STATISTICAL PHYSICS

Part 2

Theory of the Condensed State

by

E. M. LIFSHITZ and L. P. PITAEVSKIĬ
Institute of Physical Problems, U.S.S.R. Academy of Sciences

Volume 9 of *Course of Theoretical Physics*

Translated from the Russian by

J. B. SYKES and M. J. KEARSLEY

PERGAMON PRESS

OXFORD · NEW YORK · TORONTO · SYDNEY · PARIS · FRANKFURT

UK	Pergamon Press Ltd., Headington Hill Hall, Oxford OX3 0BW, England
USA	Pergamon Press Inc., Maxwell House, Fairview Park, Elmsford, New York 10523, USA
CANADA	Pergamon Press Canada Ltd., Suite 104, 150 Consumers Road, Willowdale, Ontario M2J 1P9, Canada
AUSTRALIA	Pergamon Press (Aust.) Pty. Ltd., P.O. Box 544, Potts Point, NSW 2011, Australia
FRANCE	Pergamon Press SARL, 24 rue des Ecoles, 75240 Paris, Cedex 05, France
FEDERAL REPUBLIC OF GERMANY	Pergamon Press GmbH, 6242 Kronberg-Taunus, Hammerweg 6, Federal Republic of Germany

First published in English, 1958
Second edition, revised and enlarged 1969
(Part 2), 1980
Reprinted (with corrections) 1981

British Library Cataloguing in Publication Data

Landau, Lev Davidovich
Statistical physics
Part 2: Theory of the condensed state
(Course of theoretical physics; vol. 9)
1. Mathematical physics
2. Mathematical statistics
I. Title II. Lifshitz, Evgenii Mikhailovich
III. Pitaevskiĭ, Lev Petrovich IV. Theory of the condensed state V. Series
530.1′5′95 QC21.2 78—41328

ISBN 0-08-023073-3 (Hardcover)
ISBN 0-08-023072-5 (Flexicover)

Printed in Great Britain by
A. Wheaton & Co. Ltd., Exeter

CONTENTS

NOTATION

VECTOR suffixes are denoted by Latin letters i, k, ... Spin indices are denoted by Greek letters α, β, ... Summation is implied over all repeated indices.

"4-vectors" (see the footnote to equation (13.8)) are denoted by capital letters X, P, ...

Volume element dV or d^3x.

Limit on tending to zero from above or below $+0$ or -0.

Operators are denoted by a circumflex.

Hamiltonian $\hat{H}$, $\hat{H}' = \hat{H} - \mu\hat{N}$.

Perturbation operator $\hat{V}$.

ψ operators in the Schrödinger representation $\hat{\psi}$, $\hat{\psi}^+$; in the Heisenberg representation $\hat{\Psi}$, $\hat{\Psi}^+$; in the Matsubara representaion $\hat{\Psi}^M$, $\hat{\bar{\Psi}}^M$.

Green's functions G, D.

Temperature Green's functions $\mathcal{G}$, $\mathcal{D}$.

Thermodynamic quantities are denoted as in Part 1, for example T temperature, V volume, P pressure, μ chemical potential.

Magnetic field $\mathbf{H}$; magnetic induction $\mathbf{B}$; external magnetic field $\mathfrak{H}$.

References to earlier volumes in the *Course of Theoretical Physics*:

Mechanics = Vol. 1 (*Mechanics*, third English edition, 1976).

Fields = Vol. 2 (*The Classical Theory of Fields*, fourth English edition, 1975).

QM = Vol. 3 (*Quantum Mechanics*, third English edition, 1977).

RQT = Vol. 4 (*Relativistic Quantum Theory*, first English edition, Part 1, 1971; Part 2, 1974).

Part 1 = Vol. 5 (*Statistical Physics*, Part 1, third English edition, 1980).

FM = Vol. 6 (*Fluid Mechanics*, first English edition, 1959).

ECM = Vol. 8 (*Electrodynamics of Continuous Media*, first English edition, 1960).

All are published by Pergamon Press.

PREFACE

As a brief characterization of its content, this ninth volume in the *Course of Theoretical Physics* may be said to deal with the quantum theory of the condensed state of matter. It opens with a detailed exposition of the theory of Bose and Fermi quantum liquids. This theory, set up by L. D. Landau following the experimental discoveries by P. L. Kapitza, is now an independent branch of theoretical physics. Its importance is in fact measured not so much by even the remarkable phenomena that occur in the liquid isotopes of helium as by the fact that the concepts of a quantum liquid and its spectrum are essentially the foundation for the quantum description of macroscopic bodies.

For example, a thorough understanding of the properties of metals involves treating the electrons in them as a Fermi liquid. The properties of the electron liquid are, however, complicated by the presence of the crystal lattice, and a study of the simpler case of a homogeneous isotropic liquid is a necessary preliminary step in the construction of the theory. Similarly, superconductivity in metals, which may be regarded as superfluidity of the electron liquid, is difficult to understand clearly without a previous knowledge of the simpler theory of superfluidity in a Bose liquid.

The Green's function approach is an indispensable part of the mathematical formalism of modern statistical physics. This is not only because of the convenience of calculation of Green's functions by the diagram technique, but particularly because the Green's functions directly determine the spectrum of elementary excitations in the body, and therefore constitute the language that affords the most natural description of the properties of these excitations. In the present volume, therefore, considerable attention is paid to methodological problems in the theory of Green's functions of macroscopic bodies. Although the basic ideas of the method are the same for all systems, the specific form of the diagram technique is different in different cases. It is consequently natural to develop these methods for the isotropic quantum liquids, where the essence of the procedure is seen in its purest form, without the complications arising from spatial inhomogeneity, the presence of more than one kind of particle, and so on.

For similar reasons, the microscopic theory of superconductivity is described with the simple model of an isotropic Fermi gas with weak interaction, disregarding the complications due to the presence of the crystal lattice and the Coulomb interaction.

In respect of the chapters dealing with electrons in the crystal lattice and with the theory of magnetism, we must again stress that this book is part of a course of theoretical physics and in no way attempts to be a textbook of solid state theory. Accordingly, only the most general topics are discussed here, and no reference is made to problems that involve the use of specific experimental results, nor to methods of calculation that have no evident theoretical basis. Moreover, this volume does not include the transport properties of solids, with which we intend to deal in the next and final volume of the *Course*.

Finally, this book also discusses the theory of electromagnetic fluctuations in material media and the theory of hydrodynamic fluctuations. The former was previously included in Volume 8, *Electrodynamics of Continuous Media.* Its transfer to the present volume is a consequence of the need to make use of Green's functions, whereby the entire theory can be simplified and made more convenient for application. It is also more reasonable to treat electromagnetic and hydrodynamic fluctuations in the same volume.

This is Volume 9 of the *Course of Theoretical Physics* (Part 1 of *Statistical Physics* being Volume 5). The logic of the arrangement is that the topics dealt with here are closely akin also to those in fluid mechanics (Volume 6) and macroscopic electrodynamics (Volume 8).

L. D. Landau is not among those who have actually written this book. But the reader will quickly observe how often his name occurs in it: a considerable part of the results given here are due to him, alone or with his pupils and colleagues. Our many years' association with him enables us to hope that we have accurately reflected his views on these subjects—while at the same time, of course, having regard to developments in the fifteen years since his work was so tragically terminated.

We should like to express here our thanks to A. F. Andreev, I. E. Dzyaloshinskiĭ and I. M. Lifshitz for many discussions of topics in this book. We have had great benefit from the well-known book *Quantum Field Theoretical Methods in Statistical Physics* (Pergamon, Oxford, 1965) by A. A. Abrikosov, L. P. Gor'kov and I. E. Dzyaloshinskiĭ, one of the first books in the literature of physics to deal with the new methods of statistical physics. Lastly, we are grateful to L. P. Gor'kov and Yu. L. Klimontovich for reading the book in manuscript and making a number of comments.

April 1977

E. M. Lifshitz
L. P. Pitaevskiĭ

CHAPTER I

THE NORMAL FERMI LIQUID

§ 1. Elementary excitations in a quantum Fermi liquid

AT TEMPERATURES so low that the de Broglie wavelength corresponding to the thermal motion of the atoms in a liquid becomes comparable with the distances between the atoms, the macroscopic properties of the liquid are determined by quantum effects. The theory of such quantum liquids is of considerable fundamental interest, although there exist in Nature only two such that are literally liquids, the liquid isotopes of helium He^3 and He^4 at temperatures $\sim$ 1–2°K. All other substances solidify well before quantum effects become important in them. In this connection, it may be recalled that according to classical mechanics all bodies should be solid at absolute zero (see Part 1, §64). Helium, however, because of the peculiarly weak interaction between its atoms, remains liquid down to temperatures where quantum phenomena come into effect, whereupon it need not solidify.

The calculation of the thermodynamic quantities for a macroscopic body requires a knowledge of its energy level spectrum. In a system of strongly interacting particles such as a quantum liquid, we can refer, of course, only to levels that correspond to quantum-mechanical stationary states of the whole liquid, not to states of the individual atoms. In calculating the partition function at sufficiently low temperatures, we are to take account only of the weakly excited energy levels of the liquid, lying fairly close to the ground state.

The following point is of fundamental importance for the whole theory. Any weakly excited state of a macroscopic body may be regarded, in quantum mechanics, as an assembly of separate *elementary excitations*. These behave like *quasi-particles* moving in the volume occupied by the body and possessing definite energies ε and momenta $\mathbf{p}$. The form of the function $\varepsilon(\mathbf{p})$, the *dispersion relation* for the elementary excitations, is an important characteristic of the energy spectrum of the body. It must again be emphasized that the concept of elementary excitations arises as a means of quantum-mechanical description of the collective motion of the atoms in a body, and the quasi-particles cannot be identified with the individual atoms or molecules.

There are various types of energy spectrum that can in principle occur in quantum liquids. There will be completely different macroscopic properties also, depending on the type of spectrum. We shall begin by considering a liquid with what may be called a *Fermi* spectrum. The theory of such a Fermi liquid is due to L. D. Landau (1956–1958); he derived the results given in §§1–4.†

The energy spectrum of a Fermi quantum liquid has a structure which is to some extent similar to that of an ideal Fermi gas (of particles with spin $\frac{1}{2}$). The ground state of the latter corresponds to the occupation by particles of all the states within the *Fermi sphere*, a sphere in momentum space whose radius p_F is related to the gas density N/V (number of particles per unit volume) by

$$N/V = 2 \cdot 4\pi p_F^3/3(2\pi\hbar)^3 = p_F^3/3\pi^2\hbar^3; \quad (1.1)$$

see Part 1, §57. The excited states of the gas occur when the particles pass from states of the occupied sphere to some states with $p > p_F$.

In a liquid, of course, there are no quantum states for individual particles, but to construct the spectrum of a Fermi liquid we start from the assumption that the classification of energy levels remains unchanged when the interaction between the atoms is gradually "switched on", i.e. as we go from the gas to the liquid. In this classification the role of the gas particles is taken by the elementary excitations (quasi-particles), whose number is equal to the number of atoms and which obey Fermi statistics.

It is evident that such a spectrum can occur only for a liquid of particles with half-integral spin: the state of a system of bosons (particles with integral spin) cannot be described in terms of quasi-particles obeying Fermi statistics. At the same time it must be emphasized that a spectrum of this type cannot be a universal property of all such liquids. The type of spectrum depends also on the specific nature of the interaction between atoms. This is clear from the following simple consideration: if the interaction is such that it causes the atoms to tend to associate in pairs, then in the limit we obtain a molecular liquid consisting of particles (molecules) with integral spin, for which the spectrum under consideration is certainly impossible.

Each of the quasi-particles has a definite momentum **p** (we shall return later to the question of the validity of this assertion). Let $n(\mathbf{p})$ be the momentum distribution function of the quasi-particles, normalized by the condition

$$\int n\, d\tau = N/V, \quad d\tau = d^3p/(2\pi\hbar)^3;$$

this condition will later be made more precise. The classification principle mentioned above consists in supposing that, if this function is specified, the energy E of the liquid is uniquely determined and that the ground state corresponds to a distribution function in which all states are occupied within the Fermi sphere, whose radius p_F is related to the density of the liquid by the same formula (1.1) as for an ideal gas.

† To anticipate, we may mention here for the avoidance of misunderstanding that we are referring to a non-superfluid (*normal*) Fermi liquid, such as is the liquid isotope He^3, with the reservation made in the third footnote to §54.

It is important to emphasize that the total energy E of the liquid is not simply the sum of the energies ε of the quasi-particles. In other words, E is a functional of the distribution function that does not reduce to the integral $\int n\varepsilon\, d\tau$ (as it does for an ideal gas, where the quasi-particles are the same as the actual particles and do not interact). Since the primary concept is E, the question arises how the energy of the quasi-particles is to be defined, with allowance for their interaction.

For this purpose, let us consider the change of E due to an infinitesimal change in the distribution function. It can manifestly be defined as the integral of an expression linear in the variation δn, i.e. it has the form

$$\delta E/V = \int \varepsilon(p)\delta n\, d\tau.$$

The quantity ε is the functional derivative of the energy E with respect to the distribution function. It corresponds to the change in the energy of the system when a single quasi-particle with momentum $\mathbf{p}$ is added. This quantity plays the role of the Hamiltonian function of a quasi-particle in the field of the other particles. It is also a functional of the distribution function, i.e. the form of the function $\varepsilon(\mathbf{p})$ depends on the distribution of all the particles in the liquid.

In this connection it may be noted that an elementary excitation in the type of spectrum considered may in a certain sense be treated like an atom in the self-consistent field of the other atoms. This self-consistency is, of course, not to be understood in the sense usual in quantum mechanics. Here its nature is more profound; in the Hamiltonian of the atom, not only is allowance made for the effect of the surrounding particles on the potential energy, but the dependence of the kinetic-energy operator on the momentum operator is also modified.

Hitherto we have ignored the possible spin of the quasi-particles. Since spin is a quantum-mechanical quantity, it cannot be treated classically, and we must therefore regard the distribution function as a statistical matrix with respect to the spin. The energy ε of an elementary excitation is in general not only a function of the momentum but also an operator with respect to the spin variables, which may be expressed in terms of the quasi-particle spin operator $\hat{\mathbf{s}}$. In a homogeneous isotropic liquid (not in a magnetic field and not ferromagnetic) the operator $\hat{\mathbf{s}}$ can appear in the scalar function ε only in the form of the scalars $\hat{\mathbf{s}}^2$ and $(\hat{\mathbf{s}}.\mathbf{p})^2$; the first power of the product $\hat{\mathbf{s}}.\mathbf{p}$ is inadmissible, since the spin vector is an axial vector and this product is therefore a pseudoscalar. The square $\hat{\mathbf{s}}^2 = s(s+1)$, and for spin $s = \frac{1}{2}$ the scalar $(\hat{\mathbf{s}}.\mathbf{p})^2 = \frac{1}{4}p^2$ also reduces to a constant independent of $\hat{\mathbf{s}}$. Thus in this case the energy of a quasi-particle is independent of the spin operator, and all the energy levels of the quasi-particles are doubly degenerate.

The statement that a quasi-particle has spin essentially expresses the fact that this degeneracy exists. In this sense we can say that the spin of the quasi-particles in a spectrum of the type considered is always $\frac{1}{2}$, whatever the spin of the actual particles in the liquid. For with any spin $\mathbf{s}$ other than $\frac{1}{2}$ the terms of the form $(\hat{\mathbf{s}}.\mathbf{p})^2$ would give a splitting of the $(2s+1)$-fold degenerate levels into $\frac{1}{2}(2s+1)$ doubly degenerate levels. In other words, $\frac{1}{2}(2s+1)$ different branches of the function $\varepsilon(\mathbf{p})$ would appear, each corresponding to "quasi-particles with spin $\frac{1}{2}$".

As already mentioned, when the spin of the quasi-particles is taken into account the distribution function becomes a matrix or an operator $\hat{n}(\mathbf{p})$ with respect to the spin variables. This operator may be explicitly written as an Hermitian statistical matrix $n_{\alpha\beta}(\mathbf{p})$, where α and β are spin matrix indices taking the two values $\pm\frac{1}{2}$. The diagonal matrix elements determine the numbers of quasi-particles in particular spin states. The normalization condition for the quasi-particle distribution function must therefore now be written

$$\operatorname{tr}\int \hat{n}\, d\tau \equiv \int n_{\alpha\alpha}\, d\tau = N/V, \quad d\tau = d^3p/(2\pi\hbar)^3, \tag{1.2}$$

where tr denotes the trace of the matrix with respect to the spin indices.†

The quasi-particle energy $\hat{\varepsilon}$ is in general also an operator (a matrix with respect to the spin variables). It must be defined by

$$\delta E/V = \operatorname{tr}\int \hat{\varepsilon}\, \delta\hat{n}\, d\tau \equiv \int \varepsilon_{\alpha\beta}\, \delta n_{\beta\alpha}\, d\tau. \tag{1.3}$$

If there is no spin dependence of the distribution function and the energy, so that $n_{\alpha\beta}$ and $\varepsilon_{\alpha\beta}$ reduce to unit matrices:

$$n_{\alpha\beta} = n\delta_{\alpha\beta}, \quad \varepsilon_{\alpha\beta} = \varepsilon\delta_{\alpha\beta}, \tag{1.4}$$

then the taking of the trace in (1.2) and (1.3) amounts to simply multiplying by 2:

$$2\int n\, d\tau = N/V, \quad \delta E/V = 2\int \varepsilon\, \delta n\, d\tau. \tag{1.5}$$

It is easy to see that in statistical equilibrium the quasi-particle distribution function is an ordinary Fermi distribution, the energy being represented by the quantity $\hat{\varepsilon}$ defined in (1.3). For, because the energy levels of the liquid and of the ideal Fermi gas are classified in the same manner, the entropy S of the liquid is determined by a similar combinatorial expression

$$S/V = -\operatorname{tr}\int \{\hat{n}\log\hat{n} - (1-\hat{n})\log(1-\hat{n})\}\, d\tau \tag{1.6}$$

to that for a gas (Part 1, §55). Varying this expression with the additional conditions of constant total number of particles and constant total energy,

$$\delta N/V = \operatorname{tr}\int \delta\hat{n}\, d\tau = 0, \quad \delta E/V = \operatorname{tr}\int \hat{\varepsilon}\, \delta\hat{n}\, d\tau = 0,$$

we obtain the required distribution:

$$\hat{n} = [e^{(\hat{\varepsilon}-\mu)/T} + 1]^{-1}, \tag{1.7}$$

where μ is the chemical potential of the liquid.

† Here and throughout, summation is as usual implied over repeated indices.

When the quasi-particle energy is independent of the spin, formula (1.7) signifies a similar relation between n and ε:

$$n = [e^{(\varepsilon-\mu)T}+1]^{-1}. \tag{1.8}$$

At $T = 0$, the chemical potential is equal to the limiting energy on the surface of the Fermi sphere:

$$[\mu]_{T=0} = \varepsilon_F \equiv \varepsilon(p_F). \tag{1.9}$$

It must be emphasized that, despite the formal analogy between the expression (1.8) and the ordinary Fermi distribution, it is not identical with the latter: since ε itself is a functional of n, formula (1.8) is strictly speaking a complicated implicit expression for n.

Let us now return to the assumption that a definite momentum can be assigned to each quasi-particle. The condition for this assumption to be valid is that the uncertainty in the momentum (due to the finite mean free path of the quasi-particle) should be small not only in comparison with the momentum itself but also in comparison with the width Δp of the "transitional zone" of the distribution, over which it differs appreciably from a step function:†

$$\left.\begin{aligned}\theta(\mathbf{p}) \equiv \theta(p) &= 1 \quad \text{for} \quad p < p_F,\\ &= 0 \quad \text{for} \quad p > p_F.\end{aligned}\right\} \tag{1.10}$$

It is easy to see that this condition is satisfied if the distribution $n(\mathbf{p})$ differs from (1.10) only in a small region near the surface of the Fermi sphere. For, by the Pauli principle, only quasi-particles in the transitional zone of the distribution can undergo mutual scattering, and as a result of this scattering they must enter free states in that zone. Hence the collision probability is proportional to the square of the width of the zone. Accordingly, the uncertainty in the energy and hence that in the momentum of the quasi-particle are both proportional to $(\Delta p)^2$. It is therefore clear that, when Δp is sufficiently small, the uncertainty in the momentum will be small in comparison not only with p_F but also with Δp.

Thus the method described is valid only for excited states of the liquid which are described by a quasi-particle distribution function differing from a step function in just a narrow region near the Fermi surface. In particular, for thermodynamic equilibrium distributions only sufficiently low temperatures are permissible. The (energy) width of the transitional zone of the equilibrium distribution is of the order of T. The quantum uncertainty in the energy of a quasi-particle, due to collisions, is of the order of $\hbar/\tau$, where τ is the mean free time of the quasi-particle. The condition for the theory to be applicable is therefore

$$\hbar/\tau \ll T. \tag{1.11}$$

† For future reference, it may be noted that the derivative $\theta'(p) = -\delta(p-p_F)$, since both sides give unity on integration over any range of p that includes the point $p = p_F$.

According to the preceding discussion, the time τ is inversely proportional to the squared width of the transitional zone:

$$\tau \propto T^{-2}.$$

so that (1.11) is certainly satisfied as $T \to 0$. For a liquid in which the interaction between particles is not weak, all the energy parameters are of the same order as the limiting energy ε_F; in this sense, the condition (1.11) is equivalent to $T \ll |\varepsilon_F|$.†

For almost step-function distributions (i.e. those close to the distribution for $T = 0$), as a first approximation we can replace the functional ε by its value calculated with $n(\mathbf{p}) = \theta(\mathbf{p})$. Then ε becomes a definite function of the magnitude of the momentum, and (1.7) becomes the ordinary Fermi distribution.

Near the surface of the Fermi sphere, where alone the function $\varepsilon(\mathbf{p})$ has a direct physical significance, it can then be expanded in powers of the difference $p - p_F$. We have

$$\varepsilon - \varepsilon_F \approx v_F(p - p_F), \tag{1.12}$$

where

$$\mathbf{v}_F = [\partial\varepsilon/\partial\mathbf{p}]_{p=p_F} \tag{1.13}$$

is the "velocity" of the quasi-particles on the Fermi surface. In an ideal Fermi gas, where the quasi-particles are identical with the actual particles, we have $\varepsilon = p^2/2m$, and so $v_F = p_F/m$. By analogy we can define for a Fermi liquid the quantity

$$m^* = p_F/v_F, \tag{1.14}$$

called the *effective mass* of the quasi-particle; it is positive (see the end of §2).

In terms of the quantities thus defined, the condition for the theory to be applicable may be written $T \ll v_F p_F$, and only quasi-particles with momenta p such that $|p - p_F| \ll p_F$ have any real meaning. This important fact, in particular, makes the relation (1.1) between p_F and the density of the liquid non-trivial, since its intuitive derivation (for a Fermi gas) is based on the concept of particles in states occupying the whole Fermi sphere, not just the neighbourhood of its surface.‡

The effective mass determines, in particular, the entropy S and the specific heat C of the liquid at low temperatures. These are given by the same formula as for an ideal gas (Part 1, §58), in which we need only replace the particle mass m by the effective mass m^*:

$$S = C = V\gamma T, \quad \gamma = m^* p_F/3\hbar^3 = (\tfrac{1}{3}\pi)^{2/3}(m^*/\hbar^2)(N/V)^{1/3}; \tag{1.15}$$

† For liquid He^3, however, the range of quantitative applicability of the theory is shown by experiment to be in fact limited to $T \lesssim 0.1$ °K (whereas $|\varepsilon_F| \approx 2.5$ °K).

‡ The proof of (1.1) involves the use of more complicated mathematical methods, and is given in §20 below.

because of the linear dependence on T, S and C are the same. This follows because the expression (1.6) for the entropy in terms of the distribution function is the same for a liquid and for a gas, and in the calculation of this integral only the range of momenta near p_F is important, in which the quasi-particle distribution function in the liquid and the particle distribution function in the gas are given by the same expression (1.8).†

Before the theory is further developed, the following remark should be made. Although this method of defining the quasi-particles in a Fermi liquid by exact analogy with the particles in a gas is the most convenient in systematically deriving the theory, the corresponding physical picture has the disadvantage of involving the unobservable filled Fermi sphere of quasi-particles. This could be eliminated by a formulation in which the elementary excitations occur only when $T \neq 0$. In such a picture, the elementary excitations are represented by quasi-particles outside the Fermi sphere and "holes" within it; the former are to be assigned, in the approximation corresponding to (1.12), the energy $\varepsilon = v_F(p-p_F)$, and the latter the energy $\varepsilon = v_F(p_F-p)$. The statistical distribution of each is given by the Fermi distribution formula with zero chemical potential (in accordance with the fact that the number of elementary excitations is here not constant, but is itself determined by the temperature)‡

$$n = [e^{\varepsilon/T}+1]^{-1}. \tag{1.16}$$

The elementary excitations in this picture appear or disappear only in pairs, and so the total numbers of excitations with $p > p_F$ and $p < p_F$ are always the same.

With this definition of the elementary excitations, their energy is certainly positive, being the excess of the energy of the excited level over that of the ground level of the system. The energy of the quasi-particles defined by (1.3) may be either positive or negative.

Moreover, for a liquid at zero temperature and zero pressure, the quantity $\varepsilon_F = \mu$ is certainly negative, and the values of ε close to ε_F are therefore negative also. This is clear, since, when $T = 0$ and $P = 0$, $-\mu$ is a positive quantity, the limiting value of the heat of evaporation of the liquid per particle.

† For liquid He^3 at zero pressure, $p_F/\hbar = 0.8\times 10^8\ \text{cm}^{-1}$; $m^* = 3.1\ m\,(He^3)$; p_F is found from the density of the liquid, and m^* from its specific heat.

‡ It will be recalled (cf. Part 1, §63) that under such conditions the number of quasi-particles N_{qp} is determined by the condition for thermodynamic equilibrium: the free energy F is a minimum as a function of N_{qp} for given temperature and volume: $(\partial F/\partial N_{qp})_{T,V} = 0$. This derivative is, however, just the "chemical potential of the quasi-particles"; it should not be confused with the chemical potential μ of the liquid, which is determined by the derivative of F with respect to the number of actual particles N.

§ 2. Interaction of quasi-particles

The energy of the quasi-particles, being a functional of their distribution function, varies with that function. The change of energy for a small deviation δn of the distribution function from the step function (1.10) must be

$$\delta\varepsilon_{\alpha\beta}(\mathbf{p}) = \int f_{\alpha\gamma,\beta\delta}(\mathbf{p}, \mathbf{p}')\, \delta n_{\delta\gamma}(\mathbf{p}')\, d\tau' \tag{2.1}$$

or, in a more symbolic form,

$$\delta\hat{\varepsilon}(\mathbf{p}) = \mathrm{tr}' \int \hat{f}(\mathbf{p}, \mathbf{p}')\, \delta\hat{n}(\mathbf{p}')\, d\tau',$$

where tr′ denotes the trace with respect to the pair of spin indices that correspond to the momentum $\mathbf{p}'$. The function $\hat{f}$ may be called the *interaction function* of the quasi-particles; in a Fermi gas, $\hat{f} \equiv 0$. By definition, it represents the second variational derivative of the total energy E of the liquid, and is therefore symmetrical in the variables $\mathbf{p}$, $\mathbf{p}'$ and the corresponding pairs of spin indices:

$$f_{\alpha\gamma,\beta\delta}(\mathbf{p}, \mathbf{p}') = f_{\gamma\alpha,\delta\beta}(\mathbf{p}', \mathbf{p}). \tag{2.2}$$

With the change (2.1), the energy of the quasi-particles near the surface of the Fermi sphere is given by the sum

$$\hat{\varepsilon}(\mathbf{p}) - \varepsilon_F = v_F(p - p_F) + \mathrm{tr}' \int \hat{f}(\mathbf{p}, \mathbf{p}')\, \delta\hat{n}(\mathbf{p}')\, d\tau'. \tag{2.3}$$

In particular, for thermodynamic equilibrium distributions, the second term in (2.3) gives the temperature dependence of the quasi-particle energy. The deviation $\delta\hat{n}'$ is appreciably different from zero only in a narrow band of $\mathbf{p}'$ values near the surface of the Fermi sphere, and this contains the momenta $\mathbf{p}$ of actual quasi-particles. The function $\hat{f}(\mathbf{p}, \mathbf{p}')$ in (2.1) and (2.3) can therefore be replaced in practice by its value on that surface, putting $p = p' = p_F$, so that $\hat{f}$ will depend only on the directions of the vectors $\mathbf{p}$ and $\mathbf{p}'$.

The spin dependence of the function $\hat{f}$ is due both to relativistic effects (spin–spin and spin–orbit interaction) and to the exchange interaction. The latter is the most important. When it is taken into account, the quasi-particle interaction function has (on the Fermi surface) the form

$$(p_F m^*/\pi^2\hbar^3)\hat{f}(\mathbf{p}, \mathbf{p}') = F(\vartheta) + \boldsymbol{\sigma}.\boldsymbol{\sigma}' G(\vartheta), \tag{2.4}$$

where $\boldsymbol{\sigma}$ and $\boldsymbol{\sigma}'$ are the Pauli matrices acting on the corresponding spin indices (i.e. corresponding to the variables $\mathbf{p}$ and $\mathbf{p}'$), and F and G are two functions of the angle ϑ between $\mathbf{p}$ and $\mathbf{p}'$.† The form of this expression arises from a characteristic property of the exchange interaction, which is independent of the spatial orientation of the total angular momentum of the system, so that the

† In explicit matrix form,

$$(p_F m^*/\pi^2\hbar^3) f_{\alpha\gamma,\beta\delta} = F\delta_{\alpha\beta}\delta_{\gamma\delta} + G\boldsymbol{\sigma}_{\alpha\beta}\cdot\boldsymbol{\sigma}_{\gamma\delta}. \tag{2.4a}$$

two spin operators can appear in it only as a scalar product. The functions F and G as defined by (2.4) are dimensionless. The factor separated for this purpose on the left of (2.4) is the number of quasi-particle states on the Fermi surface per unit energy interval:

$$\nu(\varepsilon_F) = [2d\tau/d\varepsilon]_{\varepsilon=\varepsilon_F} = \frac{2\cdot 4\pi p_F^2}{(2\pi\hbar)^3}\left(\frac{dp}{d\varepsilon}\right)_{p_F}$$

or

$$\nu_F = p_F^2/\pi^2\hbar^3 v_F = p_F m^*/\pi^2\hbar^3. \tag{2.5}$$

Since the trace of a Pauli matrix is zero, the second term in (2.4) vanishes when the trace tr′ is taken, and tr′ f is independent of $\boldsymbol{\sigma}$ also. This in fact also happens when the spin–orbit and spin–spin interactions are taken into account. The reason is that the scalar function tr′ $\hat{f}$ could contain the spin operator only as the product $\hat{\mathbf{s}}\cdot\mathbf{p}\times\mathbf{p}'$ of the two axial vectors $\hat{\mathbf{s}}$ and $\mathbf{p}\times\mathbf{p}'$; expressions quadratic in the components of $\hat{\mathbf{s}}$ need not be considered, since for spin $\frac{1}{2}$ they reduce to terms linear in $\hat{\mathbf{s}}$ or independent of $\hat{\mathbf{s}}$. But this product is not invariant under time reversal, and therefore cannot appear in the invariant quantity tr′ $\hat{f}$.

The following notation will be convenient:

$$f_{\alpha\gamma,\beta\gamma}(\mathbf{p}, \mathbf{p}') = \delta_{\alpha\beta} f(\mathbf{p}, \mathbf{p}'), \quad f = \tfrac{1}{2}\,\mathrm{tr}\,\mathrm{tr}'\hat{f}. \tag{2.6}$$

From the expression (2.4), we have

$$(p_F m^*/\pi^2\hbar^3) f(\vartheta) = 2F(\vartheta). \tag{2.7}$$

The quasi-particle interaction function satisfies a certain integral relation which follows from Galileo's principle of relativity. A direct consequence of this principle is that the momentum of the liquid per unit volume is equal to its mass flux density. The velocity of a quasi-particle is $\partial\varepsilon/\partial\mathbf{p}$, so that the quasi-particle flux is

$$\mathrm{tr}\int \hat{n}(\partial\hat{\varepsilon}/\partial\mathbf{p})\,d\tau.$$

Since the number of quasi-particles in the liquid is the same as the number of actual particles, it is clear that the total mass transfer by quasi-particles is found by multiplying their number flux by the actual particle mass m. Thus we obtain the equation

$$\mathrm{tr}\int \mathbf{p}\hat{n}\,d\tau = \mathrm{tr}\int m(\partial\hat{\varepsilon}/\partial\mathbf{p})\hat{n}\,d\tau. \tag{2.8}$$

Putting $n_{\alpha\beta} = n\delta_{\alpha\beta}$, $\varepsilon_{\alpha\beta} = \varepsilon\delta_{\alpha\beta}$, we vary both sides of (2.8), use (2.1), and take f from (2.6):

$$\int \mathbf{p}\,\delta n\,d\tau = m\int \frac{\partial\varepsilon}{\partial\mathbf{p}}\,\delta n\,d\tau + m\int \frac{\partial f(\mathbf{p}, \mathbf{p}')}{\partial\mathbf{p}}\,n\,\delta n'\,d\tau\,d\tau',$$

$$= m\int \frac{\partial\varepsilon}{\partial\mathbf{p}}\,\delta n\,d\tau - m\int f(\mathbf{p}, \mathbf{p}')\frac{\partial n'}{\partial\mathbf{p}'}\,\delta n\,d\tau\,d\tau',$$

where $n' \equiv n(\mathbf{p}')$; in the second integral, we have renamed the variables, and integrated by parts. Since δn is arbitrary, this gives the required relation:

$$\mathbf{p}/m = \partial\varepsilon/\partial\mathbf{p} - \int f(\mathbf{p}, \mathbf{p}')\,[\partial n(\mathbf{p}')/\partial\mathbf{p}']\,d\tau'. \tag{2.9}$$

For a step function $n(\mathbf{p}') = \theta(\mathbf{p}')$, the derivative $\partial n'/\partial\mathbf{p}'$ reduces to a delta function:

$$\partial\theta(p)/\partial\mathbf{p} = -(\mathbf{p}/p)\,\delta(p-p_F). \tag{2.10}$$

Substituting the function $\varepsilon(\mathbf{p})$ from (1.12) in (2.9), and then replacing the momentum $\mathbf{p} = p\mathbf{n}$ everywhere by the value $\mathbf{p}_F = p_F\mathbf{n}$ on the Fermi surface, and multiplying both sides of the equation by $\mathbf{p}_F$, we get the following relation between the mass m of the actual particles and the effective mass of the quasi-particles:

$$\frac{1}{m} = \frac{1}{m^*} + \frac{p_F}{(2\pi\hbar)^3}\int f(\vartheta)\cos\vartheta\,do', \tag{2.11}$$

where do' is the element of solid angle in the direction of $\mathbf{p}'$. If we substitute here the expression (2.7) for $f(\vartheta)$, this equation becomes

$$m^*/m = 1 + \overline{F(\vartheta)\cos\vartheta}, \tag{2.12}$$

where the bar denotes averaging over directions, i.e. integration over $do'/4\pi = \frac{1}{2}\sin\vartheta\,d\vartheta$.

Let us also calculate the compressibility of a Fermi liquid at absolute zero, i.e. the quantity $u^2 = \partial P/\partial\varrho$.† The density of the liquid is $\varrho = mN/V$, so that

$$u^2 = -(V^2/mN)\,\partial P/\partial V.$$

To calculate this derivative, it is convenient to express it in terms of the derivative of the chemical potential. Since the latter depends on N and V only through the ratio N/V, and for T = constant = 0 the differential $d\mu = VdP/N$, we have

$$\frac{\partial\mu}{\partial N} = -\frac{V}{N}\frac{\partial\mu}{\partial V} = -\frac{V^2}{N^2}\frac{\partial P}{\partial V},$$

and hence

$$u^2 = \frac{N}{m}\frac{\partial\mu}{\partial N}. \tag{2.13}$$

Since $\mu = \varepsilon_F$ for $T = 0$, the change $\delta\mu$ when the number of particles changes by δN is

$$\delta\mu = \int f(\mathbf{p}_F, \mathbf{p}')\,\delta n'\,d\tau' + (\partial\varepsilon_F/\partial p_F)\,\delta p_F. \tag{2.14}$$

† When $T = 0$, $S = 0$ also, and so there is no need to distinguish the isothermal and adiabatic compressibilities. The quantity u is defined by the usual expression for the velocity of sound in the liquid. It must be borne in mind, however, that at $T = 0$ ordinary sound in fact cannot be propagated in a Fermi liquid; see §4.

The first term here is the change of $\varepsilon(p_F)$ due to the change in the distribution function. The second term occurs because a change in the total number of particles also affects the value of the limiting momentum: from (1.1), $\delta N = Vp_F^2\delta p_F/\pi^2\hbar^3$. Since $\delta n'$ is appreciably different from zero only when $p' \approx p_F$, we can write, replacing f in the integral by its value on the Fermi surface,

$$\int f\,\delta n'\,d\tau' \approx \frac{1}{2}\int f\,do'\int \delta n'\frac{2\,d\tau'}{4\pi} = \frac{1}{2}4\pi\bar{f}\frac{\delta N}{4\pi V}.$$

Substituting this expression in (2.14) and putting $\partial\varepsilon_F/\partial p_F = p_F/m^*$, we obtain

$$\frac{\partial\mu}{\partial N} = \frac{\bar{f}}{2V}+\frac{\pi^2\hbar^3}{p_F m^* V}. \tag{2.15}$$

Finally, with $1/m^*$ from (2.11) and again using (1.1), we have

$$u^2 = \frac{p_F^2}{3m^2}+\frac{1}{3m}\left(\frac{p_F}{2\pi\hbar}\right)^3\int f(\vartheta)\,(1-\cos\vartheta)\,do'. \tag{2.16}$$

With $f(\vartheta)$ from (2.7), and using (2.12), we can put this expression in the form

$$u^2 = \frac{p_F^2}{3mm^*}[1+\overline{F(\vartheta)}]. \tag{2.17}$$

The function f must satisfy certain conditions that result from the requirement of stability of the ground state of the liquid. This state corresponds to occupation of all quasi-particle states within the Fermi sphere, and its energy must be a minimum with respect to any small deformation of the sphere. We shall not give the calculations in full, but only the final result,† which may be conveniently expressed by expanding the functions $F(\vartheta)$ and $G(\vartheta)$ from (2.4) in Legendre polynomials:

$$F(\vartheta) = \sum_l (2l+1)\,F_l P_l(\cos\vartheta), \quad G(\vartheta) = \sum_l (2l+1)\,G_l P_l(\cos\vartheta); \tag{2.18}$$

with this definition, the coefficients F_l and G_l are the mean values of the products FP_l and GP_l. Then the stability conditions are the inequalities

$$F_l+1 > 0, \tag{2.19}$$

$$G_l+1 > 0. \tag{2.20}$$

A comparison of (2.19) for $l = 1$ with the expression (2.12) for the effective mass shows that the latter is positive. The condition (2.19) for $l = 0$ ensures that (2.17) is positive.‡

† See I. Ya. Pomeranchuk, *Soviet Physics JETP* **8**, 361, 1959.

‡ For $l = 1$, we also have the inequality $F_1 > G_1$, as shown by A. J. Leggett, *Annals of Physics* **46**, 76, 1968.

§ 3. Magnetic susceptibility of a Fermi liquid

A quasi-particle with non-zero spin has in general a magnetic moment also. For spin $\frac{1}{2}$, the operator of this moment is $\beta\boldsymbol{\sigma}$ (the z-component of the magnetic moment is $\pm\beta$). The constant $2\beta/\hbar$ which gives the ratio of the magnetic moment of the quasi-particle to its angular momentum $\frac{1}{2}\hbar$ is equal to the corresponding constant for actual particles; clearly the value of this ratio is unchanged, whichever way the particle spins are added to the quasi-particle spin.

The existence of the magnetic moment of a quasi-particle leads in turn to to a paramagnetism of the liquid. The corresponding magnetic susceptibility may be calculated as follows.

For a "free" quasi-particle, the operator of its additional energy in a magnetic field $\mathbf{H}$ would be $-\beta\boldsymbol{\sigma}.\mathbf{H}$. In a Fermi liquid, however, we must take it into account that the interaction of the quasi-particles causes the energy of each of them to change, because of the changed distribution function in the magnetic field. In calculating the magnetic susceptibility, we must therefore write the quasi-particle energy change operator as

$$\delta\hat{\varepsilon} = -\beta\boldsymbol{\sigma}.\mathbf{H}+\mathrm{tr}' \int \hat{f}\,\delta\hat{n}'\,d\tau'. \tag{3.1}$$

The change in the distribution function is given in terms of $\delta\hat{\varepsilon}$ by $\delta\hat{n} = (\partial n/\partial\varepsilon)\,\delta\hat{\varepsilon}$;† we thus have

$$\delta\hat{\varepsilon}(\mathbf{p}) = -\beta\boldsymbol{\sigma}.\mathbf{H}+\mathrm{tr}' \int \hat{f}(\mathbf{p},\mathbf{p}')\,(dn'/d\varepsilon')\,\delta\hat{\varepsilon}(\mathbf{p}')\,d\tau'. \tag{3.2}$$

We shall need the solution of this equation only on the surface of the Fermi sphere, and seek it in the form

$$\delta\hat{\varepsilon} = -\tfrac{1}{2}\beta g\boldsymbol{\sigma}.\mathbf{H}, \tag{3.3}$$

where g is a constant. For a step function $n(\mathbf{p}') = \theta(p')$, we have

$$dn'/d\varepsilon' = -\delta(\varepsilon'-\varepsilon_F),$$

so that the integration over $dp' = d\varepsilon'/v_F$ reduces to taking the value of the integrand on the Fermi surface. Substituting $\hat{f}$ from (2.4) and noting that the Pauli matrices satisfy

$$\mathrm{tr}\,\boldsymbol{\sigma} = 0,\quad \mathrm{tr}'(\boldsymbol{\sigma}.\boldsymbol{\sigma}')\,\boldsymbol{\sigma}' = \tfrac{1}{3}\boldsymbol{\sigma}\,\mathrm{tr}'\,\boldsymbol{\sigma}'.\boldsymbol{\sigma}' = 2\boldsymbol{\sigma},$$

we find

$$g = 2-g\overline{G(\vartheta)},$$

or

$$g = 2/[1+\overline{G(\vartheta)}], \tag{3.4}$$

where the bar again denotes averaging over directions, as in (2.12).

† In calculating the field-dependent increment δn, we may neglect the change in the chemical potential. The change in the macroscopic quantity μ in an isotropic liquid can only be quadratic in the field $\mathbf{H}$ (which is assumed to be small in the calculation of the susceptibility), whereas $\delta\hat{\varepsilon}$ is of the first order in the field. Since the magnetic susceptibility of the liquid is small, we need not distinguish between the field and the induction in it.

The susceptibility χ is determined from the expression for the magnetic moment per unit volume of the liquid:

$$\chi\mathbf{H} = \beta \operatorname{tr} \int \boldsymbol{\sigma}\, \delta\hat{n}\, d\tau = \beta \operatorname{tr} \int \boldsymbol{\sigma}\delta\hat{\varepsilon}(\partial n/\partial\varepsilon)\, d\tau$$

or, after integration with the step function $n(\mathbf{p})$,

$$\chi\mathbf{H} = -\beta \frac{p_F m^*}{2\pi^2\hbar^3} \operatorname{tr} \boldsymbol{\sigma}\delta\hat{\varepsilon}(p_F).$$

Finally, substituting (3.3) and (3.4), and noting that $\operatorname{tr}(\boldsymbol{\sigma}.\mathbf{H})\boldsymbol{\sigma} = 2\mathbf{H}$, we find

$$\chi = \frac{\beta^2 p_F m^*}{\pi^2\hbar^3(1+\bar{G})} = \frac{3\gamma\beta^2}{\pi^2(1+\bar{G})}, \tag{3.5}$$

where γ is the coefficient in the linear specific heat law (1.15). The expression $\chi = 3\gamma\beta^2/\pi^2$ gives the susceptibility of a denegerate Fermi gas of particles with magnetic moment β; see Part 1, (59.5). The factor $1/(1+\bar{G})$ represents the difference between a Fermi liquid and a Fermi gas.†

The stability condition (2.20) with $l = 0$ is the same as the condition $\chi > 0$.

§ 4. Zero sound

Non-equilibrium states of a Fermi liquid are described by quasi-particle distribution functions that depend not only on the momenta but also on the coordinates and time. These functions $\hat{n}(\mathbf{p}, \mathbf{r}, t)$ satisfy a transport equation

$$d\hat{n}/dt = I(\hat{n}), \tag{4.1}$$

where $I(\hat{n})$ is the *collision integral*, giving the change in the number of quasi-particles in a given element of phase volume because of collisions between them.‡

The total time derivative in (4.1) includes both the explicit dependence of $\hat{n}$ on t and the implicit dependence due to the change in the coordinates, momentum and spin variables of the quasi-particle in accordance with its equations of motion. The distinctive feature of the Fermi liquid is that, since the quasi-particle energy is a functional of the distribution function, in an inhomogeneous liquid, ε as well as $\hat{n}$ depends on the coordinates.

For distributions $\hat{n}$ that differ only slightly from the equilibrium distribution n_0, we write

$$\hat{n}(\mathbf{p}, \mathbf{r}, t) = n_0(\mathbf{p}) + \delta\hat{n}(\mathbf{p}, \mathbf{r}, t). \tag{4.2}$$

† For He^3, $\bar{G} \approx -2/3$.

‡ This section assumes familiarity with the transport equation and in that respect goes outside the scope of the book. However, the theory of Fermi liquids would be incompletely formulated without the transport equation (and its application in §§4 and 5). We shall here need only the equation without the collision integral; problems involving the specific form of that integral will be discussed in another volume which deals with physical kinetics.

The quasi-particle energy is then $\hat{\varepsilon} = \varepsilon_0 + \delta\hat{\varepsilon}$, where ε_0 is the energy corresponding to the equilibrium distribution, and $\delta\hat{\varepsilon}$ is given by (2.1), so that

$$\frac{\partial \hat{\varepsilon}}{\partial \mathbf{r}} = \frac{\partial \delta\hat{\varepsilon}}{\partial \mathbf{r}} = \mathrm{tr}' \int \hat{f}(\mathbf{p}, \mathbf{p}') \frac{\partial \delta\hat{n}(\mathbf{p}')}{\partial \mathbf{r}} \, d\tau'. \tag{4.3}$$

If there is no external magnetic field, ε_0 and n_0 are independent of the spin.

The explicit time-dependence of $\hat{n}$ gives a term in $d\hat{n}/dt$

$$\partial \hat{n}/\partial t = \partial \delta\hat{n}/\partial t.$$

The dependence through the coordinates and momentum gives terms

$$\frac{\partial n}{\partial \mathbf{r}} \cdot \hat{\dot{\mathbf{r}}} + \frac{\partial n}{\partial \mathbf{p}} \cdot \hat{\dot{\mathbf{p}}}.$$

The quasi-particle energy $\hat{\varepsilon}$ plays the role of the Hamiltonian. From Hamilton's equations,

$$\hat{\dot{\mathbf{r}}} = \partial \hat{\varepsilon}/\partial \mathbf{p}, \quad \hat{\dot{\mathbf{p}}} = -\partial \hat{\varepsilon}/\partial \mathbf{r}.$$

Hence we have, as far as the terms of the first order in $\delta\hat{n}$,

$$\frac{\partial \delta\hat{n}}{\partial \mathbf{r}} \cdot \frac{\partial \varepsilon_0}{\partial \mathbf{p}} - \frac{\partial n_0}{\partial \mathbf{p}} \cdot \frac{\partial \delta\hat{\varepsilon}}{\partial \mathbf{r}}.$$

Finally, the time variation of the function $\hat{n}$ as an operator with regard to the spin variables is given, according to the general rules of quantum mechanics, by the commutator

$$(i/\hbar)\,[\hat{\varepsilon}, \hat{n}]. \tag{4.4}$$

However, when n_0 and ε_0 are independent of the spin, there are no terms of the first order in $\delta\hat{n}$ in this commutator.

Collecting the various terms, we obtain the equation

$$\frac{\partial \delta\hat{n}}{\partial t} + \frac{\partial \varepsilon_0}{\partial \mathbf{p}} \cdot \frac{\partial \delta\hat{n}}{\partial \mathbf{r}} - \frac{\partial \delta\hat{\varepsilon}}{\partial \mathbf{r}} \cdot \frac{\partial n_0}{\partial \mathbf{p}} = I(\hat{n}). \tag{4.5}$$

Before going on to apply the transport equation, let us discuss the conditions for it to be valid. By using the equations classical with regard to coordinates and momentum, we have assumed the motion of the quasi-particles to be quasi-classical; essentially the same assumption already underlies the description of the liquid by a distribution function that depends on both the coordinates and the momenta of the quasi-particles. The condition for quasi-classical motion is that the quasi-particle de Broglie wavelength $\hbar/p_F$ be small compared with the characteristic length L over which n varies considerably. Using instead of L the "wave number" of the inhomogeneity, $k \sim 1/L$, we can write this condition as†

$$\hbar k \ll p_F. \tag{4.6}$$

† According to the definition (1.1), $\hbar/p_F$ is of the order of the interatomic distances, so that the condition (4.6) is very weak.

The frequency ω of the distribution function variation that is established for a given k is of the order of $v_F k$, and automatically satisfies the condition

$$\hbar\omega \ll \varepsilon_F. \tag{4.7}$$

There may be any relation between $\hbar\omega$ and the temperature T. If $\hbar\omega \gg T$, the width of the transitional zone of the distribution function is $\hbar\omega$; then (4.7) is the condition necessary for the entire theory to be valid, ensuring that the quantum uncertainty in the quasi-particle energy (due to their collisions) is small compared with $\hbar\omega$.

Let us now apply the transport equation to investigate vibrational motion in a Fermi liquid.

At low (but non-zero) temperatures, collisions occur between quasi-particles in a Fermi liquid, and the mean free time $\tau \propto T^{-2}$. The nature of the waves propagated in the liquid essentially depends on the value of $\omega\tau$.

When $\omega\tau \ll 1$ (which is effectively the condition for the quasi-particle mean free path l to be small compared with the wavelength λ), the collisions are able to bring about thermodynamic equilibrium in each volume element (small compared with λ) in the liquid. This means that we have ordinary hydrodynamical sound waves propagated with velocity $u = \sqrt{(\partial P/\partial \varrho)}$. The absorption of sound waves is small when $\omega\tau \ll 1$, but increases with $\omega\tau$, and for $\omega\tau \sim 1$ becomes very strong, so that the propagation of sound waves becomes impossible.†

When $\omega\tau$ increases further to $\omega\tau \gg 1$, wave propagation again becomes possible in the Fermi liquid, but the waves have a different physical character. In these vibrations, collisions of quasi-particles are unimportant, and thermodynamic equilibrium is not established in each volume element. The process may be regarded as occurring at absolute zero of temperature. These waves are called *zero sound*.

According to the above discussion, the collision integral in the transport equation can be omitted when $\omega\tau \gg 1$; then

$$\frac{\partial \delta \hat{n}}{\partial t} + \mathbf{v} \cdot \frac{\partial \delta \hat{n}}{\partial \mathbf{r}} - \frac{\partial n_0}{\partial \mathbf{p}} \cdot \frac{\partial \delta \hat{\varepsilon}}{\partial \mathbf{r}} = 0, \tag{4.8}$$

where $\mathbf{v} = \partial \varepsilon / \partial \mathbf{p}$ is the quasi-particle velocity calculated from the unperturbed energy ε ($\mathbf{v} = v_F \mathbf{n}$, where $\mathbf{n}$ is a unit vector in the direction of $\mathbf{p}$); the suffix 0 is omitted from ε here and henceforward.

When $T = 0$, the equilibrium distribution function n_0 is a step function $\theta(p)$ cut off at the limiting momentum $p = p_F$. Its derivative is

$$\partial n_0/\partial \mathbf{p} = -\mathbf{n}\delta(p - p_F) = -\mathbf{v}\delta(\varepsilon - \varepsilon_F).$$

† When $\omega\tau \ll 1$, the sound absorption coefficient $\gamma \sim \omega^2 \eta/\varrho u^3$, where η is the viscosity of the liquid. In order of magnitude, $u \sim v_F$, $\eta/\varrho \sim v_F l \sim v_F^2 \tau$, where v_F is the quasi-particle velocity (independent of the temperature), so that $\eta \propto T^{-2}$ (I. Ya. Pomeranchuk 1950). Then $\gamma u/\omega \sim \omega\tau \propto \omega T^2$.

Assuming that the time and coordinate dependence of $\delta\hat{n}$ in the wave is given by the factor $\exp[i(\mathbf{k}.\mathbf{r}-\omega t)]$, we shall seek the solution of the transport equation in the form

$$\delta\hat{n} = \delta(\varepsilon-\varepsilon_F)\hat{\nu}(\mathbf{n})\, e^{i(\mathbf{k}.\mathbf{r}-\omega t)}. \tag{4.9}$$

Then (4.8), with $\partial\delta\hat{\varepsilon}/\partial\mathbf{r}$ from (4.3), becomes

$$(\omega - v_F\mathbf{n}.\mathbf{k})\,\hat{\nu}(\mathbf{n}) = \mathbf{n}.\mathbf{k}\frac{p_F^2}{(2\pi\hbar)^3}\,\mathrm{tr}'\int \hat{f}(\mathbf{n},\mathbf{n}')\hat{\nu}(\mathbf{n}')\,do, \tag{4.10}$$

where $\mathbf{n}$ and $\mathbf{n}'$ are unit vectors in the directions of $\mathbf{p}$ and $\mathbf{p}'$, and the integration is over the directions of $\mathbf{n}'$.

Let us consider (zero sound) vibrations which do not affect the spin properties of the liquid. This means that both the equilibrium distribution function and also its "perturbation" δn are independent of the spin variables. In such a wave, the change in the distribution function during the vibrations amounts to a deformation of the limiting Fermi surface (a sphere in the unperturbed distribution), which remains a sharp boundary between the occupied and unoccupied quasi-particle states. The function $\nu(\mathbf{n})$ is the displacement (in units of energy) of this surface in a given direction $\mathbf{n}$.

Since $\nu(\mathbf{n}')$ is independent of the spin variables, the operation tr′ in (4.10) applies only to $\hat{f}$. Writing $\hat{f}$ in the form (2.4), we have $\mathrm{tr}'\hat{f} = (2\pi^2\hbar^3/p_F m^*)F(\vartheta)$. Thus the operator $\boldsymbol{\sigma}$ no longer appears in the equation, which now becomes

$$(\omega-\mathbf{k}.\mathbf{v})\,\nu(\mathbf{n}) = \mathbf{k}.\mathbf{v}\int F(\vartheta)\,\nu(\mathbf{n}')\,do'/4\pi. \tag{4.11}$$

We take the direction of $\mathbf{k}$ as the polar axis, and define the direction of $\mathbf{n}$ by angles θ and ϕ. Introducing the wave propagation velocity $u_0 = \omega/k$ and the notation $s = u_0/v_F$, we can write the equation in the final form

$$(s-\cos\theta)\,\nu(\theta,\phi) = \cos\theta\int F(\vartheta)\,\nu(\theta',\phi')\,do'/4\pi. \tag{4.12}$$

This integral equation determines, in principle, the wave propagation velocity and the function $\nu(\mathbf{n}')$ in the waves. We see at once that, for undamped vibrations (the only ones considered here), s must exceed unity, i.e.

$$u_0 > v_F. \tag{4.13}$$

The origin of this inequality can be understood if we rewrite (4.12) as

$$\tilde{\nu}(\theta,\phi) = \cos\theta\int F(\vartheta)\frac{\tilde{\nu}(\theta',\phi')}{s-\cos\theta'}\frac{do'}{4\pi},$$

where ν has been replaced by another unknown function $\tilde{\nu} = (s-\cos\theta)\nu$. When $s = \omega/kv_F < 1$, the integrand has a pole at $\cos\theta' = s$, and in order to make the integral meaningful this pole in the plane of the complex variable $\cos\theta'$ must be avoided by some definite rule. This adds an imaginary part to the integral; the frequency ω therefore also acquires an imaginary part (for a given real k), and the wave is damped. The physical significance of the equation $\cos\theta = u_0/v_F$, corresponding to the pole, is that this is the condition for the quasi-particles to emit Cherenkov waves of zero sound.†

As an example, let us consider the case where $F(\vartheta)$ is a constant F_0. The integral on the right of (4.12) is then independent of the angles θ and ϕ, and the required function ν is therefore

$$\nu = \text{constant} \times \frac{\cos\theta}{s-\cos\theta}. \tag{4.14}$$

The Fermi surface thus becomes a surface of revolution elongated in the forward direction of wave propagation and flattened in the opposite direction. This anisotropy is a consequence of the non-equilibrium state of the liquid in each of its volume elements: in equilibrium, all properties of the liquid must be isotropic, and the Fermi surface must therefore be spherical. For comparison, it may be mentioned that an ordinary sound wave corresponds to a spherical Fermi surface with oscillating radius (the limiting momentum p_F varies with the density of the liquid), shifted as a whole by an amount depending on the velocity of the liquid in the wave; the corresponding function ν is $\nu = \delta p_F + \text{constant} \times \cos\theta$.

To find the zero sound wave propagation velocity u_0, we substitute (4.14) in (4.12):

$$F_0 \int_0^{\pi} \frac{\cos\theta}{s-\cos\theta} \frac{2\pi \sin\theta \, d\theta}{4\pi} = 1.$$

On integrating, we get an equation which implicitly determines u_0 for a given value of F_0:

$$\frac{1}{2} s \log \frac{s+1}{s-1} - 1 = 1/F_0. \tag{4.15}$$

The function on the left decreases from infinity to zero when s varies from 1 to ∞, and is always positive. Hence it follows that the waves concerned can exist only when $F_0 > 0$. It should be emphasized that the possibility of propagation of zero sound thus depends on the properties of the interaction of the quasi-particles in the Fermi liquid.

When $F_0 \to 0$, (4.15) shows that s tends to unity:

$$s-1 \approx \frac{2}{e^2} e^{-2/F_0}. \tag{4.16}$$

† This is called *Landau damping*; it will be discussed in detail in connection with plasma oscillations, in the last volume of the course. The rule for avoiding the pole in the integral is given by replacing ω by $\omega + i0$ (i.e. $s \to s+i0$); this signifies that the perturbation is made finite at all previous times (including $t \to -\infty$).

This case has more general significance than (4.15) (in which it is assumed that F is a constant $\equiv F_0$): it corresponds to zero sound in an almost ideal Fermi gas for any function $F(\vartheta)$. An almost ideal gas has $F(\vartheta)$ small in magnitude. It is seen from (4.12) that s is then close to 1, and ν is appreciably different from zero only for small angles θ. Hence, considering only the range of small angles, we can replace $F(\vartheta)$ on the right of (4.12) by its value when $\vartheta = 0$ (which corresponds to $\theta = \theta' = 0$). We then return to (4.14) and (4.16), with the constant F_0 replaced by $F(0)$.† In a slightly non-ideal gas, the velocity of zero sound exceeds that of ordinary sound by a factor $\sqrt{3}$: for the former $u_0 \approx v_F$, and for the latter (2.17) gives (with $\bar{F}$ neglected and $m \approx m^*$) $u^2 \approx \approx p_F^2/3m^{*2} = v_F^2/3$.

In the general case of an arbitrary function $F(\vartheta)$, the solution of (4.12) is not unique. The equation, in principle, allows the existence of various types of zero sound differing in the angular dependence of the amplitude $\nu(\theta, \phi)$ and propagated at various velocities. As well as the axially symmetrical solutions $\nu(\theta)$, there can also exist asymmetric solutions in which ν contains azimuthal factors $e^{\pm im\phi}$ with integral m (see Problem). For all such solutions, the integral $\int \nu \, do = 0$, i.e. the volume within the Fermi surface is fixed. This means that the vibrations do not alter the density of the liquid.

The possibility of wave propagation in a Fermi liquid at absolute zero implies that its energy spectrum may contain a branch corresponding to elementary excitations with momentum $\mathbf{p} = \hbar\mathbf{k}$ and energy $\varepsilon = \hbar\omega = u_0 p$, which are "quanta of zero sound". The fact that zero sound (with any $\mathbf{k}$) can have an arbitrary (small) intensity means, in terms of the elementary excitations, that these can occupy their quantum states in any numbers; that is, they obey Bose statistics and form what is called the *Bose branch* of the spectrum of the Fermi liquid. It must be stressed, however, that in the Landau theory it would be improper to apply the corrections, corresponding to this branch, to the thermodynamic quantities for the Fermi liquid, since these contain higher powers of the temperature (T^3 in the specific heat) than even the first corrections to the approximate theory given above.

The problem of the absorption of zero sound requires a consideration of the collisions of quasi-particles, and is outside the scope of this book.

PROBLEM

Find the velocity of propagation of asymmetric waves of zero sound when $F = F_0 + F_1 \cos\vartheta$.

SOLUTION. When

$$F = F_0 + F_1[\cos\theta\cos\theta' + \sin\theta\sin\theta'\cos(\phi' - \phi)],$$

† Vibrations corresponding to zero sound in a slightly non-ideal Fermi gas were first discussed by Yu. L. Klimontovich and V. P. Silin (1952).

there can exist solutions with $\nu \propto e^{\pm i\phi}$: putting $\nu = f(\theta)e^{i\phi}$, substituting in (4.12) and integrating over ϕ', we obtain

$$(s-\cos\theta)f = \tfrac{1}{4}F_1 \cos\theta \sin\theta \int_0^\pi \sin^2\theta' f(\theta')\, d\theta'.$$

Hence

$$\nu = \text{constant} \times \frac{\sin\theta\cos\theta}{s-\cos\theta} e^{i\phi}.$$

Substituting this expression back into the equation, we get

$$\int_0^\pi \frac{\sin^3\theta\cos\theta}{s-\cos\theta}\, d\theta = 4/F_1,$$

which gives the dependence of the velocity of propagation on F_1. The integral on the left is a monotonically decreasing function of s. Its greatest value therefore occurs when $s = 1$. Calculating the integral for $s = 1$, we find that an asymmetric wave of the type considered can be propagated if $F_1 > 6$.†

§ 5. Spin waves in a Fermi liquid

As well as the spin-independent solutions $\nu(\mathbf{n})$ considered in §4, (4.10) has solutions of the form

$$\hat{\nu} = \boldsymbol{\sigma}.\boldsymbol{\mu}(\mathbf{n}), \tag{5.1}$$

in which the variation of the quasi-particle distribution function depends on the spin component. These may be called *spin waves*.

Substituting (5.1) in (4.10), again taking $\hat{f}$ in the form (2.4), and noting that $\text{tr}'\, \boldsymbol{\sigma}'(\boldsymbol{\sigma}.\boldsymbol{\sigma}') = 2\boldsymbol{\sigma}$, we get (after cancelling $\boldsymbol{\sigma}$)

$$(s-\cos\theta)\, \boldsymbol{\mu}(\theta, \phi) = \cos\theta \int G(\vartheta)\, \boldsymbol{\mu}(\theta', \phi')\, do'/4\pi. \tag{5.2}$$

Thus, for each component of the vector $\boldsymbol{\mu}$, we get an equation that differs from (4.12) only in that F is replaced by G. Hence the subsequent calculations in §4 are applicable to spin waves also.‡

Spin waves of another kind can be propagated in a Fermi liquid when a magnetic field is present (V. P. Silin 1958). Here we shall consider only vibrations with $\mathbf{k} = 0$, in which $\delta\hat{n}$ is independent of the coordinates.

When a magnetic field $\mathbf{H}$ is present, even the quasi-particle energy and distribution function "unperturbed" by the vibrations are spin-dependent. These dependences are interrelated, and are given by (see §3)

$$\hat{\varepsilon}_0 = \varepsilon_0(\mathbf{p}) - \beta_1 \boldsymbol{\sigma}.\mathbf{H}, \quad \beta_1 = \beta/(1+\overline{G}), \tag{5.3}$$

$$\begin{aligned} \hat{n}_0 &= n_0(\mathbf{p}) - (dn_0/d\varepsilon)\beta_1 \boldsymbol{\sigma}.\mathbf{H} \\ &= n_0(\mathbf{p}) + \delta(\varepsilon - \varepsilon_F)\beta_1 \boldsymbol{\sigma}.\mathbf{H}, \end{aligned} \tag{5.4}$$

† For liquid He^3, F_0 and F_1 can be calculated from the known values of m^* and u^2 by means of (2.12) and (2.17): $F_0 = 10.8$, $F_1 = 6.3$ (at zero pressure).

‡ In liquid He^3, $G_0 = \overline{G(\vartheta)} < 0$; see the second footnote to §3. Such waves therefore cannot be propagated in it.

where $\varepsilon_0(\mathbf{p})$ is the energy in the absence of the field; the suffix 0 again indicates that these expressions relate to the equilibrium liquid.

We again seek the small variable part of the distribution function in the wave in the form

$$\delta\hat{n} = \delta(\varepsilon - \varepsilon_F)\,\boldsymbol{\sigma}.\boldsymbol{\mu}(\mathbf{n})e^{-i\omega t}.$$

The corresponding change in the quasi-particle energy is

$$\delta\hat{\varepsilon} = \boldsymbol{\sigma}.\int \boldsymbol{\mu}(\mathbf{n}')\,G(\vartheta)\,\frac{do'}{4\pi}.e^{-i\omega t}.$$

In the transport equation, we must now take into account the term (4.4) containing the commutator $[\hat{\varepsilon}, \hat{n}]$; for distributions independent of the coordinates, it becomes

$$\frac{\partial\delta\hat{n}}{\partial t} + \frac{i}{\hbar}[\hat{\varepsilon}, \hat{n}] = 0. \tag{5.5}$$

As far as terms linear in $\delta\hat{n}$ we have

$$[\hat{\varepsilon}, \hat{n}] = -\beta_1[\boldsymbol{\sigma}.\mathbf{H}, \delta\hat{n}] + \beta_1\delta(\varepsilon - \varepsilon_F)\,[\delta\hat{\varepsilon}, \boldsymbol{\sigma}.\mathbf{H}].$$

The commutators are given by the formula

$$[\boldsymbol{\sigma}.\mathbf{a}, \boldsymbol{\sigma}.\mathbf{b}] = 2i\boldsymbol{\sigma}.\mathbf{a}\times\mathbf{b},$$

where **a** and **b** are any vectors; see *QM* (55.10). The transport equation thus becomes

$$i\omega\boldsymbol{\mu}(\mathbf{n}) = (2\beta_1/\hbar)\,\mathbf{H}\times\boldsymbol{\rho}(\mathbf{n}), \tag{5.6}$$

where

$$\boldsymbol{\rho}(\mathbf{n}) = \boldsymbol{\mu}(\mathbf{n}) + \int \boldsymbol{\mu}(\mathbf{n}')\,G(\vartheta)\,do'/4\pi. \tag{5.7}$$

In the general case, the solution of (5.6) can be expanded as a series of spherical harmonics $Y_{lm}(\theta, \phi)$, with the polar axis along **H**. Each term in the expansion represents a particular type of vibration with its frequency ω_{lm}.

The first frequency ω_{00} corresponds to vibrations with $\boldsymbol{\mu}$ = constant; then $\boldsymbol{\rho} = \boldsymbol{\mu}(1+\bar{G})$, and (5.6) becomes

$$i\omega_{00}\boldsymbol{\mu} = (2\beta/\hbar)\,\mathbf{H}\times\boldsymbol{\mu};$$

the vibrations are transverse to the field ($\boldsymbol{\mu}\perp\mathbf{H}$). Writing the equation in components in the plane perpendicular to **H** and taking the determinant, we find the frequency

$$\omega_{00} = 2\beta H/\hbar. \tag{5.8}$$

Here β is the magnetic moment of a particle (actual) in the liquid. Thus ω_{00} is independent of the specific properties of the liquid. The values of all the other frequencies ω_{lm}, however, depend on the specific form of the function $G(\vartheta)$.

§ 6. A degenerate almost ideal Fermi gas with repulsion between the particles

The problem of the thermodynamic properties of an "almost ideal" degenerate gas has no direct physical significance, since the gases that actually exist in Nature condense at temperatures near absolute zero. Nevertheless, in view of the considerable methodological interest of this problem, there is value in discussing it for a hypothetical model of a gas whose particles interact in such a way that the gas cannot condense.

The condition for the gas to be almost ideal is that the range r_0 of the molecular forces be small compared with the mean distance $l \sim (V/N)^{1/3}$ between the particles. As well as the condition $r_0 \ll l$, the inequality

$$pr_0/\hbar \ll 1 \tag{6.1}$$

is valid for the particle momenta p: in a degenerate Fermi gas, the limiting momentum p_F is estimated from (1.1), which gives $p_F/\hbar \sim (N/V)^{1/3} \ll 1/r_0$.

We shall consider here only a pair interaction between particles, and assume for simplicity that the interaction $U(r)$ is independent of the particle spins. Our aim is to calculate the leading terms in the expansion of the thermodynamic quantities in powers of the ratio r_0/l, by means of quantum-mechanical perturbation theory. The difficulty is that, because of the rapid increase of the interaction energy at small distances between the particles, perturbation theory (the "Born approximation") is in fact not applicable to particle collisions. This difficulty can, however, be circumvented in the following way.

In the limiting case of "slow" collisions (as for instance when the condition (6.1) holds), the mutual scattering amplitude of particles with mass m tends to a constant limit $-a$, which in the Born approximation (see *QM*, (126.13)) is

$$-a = -mU_0/4\pi\hbar^2, \quad U_0 = \int U(r)\, d^3x; \tag{6.2}$$

this limit corresponds to the s state of the pair of particles (with spin $\frac{1}{2}$). The constant a is called the *scattering length*.† Since this quantity entirely determines the properties of the collisions, it must also determine the thermodynamic properties of the gas.

This leads to the possibility of applying a procedure known as *renormalization*. We formally replace the true energy $U(r)$ by a different function having the same value of a but such that perturbation theory can be used. So long as (i.e. in an approximation such that) the final result of the calculations contains U only in the scattering amplitude, it will be the same as the result that would be given by the actual interaction.

† The expression (6.2) takes no account of the quantum-mechanical identity of the particles. In the limit of slow collisions of identical spin-$\frac{1}{2}$ particles, scattering occurs only for antiparallel spins, and the differential cross-section for scattering into the solid angle do (in the centre-of-mass system) is $d\sigma = 4a^2\, do$; the total cross-section is obtained by integrating $d\sigma$ over a hemisphere, and is $\sigma = 8\pi a^2$ (see *QM*, §137).

The range of the actual interaction is in general the same in order of magnitude as the scattering length a. For the fictitious field $U(r)$ which serves an auxiliary purpose, the condition for the Born approximation to be valid is $a \ll r_0$. The actual small expansion parameter in this theory is, of course, $ap_F/\hbar$.

We shall need the relation between U_0 and a not only in the first approximation (6.2), but also in the second Born approximation. To find this, we recall that, if the transition probability of the system under the action of a constant perturbation $\hat{V}$ is given in the first approximation by the matrix element V_{00}, then in the second approximation V_{00} is replaced by

$$V_{00}+\sum_n{}' \frac{V_{0n}V_{n0}}{E_0-E_n},$$

where the summation is over the states (with $n \neq 0$) of the unperturbed system (see *QM*, §43). In the present case we have a system of two colliding particles, and the perturbation is their interaction $U(r)$. The perturbation matrix elements for transitions in which the particle momenta $\mathbf{p}_1$ and $\mathbf{p}_2$ become $\mathbf{p}_1'$ and $\mathbf{p}_2'$ (with $\mathbf{p}_1+\mathbf{p}_2 = \mathbf{p}_1'+\mathbf{p}_2'$) are

$$\langle \mathbf{p}_1'\alpha_1, \mathbf{p}_2'\alpha_2 | U | \mathbf{p}_1\alpha_1, \mathbf{p}_2\alpha_2\rangle = \frac{1}{V}\int U(r)\, e^{-i\mathbf{p}.\mathbf{r}/\hbar}\, d^3x, \tag{6.3}$$

where $\mathbf{p} = \mathbf{p}_2'-\mathbf{p}_2 = -(\mathbf{p}_1'-\mathbf{p}_1)$; since the interaction is independent of the spins, the particle spin components α_1 and α_2 are unaltered by the collision. The matrix element for zero momenta U_0/V plays the role of V_{00}. Thus, in changing from the first to the second approximation, we must replace U_0 by

$$U_0+\frac{1}{V}\sum_{\mathbf{p}_1'}\left[\frac{p_1^2+p_2^2-p_1'^2-p_2'^2}{2m}\right]^{-1} \left|\int Ue^{-i\mathbf{p}.\mathbf{r}/\hbar}\, d^3x\right|^2;$$

the summation is for given $\mathbf{p}_1$ and $\mathbf{p}_2$, over $\mathbf{p}_1' \neq \mathbf{p}_1, \mathbf{p}_2$. Since in our case the particle momenta are assumed small, in all the important terms in the sum we can replace the matrix elements by their values at $p = 0$. We then get the following expression for the scattering length:†

$$a = \frac{m}{4\pi\hbar^2}\left[U_0+\frac{U_0^2}{V}\sum_{\mathbf{p}_1'}\frac{2m}{p_1^2+p_2^2-p_1'^2-p_2'^2}\right]. \tag{6.4}$$

Hence, with the same accuracy,

$$U_0 = \frac{4\pi\hbar^2 a}{m}\left[1-\frac{4\pi\hbar^2 a}{mV}\sum_{\mathbf{p}_1'}\frac{2m}{p_1^2+p_2^2-p_1'^2-p_2'^2}\right]. \tag{6.5}$$

† In all the intermediate formulae we write the sums over discrete values of the particle momenta with the particles in a finite volume V; in the final calculation the summation is replaced, as usual, by integration over $Vd^3p/(2\pi\hbar)^3$.

The divergence of the sum in (6.4) for large $\mathbf{p}_1'$ and $\mathbf{p}_2'$ is due to the replacement of all matrix elements by constants and is not important, since, when this expression is later used to calculate the energy of the system, a convergent expression is still obtained, in which large momenta are not significant. We take a to be the scattering length of slow particles, which is independent of their energy. Formula (6.4) seems at first sight to depend on the momenta $\mathbf{p}_1$ and $\mathbf{p}_2$, but in fact this dependence is restricted to the imaginary part of the scattering amplitude (which exists when the method of summation is appropriately defined; cf. *QM*, (130.9)), which need not be considered, since we know that the final result will be real. This topic will be resumed in §21.

In the present section, we shall consider the model of a Fermi gas with a repulsive interaction between the particles; for such an interaction, $a > 0$. In this case, the gas has an energy spectrum of the Fermi type described in §§1 and 2.

The Hamiltonian of a system of particles (with spin $\frac{1}{2}$) having a pair interaction is, in the second quantization method,

$$\hat{H} = \sum_{\mathbf{p},\alpha} \frac{p^2}{2m} \hat{a}_{\mathbf{p}\alpha}^+ \hat{a}_{\mathbf{p}\alpha} + \frac{1}{2} \sum \langle \mathbf{p}_1'\alpha_1, \mathbf{p}_2'\alpha_2 | U | \mathbf{p}_1\alpha_1, \mathbf{p}_2\alpha_2 \rangle \hat{a}_{\mathbf{p}_1'\alpha_1}^+ \hat{a}_{\mathbf{p}_2'\alpha_2}^+ \hat{a}_{\mathbf{p}_2\alpha_2} \hat{a}_{\mathbf{p}_1\alpha_1}; \quad (6.6)$$

see *QM*, §64. Here $\hat{a}_{\mathbf{p}\alpha}^+$ and $\hat{a}_{\mathbf{p}\alpha}$ are creation and annihilation operators for a free particle with momentum $\mathbf{p}$ and spin component $\alpha\,(= \pm\frac{1}{2})$. The first term in (6.6) corresponds to the kinetic energy of the particles, and the second term to their potential energy; in the latter, the summation is over all values of the momenta and spin components, subject to the conservation of momentum in the collisions.

In accordance with the assumption that the particle momenta are small, we again replace the matrix elements in (6.6) by their values for zero momenta: $\langle 0\alpha_1, 0\alpha_2 | U | 0\alpha_1, 0\alpha_2 \rangle = U_0/V$. Next we note that, since the operators $\hat{a}_{\mathbf{p}_1\alpha_1}$ and $\hat{a}_{\mathbf{p}_2\alpha_2}$ anticommute in Fermi statistics, their product is antisymmetrical with respect to the interchange of suffixes; the same applies to the products $\hat{a}_{\mathbf{p}_1'\alpha_1}^+ \hat{a}_{\mathbf{p}_2'\alpha_2}^+$. In consequence, all terms cancel in the second sum in (6.6) that contain pairs of equal suffixes α_1, α_2 (physically, this occurs because of the fact already mentioned that, in the limit of slow collisions, only particles with opposite spins can scatter each other).

The Hamiltonian of the system thus becomes

$$\hat{H} = \sum_{\mathbf{p},\alpha} \frac{p^2}{2m} \hat{a}_{\mathbf{p}\alpha}^+ \hat{a}_{\mathbf{p}\alpha} + \frac{U_0}{V} \sum_{\mathbf{p}_1, \mathbf{p}_2, \mathbf{p}_1'} \hat{a}_{1+}'^{+} \hat{a}_{2-}'^{+} \hat{a}_{2-} \hat{a}_{1+}, \quad (6.7)$$

where $\hat{a}_{1+} \equiv \hat{a}_{\mathbf{p}_1+}$, $\hat{a}_{1+}' \equiv \hat{a}_{\mathbf{p}_1'+}$, etc., and the suffixes + and − here and henceforward replace $+\frac{1}{2}$ and $-\frac{1}{2}$.

The eigenvalues of this Hamiltonian are calculated by ordinary perturbation theory; the second term in (6.6) is treated as a small correction to the first term. The first term is diagonal, and its eigenvalues are

$$E^{(0)} = \sum_{\mathbf{p},\,\alpha} (p^2/2m)\, n_{\mathbf{p}\alpha}, \tag{6.8}$$

where $n_{\mathbf{p}\alpha}$ are the occupation numbers of the states $\mathbf{p}$, α.†

The first-order correction is given by the diagonal matrix elements of the interaction energy:

$$E_1^{(1)} = \frac{U_0}{V} \sum_{\mathbf{p}_1,\,\mathbf{p}_2} n_{1+} n_{2-}, \tag{6.9}$$

where $n_{1+} \equiv n_{\mathbf{p}_1+}$ etc.

To find the second-order correction, we use the known formula of perturbation theory,

$$E_n^{(2)} = \sum_m{}' \frac{|V_{nm}|^2}{E_n - E_m},$$

where the suffixes n and m label the states of the unperturbed system. A simple calculation (with the known matrix elements of the operators $\hat{a}_{\mathbf{p}\alpha}$ and $\hat{a}^+_{\mathbf{p}\alpha}$) gives

$$\frac{U_0^2}{V^2} \sum_{\mathbf{p}_1,\,\mathbf{p}_2,\,\mathbf{p}_1'} \frac{n_{1+} n_{2-} (1-n'_{1+})(1-n'_{2-})}{(p_1^2 + p_2^2 - p_1'^2 - p_2'^2)/2m}. \tag{6.10}$$

The structure of this expression is very clear: the squared matrix element of the transition $\mathbf{p}_1, \mathbf{p}_2 \to \mathbf{p}_1', \mathbf{p}_2'$ is proportional to the occupation numbers of the states $\mathbf{p}_1$, $\mathbf{p}_2$, and to the numbers of unoccupied positions in the states $\mathbf{p}_1'$, $\mathbf{p}_2'$.

The integral U_0 in (6.9) and (6.10) must be expressed in terms of a real physical quantity, the scattering amplitude $-a$. In the second-order terms this can be done from (6.2); in the first-order terms, the more exact formula (6.5) is needed. After these substitutions, we find as the correction of the first order in a

$$E^{(1)} = \frac{g}{V} \sum_{\mathbf{p}_1,\,\mathbf{p}_2} n_{1+} n_{2-} \tag{6.11}$$

and as the second-order correction

$$E^{(2)} = \frac{2mg^2}{V^2} \sum_{\mathbf{p}_1,\,\mathbf{p}_2,\,\mathbf{p}_1'} \frac{n_{1+} n_{2-} [(1-n'_{1+})(1-n'_{2-}) - 1]}{p_1^2 + p_2^2 - p_1'^2 - p_2'^2};$$

for brevity, we use in the intermediate formulae the "coupling constant" of the gas particles‡ $g = 4\pi\hbar^2 a/m$. In expanding the expression in the numerator,

† By assuming that the particles have definite values of the spin component, we assume that the statistical matrix $n_{\alpha\beta}(\mathbf{p})$ is also reduced to diagonal form; the functions $n_\alpha(\mathbf{p})$ with $\alpha = \pm\frac{1}{2}$ are then its diagonal elements.

‡ After the renormalization of the scattering amplitude, this quantity is no longer equal to the constant U_0 in (6.2).

we note that the terms with products of four n cancel, because their numerators are symmetrical and their denominators antisymmetrical with respect to the interchange of $\mathbf{p}_1$, $\mathbf{p}_2$ and $\mathbf{p}_1'$, $\mathbf{p}_2'$; and the summation over these variables is symmetrical. The final result is

$$E^{(2)} = -\frac{2mg^2}{V^2} \sum_{\mathbf{p}_1, \mathbf{p}_2, \mathbf{p}_1'} \frac{n_{1+}n_{2-}(n_{1+}'+n_{2-}')}{p_1^2+p_2^2-p_1'^2-p_2'^2}. \tag{6.12}$$

This sum (in which all the $n_{\mathbf{p}\alpha} \to 0$ as $\mathbf{p} \to \infty$) is convergent.

From these formulae we can calculate, first of all, the energy of the ground state. To do so, we must put all the $n_{\mathbf{p}\alpha}$ equal to unity within the Fermi sphere ($p < p_F = \hbar(3\pi^2 N/V)^{1/3}$) and zero outside. Here it should be noted that, although in the original Hamiltonian the eigenvalues of the operator products $\hat{a}_{\mathbf{p}\alpha}^+\hat{a}_{\mathbf{p}\alpha}$ give the occupation numbers of the states of the gas particles themselves, after diagonalizing the Hamiltonian by means of perturbation theory we are concerned with the quasi-particle distribution function (denoted, as in previous sections, by $n_{\mathbf{p}\alpha}$).

Since $\sum n_{\mathbf{p}+} = \sum n_{\mathbf{p}-} = \frac{1}{2}N$, we find from (6.11) the first-order correction

$$E_0^{(1)} = gN^2/4V.$$

In (6.12) we replace the summation over three momenta, together with the condition $\mathbf{p}_1+\mathbf{p}_2 = \mathbf{p}_1'+\mathbf{p}_2'$, by integration over

$$\frac{V^3}{(2\pi\hbar)^9}\,\delta(\mathbf{p}_1+\mathbf{p}_2-\mathbf{p}_1'-\mathbf{p}_2')\,d^3p_1\,d^3p_2\,d^3p_1'\,dp_2',$$

so that

$$E_0^{(2)} = -\frac{4mg^2V}{(2\pi\hbar)^9}\int \frac{\delta(\mathbf{p}_1+\mathbf{p}_2-\mathbf{p}_1'-\mathbf{p}_2')}{p_1^2+p_2^2-p_1'^2-p_2'^2}\,d^3p_1\,d^3p_2\,d^3p_1'\,d^3p_2',$$

the integration being taken over the range $p_1, p_2, p_1' \leq p_F$. The calculation of the integral† gives the following final result for the energy of the ground state:

$$E_0 = N\frac{3p_F^2}{10m}\left[1+\frac{10}{9\pi}\frac{p_Fa}{\hbar}+\frac{4(11-2\log 2)}{21\pi^2}\left(\frac{p_Fa}{\hbar}\right)^2\right], \tag{6.13}$$

where the coefficient of the square bracket is the energy of an ideal Fermi gas (K. Huang and C. N. Yang 1957).

The chemical potential of the gas at absolute zero is given by the derivative $\mu = (\partial E_0/\partial N)_V$. Expressed in terms of the limiting momentum p_F, it is

$$\mu = \frac{p_F^2}{2m}\left[1+\frac{4}{3\pi}\frac{p_Fa}{\hbar}+\frac{4(11-2\log 2)}{15\pi^2}\left(\frac{p_Fa}{\hbar}\right)^2\right]. \tag{6.14}$$

† In practice, it is simpler to proceed in a different order, beginning with the calculation of the function f (see below).

According to the general ideas of the Landau theory, the spectrum of elementary excitations $\varepsilon(\mathbf{p})$ and the quasi-particle interaction function $f_{\alpha\alpha'}(\mathbf{p}, \mathbf{p}')$ are determined by the first and second variations of the total energy with respect to the quasi-particle distribution function.† If E is written as a discrete sum over $\mathbf{p}$ and α, we have by definition

$$\delta E = \sum_{\mathbf{p}, \alpha} \varepsilon_\alpha(\mathbf{p})\, \delta n_{\mathbf{p}\alpha} + \frac{1}{2V} \sum_{\mathbf{p}, \alpha, \mathbf{p}', \alpha'} f_{\alpha\alpha'}(\mathbf{p}, \mathbf{p}')\, \delta n_{\mathbf{p}\alpha}\, \delta n_{\mathbf{p}'\alpha'}, \qquad (6.15)$$

(after differentiation of the energy, $n_{\mathbf{p}\alpha}$ is to be replaced by unity within the Fermi sphere and zero outside). There is, however, no need to calculate in this way the effective mass m^* of the quasi-particles, since it can be found more simply (see below).

To calculate the function $f_{\alpha\alpha'}(\mathbf{p}, \mathbf{p}')$ (on the Fermi surface), we twice differentiate the sum of the expressions (6.11) and (6.12), and then put $p = p' = p_F$. After making this simple calculation and changing from summation to integration, we have

$$f_{+-}(\mathbf{p}, \mathbf{p}') = g - \frac{4mg^2}{(2\pi\hbar)^3} \int \left\{ \frac{\delta(\mathbf{p}+\mathbf{p}'-\mathbf{p}_1-\mathbf{p}_2)}{2p_F^2-p_1^2-p_2^2} + \frac{\delta(\mathbf{p}+\mathbf{p}_1-\mathbf{p}'-\mathbf{p}_2)+\delta(\mathbf{p}'+\mathbf{p}_1-\mathbf{p}-\mathbf{p}_2)}{2(p_1^2-p_2^2)} \right\} d^3p_1\, d^3p_2,$$

$$f_{++}(\mathbf{p}, \mathbf{p}') = f_{--}(\mathbf{p}, \mathbf{p}') = \frac{2mg^2}{(2\pi\hbar)^3} \int \frac{\delta(\mathbf{p}+\mathbf{p}_1-\mathbf{p}'-\mathbf{p}_2)+\delta(\mathbf{p}'+\mathbf{p}_1-\mathbf{p}-\mathbf{p}_2)}{p_1^2-p_2^2}\, d^3p_1\, d^3p_2.$$

The integration in these formulae is comparatively simple, because of the lower multiplicity of the integrals.

The final result is to be put in the form (2.4), which is independent of the choice of the spin quantization axis. In this form it is

$$f_{\alpha\gamma,\beta\delta} = \frac{2\pi a\hbar^2}{m} \left\{ \left[1 + \frac{2ap_F}{\pi\hbar} \left(2 + \frac{\cos\vartheta}{2\sin\frac{1}{2}\vartheta} \log \frac{1+\sin\frac{1}{2}\vartheta}{1-\sin\frac{1}{2}\vartheta} \right) \right] \delta_{\alpha\beta}\delta_{\gamma\delta} - \left[1 + \frac{2ap_F}{\pi\hbar^2} \left(1 - \tfrac{1}{2}\sin\tfrac{1}{2}\vartheta \log \frac{1+\sin\frac{1}{2}\vartheta}{1-\sin\frac{1}{2}\vartheta} \right) \right] \boldsymbol{\sigma}_{\alpha\beta}\cdot\boldsymbol{\sigma}_{\gamma\delta} \right\}, \qquad (6.16)$$

where ϑ is the angle between the vectors $\mathbf{p}_F$ and $\mathbf{p}'_F$ (A. A. Abrikosov and I. M. Khalatnikov 1957).‡

† The matrix $f_{\alpha\alpha'}(\mathbf{p},\mathbf{p}')$ in this section is made up of the elements of the matrix $f_{\alpha\gamma,\beta\delta}(\mathbf{p},\mathbf{p}')$ that are diagonal in two pairs of suffixes (α, β and γ, δ).

‡ The function (6.16) becomes logarithmically infinite at $\vartheta = \pi$. This is because of the approximations made. A more exact analysis shows that, although $\vartheta = \pi$ is indeed a singularity of the function, the latter is zero there, not infinite; see the third footnote to §54. The invalidity of (6.16) near $\vartheta = \pi$ is unimportant in subsequent applications, which invol ve integrals convergent at this point.

The effective mass of the quasi-particles is found from this by integration as in (2.12):

$$\frac{m^*}{m} = 1+\frac{8}{15\pi^2}(7\log 2-1)\left(\frac{ap_F}{\hbar}\right)^2. \tag{6.17}$$

Formula (2.17) gives the velocity of sound in the gas:

$$u^2 = \frac{p_F^2}{3m^2}\left[1+\frac{2}{\pi}\frac{ap_F}{\hbar}+\frac{8(11-2\log 2)}{15\pi^2}\left(\frac{ap_F}{\hbar}\right)^2\right]. \tag{6.18}$$

Then, integrating u^2m/N (expressed in terms of N/V instead of p_F) with respect to N, we find from (2.13) the chemical potential of the gas, and a further integration with respect to N gives the expression (6.13) for the energy of the ground state.

Formula (6.13) represents the first terms in an expansion of the gas energy in powers of the "gaseousness parameter" $\eta = p_F a/\hbar \sim a(N/V)^{1/3}$. By similar but considerably more laborious calculations, we could derive some further terms in the expansion. The reason is that, in a Fermi gas, triple collisions contribute to the energy only in a fairly high approximation. Of three colliding particles, at least two have the same spin component; the coordinate wave function of the system must then be antisymmetric with respect to these two particles. Thus the orbital angular momentum of the relative motion of these particles is at least 1 (p state). The corresponding wave function contains an extra power of $p/\hbar$ in comparison with the s-state wave function (see *QM*, §33), and therefore the probability of such a collision contains an extra factor p^2, i.e. is reduced by a factor $\sim (pa/\hbar)^2 \sim \eta^2$ in comparison with that of a "head-on" collision of particles not obeying the Pauli principle. In consequence, triple collisions contribute to the energy only in terms containing the volume as $V^{-2}V^{-2/3}$. In other words, all terms in the expansion of the energy up to those of order $N(p_F^2/m)\eta^5$ inclusive, i.e. three more beyond those shown in (6.13), are expressed in terms of the characteristics of pair collisions only. However, these characteristics will include not only the amplitude of s-wave scattering for slow collisions, as in (6.13), but also its derivatives with respect to the energy, and the amplitude of p-wave scattering.

CHAPTER II

GREEN'S FUNCTIONS IN A FERMI SYSTEM AT $T = 0$

§ 7. Green's functions in a macroscopic system

THE method used in § 6 becomes laborious and in practice unusable in the higher orders of perturbation theory. This disadvantage is the more important in that the interaction between particles in actual physical problems is certainly not weak, and so, to ascertain the various general properties of macroscopic systems, we have to consider infinite sequences of terms in the perturbation-theory series. To overcome such difficulties, we can use a mathematical formalism similar to the one in quantum field theory.

The specific form of this treatment depends essentially on the nature of the macroscopic system to which it is to be applied. The subsequent sections of this chapter deal with the development of the formalism for a Fermi liquid at absolute zero.† The purpose of the exposition is not only the practical application of the method to such a system, but also to show how the formalism itself is constructed.

The starting-point is the second-quantized ψ operators, whose properties are known from quantum mechanics (see *QM*, §§64, 65). Here we shall need them in the Heisenberg representation, in which they depend explicitly on the time. We therefore begin by establishing some properties of the ψ operators in that representation.

We shall consider systems of spin-$\frac{1}{2}$ particles. Accordingly, the ψ operators must be given a suffix that indicates the value of the spin component and takes the values $\pm\frac{1}{2}$; these suffixes will again be written as Greek letters, and summation over repeated suffixes is implied.

By the general rule (see *QM*, § 13), the operator $\hat{f}(t)$ of any physical quantity in the Heisenberg representation is expressed in terms of the time-independent (Schrödinger) operator $\hat{f}$ of the same quantity by‡

$$\hat{f}(t) = e^{i\hat{H}t}\hat{f}e^{-i\hat{H}t},$$

where $\hat{H}$ is the Hamiltonian of the system.

† The systematic construction of this formalism is due to V. M. Galitskiĭ and A. B. Migdal (1958).

‡ In order to simplify the formulae, we shall often use units such that the quantum constant $\hbar = 1$ (so that the momentum and energy have dimensions of reciprocal length and recipro-

Here, however, it will be appropriate to modify this definition somewhat. The reason is that in quantum statistics it is more convenient to consider the states of the system not for a specified number N of particles in it but for a specified chemical potential μ. The ground state of the system, in which it is found at $T = 0$, can then be defined as the state having the lowest eigenvalue of the operator

$$\hat{H}' = \hat{H} - \mu\hat{N}, \tag{7.1}$$

and not of $\hat{H}$ as when N is specified: the probability that the system is (for a specified value of μ) in a state with energy E_n and number of particles N_n is

$$w \propto \exp\left(-\frac{E_n - \mu N_n}{T}\right) = \exp\left(-\frac{E_n'}{T}\right);$$

see Part 1, (35.1). Here E_n' are the eigenvalues of the operator $\hat{H}'$. We see that at $T = 0$ only the state with the lowest E_n' remains.†

Thus we define the Heisenberg ψ operators by the formulae

$$\left.\begin{aligned} \hat{\Psi}_\alpha(t, \mathbf{r}) &= e^{i\hat{H}'t}\hat{\psi}_\alpha(\mathbf{r})\, e^{-i\hat{H}'t}, \\ \hat{\Psi}_\alpha^+(t, \mathbf{r}) &= e^{i\hat{H}'t}\hat{\psi}_\alpha^+(\mathbf{r})\, e^{-i\hat{H}'t}. \end{aligned}\right\} \tag{7.2}$$

The Heisenberg ψ operators will be denoted by the capital letter $\hat{\Psi}$, and the Schrödinger ψ operators by $\hat{\psi}$.

The Schrödinger ψ operators obey the familiar commutation rules. The commutators of the Heisenberg operators taken at different times t and t' cannot be calculated in a general form, however. When $t = t'$, the commutation rules are the same as for the Schrödinger operators. Thus, from the rule

$$\hat{\psi}_\alpha(\mathbf{r})\hat{\psi}_\beta^+(\mathbf{r}') + \hat{\psi}_\beta^+(\mathbf{r}')\,\hat{\psi}_\alpha(\mathbf{r}) = \delta_{\alpha\beta}\,\delta(\mathbf{r}-\mathbf{r}')$$

we have the corresponding rule

$$\begin{aligned} &\hat{\Psi}_\alpha(t, \mathbf{r})\,\hat{\Psi}_\beta^+(t, \mathbf{r}') + \hat{\Psi}_\beta^+(t, \mathbf{r}')\,\hat{\Psi}_\alpha(t, \mathbf{r}) \\ &\quad = e^{i\hat{H}'t}[\hat{\psi}_\alpha(\mathbf{r})\,\hat{\psi}_\beta^+(\mathbf{r}') + \hat{\psi}_\beta^+(\mathbf{r}')\,\hat{\psi}_\alpha(\mathbf{r})]\, e^{-i\hat{H}'t} \\ &\quad = \delta_{\alpha\beta}\delta(\mathbf{r}-\mathbf{r}'). \end{aligned} \tag{7.3}$$

Similarly,

$$\left.\begin{aligned} \hat{\Psi}_\alpha(t, \mathbf{r})\,\hat{\Psi}_\beta(t, \mathbf{r}') + \hat{\Psi}_\beta(t, \mathbf{r}')\,\hat{\Psi}_\alpha(t, \mathbf{r}) &= 0, \\ \hat{\Psi}_\alpha^+(t, \mathbf{r})\,\hat{\Psi}_\beta^+(t, \mathbf{r}') + \hat{\Psi}_\beta(t, \mathbf{r}')\,\hat{\Psi}_\alpha^+(t, \mathbf{r}) &= 0. \end{aligned}\right\} \tag{7.4}$$

---

cal time respectively). To change from these to ordinary units, all momenta $\mathbf{p}$ and energies E in the formulae are to be replaced by $\mathbf{p}/\hbar$ and $E/\hbar$. Such units will, in particular, be used in the present chapter.

† The term 'Hamiltonian' will be used for both $\hat{H}$ and $\hat{H}'$.

Differentiating the definition (7.2) with respect to time, we find that the Heisenberg ψ operator satisfies the equation

$$-i\frac{\partial}{\partial t}\hat{\Psi}_\alpha(t, \mathbf{r}) = \hat{H}'\hat{\Psi}_\alpha(t,\mathbf{r}) - \hat{\Psi}_\alpha(t,\mathbf{r})\hat{H}'; \tag{7.5}$$

cf. *QM*, (13.7).

The Heisenberg and Schrödinger representations are identical for the operator of any conserved quantity (i.e. an operator that commutes with the Hamiltonian). This is true, in particular, of the Hamiltonian itself, and of the particle number operator, which also of course belongs to a conserved quantity. The expressions for these operators in terms of Schrödinger and Heisenberg ψ operators are the same. For example, the particle number operator is

$$\begin{aligned}\hat{N} &= \int \hat{\psi}^+_\alpha(\mathbf{r})\,\hat{\psi}_\alpha(\mathbf{r})\,d^3x \\ &= \int \hat{\Psi}^+_\alpha(t, \mathbf{r})\,\hat{\Psi}_\alpha(t, \mathbf{r})\,d^3x.\end{aligned} \tag{7.6}$$

The Hamiltonian of a system of interacting particles is

$$\left.\begin{aligned}\hat{H}' &= \hat{H}'^{(0)} + \hat{V}^{(1)} + \hat{V}^{(2)} + \ldots, \\ \hat{H}'^{(0)} &= -\frac{1}{2m}\int \hat{\Psi}^+_\alpha(t, \mathbf{r})\,\triangle\hat{\Psi}_\alpha(t, \mathbf{r})\,d^3x - \mu\hat{N}, \\ \hat{V}^{(1)} &= \int \hat{\Psi}^+_\alpha(t, \mathbf{r})\,U^{(1)}(\mathbf{r})\,\hat{\Psi}_\alpha(t, \mathbf{r})\,d^3x, \\ \hat{V}^{(2)} &= \tfrac{1}{2}\int \hat{\Psi}^+_\beta(t, \mathbf{r})\,\hat{\Psi}^+_\alpha(t, \mathbf{r}')\,U^{(2)}(\mathbf{r}-\mathbf{r}')\,\hat{\Psi}_\alpha(t, \mathbf{r}')\,\hat{\Psi}_\beta(t, \mathbf{r})\,d^3x\,d^3x',\end{aligned}\right\} \tag{7.7}$$

where $\hat{H}'^{(0)}$ is the Hamiltonian of a system of free particles; $\hat{V}^{(1)}$ is the operator of their interaction with the external field $U^{(1)}(\mathbf{r})$; $\hat{V}^{(2)}$ the operator of their pair interaction, $U^{(2)}(\mathbf{r}-\mathbf{r}')$ being the interaction energy of two particles. The omitted terms represent triple etc. interactions; cf. *QM*, (64.25). For simplicity, all interactions are assumed to be independent of the spins of the particles.

The commutator of $\hat{H}'$ and $\hat{\Psi}_\alpha$ in (7.5) is calculated by means of the rules (7.3) and (7.4); the delta functions that appear are removed by integration. We thus obtain a "Schrödinger equation" for $\hat{\Psi}_\alpha(t, \mathbf{r})$, in the form

$$\begin{aligned}i\frac{\partial}{\partial t}\hat{\Psi}_\alpha(t, \mathbf{r}) &= \left(-\frac{1}{2m}\triangle - \mu + U^{(1)}(\mathbf{r})\right)\hat{\Psi}_\alpha(t, \mathbf{r}) \\ &\quad + \int \hat{\Psi}^+_\beta(t, \mathbf{r}')\,U^{(2)}(\mathbf{r}-\mathbf{r}')\,\hat{\Psi}_\beta(t, \mathbf{r}')\,d^3x'.\hat{\Psi}_\alpha(t, \mathbf{r}) + \ldots.\end{aligned} \tag{7.8}$$

The concept of the *Green's function* for a macroscopic system is fundamental in the method described here. This function is defined by†

$$G_{\alpha\beta}(X_1, X_2) = -i\langle \mathrm{T}\hat{\Psi}_\alpha(X_1)\,\hat{\Psi}^+_\beta(X_2)\rangle. \tag{7.9}$$

Here and and below, X denotes for brevity the time t together with the position vector $\mathbf{r}$. The angle brackets $\langle\ldots\rangle$ denote averaging with respect to the ground

† This definition is analogous to that of the exact Green's functions (propagators) in quantum electrodynamics (cf. *RQT*, §§100, 102).

state of the system, instead of the more cumbersome notation $\langle 0|\ldots|0\rangle$ for the diagonal matrix element. The symbol T denotes the chronological product: the operators following it are to be arranged from right to left in order of increasing times t_1, t_2. For fermions, the interchange of a pair of ψ operators (as compared with their arrangement in the original writing of the product) must change the sign of the product. Explicitly,

$$G_{\alpha\beta}(X_1, X_2) = \begin{Bmatrix} -i\langle\hat{\Psi}_\alpha(X_1)\hat{\Psi}^+_\beta(X_2)\rangle & \text{for} & t_1 > t_2, \\ i\langle\hat{\Psi}^+_\beta(X_2)\hat{\Psi}_\alpha(X_1)\rangle & \text{for} & t_1 < t_2. \end{Bmatrix} \tag{7.10}$$

There are some obvious properties of the Green's function. If the system is not ferromagnetic and not in an external field, the spin dependence of the Green's function reduces to a unit matrix:

$$G_{\alpha\beta}(X_1, X_2) = \delta_{\alpha\beta}G(X_1, X_2); \tag{7.11}$$

any other dependence would distinguish a particular direction in space, the z-axis of spin quantization.† Since time is homogeneous, t_1 and t_2 appear in the Green's function only as the difference $t = t_1 - t_2$. If also the system is microscopically homogeneous in space, the coordinates of the two points appear only as the difference $\mathbf{r} = \mathbf{r}_1 - \mathbf{r}_2$. In other words, for this case we have

$$G_{\alpha\beta}(X_1, X_2) = \delta_{\alpha\beta}G(X), \quad X = X_1 - X_2. \tag{7.12}$$

It must be emphasized that microscopic homogeneity means that the body is assumed homogeneous not only as regards its mean (macroscopic) density but also as regards the probability density of various (microscopic) positions of its particles in space. Liquids and gases have this property (but solid crystals do not). Their isotropy has the result that $G(t, \mathbf{r}) = G(t, -\mathbf{r})$. In this connection, let us note once again that the function $G(t, \mathbf{r})$, by its definition, is certainly not an even function of t. The order of t_1 and t_2 in the difference $t = t_1 - t_2$ is for that reason significant.

The coordinate density matrix of a particle in the system is defined as the mean value

$$\varrho_{\alpha\beta}(\mathbf{r}_1, \mathbf{r}_2) = \frac{1}{N}\langle\hat{\Psi}^+_\beta(t, \mathbf{r}_2)\hat{\Psi}_\alpha(t, \mathbf{r}_1)\rangle. \tag{7.13}$$

From a knowledge of this matrix we can find the mean value of any quantity pertaining to an individual particle. Let $\hat{F}_{\alpha\beta}$ be some "one-particle" operator, i.e. an operator of the form

$$\hat{F}_{\alpha\beta} = \sum_a \hat{f}^{(a)}_{\alpha\beta}, \tag{7.14}$$

where $\hat{f}^{(a)}_{\alpha\beta}$ is an operator acting on the coordinates and spin of only one (the ath) particle, and the summation is over all particles in the system. In the

† This statement needs elucidation. The spin components $\hat{\Psi}_\alpha$ form a contravariant spinor of rank one (and in this sense it would be more correct to raise the index: $\hat{\Psi}^\alpha$). The components $\hat{\Psi}^+_\beta$ form a covariant spinor. Thus $G_{\alpha\beta}$ is a mixed spinor of rank two, and $\delta_{\alpha\beta}$ is a unit spinor of this kind.

second-quantization formalism, such an operator is written (in the Heisenberg representation) as

$$\hat{F}_{\alpha\beta}(t) = \int \hat{\Psi}^{+}_{\alpha}(t, \mathbf{r}) f_{\beta\gamma} \hat{\Psi}_{\gamma}(t, \mathbf{r})\, d^3x; \tag{7.15}$$

cf. *QM*, (64.23). Hence it is clear that the mean value of F can be expressed in terms of the density matrix as

$$\langle F \rangle = N\langle f \rangle = \int [\hat{f}^{(1)}_{\beta\alpha} \varrho_{\beta\alpha}(\mathbf{r}_1, \mathbf{r}_2)]_{\mathbf{r}_1=\mathbf{r}_2}\, d^3x_1, \tag{7.16}$$

where $\hat{f}^{(1)}_{\alpha\beta}$ is an operator acting on the coordinates $\mathbf{r}_1$ (we put $\mathbf{r}_2 = \mathbf{r}_1$ after applying this operator but before integrating).

According to (7.10), the density matrix can be expressed in terms of the Green's function:

$$\varrho_{\alpha\beta}(\mathbf{r}_1, \mathbf{r}_2) = -\frac{i}{N} G_{\alpha\beta}(t_1, \mathbf{r}_1;\ t_1+0, \mathbf{r}_2). \tag{7.17}$$

Here, and everywhere henceforward, writing the argument of the function as t_1+0 signifies taking the limit as it tends to t_1 from above. This ensures the correct arrangement of the ψ operators, as in the product (7.13).

For a microscopically homogeneous system, the density matrix depends only on the difference $\mathbf{r} = \mathbf{r}_1 - \mathbf{r}_2$, and if there is no spin dependence, $\varrho_{\alpha\beta} = \delta_{\alpha\beta}\varrho$, with

$$\varrho(\mathbf{r}) = -\frac{i}{N} G(t = -0, \mathbf{r}); \tag{7.18}$$

here $G_{\alpha\beta}(X_1, X_2)$ has been replaced by $G(X_1 - X_2) \equiv G(X)$ in accordance with (7.12). With $\mathbf{r}_1 = \mathbf{r}_2$, after taking the trace with respect to the spin variables, the operator product in (7.13) becomes $\hat{\Psi}^{+}_{\alpha}\hat{\Psi}_{\alpha}$, the operator of the particle number density in the system. The mean density of the body is therefore

$$N/V = 2N\varrho(0) = -2iG(t = -0, \mathbf{r} = 0), \tag{7.19}$$

where t tends to zero from below. This equation relates the chemical potential μ at $T = 0$ (on which G depends as a parameter) to the particle number density N/V.

The Fourier expansion of the function $\varrho(\mathbf{r}_1, \mathbf{r}_2)$ determines the momentum distribution of the particles:†

$$\begin{aligned} N(\mathbf{p}) &= N \int \varrho(\mathbf{r}_1, \mathbf{r}_2)\, e^{-i\mathbf{p}\cdot(\mathbf{r}_1-\mathbf{r}_2)}\, d^3(x_1 - x_2) \\ &= -i \int [G(t, \mathbf{r})]_{t=-0}\, e^{-i\mathbf{p}\cdot\mathbf{r}}\, d^3x. \end{aligned} \tag{7.20}$$

† The one-particle density matrix is (see *QM*, §14) the integral

$$\varrho(\mathbf{r}_1, \mathbf{r}_2) = \int \Psi^*(\mathbf{r}_2, q)\, \Psi(\mathbf{r}_1, q)\, dq,$$

where $\Psi(\mathbf{r}, q)$ is the wave function of the system as a whole, $\mathbf{r}$ denoting the position vector of one particle and q the set of coordinates of all the other particles, with integration over these. The Fourier components of the density matrix are equal to

$$\int \left| \int \Psi(\mathbf{r}, q) e^{i\mathbf{p}\cdot\mathbf{r}}\, d^3x \right|^2 dq,$$

and this gives its relation to the particle momentum distribution.

This is the number of particles (per unit volume) with a specified value of the spin component and with momenta in the range $d^3p/(2\pi)^3$. Here we are referring to actual particles, not to quasi-particles (which have not yet made an appearance in the formalism being described). The notation $N(\mathbf{p})$ is used in contrast to the quasi-particle distribution function $n(\mathbf{p})$.

We shall usually be concerned with the Green's function in the momentum representation, defined as the component of the Fourier expansion of $G(t, \mathbf{r})$ with respect to t and $\mathbf{r}$:

$$G(t, \mathbf{r}) = \int G(\omega, \mathbf{p}) e^{i(\mathbf{p}.\mathbf{r}-\omega t)}\, d\omega\, d^3p/(2\pi)^4, \tag{7.21}$$

$$G(\omega, \mathbf{p}) = \int G(t, \mathbf{r})\, e^{-i(\mathbf{p}.\mathbf{r}-\omega t)}\, dt\, d^3x. \tag{7.22}$$

The particle momentum distribution is expressed in terms of this function by

$$N(\mathbf{p}) = -i \lim_{t\to -0} \int_{-\infty}^{\infty} G(\omega, \mathbf{p}) e^{-i\omega t} \frac{d\omega}{2\pi}, \tag{7.23}$$

which is found by substituting (7.21) in (7.20). It is normalized by

$$-2i \lim_{t\to -0} \int G(\omega, \mathbf{p})\, e^{-i\omega t} \frac{d\omega\, d^3p}{(2\pi)^4} = \frac{N}{V}, \tag{7.24}$$

which is the condition (7.19) in the momentum representation. Thus the distribution $N(\mathbf{p})$ automatically has the correct normalization

$$2 \int N(\mathbf{p})\, d^3p/(2\pi)^3 = N/V.$$

The limit in which the integrals (7.23) and (7.24) are taken is equivalent to a particular contour rule in the plane of the complex variable ω. The presence of the factor $e^{-i\omega t}$ with $t < 0$ allows the path of integration (the real axis) to be closed by an infinite semicircle in the upper half-plane of ω, so that the integral is determined by the residues of $G(\omega, \mathbf{p})$ at its poles in that half-plane.

§ 8. Determination of the energy spectrum from the Green's function

For a microscopically homogeneous system, it is easy to determine the time and coordinate dependence of the matrix elements of the Heisenberg ψ operator with respect to stationary states having definite values of the energy and momentum.

The time dependence is given by the usual exponential factor:

$$\langle n \,|\, \hat{\Psi}_\alpha(t, \mathbf{r}) \,|\, m\rangle = e^{i\omega_{nm}t} \langle n \,|\, \hat{\psi}_\alpha(\mathbf{r}) \,|\, m\rangle, \tag{8.1}$$

but, since the Heisenberg ψ operator is defined by means of the Hamiltonian $\hat{H}'$, we have

$$\begin{aligned}\omega_{nm} &= E'_n - E'_m \\ &= E_n - E_m - \mu(N_n - N_m).\end{aligned}$$

According to the general properties of ψ operators, $\hat{\Psi}$ decreases (and $\hat{\Psi}^+$ increases) by one the number of particles in the system. Hence $N_n = N_m - 1$ in the matrix element (8.1), so that

$$\omega_{nm} = E_n(N) - E_m(N+1) + \mu, \tag{8.2}$$

where the arguments are the numbers of particles in the corresponding states.

To determine the dependence on the coordinates, we note that, since the system is homogeneous, a displacement relative to the system through an arbitrary distance $\mathbf{r}$ cannot alter the matrix elements of its ψ operators. This does not mean, however, that the matrix elements are independent of the coordinates. The reason is that the difference between $\psi_{nm}(\mathbf{r})$ and the value $\psi_{nm}(0)$ at some specified point $\mathbf{r} = 0$ is due to two causes: the displacement through $\mathbf{r}$ relative to the system itself, and the movement of the point of observation to a different position, which also changes the phases of the wave functions. In order to exclude this latter change, we shift the system through $-\mathbf{r}$, i.e. apply to its wave functions the parallel-translation operator

$$\hat{T}(-\mathbf{r}) = e^{-i\mathbf{r}.\hat{\mathbf{P}}},$$

where $\hat{\mathbf{P}}$ is the operator of the total momentum of the system; see *QM*, (15.13). These operations return the point of observation to its original position, but it remains shifted by $\mathbf{r}$ relative to the system. The invariance of the matrix elements under this transformation is expressed by

$$\langle n \,|\, \hat{\psi}_\alpha(0) \,|\, m\rangle = \langle n \,|\, e^{i\mathbf{r}.\hat{\mathbf{P}}}\hat{\psi}_\alpha(\mathbf{r})\, e^{-i\mathbf{r}.\hat{\mathbf{P}}} \,|\, m\rangle. \tag{8.3}$$

If the system has definite momenta $\mathbf{P}_n$ and $\mathbf{P}_m$ in the states n and m, then

$$\langle n \,|\, \hat{\psi}_\alpha(0) \,|\, m\rangle = e^{i\mathbf{k}_{nm}\cdot\mathbf{r}}\,\langle n \,|\, \hat{\psi}_\alpha(\mathbf{r}) \,|\, m\rangle,$$

whence

$$\left.\begin{aligned} \langle n \,|\, \hat{\Psi}_\alpha(t, \mathbf{r}) \,|\, m\rangle &= e^{i(\omega_{nm}t - \mathbf{k}_{nm}\cdot\mathbf{r})}\langle n \,|\, \hat{\psi}_\alpha(0) \,|\, m\rangle, \\ \langle n \,|\, \hat{\Psi}^+_\alpha(t, \mathbf{r}) \,|\, m\rangle &= \langle m \,|\, \hat{\Psi}_\alpha(t, \mathbf{r}) \,|\, n\rangle^*, \end{aligned}\right\} \tag{8.4}$$

where $\mathbf{k}_{nm} = \mathbf{P}_n - \mathbf{P}_m$.

Using these formulae, we can deduce an important expansion of the Green's function in momentum space, which clarifies its physical significance.

Because of the "discontinuous" definition of the function $G(t, \mathbf{r})$, in calculating $G(\omega, \mathbf{p})$ we must separate the integral over t in (7.22) into two, from $-\infty$ to 0 and from 0 to ∞. In the second (i.e. when $t = t_1 - t_2 > 0$), we expand the definition (7.10) by the matrix multiplication rule and find

$$G(t, \mathbf{r}) = \tfrac{1}{2} iG_{\alpha\alpha} = -\tfrac{1}{2} i \sum_m \langle 0 \,|\, \hat{\Psi}_\alpha(X_1) \,|\, m\rangle \langle m \,|\, \hat{\Psi}^+_\alpha(X_2) \,|\, 0\rangle,$$

with summation over all quantum states of the system. Substituting (8.4) and noting that $\mathbf{P}_0 = 0$ in the ground state, we have

$$G(t, \mathbf{r}) = -\tfrac{1}{2}i \sum_m |\langle 0|\hat{\psi}_\alpha(0)|m\rangle|^2\, e^{i(\omega_{0m}t + \mathbf{P}_m\cdot\mathbf{r})}, \tag{8.5}$$

where $\omega_{0m} = E_0(N) - E_m(N+1) + \mu$.

The spatial integration in (7.22), with $G(t, \mathbf{r})$ from (8.5), gives the delta function $\delta(\mathbf{p}-\mathbf{P}_m)$ in each term of the sum. In the integration over t (> 0), to ensure convergence, we must add to ω an infinitesimal positive imaginary part, i.e. replace ω by $\omega+i0$.† Then

$$\int_0^\infty\!\!\int G(t, \mathbf{r})\, e^{i(\omega t-\mathbf{p}\cdot\mathbf{r})}\, d^3x\, dt = \frac{1}{2}(2\pi)^3 \sum_m |\langle 0|\hat{\psi}_\alpha(0)|m\rangle|^2 \frac{\delta(\mathbf{p}-\mathbf{P}_m)}{\omega+\omega_{0m}+i0}.$$

The integral over t from $-\infty$ to 0 is calculated similarly. For $t < 0$ we have instead of (8.5)

$$G(t, \mathbf{r}) = \tfrac{1}{2}i \sum_m |\langle m\,|\,\hat{\psi}_\alpha(0)\,|\,0\rangle|^2\, e^{i(\omega_{m0}t-\mathbf{P}_m\cdot\mathbf{r})}, \tag{8.6}$$

where $\omega_{m0} = E_m(N-1)-E_0(N)+\mu$. Now, calculating the integral from $-\infty$ to 0 and adding it to the other, we obtain

$$G(\omega, \mathbf{p}) = \frac{1}{2}(2\pi)^3 \sum_m \left\{ \frac{A_m\delta(\mathbf{p}-\mathbf{P}_m)}{\omega+\mu+E_0(N)-E_m(N+1)+i0} + \right.$$
$$\left. + \frac{B_m\delta(\mathbf{p}+\mathbf{P}_m)}{\omega+\mu+E_m(N-1)-E_0(N)-i0} \right\}, \tag{8.7}$$

with the notation

$$A_m = |\langle 0\,|\,\hat{\psi}_\alpha(0)\,|\,m\rangle|^2, \quad B_m = |\langle m\,|\,\hat{\psi}_\alpha(0)\,|\,0\rangle|^2. \tag{8.8}$$

This is the required expansion.‡

We shall use the notation

$$\varepsilon_m^{(+)} = E_m(N+1)-E_0(N), \quad \varepsilon_m^{(-)} = E_0(N)-E_m(N-1) \tag{8.9}$$

for the excitation energies given by the differences between the excited level of the system with a particular number of particles and the ground state of the system with one particle more or fewer. The superscripts (+) and (−) indicate the inequalities

$$\varepsilon_m^{(+)} > \mu, \quad \varepsilon_m^{(-)} < \mu. \tag{8.10}$$

For, since $E_0(N+1)-E_0(N) \approx \partial E_0/\partial N = \mu$, the chemical potential at $T = 0$, we can write, for instance,

$$\begin{aligned}\varepsilon_m^{(+)} &= E_m(N+1)-E_0(N+1)+E_0(N+1)-E_0(N)\\ &\approx [E_m(N+1)-E_0(N+1)]+\mu.\end{aligned}$$

† This procedure is analogous to the method of calculating Green's functions in quantum electrodynamics (cf. *RQT*, §76).

‡ The corresponding expansion in quantum field theory is the Källén–Lehmann expansion (cf. *RQT*, §§101 and 108).

The difference in the square brackets (where both energies relate to systems with the same number of particles) is positive by the definition of the ground state; hence $\varepsilon_m^{(+)} > \mu$. The significance of the definition (8.9) will be discussed again below.

The displacement of the poles of the terms in the sum (as functions of ω), expressed by the terms $\pm i0$ in their denominators, is equivalent to the presence of delta-function imaginary parts according to†

$$\frac{1}{x \pm i0} = \mathrm{P}\frac{1}{x} \mp i\pi\delta(x). \tag{8.11}$$

Applying this to (8.7), we find as the real part of the Green's function

$$\text{re } G(\omega, \mathbf{p}) = 4\pi^3 \sum_m \mathrm{P}\left[\frac{A_m\delta(\mathbf{p}-\mathbf{P}_m)}{\omega+\mu-\varepsilon_m^{(+)}} + \frac{B_m\delta(\mathbf{p}+\mathbf{P}_m)}{\omega+\mu-\varepsilon_m^{(-)}}\right], \tag{8.12}$$

and its imaginary part (since each difference $\varepsilon_m^{(+)} - \mu > 0$, and each difference $\varepsilon_m^{(-)} - \mu < 0$)

$$\text{im } G(\omega, \mathbf{p}) = \begin{cases} -4\pi^4 \sum_m A_m\delta(\mathbf{p}-\mathbf{P}_m)\,\delta(\omega+\mu-\varepsilon_m^{(+)}) & \text{for } \omega > 0, \\ 4\pi^4 \sum_m B_m\delta(\mathbf{p}+\mathbf{P}_m)\,\delta(\omega+\mu-\varepsilon_m^{(-)}) & \text{for } \omega < 0. \end{cases} \tag{8.13}$$

Hence we always have

$$\text{sgn im } G(\omega, \mathbf{p}) = -\text{sgn } \omega. \tag{8.14}$$

We may also notice the asymptotic behaviour of the function $G(\omega, \mathbf{p})$ as $\omega \to \infty$. From (8.7),

$$G(\omega, \mathbf{p}) \approx \frac{4\pi^3}{\omega} \sum_m [A_m\delta(\mathbf{p}-\mathbf{P}_m) + B_m\delta(\mathbf{p}+\mathbf{P}_m)].$$

The coefficient of $1/\omega$ is easily seen to be the Fourier component with respect to $\mathbf{r}_1 - \mathbf{r}_2$ of

$$\tfrac{1}{2}\{\hat{\Psi}_\alpha(t, \mathbf{r}_1)\hat{\Psi}_\alpha^+(t, \mathbf{r}_2) + \hat{\Psi}_\alpha^+(t, \mathbf{r}_2)\hat{\Psi}_\alpha(t, \mathbf{r}_1)\} = \delta(\mathbf{r}_1 - \mathbf{r}_2),$$

i.e. unity. Thus

$$G(\omega, \mathbf{p}) \to 1/\omega \quad \text{as} \quad |\omega| \to \infty. \tag{8.15}$$

† See *QM* (43.10). The symbol P denotes that in the integration of expressions of the form $f(x)/(x \pm i0)$ the integral is to be taken as a principal value:

$$\int_{-\infty}^{\infty} \frac{f(x)}{x \pm i0} dx = \mathrm{P}\int_{-\infty}^{\infty} \frac{f(x)}{x} dx \mp i\pi f(0).$$

The second term comes from the passage round the pole $x = -i0$ or $x = i0$ along a semicircle above or below it respectively.

The chief property of the Green's function in the momentum representation is that its poles can only be at the points $\omega = \varepsilon_m - \mu$, where ε_m are the discrete excitation energies of the system, defined as shown above. Each of these energies corresponds to a definite value of the momentum $\mathbf{P}_m$ of the system, as is evident from the presence of a corresponding delta function in each pole term of the Green's function.

We are interested here, however, in the Green's function of a macroscopic body. This means that we are considering the limit in which the volume V and the number of particles N tend to infinity (for a fixed finite value of the ratio N/V). In this limit, the separations between the levels in the system tend to zero, the poles of the function $G(\omega, \mathbf{p})$ merge, and we can say only that this function has an imaginary part for values of $\omega + \mu$ in the continuous range of possible values of the excitation energy of the system. Excitations in which the whole momentum $\mathbf{p}$ of the macroscopic system can be ascribed to one quasi-particle with a definite dispersion relation $\varepsilon(\mathbf{p})$ (in the ground state of the system, $\mathbf{p} = 0$) form an exception; such values correspond to isolated poles of the Green's function.

If the momentum $\mathbf{p}$ is made up of the momenta of more than one quasi-particle, the energy of the system is not uniquely determined by the value of $\mathbf{p}$: a given momentum of the system can be composed in various ways of quasi-particle momenta, with the total energy of the quasi-particles covering a continuous range of values; the pole is removed by integration over all such states.

Thus the quasi-particle dispersion relation is determined by the equation

$$G^{-1}(\varepsilon - \mu, \mathbf{p}) = 0 \tag{8.16}$$

(V. L. Bonch-Bruevich 1955).

It should be emphasized that the definition of the excitation energy as in (8.9) in fact corresponds to the definition of the quasi-particle energy in the Landau theory: the difference $\varepsilon_m^{(+)}$ is the change in the energy of the system when one particle is added to it, and if the whole of this change is ascribed to one quasi-particle we have ε defined in accordance with (1.3). Similarly, $-\varepsilon_m^{(-)}$ is the change in energy when one particle is removed, and so $\varepsilon_m^{(-)}$ is the energy of the quasi-particle removed. It is therefore natural that $\varepsilon_m^{(-)} < \mu$, since in the Landau theory a quasi-particle can be removed only from within the Fermi sphere.†

Since all the excited states that occur in the expansion (8.7) are obtained from the ground state by adding or removing one particle with spin $\frac{1}{2}$, it is clear that, for a system of fermions, the poles of the Green's function determine only the spectrum of Fermi-type elementary excitations. It will be shown in §18 how the Bose branch is determined.

† It should be noted that the excited level E_m of the system appears with the negative sign in the definition of the quasi-particle energy $\varepsilon_m^{(-)}$. This is the reason why the momentum of these quasi-particles $\mathbf{p} = -\mathbf{P}_m$, as is seen from the delta function $\delta(\mathbf{p} + \mathbf{P}_m)$ in the corresponding terms of the expansion (8.7).

The description of the spectrum of a macroscopic system by means of the concept of quasi-particles with a definite dependence of ε on $\mathbf{p}$ is an approximate one, whose accuracy diminishes with increasing $|\varepsilon-\mu|$. The departure from the picture of independent quasi-particles is shown by the shift of the Green's function pole into the complex domain, $\varepsilon(\mathbf{p})$ becoming complex. According to the general principles of quantum mechanics (see *QM*, §134), complex energy levels signify a finite lifetime τ of the excited state of the system: $\tau \sim 1/|\text{im}\,\varepsilon|$. The quantity im ε itself represents the degree of "broadening" of the quasi-particle energy values (the level width). Of course, this treatment is meaningful only if the imaginary part is sufficiently small, $|\text{im}\,\varepsilon| \ll |\varepsilon-\mu|$. As explained in §1, this condition is in fact satisfied for weakly excited states of the system, since $|\text{im}\,\varepsilon| \sim 1/\tau \propto (p-p_F)^2$, whereas $\text{re}(\varepsilon-\mu) \propto |p-p_F|$.

The required sign of im ε is ensured by the fixed sign of the imaginary part of the Green's function: near its pole, this function has the form

$$G(\omega, \mathbf{p}) \approx Z/[\omega+\mu-\varepsilon(\mathbf{p})], \tag{8.17}$$

and the constant $Z > 0$, as follows from the fact that the coefficients A_m and B_m in the expansion (8.7) are positive; Z is often called the *renormalization constant* (by analogy with quantum electrodynamics). The imaginary part of the Green's function is

$$\text{im}\, G \approx Z\, \text{im}\, \varepsilon/|\omega+\mu-\varepsilon|^2.$$

Since this expression relates to values of $\omega \approx \varepsilon-\mu$, we find, on comparing its sign with the rule (8.14), that

$$\left.\begin{aligned} \text{im}\,\varepsilon < 0 \quad &\text{when} \quad \text{re}\,\varepsilon > \mu, \\ \text{im}\,\varepsilon > 0 \quad &\text{when} \quad \text{re}\,\varepsilon < \mu, \end{aligned}\right\} \tag{8.18}$$

as it should be: this sign of im ε corresponds in both cases ($\varepsilon_m^{(+)}$ and $\varepsilon_m^{(-)}$ in (8.9)) to the correct negative imaginary increment to the energy E_m of the excited state.

The analytical properties of the Green's function will be further discussed in §36, where this question will be considered for the general case of arbitrary temperatures.

§ 9. Green's function of an ideal Fermi gas

To illustrate the general relations given in §8, let us calculate the Green's function of an ideal gas.

The Schrödinger ψ operators can always be written as an expansion

$$\hat{\psi}_\alpha(\mathbf{r}) = \sum_{\mathbf{p},\sigma} \hat{a}_{\mathbf{p}\sigma}\psi_{\mathbf{p}\alpha}(\mathbf{r}, \sigma) \tag{9.1}$$

in terms of a complete set of functions $\psi_{\mathbf{p}\alpha}$, the spinor wave functions of a free particle with momentum $\mathbf{p}$ (and spin projection σ), i.e. in plane waves

$$\psi_{\mathbf{p}\alpha} = \frac{u_\alpha(\sigma)}{\sqrt{V}} e^{i\mathbf{p}\cdot\mathbf{r}}, \tag{9.2}$$

where u_α is the spinor amplitude normalized by the condition $u_\alpha u_\alpha^* = 1$; this choice of the functions $\psi_{\mathbf{p}\alpha}$ has no connection with the actual interaction of the particles in the system.

For a system of non-interacting particles, the Heisenberg ψ operator can also be written in an explicit form. In this case, the change from the Schrödinger to the Heisenberg representation consists in placing in each term of the sum in (9.1) the corresponding time factor:

$$\hat{\Psi}_\alpha(t, \mathbf{r}) = \sum_{\mathbf{p},\,\sigma} \hat{a}_{\mathbf{p}\sigma}\psi_{\mathbf{p}\alpha}(\mathbf{r}, \sigma) \exp\left[-i\left(\frac{p^2}{2m} - \mu\right)t\right]. \tag{9.3}$$

This is easily seen if we note that the matrix elements of the Heisenberg operator for every transition $i \to f$ must contain factors $\exp[-i(E_i' - E_f')t]$, where E_i' and E_f' are the energies of the initial and final states (in this case, eigenvalues of the Hamiltonian $\hat{H}' = \hat{H} - \mu\hat{N}$). For a transition with decrease in the number of particles in the state $\mathbf{p}$, α by one, the difference $E_i' - E_f' = p^2/2m - \mu$, so that the condition stated is satisfied.

However, instead of directly calculating the Green's function by means of (9.3) from the definition (7.10), it is more convenient to begin by converting this definition into an equivalent differential equation. To do so, we differentiate $G_{\alpha\beta}(X_1 - X_2)$ with respect to t_1. It is necessary to take account of the discontinuity of this function at $t_1 = t_2$: according to the definition (7.10), the amount of the discontinuity is

$$\begin{aligned}[G_{\alpha\beta}] &\equiv [G_{\alpha\beta}]_{t_1 = t_2 + 0} - [G_{\alpha\beta}]_{t_1 = t_2 - 0} \\ &= -i\langle \hat{\Psi}_\alpha(t_1, \mathbf{r}_1)\hat{\Psi}_\beta^+(t_1, \mathbf{r}_2) + \hat{\Psi}_\beta^+(t_1, \mathbf{r}_2)\hat{\Psi}_\alpha(t_1, \mathbf{r}_1)\rangle\end{aligned}$$

or, from (7.3),†

$$[G_{\alpha\beta}] = -i\delta_{\alpha\beta}\delta(\mathbf{r}_1 - \mathbf{r}_2). \tag{9.4}$$

The presence of the discontinuity gives rise to a term $[G_{\alpha\beta}]\delta(t_1 - t_2)$ on differentiation. Hence

$$\frac{\partial}{\partial t_1} G_{\alpha\beta} = -i\left\langle \mathrm{T}\frac{\partial \hat{\Psi}_\alpha(X_1)}{\partial t_1}\hat{\Psi}_\beta^+(X_2)\right\rangle - i\delta_{\alpha\beta}\delta(\mathbf{r}_1 - \mathbf{r}_2)\,\delta(t_1 - t_2). \tag{9.5}$$

For a system of free particles, the Heisenberg ψ operator satisfies the equation

$$i\frac{\partial \hat{\Psi}_\alpha}{\partial t} = -\frac{1}{2m}\triangle\hat{\Psi}_\alpha - \mu\hat{\Psi}_\alpha;$$

† It must be emphasized that the magnitude of the discontinuity does not depend on the interaction of the particles.

cf. (7.8). Substituting this derivative in (9.5) and again using the definition (7.10), we get as the equation for the Green's function

$$\left(i\frac{\partial}{\partial t}+\frac{\triangle}{2m}+\mu\right)G^{(0)}(t, \mathbf{r}) = \delta(t)\,\delta(\mathbf{r}), \tag{9.6}$$

where we have put $G^{(0)}_{\alpha\beta} = \delta_{\alpha\beta}G^{(0)}$; the superscript (0) to G indicates that there is no interaction between the particles.

This equation has the Fourier transform

$$\left(\omega-\frac{p^2}{2m}+\mu\right)G^{(0)}(\omega, \mathbf{p}) = 1.$$

In determining the Green's function from this, we must add to ω an infinitesimal imaginary part in such a way that the imaginary part of G has the correct sign in accordance with (8.14): .

$$G^{(0)}(\omega, \mathbf{p}) = \left[\omega-\frac{p^2}{2m}+\mu+i0.\,\mathrm{sgn}\,\omega\right]^{-1}. \tag{9.7}$$

The pole of this expression is at $\omega+\mu = \varepsilon(\mathbf{p}) = p^2/2m$, in accordance with the fact that in an ideal gas the quasi-particles are the same as the actual particles. The chemical potential of an ideal Fermi gas is $\mu = p_F^2/2m$. For weakly excited states, p is close to p_F, so that we can put $p^2/2m \approx \mu+v_F(p-p_F)$, where $v_F = p_F/m$, and write the Green's function for such states in the form

$$G^{(0)}(\omega, \mathbf{p}) = [\omega-v_F(p-p_F)+i0.\,\mathrm{sgn}\,\omega]^{-1}. \tag{9.8}$$

In all integrations involving the function $G^{(0)}$, the presence of the infinitesimal imaginary part in its denominator is important only near the pole, when $\omega \approx v_F(p-p_F)$. In this sense, sgn ω in (9.7) may be replaced by sgn $(p-p_F)$, and $G^{(0)}$ written as

$$G^{(0)}(\omega, \mathbf{p}) = [\omega^2-p^2/2m+\mu+i0.\,\mathrm{sgn}\,(p-p_F)]^{-1}. \tag{9.9}$$

This change is important in that $G^{(0)}$ in the form (9.9) is a single function of the complex variable ω, analytic throughout the plane, and the methods of the theory of analytic functions can be used to calculate the integrals.

For instance, to calculate the integral (7.23) (the particle momentum distribution) for a non-zero negative t, we close the contour of integration (the real ω-axis) by an infinite semicircle in the upper half-plane (and can then put $t = 0$). The integral

$$N(\mathbf{p}) = -\frac{i}{2\pi}\int\frac{d\omega}{\omega-p^2/2m+\mu+i0.\,\mathrm{sgn}\,(p-p_F)}$$

is now determined by the residue of the integrand at the pole in the upper half-plane. When $p > p_F$ there is no such pole, and $N(\mathbf{p}) = 0$. If $p < p_F$, however, we find $N(\mathbf{p}) = 1$, as it should be for the ground state of an ideal Fermi gas.

§ 10. Particle momentum distribution in a Fermi liquid

The Green's function of a Fermi liquid cannot, of course, be calculated in a general form as was done for a Fermi gas. But the statement that a Fermi liquid has a spectrum of the type described in §1 implies that its Green's function has a pole at

$$\omega = \varepsilon(\mathbf{p}) - \mu \approx v_F(p - p_F), \quad v_F = p_F/m^*. \tag{10.1}$$

It can therefore be written as

$$G(\omega, \mathbf{p}) = \frac{Z}{\omega - v_F(p - p_F) + i0 . \operatorname{sgn} \omega} + g(\omega, \mathbf{p}), \tag{10.2}$$

where $g(\omega, \mathbf{p})$ is a function finite at the point (10.1). As already noted in connection with (8.17), the coefficient Z (the residue of G at the pole) is positive.

An interesting conclusion can be drawn from (10.2) about the nature of the particle (not quasi-particle) momentum distribution in the liquid. We calculate the difference between the values of the distribution function $N(\mathbf{p})$ (which in practice depends only on the magnitude p) on the two sides of the surface of the Fermi sphere, i.e. the limit of the difference $N(p_F - q) - N(p_F + q)$ as $q \to +0$.

The distribution $N(\mathbf{p})$ is expressed in terms of the Green's function by the integral (7.23). Since $g(\omega, \mathbf{p})$ is finite, it is evident that the difference between the integrals of g tends to zero with q. It is therefore sufficient to consider the difference between the integrals of the pole terms in (10.2). Since, in this integration, the term $i0$ in the denominator is important only near the pole, we can, as already mentioned in §9, replace sgn ω by $\operatorname{sgn}(p - p_F)$. Then

$$N(p_F - q) - N(p_F + q) = -i \int_{-\infty}^{\infty} \left\{ \frac{Z}{\omega + v_F q - i0} - \frac{Z}{\omega - v_F q + i0} \right\} \frac{d\omega}{2\pi};$$

since this integral of the difference converges, the factor $e^{-i\omega t}$ in it, with $t = -0$, may be omitted. Now, closing the contour of integration by an infinite semicircle in either half-plane, we find that the whole integral is equal to Z, and independent of q. Thus

$$N(p_F - 0) - N(p_F + 0) = Z \tag{10.3}$$

(A. B. Migdal 1957).

It has been mentioned above that $Z > 0$. Since $N(\mathbf{p}) \leq 1$, it follows from (10.3) that

$$0 < Z \leq 1; \tag{10.4}$$

the value $Z = 1$ is reached only in the limit of an ideal gas.

The particle momentum distribution in a Fermi liquid at $T = 0$ therefore has, as in a gas, a discontinuity on the surface of the Fermi sphere, decreasing towards the outside. Unlike the gas case, however, the magnitude of the discontinuity is less than unity, and $N(\mathbf{p})$ remains non-zero for $p > p_F$, as shown in Fig. 1 by the continuous curve; the broken curve corresponds to a gas.

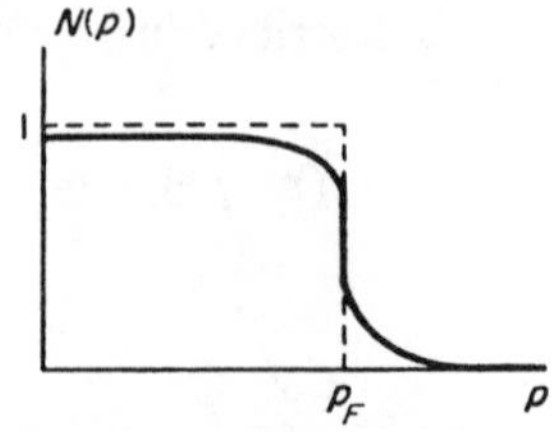

FIG. 1.

§ 11. Calculation of thermodynamic quantities from the Green's function

A knowledge of the Green's function of a system is sufficient to describe its thermodynamic properties. When $T = 0$, these properties are expressed by the dependence of the energy of the system (which is the ground-state energy E_0) on the density N/V.

When the quasi-particle dispersion relation $\varepsilon(p)$ has been determined (by solving equation (8.16)), this dependence can be found by using the fact that

$$\varepsilon(p_F) = \mu. \tag{11.1}$$

Since the dependence of p_F on N/V is known, from (1.1):

$$p_F = (3\pi^2)^{1/3}(N/V)^{1/3}, \tag{11.2}$$

equation (11.1) determines the function $\mu(N/V)$ (though in an implicit form, since the dispersion relation $\varepsilon(p)$ in general contains μ as a parameter). At $T = 0$ (and therefore $S = 0$), the chemical potential $\mu = (\partial E_0/\partial N)_V$; integration of this gives the required energy

$$E_0 = \int_0^N \mu(N/V)\,dN; \tag{11.3}$$

when $N = 0$, $E_0 = 0$, of course.

Another way of describing the thermodynamic properties at $T = 0$ is to calculate the thermodynamic potential Ω. According to the general definition (see Part 1, §24), this potential $\Omega = E - TS - \mu N = -PV$, and its differential $d\Omega = -SdT - Nd\mu$; when $T = 0$, $S = 0$ also and these expressions reduce to

$$\Omega = E - \mu N, \tag{11.4}$$

$$d\Omega = -Nd\mu. \tag{11.5}$$

The significance of the potential Ω is that it describes the properties of the system at constant V.

The simplest method of expressing Ω in terms of the Green's function is to use the relation (7.24) between N/V and G. Substituting N from (7.24) in (11.5) and integrating with respect to μ (with V constant), we obtain

$$\Omega(\mu) = 2iV \int_0^\mu d\mu \,.\, \lim_{t\to-0} \int G(\omega, \mathbf{p}) e^{-i\omega t} \frac{d^3p\, d\omega}{(2\pi)^4}, \tag{11.6}$$

since again $\Omega = 0$ when $\mu = 0$.

§ 12. Ψ operators in the interaction representation

The Green's function for a system of interacting particles cannot, of course, be calculated in a general form. There is, however, a mathematical technique (similar to the diagram technique in quantum field theory) whereby it can be calculated as a power series in the particle interaction energy, each term being expressed by means of the Green's functions of a system of free particles and the interaction operator.

We shall use, as well as the Heisenberg representation, a representation of operators in which their time dependence is given not by the actual Hamiltonian of the system

$$\hat{H}' = \hat{H}'^{(0)} + \hat{V} = \hat{H}^{(0)} - \mu\hat{N} + \hat{V}$$

(where $\hat{V}$ is the interaction operator) but by the free-particle Hamiltonian $\hat{H}'(0)$:

$$\hat{\Psi}_0(t, \mathbf{r}) = \exp(i\hat{H}'^{(0)}t)\, \hat{\psi}(\mathbf{r}) \exp(-i\hat{H}'^{(0)}t). \tag{12.1}$$

The operators and wave functions in this *interaction representation* will be distinguished by the suffix 0. By expressing the Green's function in terms of the operators $\hat{\Psi}_0$ (instead of the Heisenberg operators $\hat{\Psi}$) we take the first step towards the objective of expressing G in terms of $G^{(0)}$ and V.

In this section, Φ or ϕ will denote wave functions in "occupation number space" (in contrast to the coordinate wave functions Ψ or ψ); these functions are acted on by second-quantized operators. Let ϕ be such a function in the Schrödinger representation; its time dependence is given by the wave equation

$$i\partial\phi/\partial t = (\hat{H}'^{(0)} + \hat{V})\phi. \tag{12.2}$$

In the Heisenberg representation, where the whole of the time dependence is transferred to the operators, the wave function Φ of the system is a constant, independent of time. In the interaction representation, however, the wave function Φ_0 is time-dependent, but only because of the interaction of the particles in the system, and is given by

$$i\partial\Phi_0(t)/\partial t = \hat{V}_0(t)\Phi_0(t), \tag{12.3}$$

where

$$\hat{V}_0(t) = \exp(i\hat{H}'^{(0)}t)\, \hat{V} \exp(-i\hat{H}'^{(0)}t) \tag{12.4}$$

is the interaction operator in that representation; for operators having the form (7.6), (7.7), the change to the new representation is obtained by simply substituting $\hat{\Psi}_0$ for $\hat{\Psi}$. Equation (12.3) is easily derived, since the transformation of operators by (12.1) corresponds to transformation of the wave functions according to

$$\Phi_0 = \exp(i\hat{H}'^{(0)}t)\phi; \tag{12.5}$$

see *QM*, §12. Differentiating this and using (12.2), we get (12.3).†

From (12.3), the values of $\Phi_0(t)$ at two successive instants are related by

$$\begin{aligned}\Phi_0(t+\delta t) &= [1-i\delta t.\hat{V}_0(t)]\Phi_0(t)\\ &= \exp\{-i\delta t.\hat{V}_0(t)\}\Phi_0(t).\end{aligned}$$

Accordingly, the value of Φ_0 at any instant t can be expressed in terms of its value at some initial instant t_0 ($<t$) by

$$\Phi_0(t) = \hat{S}(t, t_0)\Phi_0(t_0), \tag{12.6}$$

where

$$\hat{S}(t, t_0) = \prod_{t_i=t_0}^{t} \exp\{-i\delta t.\hat{V}_0(t_i)\}; \tag{12.7}$$

the factors in this product are clearly arranged from right to left in order of increasing time t_i; it is understood that we take the limit of the product over all the infinitesimal intervals δt between t_0 and t. If $V_0(t)$ were an ordinary function, this limit would reduce simply to

$$\exp\left\{-i\int_{t_0}^{t} V_0(t)\,dt\right\},$$

but this result depends on the commutativity of the factors pertaining to different instants, which is assumed in changing from the product in (12.7) to the summation in the exponent. For the operator $\hat{V}_0(t)$ there is no such commutativity, and the reduction to an ordinary integral is not possible. Instead, we can write (12.7) in the symbolic form

$$\hat{S}(t, t_0) = \mathrm{T}\exp\left\{-i\int_{t_0}^{t} \hat{V}_0(t)\,dt\right\}, \tag{12.8}$$

where T denotes the chronological ordering of the factors in the same sequence as in (12.7), i. e. with the time increasing from right to left.

The operator $\hat{S}$ is unitary ($\hat{S}^{-1} = \hat{S}^+$), and has the obvious properties

$$\left.\begin{aligned}\hat{S}(t_3, t_2)\,\hat{S}(t_2, t_1) &= \hat{S}(t_3, t_1),\\ \hat{S}^{-1}(t_2, t_1)\,\hat{S}^{-1}(t_3, t_2) &= \hat{S}^{-1}(t_3, t_1).\end{aligned}\right\} \tag{12.9}$$

† Equation (12.3) is the same as *RQT* (73.5), and the following method of solution repeats that given in *RQT* §73.

To simplify the subsequent analysis, we make the formal hypothesis (which does not affect the final results) that the interaction $\hat{V}_0(t)$ is adiabatically "switched on" between $t = -\infty$ and a finite time, and adiabatically "switched off" at $t = +\infty$. Then, as $t \to -\infty$, before the interaction begins, the wave function $\Phi_0(t)$ coincides with the Heisenberg function Φ. Putting $t_0 = -\infty$ in (12.6), we get

$$\Phi_0(t) = \hat{S}(t, -\infty)\Phi. \tag{12.10}$$

Having thus established the relation between the wave functions in the two representations, we also have the transformation rule for operators, including ψ operators:

$$\hat{\Psi} = \hat{S}^{-1}(t, -\infty)\,\hat{\Psi}_0\hat{S}(t, -\infty). \tag{12.11}$$

Since $\hat{S}$ is unitary, the operators $\hat{\Psi}^+$ are transformed in the same way.

Let us now express the Green's function in terms of ψ operators in the interaction representation.† Let $t_1 > t_2$; then

$$\begin{aligned} G_{\alpha\beta}(X_1, X_2) &= -i\langle\hat{\Psi}_\alpha(t_1)\,\hat{\Psi}_\beta^+(t_2)\rangle \\ &= -i\langle\hat{S}^{-1}(t_1, -\infty)\,\hat{\Psi}_{0\alpha}(t_1)\,\hat{S}(t_1, -\infty)\times \\ &\quad\times\hat{S}^{-1}(t_2, -\infty)\,\hat{\Psi}_{0\beta}^+(t_2)\,\hat{S}(t_2, -\infty)\rangle. \end{aligned}$$

According to (12.9),

$$\begin{aligned} \hat{S}(t_1, -\infty)\,\hat{S}^{-1}(t_2, -\infty) &= \hat{S}(t_1, t_2)\,\hat{S}(t_2, -\infty)\,\hat{S}^{-1}(t_2, -\infty) \\ &= \hat{S}(t_1, t_2), \\ \hat{S}^{-1}(t_1, -\infty) &= \hat{S}^{-1}(t_1, -\infty)\,\hat{S}^{-1}(\infty, t_1)\,\hat{S}(\infty, t_1) \\ &= \hat{S}^{-1}(\infty, -\infty)\,\hat{S}(\infty, t_1). \end{aligned}$$

Substitution in the preceding expression gives

$$G_{\alpha\beta}(X_1, X_2) = -i\langle\hat{S}^{-1}(\infty, -\infty)\,\hat{S}(\infty, t_1)\,\hat{\Psi}_{0\alpha}(t_1)\,\hat{S}(t_1, t_2)\,\hat{\Psi}_{0\beta}^+(t_2)\,\hat{S}(t_2, -\infty)\rangle.$$

Taking the operators $\hat{S}$ as the products (12.7) we see that all factors from the second onwards in the averaged expression are in chronological order from right to left, $t = -\infty$ to $t = \infty$. We can therefore write

$$G_{\alpha\beta}(X_1, X_2) = -i\langle\hat{S}^{-1}\,\mathrm{T}[\hat{\Psi}_{0\alpha}(t_1)\,\hat{\Psi}_{0\beta}^+(t_2)\,\hat{S}]\rangle, \tag{12.12}$$

with

$$\hat{S} = \hat{S}(\infty, -\infty) = \mathrm{T}\exp\left\{-i\int_{-\infty}^{\infty}\hat{V}_0(t)\,dt\right\}. \tag{12.13}$$

The calculation with $t_1 < t_2$ differs from the above only in the notation, and the final result (12.12), (12.13) is valid for any t_1 and t_2.

The transformation made does not depend on the state of the system with respect to which the averaging is done. However, if the averaging is with

† This derivation repeats the one given in *RQT*, §100.

respect to the ground state (as in (12.12)), the transformation can be carried further. To do so, we note that the adiabatic switching on or off of the interaction, like any adiabatic perturbation, cannot cause a transition with change of energy of the quantum system (see *QM*, §41). Hence a system in a non-degenerate state (such as the ground state) will remain in that state. That is, the effect of the operator $\hat{S}$ on the wave function $\Phi = \Phi_0(-\infty)$ must reduce to multiplication by a phase factor (which does not affect the state), the mean value of $\hat{S}$ in the ground state: $\hat{S}\Phi = \langle\hat{S}\rangle\Phi$. Similarly, $\Phi^*\hat{S}^{-1} = \langle\hat{S}\rangle^{-1}\Phi^*$. Thus we have finally the following formula for the Green's function in terms of operators in the interaction representation:†

$$iG_{\alpha\beta}(X_1, X_2) = \frac{1}{\langle\hat{S}\rangle}\langle \mathrm{T}[\hat{\Psi}_{0\alpha}(X_1)\,\hat{\Psi}^+_{0\beta}(X_2)\,\hat{S}]\rangle. \tag{12.14}$$

According to the meaning of this representation, the averaging in (12.14) is with respect to the ground state of a system of free particles: the properties of the operators $\hat{\Psi}_0$ are the same as those of the Heisenberg operators $\hat{\Psi}$ in the absence of interactions, and the Heisenberg wave function Φ is independent of time, so that it is the same as its value at $t = -\infty$, when there is no interaction. Hence, in particular,

$$\langle \mathrm{T}\hat{\Psi}_{0\alpha}(X_1)\,\hat{\Psi}^+_{0\beta}(X_2)\rangle = iG^{(0)}_{\alpha\beta}(X_1, X_2) \tag{12.15}$$

is the Green's function of a system of non-interacting particles.

§ 13. The diagram technique for Fermi systems

The significance of symbolic expressions such as (12.14) is that they make it easily possible to write down the successive terms in expansions in powers of $\hat{V}$. For example,

$$\langle \mathrm{T}\hat{\Psi}_{0\alpha}(X)\,\hat{\Psi}^+_{0\beta}(X')\,\hat{S}\rangle =$$
$$\sum_{n=0}^{\infty}\frac{(-i)^n}{n!}\int_{-\infty}^{\infty} dt_1\ldots\int_{-\infty}^{\infty} dt_n \langle \mathrm{T}\,\hat{\Psi}_{0\alpha}(X)\hat{\Psi}^+_{0\beta}(X')\,\hat{V}_0(t_1)\ldots\hat{V}_0(t_n)\rangle, \tag{13.1}$$

and the expression for $\langle\hat{S}\rangle$ differs from the above only in that the factors $\hat{\Psi}_{0\alpha}\hat{\Psi}^+_{0\beta}$ do not appear in the T product. As already mentioned, the operator $\hat{V}_0(t)$ in the interaction representation is found from (7.7) by replacing all the $\hat{\Psi}$ by $\hat{\Psi}_0$. The calculation of successive terms in the expansion (13.1) thus reduces to the averaging, with respect to the ground state, of the T product of various numbers of ψ operators of free particles.

† The notation in (12.14) is in a certain way conventional: although it contains the symbol T twice (once explicitly and once in the definition of S), all factors in the product must really be arranged in a single chronological sequence.

These calculations are made largely automatic by the rules of the *diagram technique*, which, however, essentially depend on the nature of the physical system considered. The technique described in this section relates to non-superfluid Fermi systems, the particles being assumed to have a spin-independent pair interaction. The corresponding interaction operator is

$$\hat{V}_0(t) = \frac{1}{2}\int \hat{\Psi}^+_{0\gamma}(t, \mathbf{r}_1)\,\hat{\Psi}^+_{0\delta}(t, \mathbf{r}_2)\,U(\mathbf{r}_1-\mathbf{r}_2)\,\hat{\Psi}_{0\delta}(t, \mathbf{r}_2)\,\hat{\Psi}_{0\gamma}(t, \mathbf{r}_1)\,d^3x_1\,d^3x_2, \quad (13.2)$$

where $U(\mathbf{r}_1-\mathbf{r}_2)$ is the interaction energy of two particles; the superscripts (2) to $\hat{V}$ and U are omitted.

The mean value of products of ψ operators is calculated by *Wick's theorem*:† the average of the product of any (even) number of operators $\hat{\Psi}$ and $\hat{\Psi}^+$ is equal to the sum of products of all possible means (contractions) of pairs of these operators. In each pair, the operators are in the same order as the original product. The sign of each term in the sum is given by the factor $(-1)^P$, where P is the number of interchanges of operators needed to bring all the averaged operators together.

Only those contractions are non-zero which contain one operator $\hat{\Psi}$ and one $\hat{\Psi}^+$: in the diagonal matrix element, all particles annihilated by the operator $\hat{\Psi}$ must be created again by $\hat{\Psi}^+$. It is therefore clear that the mean value of the product of several ψ operators can be non-zero only if it contains the same number of operators $\hat{\Psi}$ and $\hat{\Psi}^+$.

When applied to the average of the T product, Wick's theorem enables it to be expressed in terms of the means of paired T products, i.e., according to (12.15), in terms of the Green's functions of free particles. We shall do this for the first-order correction to the Green's function of a system of interacting particles.

First of all, let us note that, in expanding the expression in the numerator of (12.14) by Wick's theorem, we get, in particular, terms of the form

$$\langle \mathrm{T}\hat{\Psi}_{0\alpha}(X_1)\,\hat{\Psi}^+_{0\beta}(X_2)\rangle\,\langle\hat{S}\rangle = iG^{(0)}_{\alpha\beta}(X_1, X_2)\,\langle\hat{S}\rangle, \quad (13.3)$$

in which the pair of ψ operators that are "outside" $\hat{S}$ are contracted; the expression for $\langle\hat{S}\rangle$ contains, in each term of its expansion, only contractions of "inside" operators. The factor $\langle\hat{S}\rangle$ cancels entirely with the denominator in (12.14), and so all these terms give just the "unperturbed" Green's function $iG^{(0)}_{\alpha\beta}$.

Retaining the first two terms of the expansion in (13.1), substituting (13.2) and renaming the variables, we find

$$iG_{\alpha\beta}(X_1, X_2) \approx iG^{(0)}_{\alpha\beta} + iG^{(1)}_{\alpha\beta},$$

where

$$iG^{(1)}_{\alpha\beta} = -\tfrac{1}{2}\,i\langle \mathrm{T}\hat{\Psi}_{0\alpha}(X_1)\,\hat{\Psi}^+_{0\beta}(X_2)\times$$
$$\times\int_{-\infty}^{\infty} dt\int d^3x_3\,d^3x_4\,\hat{\Psi}^+_{0\gamma}(t, \mathbf{r}_3)\,\hat{\Psi}^+_{0\delta}(t, \mathbf{r}_4)\,U(\mathbf{r}_3-\mathbf{r}_4)\,\hat{\Psi}_{0\delta}(t, \mathbf{r}_4)\,\hat{\Psi}_{0\gamma}(t, \mathbf{r}_3)\rangle.$$

† The proof is given at the end of the section so as not to impede the discussion here.

For greater compactness in the formulae, we use the notation

$$U(X_1 - X_2) = U(\mathbf{r}_1 - \mathbf{r}_2)\,\delta(t_1 - t_2). \tag{13.4}$$

Then†

$$iG^{(1)}_{12} = -\tfrac{1}{2}i \int \langle \mathrm{T}\hat{\Psi}_1 \hat{\Psi}^+_2 \hat{\Psi}^+_3 \hat{\Psi}^+_4 \hat{\Psi}_4 \hat{\Psi}_3 \rangle\, U_{34}\, d^4X_3\, d^4X_4,$$

where $d^4X = dt\, d^3x$.

In order to average by Wick's theorem, we write out the operators separately and show all the relevant contractions:

$$\langle \Psi_1 \Psi^+_2 \Psi^+_3 \Psi^+_4 \Psi_4 \Psi_3 \rangle \to \Psi_1 \Psi^+_2 \Psi^+_3 \Psi^+_4 \Psi_4 \Psi_3 + \Psi_1 \Psi^+_2 \Psi^+_3 \Psi^+_4 \Psi_4 \Psi_3 +$$
$$+ \Psi_1 \Psi^+_2 \Psi^+_3 \Psi^+_4 \Psi_4 \Psi_3 + \Psi_1 \Psi^+_2 \Psi^+_3 \Psi^+_4 \Psi_4 \Psi_3.$$

The terms containing contractions $\Psi_1 \Psi^+_2$ have been omitted in accordance with the previous discussion. The operators contracted in pairs (joined by the loops) are to be interchanged so as to be adjacent. For instance, the first term written above denotes the product

$$\langle \mathrm{T}\hat{\Psi}_1 \hat{\Psi}^+_3 \rangle \langle \mathrm{T}\hat{\Psi}^+_2 \hat{\Psi}_4 \rangle \langle \mathrm{T}\hat{\Psi}^+_4 \hat{\Psi}_3 \rangle,$$

and the last one is

$$-\langle \mathrm{T}\hat{\Psi}_1 \hat{\Psi}^+_4 \rangle \langle \mathrm{T}\hat{\Psi}^+_2 \hat{\Psi}_4 \rangle \langle \hat{\Psi}^+_3 \hat{\Psi}_3 \rangle.$$

The contractions of products of ψ operators with different arguments are replaced according to

$$\Psi_1 \Psi^+_3 \equiv \langle \mathrm{T}\hat{\Psi}_1 \hat{\Psi}^+_3 \rangle = iG^0_{13},\quad \Psi^+_2 \Psi_4 = -iG^0_{24}, \text{ etc.}$$

Those of ψ operators with the same argument represent the spatial number density of particles in an ideal gas (denoted by $n^{(0)}$), regarded as a function of the chemical potential:‡

$$\langle \hat{\Psi}^+ \hat{\Psi} \rangle = n^{(0)}(\mu) = (2m\mu)^{3/2}/3\pi^2. \tag{13.5}$$

Thus we have

$$iG^{(1)}_{12} = \tfrac{1}{2} \int d^4X_3\, d^4X_4 U_{34} [-G^{(0)}_{13} G^{(0)}_{34} G^{(0)}_{42} - G^{(0)}_{14} G^{(0)}_{43} G^{(0)}_{32}$$
$$+ in^{(0)} G^{(0)}_{13} G^{(0)}_{32} + in^{(0)} G^{(0)}_{14} G^{(0)}_{42}].$$

† Here and below, to simplify some particularly cumbersome expressions, we omit the suffix in $\hat{\Psi}_0$ and denote by numerals 1, 2, ... the set of values of the argument X and the spin index:

$$\hat{\Psi}_1 \equiv \hat{\Psi}_\alpha(X_1), \qquad \hat{\Psi}_2 \equiv \hat{\Psi}_\beta(X_2),$$
$$G_{12} \equiv G_{\alpha\beta}(X_1, X_2), \quad U_{12} \equiv U(X_1 - X_2), \ \ldots$$

‡ Such contractions always arise from ψ operators that appear in the same interaction operator $\hat{V}$. Hence $\hat{\Psi}^+$ in such terms is always to the left of $\hat{\Psi}$.

These four terms are equal in pairs, differing only in the naming of the variables of integration X_3 and X_4. Thus the factor $\frac{1}{2}$ disappears and the first-order correction to the Green's functions has two terms:

$$iG^{(1)}_{12} = \int U_{34}[in^{(0)}G^{(0)}_{14}G^{(0)}_{42} - G^{(0)}_{13}G^{(0)}_{34}G^{(0)}_{42}]\, d^4X_3\, d^4X_4. \tag{13.6}$$

The structure of these terms is conveniently represented graphically by means of the *Feynman diagrams*

(13.7)

Here the continuous line 4 ← 2 denotes the contraction $\widehat{\Psi_4\Psi_2^+}$ (i.e the function $iG^{(0)}_{42}$); the numerals refer to the variables X_4 and X_2 on which the contracted operators depend, and the direction of the arrow corresponds to the direction from $\hat{\Psi}^+$ to $\hat{\Psi}$ in the contraction. The contraction $\widehat{\Psi^+\Psi}$ of two operators depending on the same variables (i.e. the density $n^{(0)}$) is represented by a loop – a closed continuous line. The broken line 3--4 denotes the factor U_{34}. Integration is implied over all variables shown at interior points in the diagram (points of intersection of lines). The variables (X_1 and X_2) shown at the external lines of the diagram remain free.

The first-order terms arising from (13.3) would have diagrams in two separate parts: a straight segment ($iG^{(0)}_{\alpha\beta}$), and a diagram with closed loops of continuous lines, e.g.

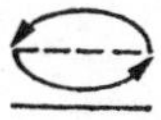

With an understanding of the method of operator contraction and the structure of the corresponding diagrams, we can see the origin of the general rule according to which, in all orders of perturbation theory, the role of the factor $\langle\hat{S}\rangle^{-1}$ in (12.14) is to take into account only "connected" diagrams with two external lines, which contain no detached loops without external lines that are unconnected to the rest of the diagram by either continuous or broken lines. Cf. *RQT*, §100, for a similar situation in quantum electrodynamics.

The cancelling of the factor $\frac{1}{2}$ in (13.6) is an instance of a general rule: it is not necessary to include (in the nth-order terms) the factor $1/n!$ from the expansion (13.1) or the factor 2^{-n} from the coefficients $\frac{1}{2}$ in (13.2). The reason is that diagrams of order n contain n broken lines i---k. The factor $1/n!$ cancels from the combination of terms differing by interchanges of pairs of numbers i, k among all n broken lines. The factor 2^{-n} cancels from the interchanges of the numbers i, k between the ends of each line.

The rules of the diagram technique will be finally formulated for the calculation of the Green's function not in the coordinate representation but directly in the momentum representation, which is the most important in physical applications.

The change to the momentum representation is made by means of the Fourier expansion (7.21), (7.22), which we write in the four-dimensional form†

$$\left.\begin{aligned} G(X) &= \int G(P)\, e^{-iPX}\, d^4P/(2\pi)^4, \\ G(P) &= \int G(X) e^{iPX}\, d^4X, \end{aligned}\right\} \tag{13.8}$$

where the "4-momentum" $P = (\omega, \mathbf{p})$, and $PX = \omega t - \mathbf{p}.\mathbf{r}$. We can similarly expand the interaction potential:

$$U(X) = \delta(t)\, U(\mathbf{r}) = \int U(Q) e^{-iQX}\, d^4Q/(2\pi)^4, \tag{13.9}$$

where $Q = (q_0, \mathbf{q})$; $U(Q)$ is the same as the three-dimensional expansion component,

$$U(Q) \equiv U(\mathbf{q}) = \int U(\mathbf{r}) e^{-i\mathbf{q}.\mathbf{r}}\, d^3x. \tag{13.10}$$

Since $U(\mathbf{r})$ is even, it is clear that $U(-\mathbf{q}) = U(\mathbf{q})$.

Let us make this expansion for the first-order correction $G^{(1)}_{12} \equiv G^{(1)}_{\alpha\beta}(X_1 - X_2)$. To do so, we multiply equation (13.6) by $\exp[iP(X_1 - X_2)]$ and integrate over $d^4(X_1 - X_2)$.

In the first term we write

$$e^{iP(X_1-X_2)} = e^{iP(X_1-X_3)}\, e^{iP(X_3-X_2)}$$

and, changing the variables of integration, obtain

$$in^{(0)} \int G^{(0)}_{\alpha\gamma}(X_1 - X_3)\, e^{iP(X_1-X_3)}\, d^4(X_1 - X_3) \times$$
$$\times G^{(0)}_{\gamma\beta}(X_3 - X_2)\, e^{iP(X_3-X_2)}\, d^4(X_3 - X_2) \int U(X_3 - X_4)\, d^4(X_3 - X_4).$$

The first two integrals give $G^{(0)}_{\alpha\gamma}(P)\, G^{(0)}_{\gamma\beta}(P)$, and the third is $U(0) = \int U(\mathbf{r})\, d^3x$, the value of $U(\mathbf{q})$ for $\mathbf{q} = 0$.

Similarly, in the second term we write

$$e^{iP(X_1-X_2)} = e^{iP(X_1-X_3)}\, e^{iP(X_3-X_4)}\, e^{iP(X_4-X_2)}$$

and, after changing to integration with respect to $X_1 - X_3$, $X_3 - X_4$, $X_4 - X_2$, obtain

$$-G^{(0)}_{\alpha\gamma}(P) \int G^{(0)}_{\gamma\delta}(X)\, U(X)\, e^{iPX}\, d^4X . G^{(0)}_{\delta\beta}(P).$$

The remaining integral is expressed in terms of the Fourier components of $G^{(0)}_{\gamma\delta}$ and U by means of the formula for the Fourier components of the product of two functions‡

$$\int f(X)\, g(X)\, e^{iPX}\, d^4X = \int f(P_1)\, g(P - P_1)\, d^4P_1/(2\pi)^4. \tag{13.11}$$

† Though using for convenience a four-dimensional terminology in the discussion and notation, we must reiterate that it is entirely unconnected with relativistic invariance.

‡ To prove this formula, we must substitute on the left the functions $f(X)$ and $g(X)$ as Fourier expansions:

$$\int f(X)\, g(X)\, e^{iPX}\, d^4X = \int f(P_1)\, g(P_2)\, e^{i(P-P_1-P_2)}\, d^4X\, d^4P_1\, d^4P_2/(2\pi)^4.$$

Thus the first-order correction to the Green's functions in the momentum representation is finally

$$iG^{(1)}_{\alpha\beta}(P) = in^{(0)}\,U(0)\,G^{(0)}_{\alpha\gamma}(P)\,G^{(0)}_{\gamma\beta}(P) -$$
$$- \int G^{(0)}_{\alpha\gamma}(P)\,G^{(0)}_{\gamma\delta}(P_1)\,G^{(0)}_{\delta\beta}(P)\,U(\mathbf{p}-\mathbf{p}_1)\,d^4P_1/(2\pi)^4. \qquad (13.12)$$

Each of the two terms in (13.12) corresponds to a particular Feynman diagram, and this equation may be written

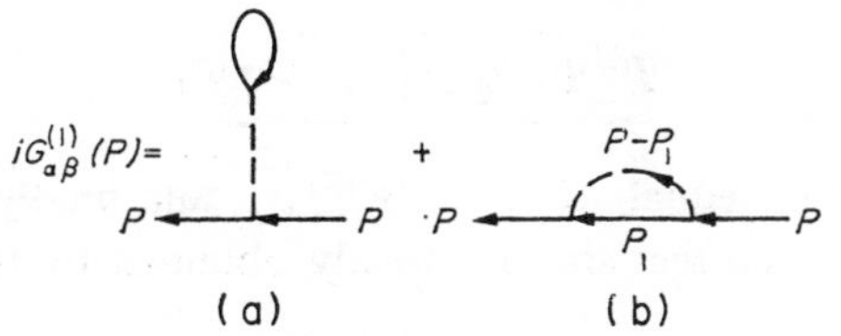

(13.13)

The points of intersection of lines are called *vertices* of the diagram. Each diagram has $2n$ vertices, where n is the order of the perturbation theory. At each vertex, two continuous lines and one broken line meet. To each continuous line is attached its "4-momentum" P in the direction shown by the arrow (and the direction of the arrows is unchanged along each continuous sequence of such lines). To each broken line is attached a 4-momentum Q; for these lines, conventionally any direction of the arrow may be chosen.† The "conservation of 4-momentum" holds at the vertices of the diagram: the sum of the 4-momentum for the ingoing lines is equal to that for the outgoing lines at each vertex. Each vertex also has a particular spin index α. Each diagram has two *external* lines (one ingoing and one outgoing), whose 4-momentum is the argument of the required Green's function $iG_{\alpha\beta}(P)$; these two lines also have the spin indices α and β of that function. The remaining lines in the diagram are called *internal* lines.

The analytical form of the terms corresponding to each diagram is deduced from the following rules:

1. Each continuous line between vertices α and β is associated with the factor $iG^{(0)}_{\alpha\beta}(P)$, and each broken line with the factor $-iU(Q)$. A closed loop with one vertex is associated with the factor $n^{(0)}(\mu)$.

The integration over d^4X is effected by the formula

$$e^{iPX}\,d^4X = (2\pi)^4\delta^4(P),$$

where the "four-dimensional" delta function $\delta^{(4)}$ is defined as the product of delta functions of the components of the "4-vector" P. The resulting factor $\delta^{(4)}(P-P_1-P_2)$ is removed by integration over d^4P_2, and we have the right-hand side of (13.11).

† The "time" components of the 4-vectors $Q = (q_0, \mathbf{q})$ are in general non-zero, but the function $U(Q)$ is independent of q_0 by the definition (13.10). The arbitrariness of the direction of the broken line arises because the function $U(-Q) = U(Q)$ is even.

2. At each vertex, the conservation of 4-momentum holds. There is integration over $d^4P/(2\pi)^4$ for the 4-momentum of internal lines that are left indeterminate. At each vertex there is summation over a pair of dummy spin indices, one from each of the adjacent $G^{(0)}$ factors.

3. The common factor of the diagram in $iG_{\alpha\beta}$ is $(-1)^L$, where L is the number of closed loops of continuous lines with more than one vertex in the diagram.

This last rule arises as follows. A closed loop with k (> 1) vertices comes from the contraction of ψ operators in the form

$$\Psi_1^+\Psi_1\Psi_2^+\Psi_2\dots\Psi_k^+\Psi_k$$

Here, the contractions equal $iG_{12}^{(0)}, \dots, iG_{k-1,k}^{(0)}$, and finally $-iG_{k1}^{(0)}$. For loops with one vertex, the correct sign is already obtained by the presence of $n^{(0)}$ from rule 1.

As an example, here is a set of diagrams giving the second-order correction to the Green's functions:

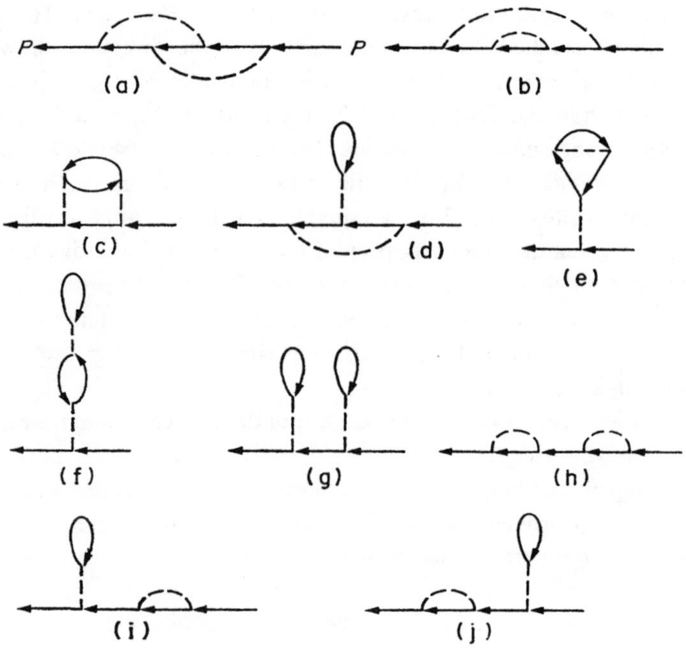

(13.14)

Lastly, let us return to Wick's theorem and prove it in the "macroscopic limit" (i.e. as $V \to \infty$ or, equivalently for a given density of the system, as $N \to \infty$), which is the only important case in statistical applications.

Let us consider, for example, the averaged product of four ψ operators, of the type

$$\langle\hat{\Psi}_{01}\hat{\Psi}_{02}\hat{\Psi}_{03}^+\hat{\Psi}_{04}^+\rangle = \frac{1}{V^2}\sum_{\mathbf{p}_1\dots\mathbf{p}_4}\langle\hat{a}_{\mathbf{p}_1}\hat{a}_{\mathbf{p}_2}\hat{a}_{\mathbf{p}_3}^+\hat{a}_{\mathbf{p}_4}^+\rangle\exp(\dots); \tag{13.15}$$

the ψ operators are in the form (9.3), and the obvious but lengthy exponents are omitted. In this sum, the only non-zero terms are those containing equal numbers of operators $\hat{a}_{\mathbf{p}}$ and $\hat{a}^+_{\mathbf{p}}$ with the same values of the momenta. They include terms in which the momenta are equal in pairs, e.g. $\mathbf{p}_1 = \mathbf{p}_4$ and $\mathbf{p}_2 = \mathbf{p}_3$. These correspond to the paired contraction

$$\Psi_{01}\Psi_{02}\Psi^+_{03}\Psi^+_{04}$$

and are expressed by a sum of the form

$$\frac{1}{V^2}\sum_{\mathbf{p}_1,\mathbf{p}_2} \langle \hat{a}_{\mathbf{p}_1}\hat{a}^+_{\mathbf{p}_1}\rangle \langle \hat{a}_{\mathbf{p}_2}\hat{a}^+_{\mathbf{p}_2}\rangle \exp(\ldots).$$

In the limit $V \to \infty$, the summation over $\mathbf{p}_1$ and $\mathbf{p}_2$ is replaced by integration over $V^2 d^3p_1 d^3p_2/(2\pi)^6$; the volume V cancels, and the expression remains finite. In the sum (13.15), terms with $\mathbf{p}_1 = \mathbf{p}_2 = \mathbf{p}_3 = \mathbf{p}_4$ are also non-zero; they form a sum

$$\frac{1}{V^2}\sum_{\mathbf{p}} \langle \hat{a}_{\mathbf{p}}\hat{a}_{\mathbf{p}}\hat{a}^+_{\mathbf{p}}\hat{a}^+_{\mathbf{p}}\rangle \exp(\ldots),$$

but after the change to integration one factor $1/V$ remains, and this expression vanishes in the limit $V \to \infty$.

This is clearly a general result: in the limit $V \to \infty$, only the results of paired contractions are non-zero in the mean value of a product of ψ operators.

In the proof given, no essential use has been made of the fact that the averaging is with respect to the ground state, and it therefore remains valid for averaging with respect to any quantum state of the system.

§ 14. The self-energy function

The rules of the diagram technique formulated in §13 have an important property: the common factor in the diagram is independent of its order. Consequently, each "figure" in the diagram has a definite analytical significance, whatever the diagram in which it appears, and can be calculated independently beforehand. In fact, we can calculate beforehand the sum of several figures having a definite number of external lines, and then substitute it as a "block" in more complex diagrams. This is one of the chief advantages of the diagram technique.

One such block, which is also of considerable independent importance, is the *self-energy function*.† In order to arrive at this concept, let us consider all the Green's function diagrams that cannot be separated into two parts joined by only one continuous line. These include, for example, the two diagrams of

† Compare the corresponding definition of the compact self-energy function in quantum electrodynamics (*RQT*, §§100, 102).

first-order perturbation theory (13.13), and the second-order diagrams (13.14 a–f). All have the same type of structure: one factor $iG^{(0)}_{\alpha\beta}$ at each end, and an internal part (a function of P), called the self-energy function. The sum of all possible such parts is called the exact or complete self-energy function or the *mass operator*; we shall denote it by $-i\Sigma_{\alpha\beta}(P)$.

All diagrams of the self-energy type give a contribution to the Green's function

$$iG^{(0)}_{\alpha\beta}(P)[-i\Sigma_{\beta\gamma}(P)]\,iG^{(0)}_{\gamma\delta}(P) = iG^{(0)}(P)\,\Sigma(P)\,G^{(0)}(P)\delta_{\alpha\delta}, \tag{14.1}$$

where we have written $G^{(0)}_{\alpha\beta} = G^{(0)}\delta_{\alpha\beta}$ and also

$$\Sigma_{\alpha\beta}(P) = \delta_{\alpha\beta}\Sigma(P). \tag{14.2}$$

The complete Green's function (represented graphically by a thick continuous line) is given by the sum of an infinite series

(14.3)

where the circles denote exact self-energy functions $-i\Sigma_{\alpha\beta}$. Each term in this series from the third onwards is a set of diagrams which can be dissected into two, three, ... parts joined by one continuous line.

If we detach from each term of the series (14.3), from the second onwards, one circle and the line to its right, the remaining series is again the complete function. Thus

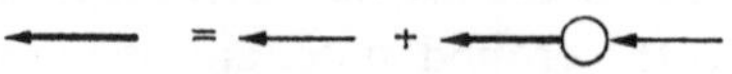

(14.4)

Analytically, this is written

$$G = G^{(0)} + G\Sigma G^{(0)} \tag{14.5}$$

or, dividing by $G^{(0)}G$,

$$\frac{1}{G(P)} = \frac{1}{G^{(0)}(P)} - \Sigma(P). \tag{14.6}$$

We note that the sign of the imaginary part of Σ is the same as that of im G, and from (8.14)

$$\operatorname{sgn}\operatorname{im}\Sigma(\omega, \mathbf{p}) = -\operatorname{sgn}\omega. \tag{14.7}$$

This follows from (14.6), since the sign of im G^{-1} is the reverse of that of im G, and from (9.7) im $[G^{(0)}]^{-1} = 0$.

Thus the calculation of G reduces to that of Σ, which requires the use of a smaller number of diagrams. The number can be still further reduced, since some of the remaining diagrams can be summed at once in a very simple form.

Let us select among all the diagrams that determine Σ (with a pair interaction between particles) those which represent various "offshoots" connected to the external lines by one broken line, and denote their sum by Σ_a. All such diagrams are present in one *skeleton diagram*† of the form

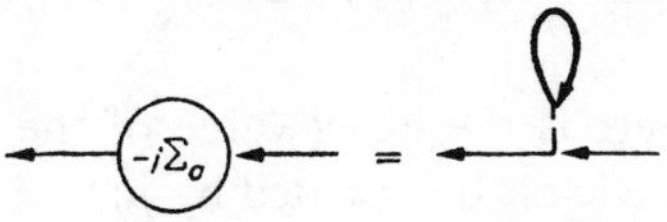

(14.8)

The remaining part of Σ is denoted by Σ_b. For example, the following diagrams of the first and second orders belong to the first class:

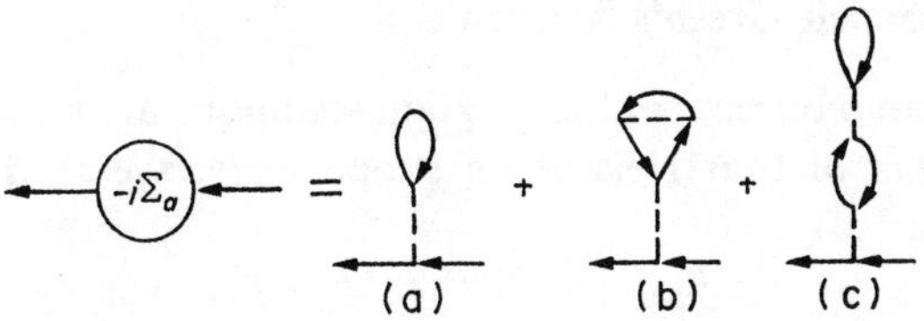

(14.9)

and to the second class:

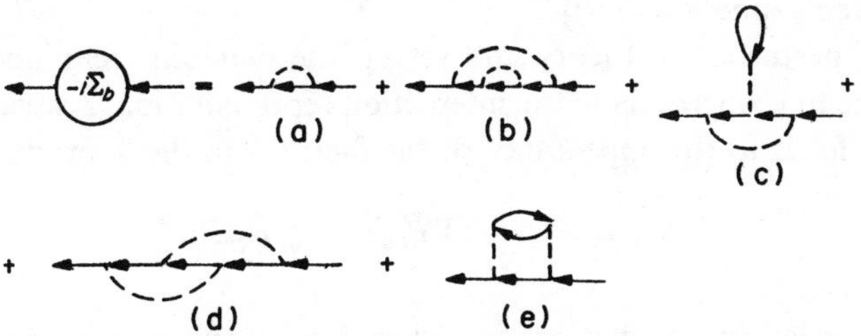

(14.10)

The thick loop in the diagram (14.8) corresponds to the exact density $n(\mu)$ of the system, just as the thin loop in (13.13a) corresponds to the ideal gas density $n^{(0)}(\mu)$. It therefore follows from the definition (14.8) that

$$-i\Sigma_a = -in(\mu)\,U(0). \tag{14.11}$$

Thus

$$\Sigma = n(\mu)\,U(0)+\Sigma_b, \tag{14.12}$$

and only the diagrams in Σ_b need be specially calculated.

The quasi-particle dispersion relation is given by (8.16). Expressing G there in terms of Σ by (14.6) and taking $G^{(0)}$ from (9.7), we obtain the equation

$$\frac{1}{G^{(0)}(\varepsilon-\mu, \mathbf{p})} = \varepsilon(\mathbf{p})-\frac{p^2}{2m} = \Sigma(\varepsilon-\mu, \mathbf{p}). \tag{14.13}$$

On the boundary of the Fermi sphere, where $p = p_F$, the energy of the quasi-particle is equal to μ. Hence we see that

$$\mu-\Sigma(0, \mathbf{p}_F) = p_F^2/2m. \tag{14.14}$$

† As in quantum field theory, skeleton diagrams are those made up of thick lines and blocks; each such diagram is equivalent to a definite infinite set of ordinary diagrams of various orders.

The dispersion relation therefore has the form (for p close to p_F)

$$\varepsilon(\mathbf{p})-\mu = \frac{p_F}{m}(p-p_F)+\Sigma(\varepsilon-\mu, \mathbf{p}_F)-\Sigma(0, \mathbf{p}_F). \tag{14.15}$$

We emphasize that p_F here is the *exact* value of the limiting momentum for a system of interacting particles. It is related by $p_F^3/3\pi^2 = n$ to the exact density $n(\mu)$, not to the approximate density $n^{(0)}$, as in (13.5).

§ 15. The two-particle Green's function

Other important concepts in the diagram technique are reached by considering the T product of four Heisenberg ψ operators, averaged with respect to the ground state:†

$$K_{34,\,12} = \langle \mathrm{T}\hat{\Psi}_3\hat{\Psi}_4\hat{\Psi}_1^+\hat{\Psi}_2^+\rangle. \tag{15.1}$$

This is called the *two-particle Green's function* (as distinct from the single-particle Green's function (7.9)).

To apply perturbation theory and set up the diagram technique, we must again change to ψ operators in the interaction representation. As with the function G, this leads to the appearance of the factor $\hat{S}$ in the T product:

$$K_{34,\,12} = \frac{1}{\langle \hat{S}\rangle}\langle \mathrm{T}\hat{\Psi}_{03}\hat{\Psi}_{04}\hat{\Psi}_{01}^+\hat{\Psi}_{02}^+\hat{S}\rangle. \tag{15.2}$$

In the zero-order approximation (i.e. when $\hat{S} = 1$) this expression becomes a sum of products of two contractions expressible in terms of $G^{(0)}$ functions:

$$K_{34,\,12}^{(0)} = G_{31}^{(0)}G_{42}^{(0)} - G_{32}^{(0)}G_{41}^{(0)}. \tag{15.3}$$

The subsequent discussion of the properties of the two-particle Green's function thus defined will be given in the momentum representation.

For a homogeneous system, $K_{34,12}$ in fact depends only on three independent differences of the arguments, for example X_3-X_2, X_4-X_2, X_1-X_2. In the momentum representation, this property has the consequence that the Fourier component with respect to all the variables $X_1, \ldots, X_4$ contains a delta function:

$$\begin{aligned}\int K_{34,\,12} \exp\{i(P_3X_3+P_4X_4-P_1X_1-P_2X_2)\}\, d^4X_1 \ldots d^4X_4 \\ = (2\pi)^4\,\delta^{(4)}(P_3+P_4-P_1-P_2)\,K_{\gamma\delta,\,\alpha\beta}(P_3, P_4; P_1, P_2).\end{aligned} \tag{15.4}$$

This is easily seen by noting that

$$\begin{aligned}&P_3X_3+P_4X_4-P_1X_1-P_2X_2 \\ &\quad = P_3(X_3-X_2)+P_4(X_4-X_2)-P_1(X_1-X_2)-X_2(P_1+P_2-P_3-P_4),\end{aligned}$$

† We are again using the simplified notation in which the suffixes 1,2,... denote the 4-coordinates together with the spin index: $X_1\alpha$, $X_2\beta$, ... ; cf. the second footnote to §13. The full notation is shown by

$$K_{34,\,12} \equiv K_{\gamma\delta,\,\alpha\beta}(X_3, X_4; X_1, X_2).$$

and changing to integration over $X_3-X_4, X_4-X_2, X_1-X_2, X_2$. It may be noted in passing that the inverse Fourier transform may be written

$$K_{34,\,12} = \int K_{\gamma\delta,\,\alpha\beta}(P_3, P_4; P_1, P_3+P_4-P_1)\times$$
$$\times \exp\{-i[P_3(X_3-X_2)+P_4(X_4-X_2)-P_1(X_1-X_2)]\}\frac{d^4P_1\, d^4P_3\, d^4P_4}{(2\pi)^{12}}. \quad (15.5)$$

The function $K_{\gamma\delta,\,\alpha\beta}(P_3, P_4; P_1, P_2)$ defined in this way will be called the two-particle Green's function in the momentum representation: its arguments are related by

$$P_1+P_2 = P_3+P_4.$$

In the zero-order approximation, in agreement with (15.3), we have

$$K^{(0)}_{\gamma\delta,\,\alpha\beta}(P_3, P_4; P_1, P_2)$$
$$= (2\pi)^4[\delta^{(4)}(P_1-P_3)\, G^{(0)}_{\gamma\alpha}(P_1)\, G^{(0)}_{\delta\beta}(P_2) - \delta^{(4)}(P_1-P_4)\, G^{(0)}_{\gamma\beta}(P_2)\, G^{(0)}_{\delta\alpha}(P_1)], \quad (15.6)$$

i.e. K reduces to a sum of two products of single-particle Green's functions.

In higher approximations of perturbation theory, terms appear which amount to corrections to these single-particle functions, together with terms that do not form products of G functions. This part of the two-particle Green's function is of independent interest. To derive it, we put K in the form

$$K_{\alpha_3\alpha_4,\,\alpha_1\alpha_2}(P_3, P_4; P_1, P_2)$$
$$= (2\pi)^4[\delta^{(4)}(P_1-P_3)\, G_{\alpha_3\alpha_1}(P_1)\, G_{\alpha_4\alpha_2}(P_2) -$$
$$- \delta^{(4)}(P_1-P_4)\, G_{\alpha_3\alpha_2}(P_2)\, G_{\alpha_4\alpha_1}(P_1)] +$$
$$+ G_{\alpha_3\beta_3}(P_3)\, G_{\alpha_4\beta_4}(P_4) i\Gamma_{\beta_3\beta_4,\,\beta_1\beta_2}(P_3, P_4; P_1, P_2)\, G_{\beta_1\alpha_1}(P_1)\, G_{\beta_2\alpha_2}(P_2). \quad (15.7)$$

The function Γ thus defined is called a *vertex function.*

According to the definition (15.1), a two-particle Green's function in the space–time representation is antisymmetric with respect to interchanges of arguments (together with the spin suffixes) in the first or second pair: 1 and 2, or 3 and 4. Hence we have the analogous symmetry property for the Green's function and the vertex function in the momentum representation:

$$\Gamma_{\gamma\delta,\,\alpha\beta}(P_3, P_4; P_1, P_2) = -\Gamma_{\delta\gamma,\,\alpha\beta}(P_4, P_3; P_1, P_2)$$
$$= -\Gamma_{\gamma\delta,\,\beta\alpha}(P_3, P_4; P_2, P_1). \quad (15.8)$$

The reason for separating the four G factors in the definition of Γ (the last term in (15.7)) becomes clear if we trace back the nature of the diagrams that arise when the expression (15.2) for the two-particle Green's function is expanded. The analysis below again assumes a pair interaction between particles.

In the zero-order approximation, the function K is assigned diagrams

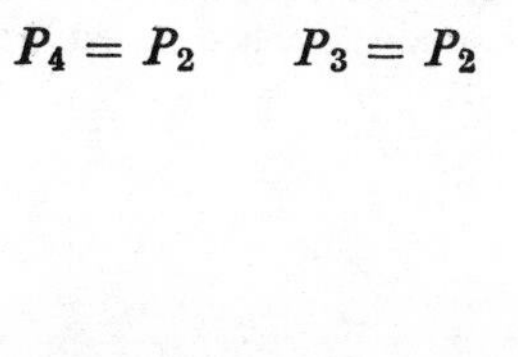

corresponding to the terms in (15.6). In first-order perturbation theory, diagrams appear of the types†

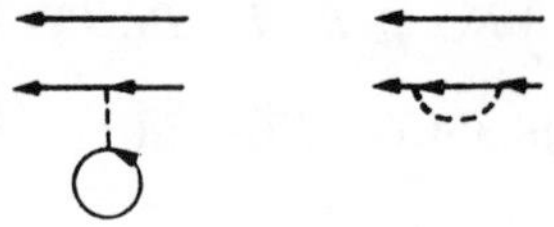

representing corrections to each of the factors in (15.6). There also appear diagrams that do not separate into two parts:

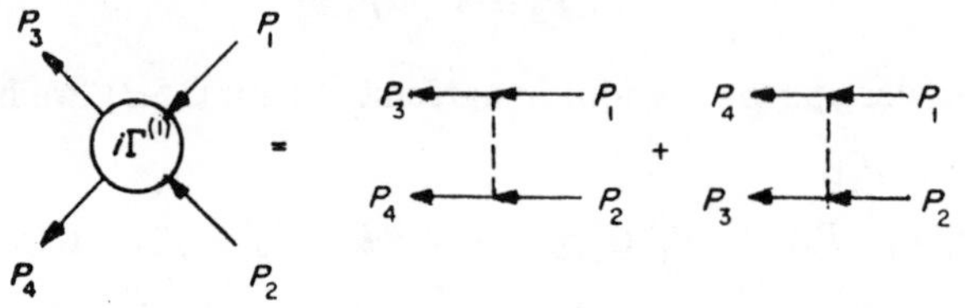

(15.9)

The four arrows $P_1, \ldots, P_4$ correspond to the four G factors in the last term in (15.7), and the internal part of the diagrams determines (in first order) the vertex function, the circle on the left of the diagram equation (15.9). Writing these diagrams in analytical form, we have

$$\Gamma^{(1)}_{\gamma\delta,\alpha\beta}(P_3, P_4; P_1, P_2) = -\delta_{\alpha\gamma}\delta_{\beta\delta}U(P_1-P_3)+\delta_{\alpha\delta}\delta_{\beta\gamma}U(P_1-P_4).$$

The diagrams of higher orders contain corrections of three types: (1) further corrections to two unconnected continuous lines, (2) corrections of self-energy type to external lines in the diagrams (15.9), (3) corrections forming a figure that replaces the broken line in the diagrams (15.9); the sum of all possible such figures gives the exact vertex function $i\Gamma$. In the graphical representation of the two-particle Green's function by a sum of skeleton diagrams,

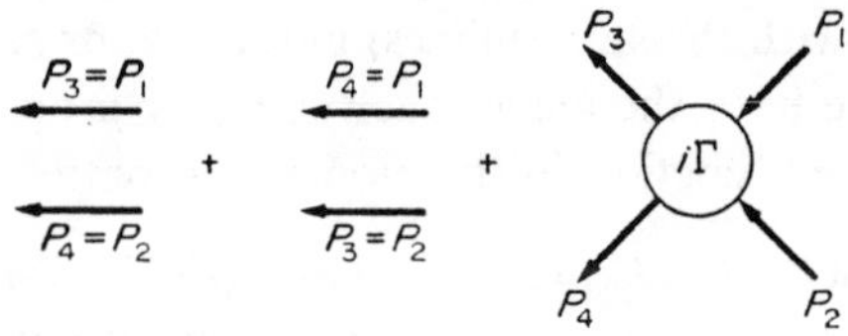

(15.10)

the thick lines represent exact G functions, and the circle conventionally represents the vertex function.

The calculation of the vertex function in various orders of perturbation theory must be made by means of the diagram technique rules formulated in §13, and diagrams with four external lines are to be considered (rather than those with two as in the calculation of G). Rule (3), which gives the sign of the whole diagram, is to be supplemented by the following point; if external lines

† As for the single-particle Green's function, the factor $\langle \hat{S} \rangle^{-1}$ in the definition (15.2) leads to the vanishing of diagrams that contain detached closed loops of continuous lines.

1 and 4, and 2 and 3, are joined by sequences of continuous lines (instead of 1 and 3, and 2 and 4), the sign of the diagram is reversed.

As an example, the following are all the diagrams that determine the vertex function in second-order perturbation theory:

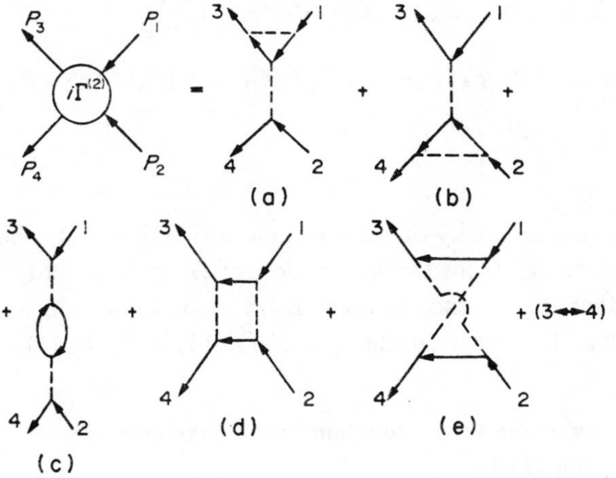

(15.11)

The self-energy function Σ and the vertex function Γ are not independent; they are related by a certain integral equation called *Dyson's equation*.†

To derive this, we use equation (9.5), which is valid (as mentioned in §9) even when the interaction of the particles is taken into account. There is a difference from the derivation in §9, however, in that the ψ operator now satisfies equation (7.8). Omitting in the latter the term containing the external field, and substituting from it the derivative $\partial\hat{\Psi}/\partial t_1$ in (9.5), we obtain

$$\left(i\frac{\partial}{\partial t_1}+\frac{\Delta_1}{2m}+\mu\right)G_{\alpha\beta}(X_1-X_2)-\delta_{\alpha\beta}\delta^{(4)}(X_1-X_2)$$
$$=-i\int\langle \mathrm{T}\hat{\Psi}^+_\gamma(X_3)\,U(X_1-X_2)\,\hat{\Psi}_\gamma(X_3)\,d^4X_3.\hat{\Psi}_\alpha(X_1)\,\hat{\Psi}^+_\beta(X_2)\rangle$$
$$=-i\int K_{\gamma\alpha,\gamma\beta}(X_3,X_1;X_3,X_2)\,U_{13}d^4X_3. \tag{15.12}$$

This equation solves the problem in principle, since K is expressed in terms of Γ by (15.7). We have only to change to the momentum representation. To do so, we multiply (15.12) by $\exp[iP(X_1-X_2)]$ and integrate over $d^4(X_1-X_2)$, taking $K_{31,32}$ in the form (15.5) and U_{13} in the form (13.9). Then the integration with respect to 4-coordinates gives delta functions, which are removed by the integration with respect to 4-momenta. The result is

$$[G^{(0)-1}(P)\,G(P)-1]\,\delta_{\alpha\beta}$$
$$=-i\int K_{\gamma\alpha,\gamma\beta}(P_3,P_4;P_3+P_4-P,P)\,U(P-P_4)\frac{d^4P_3\,d^4P_4}{(2\pi)^8}, \tag{15.13}$$

with $G^{(0)}(P)$ from (9.7).

† It is analogous to Dyson's equation in quantum electrodynamics (see *RQT*, §104).

It now remains to express K in terms of Γ. Substituting (15.7) in (15.13), we finally obtain Dyson's equation in the form

$$\delta_{\alpha\beta}[G^{(0)-1}(P)-G^{-1}(P)] = \delta_{\alpha\beta}\Sigma(P)$$
$$= U(0)\, n(\mu)\, \delta_{\alpha\beta}+i\delta_{\alpha\beta}\int U(P-P_1)\, G(P_1)\frac{d^4P_1}{(2\pi)^4}$$
$$+\int \Gamma_{\gamma\alpha,\gamma\beta}(P_3, P_4; P_3+P_4-P, P)\, G(P_3)\, G(P_4)\, G(P_3+P_4-P)\times$$
$$\times U(P-P_4)\frac{d^4P_3\, d^4P_4}{(2\pi)^8}. \qquad (15.14)$$

Here $n(\mu)$ is the exact density of the system as a function of its chemical potential; this factor comes from the integration of G by formula (7.24), together with the fact that the G function arose from a contraction in which $\hat{\Psi}^+$ is to the left of $\hat{\Psi}$. The first term on the right of (15.14) is Σ_a (14.11).

§ 16. The relation of the vertex function to the quasi-particle scattering amplitude

The mathematical formalism developed in the preceding sections makes possible a rigorous justification and fuller understanding of the significance of the fundamental relations in the Landau theory of the Fermi liquid, which have been introduced in Chapter I in a partly intuitive form. This topic will be the subject of §§16–20.†

There is a close relation between the vertex function and the mutual scattering amplitude of quasi-particles. For a better elucidation of this relation, let us consider it first in terms of the purely quantum-mechanical problem of the scattering of two particles in a vacuum.

In quantum mechanics, diagrams with four external lines (two ingoing and two outgoing) correspond to a collision of two particles; in the analytical form of the diagram, its external lines correspond to the wave function (plane wave) amplitudes of free particles (cf. *RQT*, §103). Let us see how such diagrams of different orders in fact give successive terms in the ordinary (non-relativistic) Born expansion of the scattering amplitude.

First of all, in a vacuum many of the diagrams are zero. This is most simply understood in the coordinate representation, since in a vacuum all contractions of the form $\langle\hat{\Psi}^+\hat{\Psi}\rangle$ are zero in which the annihilation operator is to the right and acts first on the vacuum state; only contractions of the form $\langle\hat{\Psi}\hat{\Psi}^+\rangle$ remain. Hence all diagrams with closed loops of continuous lines are zero, since they always contain a contraction of the form $\langle\hat{\Psi}^+\hat{\Psi}\rangle$. For the same reason, all corrections to the Green's function, i.e. to the internal continuous

† The content of §§16–18 is due to L. D. Landau (1958), and that of §§19 and 20 to L. D. Landau and L. P. Pitaevskiĭ (1959).

lines in the diagrams, are zero.† Lastly, diagrams with intersecting broken lines are zero; for example, in the diagram

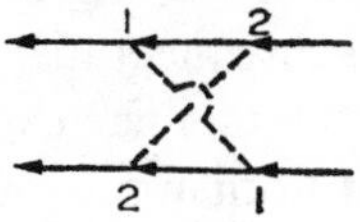

(where 1 and 2 denote the arguments t_1 and t_2) the upper internal line corresponds to the contraction $\langle\hat{\Psi}_2^+\hat{\Psi}_1\rangle = 0$ if $t_2 > t_1$, and the lower line to $\langle\hat{\Psi}_1^+\hat{\Psi}_2\rangle = 0$ if $t_2 < t_1$.

Thus, for two particles in a vacuum, there remain only the following diagrams forming a "ladder series":

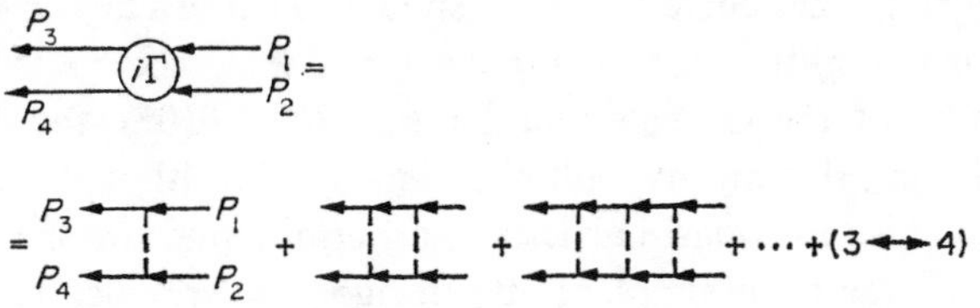

(16.1)

The internal continuous lines here correspond to the vacuum Green's functions

$$G^{(\text{vac})}(\omega, \mathbf{p}) = \left[\omega - \frac{\mathbf{p}^2}{2m} + i0\right]^{-1} \tag{16.2}$$

(formula (9.7) with $\mu = 0$). It should be noted that (because μ is absent from the denominator) the pole of this function is always in a particular (the lower) complex ω half-plane. The vanishing of the diagrams listed above occurs, from the mathematical point of view, precisely because all the poles of the integrands lie in one half-plane; the vanishing of the integrals is obvious if the path of integration is closed in the other half-plane.

The ladder series (16.1) can be summed by reducing it to an integral equation (cf. the summation of the similar series (17.3) below). If the diagrams with interchanged external lines 3 and 4 are at first omitted, this equation is equivalent to Schrödinger's equation for two particles, ignoring their identity, written in the momentum representation; see *QM* (130.9). Accordingly the vertex function is expressed in terms of the scattering amplitude f of the two particles by

$$\Gamma_{\gamma\delta,\alpha\beta}(P_3, P_4; P_1, P_2) = \delta_{\alpha\gamma}\delta_{\beta\delta}(4\pi/m)f. \tag{16.3}$$

The addition of the diagrams with interchanged external lines 3 and 4 brings about the antisymmetrization of the amplitude, as is correct for fermions.

† The vanishing of all corrections to the Green's function in the vacuum simply expresses the fact that a single particle cannot interact with anything. Here it may be recalled that the existence of vacuum corrections to the Green's function of a particle in the relativistic theory is due to the possible occurrence of virtual electron pairs or photons in intermediate states.

In the first approximation of perturbation theory, only the first diagram (16.1) and the one with interchanged external lines remain; these do not involve $G^{(\text{vac})}$. The expression for the scattering amplitude is then the usual one in the first Born approximation. The subsequent diagrams, after the integration over intermediate frequencies, give the familiar expressions for the corrections to the amplitude in the subsequent Born approximations.

In a Fermi liquid, the interaction of the colliding particles with the particles of the medium causes them to be effectively replaced by quasi-particles. All the corrections to the internal lines of the diagram resulting from this interaction are automatically taken into account by the definition of the function Γ. A further allowance must, however, be made for the corrections to the external lines. In quantum field theory (by virtue of the general requirements of a unitary scattering matrix), these corrections are shown to cause a factor $\sqrt{Z}$ to appear in the scattering amplitude for each free external line, where Z is the renormalization constant of the Green's function (see *RQT*, §107); for diagrams with four external lines, this means multiplication by Z^2. Although the proof given in *RQT* is valid also for quasi-particles in a Fermi liquid, we shall here explain the origin of this factor by simpler (but not rigorous) arguments.

The Green's function of a liquid, near its pole (the first term in (10.2)) differs from that for an ideal gas only by the factor Z. If $\hat{\Psi}$ and $\hat{\Psi}^+$ are replaced by the operators $\hat{\Psi}_{qu} = \hat{\Psi}/\sqrt{Z}$, $\hat{\Psi}^+_{qu} = \hat{\Psi}^+/\sqrt{Z}$, the Green's function $G_{qu} = G/Z$ formed from them will look exactly like that for an ideal gas, near the pole. In this sense such operators may be regarded as ψ operators of an ideal gas of quasi-particles. The two-particle Green's function determined from them is $K_{qu} = K/Z^2$, and therefore, by the definition (15.7), the vertex part $\Gamma_{qu} = \Gamma Z^2$, as required.

In the application to quasi-particles, what is of interest is the number of collisions (per unit time and liquid volume) rather than the collision cross-section. For collisions with a given change in the momenta and spin components of the particles ($\mathbf{p}_1\alpha, \mathbf{p}_2\beta \to \mathbf{p}_3\gamma, \mathbf{p}_4\delta$), this number is

$$dW = 2\pi|Z^2\Gamma_{\gamma\delta,\alpha\beta}(P_3, P_4; P_1, P_2)|^2\,\delta(\varepsilon_3+\varepsilon_4-\varepsilon_1-\varepsilon_2)\times$$
$$\times n_{\mathbf{p}_1}n_{\mathbf{p}_2}(1-n_{\mathbf{p}_3})(1-n_{\mathbf{p}_4})\,d^3p_1\,d^3p_2\,d^3p_3/(2\pi)^9, \qquad (16.4)$$

where $\mathbf{p}_1+\mathbf{p}_2 = \mathbf{p}_3+\mathbf{p}_4$, and $n_\mathbf{p}$ is the quasi-particle distribution function. The factors $n_{\mathbf{p}_1}$ and $n_{\mathbf{p}_2}$ simply express the fact that the number of collisions of quasi-particles with given initial momenta (and spin components) is proportional to the numbers of such quasi-particles per unit volume. The factors $(1-n_{\mathbf{p}_3})$ and $(1-n_{\mathbf{p}_4})$ are due to the fact that, in accordance with the Pauli principle, a collision can occur only if the final states are unoccupied.

§ 17. The vertex function for small momentum transfers

An important role in the theory of Fermi liquids is played by the vertex function with almost equal values of the pairs of variables P_1, P_3 and P_2, P_4 (we shall see, in particular, that it is closely related to the quasi-particle interaction function). Using the relation $P_1+P_2 = P_3+P_4$, we put $P_3 = P_1+K$, $P_4 = P_2-K$, and write in simplified notation

$$\Gamma_{\gamma\delta,\,\alpha\beta}(P_1+K, P_2-K; P_1, P_2) = \Gamma_{\gamma\delta,\,\alpha\beta}(K; P_1, P_2); \tag{17.1}$$

this function will be considered for small K. In terms of quasi-particle scattering processes, this means that we consider collisions with a small transfer of 4-momentum, which are close to "forward scattering".

When $K = 0$, as we shall see, the function Γ has a singularity; we shall be interested in the part of the function that contains this singularity. The origin of the singularity is easily understood from the skeleton diagram

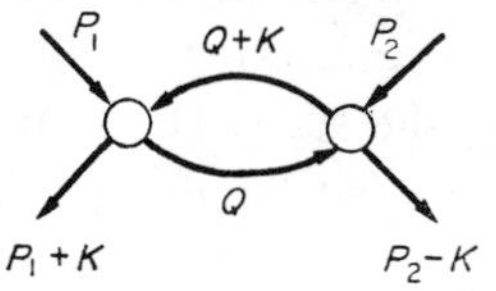

(17.2)

which includes the set of diagrams of the two-particle Green's function that can be cut between the pairs of external lines P_1, P_3 and P_2, P_4 into two parts joined by two continuous lines.† The two thick joining lines correspond to the exact one-particle Green's functions $G(Q)$ and $G(Q+K)$, with integration over the 4-momentum Q in the diagram. As $K \to 0$, the arguments of these two functions become closer, and therefore so do their poles. These may "pinch" between them the contour of integration (see below), which is the source of the singularity in the function Γ.

To calculate the exact function Γ, we must sum the whole perturbation-theory series. Since our aim is to separate the part that has a singularity when $K = 0$, we must first distinguish the contribution from all diagrams that cannot be cut through pairs of continuous lines having almost equal (differing by K) values of the 4-momentum. This part of the function Γ, which has no singularity at $K = 0$, is denoted by $\tilde{\Gamma}$; in it we can put $K = 0$, since it is a function only of the variables P_1 and P_2: $\tilde{\Gamma}_{\gamma\delta,\,\alpha\beta}(P_1, P_2)$. The "dangerous" diagrams can be classified by the numbers of pairs of lines with almost equal arguments

† For example, in second-order perturbation theory (with respect to the pair interaction), (17.2) contains the diagrams (15.11a, b, c), and (15.11e) with interchanged external lines 3 and 4.

which they contain. Thus the total vertex part Γ is represented by the following infinite ladder series of diagrams:

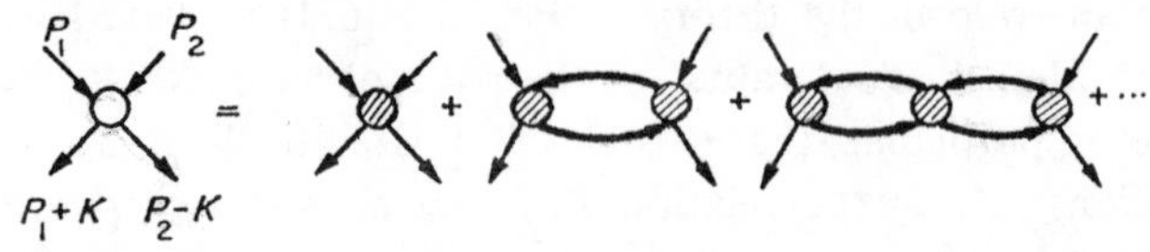

(17.3)

Here the white circle corresponds to the required $i\Gamma$, and the shaded circles represent $i\tilde{\Gamma}$. The external lines in these diagrams do not enter into the determination of Γ, and serve only to indicate the number and values of the ingoing and outgoing 4-momenta.

All the internal lines in the diagrams (17.3) are thick, i.e. they correspond to exact G functions. Here it should be emphasized that the possibility of representing Γ in the form of these skeleton diagrams (and therefore all the conclusions drawn from them) does not presuppose a pair interaction between particles, since there are no explicit broken lines, and the nature of the interaction actually affects only the internal structure of the blocks represented by circles, which is of no interest in this connection.†

The problem of summing the series (17.3) amounts to the solution of the integral equation; to derive this, we "multiply" the whole series by a further $\tilde{\Gamma}$:

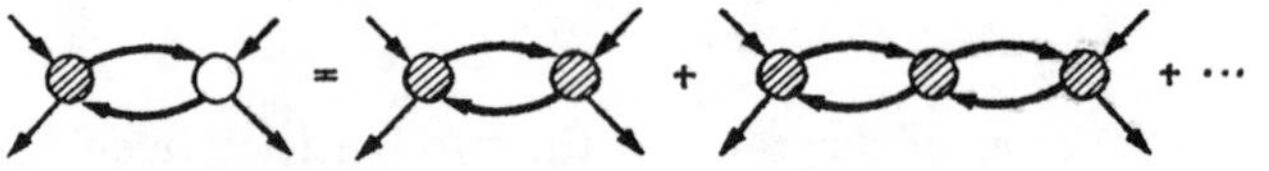

Comparison with the original series (17.3) gives the equation

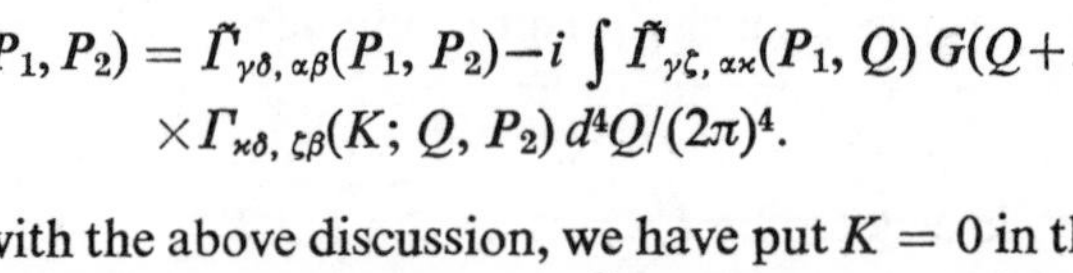

(17.4)

This diagram equation, when written in analytical form, gives the required integral equation

$$\Gamma_{\gamma\delta,\alpha\beta}(K; P_1, P_2) = \tilde{\Gamma}_{\gamma\delta,\alpha\beta}(P_1, P_2) - i \int \tilde{\Gamma}_{\gamma\zeta,\alpha\varkappa}(P_1, Q)\, G(Q+K)\, G(Q) \times$$
$$\times \Gamma_{\varkappa\delta,\zeta\beta}(K; Q, P_2)\, d^4Q/(2\pi)^4. \tag{17.5}$$

In accordance with the above discussion, we have put $K = 0$ in the functions $\tilde{\Gamma}$, used the abbreviated notation for Γ and $\tilde{\Gamma}$ previously described, and also put $G_{\alpha\beta} = G\delta_{\alpha\beta}$.

† Only such general properties as the conservation of particle number are assumed. This latter is shown by the constant difference between the number of lines going right and left at each cross-section of the diagram (equal to zero for cross-sections of the type shown in (17.3)).

To investigate this equation, let us first consider the product $G(Q+K)G(Q)$ in its kernel. As already mentioned, for small K the poles of the two factors are close together. Near these poles, the G functions are represented by the pole terms in (10.2). Denoting the components of the 4-vectors K and Q by

$$K = (\omega, \mathbf{k}), \quad Q = (q_0, \mathbf{q}), \tag{17.6}$$

we can write in this region

$$G(Q)\,G(Q+K) \approx Z^2[q_0 - v_F(q-p_F)+i\delta_1]^{-1}[q_0+\omega - v_F(|\mathbf{q}+\mathbf{k}|-p_F)+i\delta_2]^{-1}, \tag{17.7}$$

where δ_1 and δ_2 are infinitesimal increments whose signs near the poles are given by

$$\left.\begin{aligned} \operatorname{sgn}\delta_1 &= \operatorname{sgn}(q-p_F), \\ \operatorname{sgn}\delta_2 &= \operatorname{sgn}(|\mathbf{q}+\mathbf{k}|-p). \end{aligned}\right\} \tag{17.8}$$

The signs of δ_1 and δ_2 determine the position of the poles in the upper or lower half-plane of the complex variable q_0. The singularity in the kernel of the integral equation, and therefore in the solution of the equation, arises from the pinching of the contour of integration with respect to q_0 (the real axis) between the poles, for which the latter must be on opposite sides of the contour, i.e. in opposite half-planes.

Let us first suppose that $\mathbf{q}.\mathbf{k} > 0$, i.e. $\cos\theta > 0$, where θ is the angle between $\mathbf{q}$ and $\mathbf{k}$. Then $|\mathbf{q}+\mathbf{k}| > q$, and δ_1 and δ_2 have opposite signs ($\delta_1 < 0$, $\delta_2 > 0$) if $q < p_F$, $|\mathbf{q}+\mathbf{k}| > p_F$, which, in view of the smallness of k, is equivalent to the conditions

$$p_F - k\cos\theta < q < p_F. \tag{17.9}$$

In the subsequent integration with respect to q_0 in (17.5), the contour of integration may be replaced by an infinite semicircle in either the upper or the lower half-plane; the integral is then given by the residue of the integrand at the corresponding pole. Because of the narrowness of the range (17.9) when k is small, we can take $k = 0$ in the factors Γ and $\tilde{\Gamma}$ in the integrand, and similarly $q_0 \approx 0$ for the position of the poles when k and ω are small.

In other words, as regards its role in the kernel of the integral equation (17.5), the product of pole factors (17.7) is equivalent to the delta functions $A\delta(q_0)\,\delta(q-p_F)$, with a coefficient A given by

$$A = \int \frac{Z^2\,dq_0\,dq}{[q_0 - v_F(q-p_F)+i\delta_1]\,[q_0+\omega - v_F(|\mathbf{q}+\mathbf{k}|-p_F)+i\delta_2]}.$$

When q is outside the range (17.9), both poles lie in the same half-plane of complex q_0; when the contour of integration with respect to q_0 is completed in the other half-plane, we see that the integral is zero. In the range (17.9), completing the contour in one half-plane and calculating the integral from the residue at the pole in that half-plane, we find

$$A = \int \frac{2\pi i Z^2\, dq}{\omega - v_F(|\mathbf{q}+\mathbf{k}|-q)+i0},$$

where we have used the fact that $\delta_1 < 0$ and $\delta_2 > 0$ in the range (17.9). Since, by (17.9), $q \approx p_F \gg k$, we can put $|\mathbf{q}+\mathbf{k}|-q \approx k \cos\theta$, and then, with the limits given by (17.9),

$$A = 2\pi i Z^2 k \cos\theta/(\omega - kv_F\cos\theta).$$

It is easy to prove by the same method that a similar expression for A (but with the opposite sign of $i0$) is obtained when $\cos\theta < 0$ (when the integration is to be taken over the range $q > p_F$, $|\mathbf{q}+\mathbf{k}| < p_F$). Thus we have in the kernel of (17.5)

$$G(Q)\,G(Q+K) = \frac{2\pi i Z^2 \mathbf{l}.\mathbf{k}\,\delta(q_0)\,\delta(q-p_F)}{\omega - v_F\mathbf{l}.\mathbf{k}+i0.\mathrm{sgn}\,\omega}+\phi(Q), \tag{17.10}$$

where $\mathbf{l}.\mathbf{k}$ is written in place of $k\cos\theta$ ($\mathbf{l} = \mathbf{q}/q$), and the function ϕ has (when K is small) no delta-function part, and we can therefore put in it $K = 0$.

Substituting (17.10) in (17.5), we get the basic integral equation in the form

$$\begin{aligned}\Gamma_{\gamma\delta,\alpha\beta}(K; P_1, P_2) &= \tilde{\Gamma}_{\gamma\delta,\alpha\beta}(P_1, P_2)\\ &\quad - i\int \tilde{\Gamma}_{\gamma\zeta,\alpha\varkappa}(P_1, Q)\,\phi(Q)\,\Gamma_{\varkappa\delta,\zeta\beta}(K; Q, P_2)\, d^4Q/(2\pi)^4\\ &\quad + \frac{Z^2p_F^2}{(2\pi)^3}\int \tilde{\Gamma}_{\gamma\zeta,\alpha\varkappa}(P_1, Q_F)\,\Gamma_{\varkappa\delta,\zeta\beta}(K; Q_F, P_2)\frac{\mathbf{l}.\mathbf{k}\,do_l}{\omega - v_F\mathbf{l}.\mathbf{k}}.\end{aligned} \tag{17.11}$$

In the last term we have written $d^4Q = q^2\,dq\,do_l\,dq_0$ (where do_l is the element of solid angle in the direction of $\mathbf{l}$) and have removed the delta functions by integration over $dq\,dq_0$. In this term the argument Q in the functions Γ and $\tilde{\Gamma}$ is taken on the Fermi surface: $Q_F = (0, p_F\mathbf{l})$.

The factor $\mathbf{l}.\mathbf{k}/(\omega - v_F\mathbf{l}.\mathbf{k})$ in the kernel of (17.11) has a specific property: its limit as $\mathbf{k} \to 0$ and $\omega \to 0$ depends on the limit of the ratio ω/k. The solution of the equation must therefore have the same property: the limit of the function $\Gamma(K; P_1, P_2)$ as $K \to 0$ depends on the way in which ω and $\mathbf{k}$ tend to zero.

Let $\Gamma^\omega(P_1, P_2)$ denote the limit

$$\Gamma^\omega_{\gamma\delta,\alpha\beta}(P_1, P_2) = \lim_{K\to 0}\Gamma_{\gamma\delta,\alpha\beta}(K; P_1, P_2) \quad \text{for} \quad k/\omega \to 0; \tag{17.12}$$

we shall see in §18 that the quasi-particle interaction function is related to this quantity. With that method of taking the limit, the kernel of the last integral term in (17.11) is zero, and so Γ^ω satisfies the equation

$$\Gamma^\omega_{\gamma\delta,\alpha\beta}(P_1, P_2) = \tilde{\Gamma}_{\gamma\delta,\alpha\beta}(P_1, P_2) - i\int \tilde{\Gamma}_{\gamma\zeta,\alpha\varkappa}(P_1, Q)\,\phi(Q)\,\Gamma^\omega_{\varkappa\delta,\zeta\beta}(Q, P_2)\, d^4Q/(2\pi)^4. \tag{17.13}$$

Because of (15.8),

$$\Gamma^{\omega}_{\gamma\delta,\,\alpha\beta}(P_1, P_2) = \Gamma^{\omega}_{\delta\gamma,\,\beta\alpha}(P_2, P_1). \tag{17.14}$$

We can eliminate $\tilde{\Gamma}$ from the two equations (17.11) and (17.13). The result is

$$\Gamma_{\gamma\delta,\,\alpha\beta}(K; P_1, P_2) = \Gamma^{\omega}_{\gamma\delta,\,\alpha\beta}(P_1, P_2) + \frac{Z^2 p_F^2}{(2\pi)^3} \int \Gamma^{\omega}_{\gamma\zeta,\,\alpha\varkappa}(P_1, Q_F)\Gamma_{\varkappa\delta,\,\zeta\beta}(K; Q_F, P_2)\frac{\mathbf{l}.\mathbf{k}do_l}{\omega - v_F\mathbf{l}.\mathbf{k}}, \tag{17.15}$$

since, if we formally write (17.13) as $\tilde{\Gamma} = \hat{L}\Gamma^{\omega}$, (17.11) becomes

$$\hat{L}\Gamma = \tilde{\Gamma} + \frac{Z^2 p_F^2}{(2\pi)^3}\int \tilde{\Gamma}\Gamma \frac{\mathbf{l}.\mathbf{k}do_l}{\omega - v_F\mathbf{l}.\mathbf{k}}.$$

Substituting here $\tilde{\Gamma} = \hat{L}\Gamma^{\omega}$ and applying the operator $\hat{L}^{-1}$ to both sides of the equation, we get (17.15).

We now define the function Γ^k by

$$\Gamma^{k}_{\gamma\delta,\,\alpha\beta}(P_1, P_2) = \lim_{K\to 0} \Gamma_{\gamma\delta,\,\alpha\beta}(K; P_1, P_2) \quad \text{with} \quad \omega/k \to 0. \tag{17.16}$$

This function (multiplied by Z^2) is the forward scattering amplitude (i.e. that of the transition $P_1, P_2 \to P_1, P_2$), corresponding to actual physical processes undergone by quasi-particles at the Fermi surface: collisions that leave the quasi-particles on that surface are accompanied by a change of momentum without change of energy, and therefore the passage to the limit of zero momentum transfer ($\mathbf{k} \to 0$) must be made with exactly zero energy transfer ($\omega = 0$). The function Γ^{ω} defined above corresponds to the non-physical limit of "scattering" with a small energy transfer and exactly zero momentum transfer ($\mathbf{k} = 0$).

Putting $\omega = 0$ in (17.15), taking the limit $\mathbf{k} \to 0$ and multiplying both sides by Z^2, we get

$$Z^2\Gamma^{k}_{\gamma\delta,\,\alpha\beta}(P_1, P_2) = Z^2\Gamma^{\omega}_{\gamma\delta,\,\alpha\beta}(P_1, P_2) - \frac{p_F^2}{v_F(2\pi)^3}\int Z^2\Gamma^{\omega}_{\gamma\zeta,\,\alpha\varkappa}(P_1, Q_F)\, Z^2\Gamma^{k}_{\varkappa\delta,\,\zeta\beta}(Q_F, P_2)\,do_l. \tag{17.17}$$

Thus there is a general relation between the two limiting forms of the forward scattering amplitude.

The antisymmetry properties (15.8) for Γ give some information about the behaviour of Γ^k and Γ^{ω} when $P_1 \to P_2$. Putting $P_1 = P_2$ and $\alpha = \beta$ in this equation, we get

$$\Gamma_{\gamma\delta,\,\alpha\alpha}(P_1 + K, P_1 - K; P_1, P_1) = 0; \tag{17.18}$$

there is *no* summation with respect to α here.† The transition to Γ^{ω} or Γ^{k}

† When only the exchange interaction between the quasi-particle spins is taken into account, the only non-zero $\Gamma_{\gamma\delta,\,\alpha\alpha}$ are the $\Gamma_{\alpha\alpha,\,\alpha\alpha}$. This expresses the constancy of the spin vector in scattering, and may be verified directly from an expression of the type (2.4).

in this equation is to be made with caution, since in the latter functions we first put $K = 0$, but in (17.18) we first put $P_1 = P_2$.

Let K and $P_1 - P_2 \equiv S = (s_0, \mathbf{s})$ be simultaneously small. Then, as well as the diagrams (17.2), the diagrams

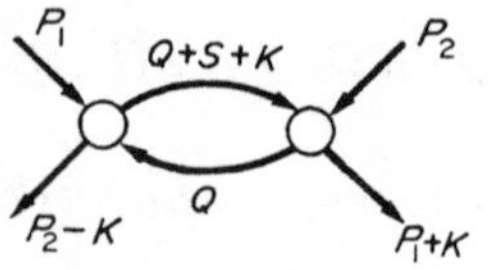

are dangerous. When K and $S \to 0$, the function $\Gamma_{\gamma\delta,\,\alpha\alpha}$ will therefore depend on the two "singular" arguments

$$x = \omega/k, \quad y = (s_0 + \omega)/|\mathbf{s}+\mathbf{k}|,$$

and (17.18) shows that this function is zero when $x = y$. We shall consider the values of Γ on the Fermi surface; then $\omega = s_0 = 0$, and so $y = 0$. Hence, in this limit, (17.18) is valid only if also $x = 0$. Thus, on the Fermi surface it is valid for Γ^k:

$$\Gamma^k_{\gamma\delta,\,\alpha\alpha}(P_1, P_1) = 0 \tag{17.19}$$

(N. D. Mermin 1967).

§ 18. The relation of the vertex function to the quasi-particle interaction function

Just as intermediate states with particle numbers $N \pm 1$ are involved in the matrix element (7.9), which determines the one-particle Green's function, so intermediate states with N, $N \pm 1$ and $N \pm 2$ particles are involved in the matrix element (15.1) of the two-particle Green's function.†

Because of the presence of intermediate states with $N \pm 1$ particles, the two-particle Green's function has poles which coincide with those of the function G, i.e. with the quasi-particle energy. The corresponding factors are, however, shown explicitly in (15.7). Hence the vertex function Γ defined by this formula only has poles corresponding to states with N or $N \pm 2$ particles. The angular momentum of these states differs by 0 or 1 from that of the ground state, and so the elementary excitations corresponding to these poles have integral spin (0 or 1) and hence obey Bose statistics. Thus the poles of the vertex function determine the Bose branches of the energy spectrum of a Fermi liquid.

† States with N particles arise with such a sequence of operators in the T product as, for example, $\hat{\Psi}_3 \hat{\Psi}_1^+ \hat{\Psi}_4 \hat{\Psi}_2^+$. States with $N+2$ particles correspond to such sequences as $\hat{\Psi}_3 \hat{\Psi}_4 \hat{\Psi}_1^+ \hat{\Psi}_2^+$.

The poles arising from the intermediate states without change of particle number correspond to elementary excitations that represent zero sound quanta. In the diagram technique, intermediate states correspond to different cross-sections that divide the diagrams into two parts between various external lines. In the present case, intermediate states without change of particle number correspond to cross-sections of the diagrams (17.3) at one of the pairs of continuous lines joining adjacent blocks $\tilde{\Gamma}$; the constancy of particle number in these states is expressed by the equal numbers of lines passing in each direction through the cross-section. The 4-momentum transfer through such a cross-section is $(Q+K)-Q = K$; accordingly, the elementary excitations without change of particle number correspond to poles of the vertex function $\Gamma(K; P_1, P_2)$ with respect to the variable K.

We have seen previously, in the derivation of (17.10), that one of the two momenta $\mathbf{q}$ and $\mathbf{q}+\mathbf{k}$ (which appear in the 4-vectors Q and $Q+K$) must be greater than the limiting momentum p_F, and the other must be less. On the other hand, in excitation from the ground state, only "particles" can be outside the Fermi sphere, and only "holes" within it. In this sense, we can say that the zero-point excitations in a Fermi liquid may be regarded as particle–hole bound states.†

Elementary excitations corresponding to intermediate states with $N\pm 2$ particles (and to the poles of the function $\Gamma(K; P_1, P_2)$ with respect to the variable P_1+P_2) could be regarded as bound states of two particles or two holes. The presence of such states would, however, lead (as will be shown in Chapter V) to superfluidity of the Fermi liquid, and this in turn necessitates a considerable change in the whole mathematical formalism of the diagram technique.

Thus, to determine the Bose branch of the energy spectrum of a non-superfluid Fermi liquid, we must examine the poles of the vertex function $\Gamma(K; P_1, P_2)$ with respect to the variable $K = (\omega, \mathbf{k})$. For each value of $\mathbf{k}$, a particular energy $\omega(\mathbf{k})$ corresponds to the pole, and the dispersion relation for these excitations is thereby determined. For weakly excited states, ω and $\mathbf{k}$ are small, so that we can use the equations derived for the function $\Gamma(K; P_1, P_2)$ in the range of small K.

Near a pole of Γ, the left-hand side and the integral on the right-hand side of (17.15) become arbitrarily large; the term $\Gamma^{\omega}(P_1, P_2)$ remains finite, and may therefore be omitted. Moreover, the variable P_2 and the suffixes β and δ are unaffected by the operations on Γ in (17.15), and so they act as unimportant parameters in that equation. Lastly, we shall consider the function Γ on the surface of the Fermi sphere, i.e. we shall put $P_1 = (0, p_F\mathbf{n})$, where $\mathbf{n}$ is a variable unit vector. From all these facts we conclude that the determination of

† In this formulation, the problem is formally very similar to that of determining the electron–positron bound state levels in quantum electrodynamics (see *RQT*, §122). In particular, equations (17.4) and (17.5) are analogous to the Bethe–Salpeter equation, *RQT* (122.10), (122.11).

the acoustic excitations in a Fermi liquid reduces to the problem of finding the eigenvalues of the integral equation

$$\chi_{\gamma\alpha}(\mathbf{n}) = \frac{Z^2 p_F^2}{(2\pi)^3} \int \Gamma^{\omega}_{\gamma\zeta,\alpha\varkappa}(\mathbf{n}, \mathbf{l})\, \chi_{\varkappa\zeta}(\mathbf{l}) \frac{\mathbf{l}.\mathbf{k}\, do_l}{\omega - v_F \mathbf{l}.\mathbf{k}}, \tag{18.1}$$

where $\chi_{\gamma\alpha}(\mathbf{n})$ is an auxiliary function.

This equation may be transformed by replacing χ by another function

$$\nu_{\gamma\alpha}(\mathbf{n}) = \frac{\mathbf{n}.\mathbf{k}}{\omega - v_F \mathbf{n}.\mathbf{k}} \chi_{\gamma\alpha}(\mathbf{n}). \tag{18.2}$$

Equation (18.1) then becomes

$$(\omega - v_F \mathbf{n}.\mathbf{k})\, \nu_{\gamma\alpha}(\mathbf{n}) = \mathbf{k}.\mathbf{n} \frac{p_F^2 Z^2}{(2\pi)^3} \int \Gamma^{\omega}_{\gamma\zeta,\alpha\varkappa}(\mathbf{n}, \mathbf{n}')\, \nu_{\varkappa\zeta}(\mathbf{n}')\, do', \tag{18.3}$$

with $\mathbf{n}'$ in place of $\mathbf{l}$.

This equation has exactly the same form as the transport equation (4.10) for the vibrations of a Fermi liquid. Comparison of the two equations gives the following correlation between the quasi-particle interaction function and the function Γ^{ω}:

$$f_{\gamma\delta,\alpha\beta}(p_F \mathbf{n}, p_F \mathbf{n}') = Z^2 \Gamma^{\omega}_{\gamma\delta,\alpha\beta}(\mathbf{n}, \mathbf{n}'). \tag{18.4}$$

This shows the relation between the function f and the properties of quasi-particle scattering.†

Equation (18.4) relates f to the non-physical scattering amplitude. We now use (17.17) to obtain an explicit relation between f and the "physical" forward scattering amplitude for quasi-particles on the Fermi surface, which we denote by

$$A_{\gamma\delta,\alpha\beta}(\mathbf{n}_1, \mathbf{n}_2) = Z^2 \Gamma^{k}_{\gamma\delta,\alpha\beta}(\mathbf{n}_1, \mathbf{n}_2). \tag{18.5}$$

The relation (17.17) on the Fermi surface is

$$A_{\gamma\delta,\alpha\beta}(\mathbf{n}_1, \mathbf{n}_2) = f_{\gamma\delta,\alpha\beta}(\mathbf{n}_1, \mathbf{n}_2) - \frac{p_F^2}{2\pi^2 v_F} \int f_{\gamma\zeta,\alpha\varkappa}(\mathbf{n}_1, \mathbf{n}')\, A_{\varkappa\delta,\zeta\beta}(\mathbf{n}', \mathbf{n}_2) \frac{do'}{4\pi}. \tag{18.6}$$

The spin dependence of the functions A and f can be expressed by means of the Pauli matrices $\boldsymbol{\sigma}$. In the general case, these functions may contain any scalar combinations of the four vectors $\mathbf{n}_1$, $\mathbf{n}_2$, $\boldsymbol{\sigma}_1$, $\boldsymbol{\sigma}_2$. If there is an exchange

† The above general proof is due to L. D. Landau (1958). For a slightly non-ideal Fermi gas, the derivation of the transport equation by summation of specific diagrams of the type (17.3) was earlier given by A. B. Migdal and V. M. Galitskiĭ (1958). For a gas, the G functions (in the zero-order approximation) contain only pole terms, and so the exclusion of non-pole terms does not arise.

interaction between the particles, the only permissible scalar products are $\mathbf{n}_1.\mathbf{n}_2$ and $\boldsymbol{\sigma}_1.\boldsymbol{\sigma}_2$. The functions A and f can then be written (as with f in (2.4)) as

$$\left.\begin{aligned}\frac{p_F^2}{\pi^2 v_F} f_{\gamma\delta,\alpha\beta}(\mathbf{n}_1, \mathbf{n}_2) &= F(\vartheta)\,\delta_{\alpha\gamma}\delta_{\beta\delta}+G(\vartheta)\,\boldsymbol{\sigma}_{\gamma\alpha}.\boldsymbol{\sigma}_{\delta\beta},\\ \frac{p_F^2}{\pi^2 v_F} A_{\gamma\delta,\alpha\beta}(\mathbf{n}_1, \mathbf{n}_2) &= B(\vartheta)\,\delta_{\alpha\gamma}\delta_{\beta\delta}+C(\vartheta)\,\boldsymbol{\sigma}_{\gamma\alpha}.\boldsymbol{\sigma}_{\delta\beta},\end{aligned}\right\} \tag{18.7}$$

where the coefficients F, G, B, C, are functions only of the angle ϑ between $\mathbf{n}_1$ and $\mathbf{n}_2$. They may be expanded in series of Legendre polynomials:

$$B(\vartheta) = \sum_{l=0}^{\infty} (2l+1)\, B_l P_l(\cos\vartheta), \;\ldots . \tag{18.8}$$

Substituting (18.7) and (18.8) in (18.6), and calculating the integral with the addition theorem for Legendre polynomials, we get

$$B_l = F_l(1-B_l), \quad C_l = G_l(1-C_l). \tag{18.9}$$

These formulae establish a simple algebraic relation between the expansion coefficients of f and A.

The stability conditions (2.19) and (2.20) give similar inequalities for the coefficients B_l and C_l:

$$B_l < 1, \quad C_l < 1. \tag{18.10}$$

Moreover, these coefficients satisfy a relation that follows from (17.19): $B(0)+ + C(0) = 0$, or

$$\sum_{l=0}^{\infty} (2l+1)\,(B_l+C_l) = 0. \tag{18.11}$$

Equations (18.9) and (18.11) together with the conditions (18.10) are sufficient to prove an interesting theorem: in every stable Fermi liquid, there is at least one branch (ordinary or spin) of axially symmetric zero sound.†

§ 19. Identities for derivatives of the Green's function

In the mathematical formalism of Green's functions, an important part is played by certain identical relations between the derivatives of these functions and the quasi-particle scattering amplitude. These relations are all derived in the same way by calculating the change in the Green's functions caused by some fictitious "external field" for which the result of its action on the system is already known.

† See N. D. Mermin, *Physical Review* **159,** 161, 1967.

First of all, then, let us calculate the change δG in the Green's function caused by an arbitrary "external field", for which the corresponding term in the Hamiltonian is

$$\delta \hat{V}^{(1)} = \int \hat{\Psi}_\alpha^+(t, \mathbf{r})\, \delta \hat{U} \hat{\Psi}_\alpha(t, \mathbf{r})\, d^3x, \tag{19.1}$$

where $\delta \hat{U}$ is some operator acting on functions of $\mathbf{r}$ (and possibly depending on the time t).

When the external field is present, the Green's function depends on the two 4-momenta P_1 and P_2. In the diagram technique, such a field is represented by a new graphical feature, an external broken line:

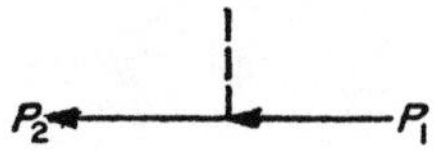

and this line is associated with a factor

$$-i\delta U(P_2, P_1) = -i \int e^{iP_2X}\, \delta \hat{U} e^{-iP_1X}\, d^4X. \tag{19.2}$$

In the first order with respect to the external field, the correction to the exact Green's function is represented by a sum of two skeleton diagrams:

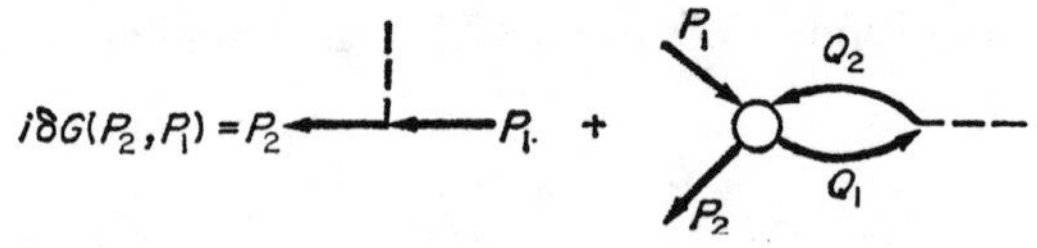

(19.3)

where all the continuous lines are thick (exact G functions) and the circle is an exact vertex function ($i\Gamma$). In analytical form, this equation is

$$\begin{aligned} \delta G_{\beta\alpha}(P_2, P_1) = {} & G_{\beta\gamma}(P_2)\, \delta U(P_2, P_1)\, G_{\gamma\alpha}(P_1) - \\ & - iG_{\beta\gamma}(P_2)\, G_{\varepsilon\alpha}(P_1) \int \Gamma_{\gamma\delta, \varepsilon\zeta}(P_2, Q_1; P_1, Q_2) \times \\ & \times \delta U(Q_2, Q_1)\, G_{\zeta\varkappa}(Q_2)\, G_{\varkappa\delta}(Q_1)\, d^4Q_1/(2\pi)^4, \end{aligned} \tag{19.4}$$

with $Q_2 + P_1 = P_2 + Q_1$.

The first two identities to be considered are due to the conservation of the number of particles in the system. In the Hamiltonian of the system, this property is expressed by the appearance of the ψ operators in pairs: one $\hat{\Psi}^+(X)$ and one $\hat{\Psi}(X)$ for each argument X.

We apply a gauge transformation to the ψ operators:

$$\hat{\Psi}_\alpha(X) = \hat{\Psi}'_\alpha(X) e^{-i\chi(X)}, \quad \hat{\Psi}_\alpha^+ = \hat{\Psi}'^+_\alpha e^{i\chi(X)}, \tag{19.5}$$

where $\chi(X)$ is a real function.† From the above-mentioned property of the

† This is analogous to the gauge transformation in quantum electrodynamics; cf. *QM* (111.2)–(111.9).

Hamiltonian, if $\hat{\Psi}$ satisfies the "Schrödinger equation" (7.8), $\hat{\Psi}'$ satisfies a similar equation with the changes

$$\triangle \to (\nabla - i\nabla\chi)^2, \quad \frac{\partial}{\partial t} \to \frac{\partial}{\partial t} - i\frac{\partial \chi}{\partial t}.$$

For an infinitesimal $\chi = \delta\chi$, this change in the equation is equivalent to adding to the Hamiltonian an "external field"

$$\delta\hat{U} = -\frac{\partial \delta\chi}{\partial t} + \frac{i}{2m}(\triangle\delta\chi + 2\nabla\delta\chi \cdot \nabla).$$

In particular, if

$$\delta\chi(X) = \mathrm{re}\, \chi_0 e^{-iKX}, \quad K = (\omega, \mathbf{k}),$$

where the symbol re can in fact be omitted, since the subsequent operations are linear, we have

$$\delta U(P_2, P_1) = i(2\pi)^4 \chi_0 \delta^{(4)}(P_2 - P_1 - K)\left\{\omega - \frac{1}{2m}\mathbf{k} \cdot (\mathbf{p}_1 + \mathbf{p}_2)\right\}. \tag{19.6}$$

On the other hand, the Green's function constructed from the ψ operators

$$\hat{\Psi}'_\alpha = \hat{\Psi}_\alpha(1 + i\delta\chi), \quad \hat{\Psi}'^+_\alpha = \hat{\Psi}^+_\alpha(1 - i\delta\chi)$$

differs from that constructed from $\hat{\Psi}$ and $\hat{\Psi}^+$ by

$$\delta G_{\alpha\beta}(X_1, X_2) = iG_{\alpha\beta}(X_1 - X_2)[\delta\chi(X_1) - \delta\chi(X_2)]$$

or, in Fourier components,

$$\begin{aligned} \delta G_{\alpha\beta}(P_2, P_1) &= \int \delta G_{\alpha\beta}(X_1, X_2)\, e^{i(P_2X_1 - P_1X_2)}\, d^4X_1\, d^4X_2 \\ &= i[G_{\alpha\beta}(P_1) - G_{\alpha\beta}(P_2)]\, \delta\chi(P_2 - P_1), \end{aligned} \tag{19.7}$$

where

$$\delta\chi(P) = \int \delta\chi(X)\, e^{iPX}\, d^4X = (2\pi)^4 \chi_0 \delta^{(4)}(P - K).$$

Thus the same change $\delta G_{\alpha\beta}$ has been expressed in the two forms (19.7) and (19.4), where δU is to be substituted from (19.6). Equating these two expressions, we get, after putting $G_{\alpha\beta} = G\delta_{\alpha\beta}$ and renaming some of the variables,

$$\begin{aligned} \delta_{\alpha\beta}[G(P+K) - G(P)] = G(P+K)\, G(P)\Bigg\{&\left[-\omega + \frac{\mathbf{k} \cdot (2\mathbf{p} + \mathbf{k})}{2m}\right]\delta_{\alpha\beta} + \\ &+ i\int \Gamma_{\beta\delta,\alpha\delta}(K; P, Q)\, G(Q)\, G(Q-K)\left[\omega - \frac{\mathbf{k} \cdot (2\mathbf{q} - \mathbf{k})}{2m}\right]\frac{d^4Q}{(2\pi)^4}\Bigg\}. \end{aligned}$$

The required identities are obtained by taking the limit of this equation as $\omega \to 0$, $\mathbf{k} \to 0$; then

$$G(P+K) - G(P) \to \omega\frac{\partial G}{\partial p_0} + \mathbf{k} \cdot \frac{\partial G}{\partial \mathbf{p}}, \tag{19.8}$$

where $P = (p_0, \mathbf{p})$. Taking the limit with the condition $k/\omega \to 0$, we get the first identity:

$$\delta_{\alpha\beta}\frac{\partial G(P)}{\partial p_0} = -\{G^2(P)\}_\omega \left[\delta_{\alpha\beta} - i\int \Gamma^\omega_{\beta\delta,\alpha\delta}(P, Q)\{G^2(Q)\}_\omega \frac{d^4Q}{(2\pi)^4}\right], \tag{19.9}$$

with the notation

$$\{G^2(P)\}_\omega = \lim_{\omega,\,\mathbf{k}\to 0} G(P)G(P+K), \qquad k/\omega \to 0. \tag{19.10}$$

Similarly, taking the limit with the condition $\omega/k \to 0$, we get a second identity:

$$\delta_{\alpha\beta}\frac{\partial G}{\partial \mathbf{p}} = \{G^2(P)\}_k \left[\frac{\mathbf{p}}{m}\delta_{\alpha\beta} - i\int \Gamma^k_{\beta\delta,\alpha\delta}(P, Q)\frac{\mathbf{q}}{m}\{G^2(Q)\}_k \frac{d^4Q}{(2\pi)^4}\right] \tag{19.11}$$

with the corresponding notation $\{G^2(P)\}_k$.

Next, let us consider the change in the Green's function when a constant field

$$\delta\hat{U} = \delta U(\mathbf{r}) = U_0 e^{i\mathbf{k}\cdot\mathbf{r}} \tag{19.12}$$

is applied to the system. When $\mathbf{k} \to 0$, this field varies slowly in space, and so its influence on the system can be treated macroscopically. According to the thermodynamic condition of equilibrium in an external field, we must have $\mu + \delta U =$ constant (see Part 1, §25); when $\mathbf{k} \to 0$, this means that the chemical potential μ changes by the small amount $-U_0$. The corresponding change in the Green's function is

$$\delta G_{\alpha\beta}(X_1, X_2) = -U_0\delta_{\alpha\beta}\,\partial G(X_1 - X_2)/\partial\mu,$$

and its Fourier component, defined as in (19.7), is

$$\delta G_{\alpha\beta}(P_2, P_1) = -(2\pi)^{(4)}\,\delta^{(4)}(P_2 - P_1)U_0\delta_{\alpha\beta}\partial G(P_1)/\partial\mu.$$

The same change in the Green's function can also be calculated from equation (19.4), this time with

$$\delta U(P_2, P_1) = (2\pi)^4\, U_0\delta^{(4)}(P_2 - P_1 - K) \qquad (K = 0, \mathbf{k}).$$

The passage to the limit $\mathbf{k} \to 0$ in this case (constant field, $\omega \equiv 0$) corresponds to the case $\omega/k \to 0$. This gives the identity

$$\delta_{\alpha\beta}\frac{\partial G(P)}{\partial \mu} = -\{G^2(P)\}_k \left[\delta_{\alpha\beta} - i\int \Gamma^k_{\beta\delta,\alpha\delta}(P, Q)\{G^2(Q)\}_k \frac{d^4Q}{(2\pi)^4}\right]. \tag{19.13}$$

Lastly, one more identity results from the Galilean invariance of the system. To derive it, let us consider the liquid in a system of coordinates moving with a small velocity $\delta\mathbf{w}(t) = \mathbf{w}_0 e^{-i\omega t}$ that varies slowly with time. The change to such coordinates is equivalent to the imposition of an external field whose operator is†

$$\delta\hat{U} = -\delta\mathbf{w}\,.\,\hat{\mathbf{p}} = i\delta\mathbf{w}\,.\,\nabla \tag{19.14}$$

or, in the momentum representation,

$$\delta U(P_2, P_1) = -\mathbf{p}_1\cdot\mathbf{w}_0(2\pi)^4\,\delta^{(4)}(P_2-P_1-K), \qquad K = (\omega, 0).$$

This expression is to be substituted in (19.4), and then the limit $\omega \to 0$ is to be taken.

When $\omega \to 0$, we have a Galilean transformation from one inertial frame of reference to another moving with constant velocity $\delta\mathbf{w}$. If there is in the liquid an elementary excitation with energy $\varepsilon(\mathbf{p})$, its energy in the frame moving relative to the liquid with velocity $\delta\mathbf{w}$ is $\varepsilon-\mathbf{p}.\delta\mathbf{w}$.‡ Hence, in the new frame, the frequency p_0 must appear in the function $G(P)$ as $p_0+\mathbf{p}.\delta\mathbf{w}$ (so that the pole of the function is shifted by $-\mathbf{p}.\delta\mathbf{w}$). Then

$$\delta G = \mathbf{p}\,.\,\delta\mathbf{w}\,\partial G/\partial p_0,$$

and we arrive at the identity

$$\delta_{\alpha\beta}\mathbf{p}\frac{\partial G(P)}{\partial p_0} = -\{G^2(P)\}_\omega\left\{\delta_{\alpha\beta}\mathbf{p}-i\int\Gamma^{\omega}_{\beta\delta,\alpha\delta}(P, Q)\,\mathbf{q}\{G^2(Q)\}_\omega\frac{d^4Q}{(2\pi)^4}\right\}. \tag{19.15}$$

We shall need to use these identities, in particular, for values of the free variable $P = (p_0, \mathbf{p})$ on the Fermi surface: $P_F = (0, \mathbf{p}_F)$. Transferring the factor $G^2(P)$ from the right-hand to the left-hand sides, we replace the derivatives of $G(P)$ there by those of $G^{-1}(P)$; the way in which the limit $K \to 0$ is taken in $G(P)G(P+K)$ is unimportant.

Near the Fermi surface, the Green's function is determined by the pole term, so that

$$G^{-1}(P) = \frac{1}{Z}[p_0 - v_F(p-p_F)].$$

Hence, on the surface itself,

$$\frac{\partial G^{-1}}{\partial p_0} = \frac{1}{Z}, \qquad \frac{\partial G^{-1}}{\partial \mu} = \frac{v_F}{Z}\frac{dp_F}{d\mu}.$$

† In the classical Lagrangian of a free particle, $L = \frac{1}{2}mv^2$, the change to moving coordinates is effected by substituting $\mathbf{v}\to\mathbf{v}+\delta\mathbf{w}$, and gives an increment $\delta L = m\mathbf{v}.\delta\mathbf{w}$ which is small if $\delta\mathbf{w}$ is small. Accordingly (cf. *Mechanics* (40.7)) the increment of the Hamilton's function is $\delta H = -\mathbf{p}.\delta\mathbf{w}$, and in quantum mechanics this corresponds to the operator (19.14).

‡ See the more detailed discussion in §23.

Consequently, for example, the identities (19.9) and (19.13), on the Fermi surface, take the form

$$i\int \Gamma^{\omega}_{\beta\delta,\,\alpha\delta}(P_F, Q)\{G^2(Q)\}_{\omega}\frac{d^4Q}{(2\pi)^4} = \left(1-\frac{1}{Z}\right)\delta_{\alpha\beta}, \tag{19.16}$$

$$i\int \Gamma^{k}_{\beta\delta,\,\alpha\delta}(P_F, Q)\{G^2(Q)\}_{k}\frac{d^4Q}{(2\pi)^4} = \left(1-\frac{v_F}{Z}\frac{dp_F}{d\mu}\right)\delta_{\alpha\beta}. \tag{19.17}$$

§ 20. Derivation of the relation between the limiting momentum and the density

The relations derived in the preceding sections provide a consistent proof of the fundamental proposition in the Landau theory of the Fermi liquid: the relation between the limiting momentum p_F and the density N/V of the liquid is given by the same formula (1.1) as for an ideal gas.

The idea of the proof is to calculate independently the changes in N and p_F due to an infinitesimal change in the chemical potential μ, and then to compare them.

According to (7.24), the total number of particles in a given volume V, as a function of the chemical potential, is given by the integral

$$N = -2iV\lim_{t\to -0}\int G(P)\,e^{-ip_0t}\frac{d^4P}{(2\pi)^4}, \qquad P = (p_0, \mathbf{p}). \tag{20.1}$$

Hence the derivative

$$\frac{1}{V}\frac{dN}{d\mu} = -2i\int\frac{\partial G(P)}{\partial\mu}\frac{d^4P}{(2\pi)^4}. \tag{20.2}$$

Since this integral converges for large p_0 ($\partial G/\partial\mu \propto 1/p_0^2$ when $|p_0| \to \infty$), the factor e^{-ip_0t} may be omitted from the integrand. After substitution of $\partial G/\partial\mu$ from the identity (19.13) summed over $\alpha = \beta$, we find

$$\frac{1}{V}\frac{dN}{d\mu} = -2i\int\{G^2(P)\}_k\frac{d^4P}{(2\pi)^4} + \int\{G^2(P)\}_k\Gamma^k(P, Q)\{G^2(Q)\}_k\frac{d^4P\,d^4Q}{(2\pi)^8},$$

where we have put $\Gamma = \Gamma_{\alpha\gamma,\,\alpha\gamma}$ for brevity. The object of the calculation is now to express the right-hand side of this equation in terms of an integral over the Fermi surface only.

First, we replace Γ^k in the second integral from (17.17), with S_F in place of Q_F:

$$\frac{1}{V}\frac{dN}{d\mu} = 2i\int\{G^2(P)\}_k\frac{d^4P}{(2\pi)^4} + \int\{G^2(P)\}_k\Gamma^{\omega}(P, Q)\{G^2(Q)\}_k\frac{d^4P\,d^4Q}{(2\pi)^8} -$$

$$-\frac{p_F^2Z^2}{v_F(2\pi)^3}\int\{G^2(P)\}_k\Gamma^{\omega}_{\alpha\zeta,\,\alpha\varkappa}(P, S_F)\Gamma^{k}_{\varkappa\gamma,\,\zeta\gamma}(S_F, Q)\{G^2(Q)\}_k\frac{d^4P\,d^4Q\,do_S}{(2\pi)^8}. \tag{20.3}$$

We begin by transforming the last term. In the integrand, only the last two factors depend on Q; the integral of these over d^4Q is given (on the Fermi surface, $S = S_F$) by (19.17), and this term therefore becomes

$$i\frac{p_F^2 Z^2}{v_F(2\pi)^3}\int \{G^2(P)\}_k \Gamma^\omega(P, S_F)\frac{d^4P\, do_S}{(2\pi)^4}\left(1-\frac{v_F}{Z}\frac{dp_F}{d\mu}\right).$$

Next, we note that in the integration over d^4P the limiting values of $G(P)\,G(P+$ $+K)$ are to be taken in the form (17.10); hence $\{G^2(P)\}_\omega = \phi(P)$, and

$$\{G^2(P)\}_k = \{G^2(P)\}_\omega - \frac{2\pi i Z^2}{v_F}\delta(p_0)\,\delta(p-p_F). \tag{20.4}$$

This gives

$$i\frac{p_F^2 Z^2}{v_F(2\pi)^3}\left(1-\frac{v_F}{Z}\frac{dp_F}{d\mu}\right)\left\{\iint \{G^2(P)\}_\omega \Gamma^\omega(P, S_F)\frac{d^4P\, do_S}{(2\pi)^4} - 8\pi i\bar{F}\right\},$$

where, in accordance with (18.4), we have used the quasi-particle interaction function and expressed $f_{\alpha\xi,\alpha\xi}$ in terms of $F(\vartheta)$ by (2.6), (2.7); the bar over F denotes integration over $do/4\pi$. The remaining integral over d^4P is given by (19.16), and the integration over do_S gives a further factor of 4π. Thus the third term in (20.3) is equal to

$$-\frac{p_F^2 Z^2}{v_F \pi^2}\left(\frac{v_F}{Z}\frac{dp_F}{d\mu}-1\right)\left\{1-\frac{1}{Z}+\bar{F}\right\}. \tag{20.5}$$

The second term in (20.3) is transformed similarly: the quantities $\{G^2(P)\}_k$ and $\{G^2(Q)\}_k$ are expressed in terms of $\{G^2(P)\}_\omega$ and $\{G^2(Q)\}_\omega$ by (20.4), and the identities (19.9) and (19.16) are then used. This term is then found to be equal to

$$-2i\int \frac{\partial G}{\partial p_0}\frac{d^4P}{(2\pi)^4} - 2i\int \{G^2(P)\}_\omega \frac{d^4P}{(2\pi)^4} + \frac{p_F^2 Z^2}{v_F \pi^2}\left\{2\left(\frac{1}{Z}-1\right)-\bar{F}\right\}. \tag{20.6}$$

The first integral gives zero on integration with respect to p_0, since $G \to 0$ when $p_0 \to \pm\infty$.

Lastly, the first term in (20.3), with the substitution of (20.4), becomes

$$2i\int \{G^2(p)\}_\omega \frac{d^4P}{(2\pi)^4} + \frac{p_F^2 Z^2}{v_F \pi^2}. \tag{20.7}$$

Adding the contributions (20.5)–(20.7), we get

$$\frac{1}{V}\frac{dN}{d\mu} = \frac{p_F^2}{\pi^2}\frac{dp_F}{d\mu} + \frac{p_F^2 Z}{\pi^2 v_F}\left\{1-\frac{dp_F}{d\mu}v_F(1+\bar{F})\right\}. \tag{20.8}$$

On the other hand, by putting

$$\delta n' = (\partial n'/\partial p_F)\,\delta p_F = \delta(p'-p_F)\,\delta p_F$$

in (2.14), we easily find

$$d\mu/dp_F = v_F(1+\bar{F}). \tag{20.9}$$

It should be emphasized that, in the derivation of (2.14), no specific dependence of p_F on N/V has been assumed, and we can therefore use this relation to find that dependence; equation (20.9) can, of course, also be obtained with the aid of the same relations for the vertex functions as were used in deriving (20.8).†

From this equation we see that the quantity in the braces in (20.8) is zero, and so

$$\frac{d}{d\mu}\left(\frac{N}{V}\right) = \frac{p_F^2}{\pi^2}\frac{dp_F}{d\mu} = \frac{d}{d\mu}\left[\frac{8\pi p_F^3}{3(2\pi)^3}\right]. \tag{20.10}$$

When $N/V \to 0$ we have a gas, and in this limit the dependence of p_F on N/V must therefore be the same as for a gas. This determines the constant in the integration of (20.10), and we have finally the required relation (1.1):

$$N/V = 8\pi p_F^3/3(2\pi)^3.$$

§ 21. Green's function of an almost ideal Fermi gas

To illustrate the application of the diagram technique, we shall calculate in this section the Green's function of an almost ideal Fermi gas in the model discussed in §6 by means of ordinary perturbation theory (V. M. Galitskiĭ 1958). The gas, it will be recalled, has repulsion between the particles, and the device described in §6 allows us to apply perturbation theory to this interaction, provided that the final result involves only the scattering amplitude.

As shown in §14, the determination of the Green's function reduces to the calculation of the self-energy function $\Sigma_{\alpha\beta}(P)$. In first- and second-order perturbation theory, it is given by the set of diagrams (14.9) and (14.10). These may be put in the form

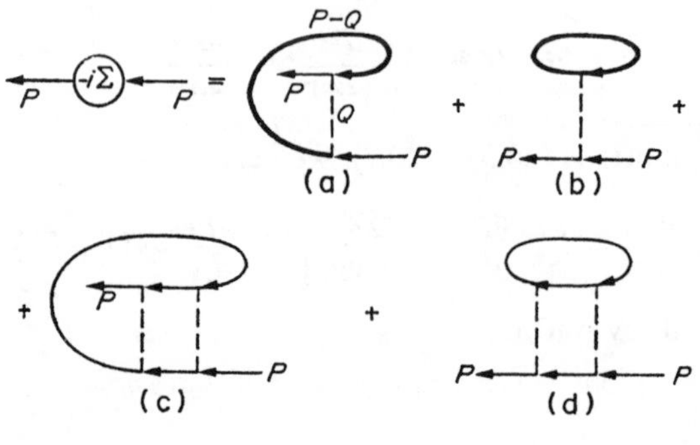

(21.1)

† Formula (2.11) for the effective mass can be deduced from the relation (17.17) and the identities (19.11) and (19.15).

The diagrams (21.1a, b) include the first-order diagrams (14.10a) and (14.9a) and the second-order diagrams (14.10b, c) and (14.9b, c); the latter differ from the former only by corrections to the internal continuous line. These lines are shown thick in (21.1a, b) and must therefore be correlated not with the ideal-gas Green's functions $G^{(0)}$ but with the functions G corrected as far as first-order terms. The diagrams (21.1c, d) are the second-order diagrams (14.10d, e). All the diagrams have been subjected to a deformation in order to clarify their structure; they are the first terms in a "ladder" series of diagrams with four external lines, in each of which a pair of external lines have been "short-circuited" in two different ways.

We first calculate the diagram (21.1a). Its analytical expression is

$$\left.\begin{aligned} [-i\Sigma(P)]_a &= \int U(Q)\,G(P-Q)\,d^4Q/(2\pi)^4, \\ Q = (q_0, \mathbf{q}), &\qquad P = (\omega, \mathbf{p}); \end{aligned}\right\} \tag{21.2}$$

the common factor $\delta_{\alpha\beta}$ is omitted. We first integrate over q_0. Since the factor $U(Q) \equiv U(\mathbf{q})$ does not depend on q_0, and $G \propto 1/q_0$ when $|q_0| \to \infty$, the manner of integration has to be more precisely described. For this, we must go back to the origin of the diagram (21.1a), and note that the continuous line there corresponds to the contraction of a pair of ψ operators arising from one operator $\hat{V}$. This means that $\hat{\Psi}$ and $\hat{\Psi}^+$ are taken at the same instant, and $\hat{\Psi}^+$ is to the left of $\hat{\Psi}$ in the contraction. That is to say, in the coordinate representation the G function occurring is taken for $t = t_1 - t_2 \to -0$. In the momentum representation, this means including a factor $\exp(-iq_0 t)$ in the integrand of (21.2) and taking the limit as $t \to -0$. Now using (7.23), we find

$$[-i\Sigma]_a = i\int U(\mathbf{q})\,N(\mathbf{p}-\mathbf{q})\,d^3q/(2\pi)^3, \tag{21.3}$$

where $N(\mathbf{p})$ is the particle distribution function.

The Fourier component $U(\mathbf{q})$ depends markedly on $\mathbf{q}$ only when $q \gtrsim 1/r_0$, where r_0 is the range of action of the field $U(r)$; these values are certainly large (for a rarefied gas) in comparison with p_F. If we consider only values $|p-p_F| \ll 1/r_0$, then for these values of $\mathbf{q}$ we have $N(\mathbf{p}-\mathbf{q}) \approx 0$. Hence $U(\mathbf{q})$ in (21.3) may be replaced by $U(0)$ and taken outside the integral.† The remaining integral is half (because of the specified value of the spin component) the gas density $n(\mu)$:

$$[\Sigma]_a = -\tfrac{1}{2}n(\mu)\,U(0).$$

The diagram (21.1b), with the closed continuous line, gives $[\Sigma]_b = n(\mu)\,U(0)$. Thus the contribution to Σ from the two diagrams is

$$[\Sigma]_{a,b} = \tfrac{1}{2}n(\mu)\,U(0) = (2\pi/m)\,n(\mu)\,a, \tag{21.4}$$

where a is the scattering length defined by (6.2).

† The resulting error is easily seen to be of the relative order of magnitude $\sim (p_F r_0)^2$, and therefore has no effect even on the terms of the next order in $p_F r_0$.

The expression (21.4) includes, in particular, the whole of the first-order effect. In this approximation, $n(\mu)$ is to be understood as the ideal gas density $n^{(0)}(\mu)$, so that

$$\Sigma^{(1)} \equiv [\Sigma]^{(1)}_{a,b} = (2\pi/m)\, n^{(0)}(\mu)\, a. \tag{21.5}$$

For the subsequent calculation we define an auxiliary function F given by the ladder diagrams

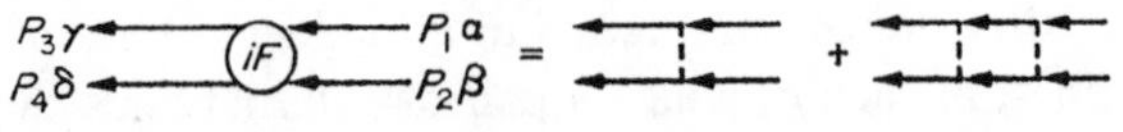

(21.6)

(as usual, $P_1+P_2 = P_3+P_4$). In analytical form,

$$iF_{\gamma\delta,\alpha\beta}(P_3, P_4; P_1, P_2) = i\delta_{\alpha\gamma}\delta_{\beta\delta}(F^{(1)}+F^{(2)}), \tag{21.7}$$

where

$$iF^{(1)} = -iU(P_3-P_1), \tag{21.8}$$

$$iF^{(2)} = \int G^{(0)}(P')\, U(P_1-P')\, G^{(0)}(P_1+P_2-P')\, U(P'-P_3)\, d^4P'/(2\pi)^4. \tag{21.9}$$

Expanding both diagrams (21.1c), (21.1d) and expressing them in terms of $F^{(2)}$, we obtain

$$\begin{aligned}[-i\Sigma(P)]_{c,d} = &-\int G^{(0)}(Q)\, F^{(2)}(P, Q; Q, P)\, d^4Q/(2\pi)^4 \\ &+2\int G^{(0)}(Q)\, F^{(2)}(P, Q; P, Q)\, d^4Q/(2\pi)^4;\end{aligned} \tag{21.10}$$

the same integrals with $F^{(1)}$ in place of $F^{(2)}$ give (21.5). The difference in sign between the two integrals is due to the presence of the closed loop in the diagram (21.1d); the delta factors in the first diagram give $\delta_{\alpha\gamma}\delta_{\gamma\beta} = \delta_{\alpha\beta}$, and those in the second diagram $\delta_{\alpha\beta}\delta_{\gamma\gamma} = 2\delta_{\alpha\beta}$.

Let us now calculate $F^{(2)}$. Since $U(Q)$ is independent of q_0, the integration with respect to p_0' reduces to

$$\int_{-\infty}^{\infty} G^{(0)}(P')\, G^{(0)}(P_1+P_2-P')\, dp_0'/2\pi.$$

Substituting here $G^{(0)}$ from (9.9) and using the convergence of the integral for $|p_0'| \to \infty$, we close the contour of integration by an infinite semicircle in one half-plane of the complex variable p_0'; the integral is zero unless the poles of the two functions $G^{(0)}$ lie in different half-planes, i.e.

$$\operatorname{sgn}(p'-p_F) = \operatorname{sgn}(|\mathbf{p}_1+\mathbf{p}_2-\mathbf{p}'|-p_F). \tag{21.11}$$

The result is

$$F^{(2)}(P_3, P_4; P_1, P_2)$$
$$= -\int \frac{U(\mathbf{p}_1-\mathbf{p}')\, U(\mathbf{p}'-\mathbf{p}_3)\operatorname{sgn}(p'-p_F)}{\omega_1+\omega_2+2\mu-\dfrac{1}{2m}[\mathbf{p}'^2+(\mathbf{p}_1+\mathbf{p}_2-\mathbf{p}')^2]+i0.\operatorname{sgn}(p'-p_F)}\frac{d^3p'}{(2\pi)^3}, \tag{21.12}$$

where $\omega_1 \equiv p_{10}$, $\omega_2 \equiv p_{20}$. In order to satisfy the condition (21.11) automatically, we must substitute in the numerator of the integrand

$$\operatorname{sgn}(p'-p_F) \to 1-\theta(\mathbf{p}')-\theta(\mathbf{p}_1+\mathbf{p}_2-\mathbf{p}'),$$

where $\theta(\mathbf{p})$ is the step function (1.10).

We have seen in §16 that a sequence of ladder diagrams determines (in a vacuum) the mutual scattering amplitude of two particles. Hence the expression (21.12) contains the correction to the first-order terms in the scattering amplitude. This correction can be taken into account by substituting in $F^{(1)}$ (21.8)

$$U(\mathbf{p}_3-\mathbf{p}_1) \to -(4\pi/m)\,\mathrm{re}\, f(\mathbf{p}_3, \mathbf{p}_1).$$

where f is the scattering amplitude† in a vacuum, correct to the second order, and at the same time subtracting from the expression $F^{(2)}$ (21.12) the real part of its value in a vacuum, i.e. for $p_F = 0$, $\mu = 0$, and the values $\omega_1 = p_1^2/2m$, $\omega_2 = p_2^2/2m$ corresponding to the energies of two real colliding particles (the "physical" external lines of the diagrams). We can then replace $-\mathrm{re}\, f$ by the value for zero energy, i.e. the scattering length a.‡ We thus have

$$F^{(2)}(P_3, P_4; P_1, P_2)$$
$$= -\left(\frac{4\pi a}{m}\right)^2 \int\left\{\frac{1-\theta(\mathbf{p}')-\theta(\mathbf{p}_1+\mathbf{p}_2-\mathbf{p}')}{\omega_1+\omega_2+2\mu-\frac{1}{2m}[p'^2+(\mathbf{p}_1+\mathbf{p}_2-\mathbf{p}')^2]+i0.\operatorname{sgn}(p'-p_F)}\right.$$
$$\left.-\mathrm{P}\frac{2m}{p_1^2+p_2^2-p'^2-(\mathbf{p}_1+\mathbf{p}_2-\mathbf{p}')^2}\right\}\frac{d^3p'}{(2\pi)^3}. \qquad (21.13)$$

The symbol P in the second term means that the integral is taken as a principal value; this is the result of separating the real part of the integral by means of the rule (8.11).

Since the expression (21.13) is symmetrical in P_1 and P_2, the two integrals in (21.10) are the same, and

$$[-i\Sigma(P)]_{c,d} = \int G^{(0)}(Q)\, F^{(2)}(P, Q; P, Q)\, d^4Q/(2\pi)^4.$$

When the first term from (21.13) is substituted, the integral with respect to q_0 is non-zero if

$$\operatorname{sgn}(p'-p_F) = -\operatorname{sgn}(q-p_F), \qquad (21.14)$$

† Not to be confused with the quasi-particle interaction function.

‡ This replacement could not be made in (21.12), since it would cause the integral to diverge for large $\mathbf{p}'$. After the subtraction mentioned, the integral converges (for $p' \sim p_F$) even with this replacement, which is therefore feasible. The subtraction of only the real part of the integral (and accordingly the replacement of U by $\mathrm{re}\, f$) is done in order to avoid a difficulty concerning the imaginary part of the scattering amplitude. The reason is that, for small momenta, $\mathrm{re}\, f$ is expanded in even powers of the momentum and $\mathrm{im}\, f$ in odd powers (see *QM*, §132). Hence the inclusion of the momentum dependence of f would lead to corrections of relative order $(p_F a)^2$, which are negligible. The substitution $U \to -4\pi f/m$, however, would mean taking into account the imaginary part of f, which brings in corrections of relative order $p_F a$.

so that the two poles of the integrand are again in different half-planes of q_0. When the second term from (21.13) is substituted, only the factor $G_0(Q)$ depends on q_0; the integration with respect to q_0 is carried out by means of formula (7.23) and gives $N^{(0)}(\mathbf{q})$, the particle distribution function in an ideal gas, i.e. the step function $\theta(\mathbf{q})$. The result, when the contributions from all diagrams (21.1a)–(21.1d) are collected, is

$$\Sigma(\omega, \mathbf{p}) = (2\pi/m)\, n(\mu)\, a + \Sigma^{(2)}(\omega, \mathbf{p}), \tag{21.15}$$

where

$$\Sigma^{(2)}(\omega, \mathbf{p}) = \left(\frac{4\pi a}{m}\right)^2 \int \left\{ \frac{[1-\theta(\mathbf{p}')-\theta(\mathbf{p}+\mathbf{q}-\mathbf{p}')][\theta(\mathbf{q})-\theta(\mathbf{p}')]}{\omega+\mu+\dfrac{1}{2m}[q^2-p'^2-(\mathbf{p}+\mathbf{q}-\mathbf{p}')^2]+i0.\operatorname{sgn}(p'-p_F)} - \mathrm{P}\frac{2m\theta(\mathbf{q})}{p^2+q^2-p'^2+(\mathbf{p}+\mathbf{q}-\mathbf{p}')^2} \right\} \frac{d^3q\, d^3p'}{(2\pi)^6}; \tag{21.16}$$

the factor $\theta(\mathbf{q})-\theta(\mathbf{p}')$ in the numerator of the first term in the integrand replaces $-\operatorname{sgn}(q-p_F)$ with the condition (21.14).

First, we note that Σ has an imaginary part. It is separated from (21.16) by means of the rule (8.11), and is

$$\operatorname{im}\Sigma(\omega, \mathbf{p}) = -\left(\frac{4\pi a}{m}\right)^2 \pi \int \{\theta(\mathbf{q})[1-\theta(\mathbf{p}')][1-\theta(\mathbf{p}+\mathbf{q}-\mathbf{p}')] + [1-\theta(\mathbf{q})]\,\theta(\mathbf{p}')\,\theta(\mathbf{p}+\mathbf{q}-\mathbf{p}')\}\,\delta\left[\omega+\mu+\frac{1}{2m}(q^2-p'^2-(\mathbf{p}+\mathbf{q}-\mathbf{p}')^2)\right]\frac{d^3q\,d^3p'}{(2\pi)^6}; \tag{21.17}$$

the expression in the braces is transformed using the fact that $\theta^2(\mathbf{p}) \equiv \theta(\mathbf{p})$.

The quasi-particle energy spectrum is calculated, according to (14.13), as

$$\varepsilon(\mathbf{p}) = \frac{p^2}{2m} + \frac{2\pi}{m} n(\mu)\, a + \Sigma^{(2)}\left(\frac{p^2}{2m} - \mu, \mathbf{p}\right); \tag{21.18}$$

in $\Sigma^{(2)}$ we can put $\varepsilon \approx p^2/2m$ with the necessary accuracy. The fact that Σ is complex means that the excitations are damped ($\operatorname{im}\varepsilon \neq 0$).

The presence of this damping expresses the instability of quasi-particles due to the possibility of their actual decay. A quasi-particle may lose part of its energy and so give rise to a pair of quasi-particles (particle and hole). Let us consider, for example, the first term in the braces in the integrand of (21.17). From the properties of the step function, this term is non-zero if

$$p' > p_F, \quad |\mathbf{q}+\mathbf{p}-\mathbf{p}'| > p_F, \quad q < p_F.$$

These inequalities correspond to a process in which a quasi-particle with initial momentum $\mathbf{p}(p > p_F)$ enters a state $\mathbf{p}'$ $(p > p' > p_F)$, and the momentum $\mathbf{p}-\mathbf{p}'$ is transmitted to a particle within the Fermi sphere (momentum $q < p_F$), which is excited to a state with momentum $\mathbf{q}+\mathbf{p}-\mathbf{p}'$ outside the Fermi sphere; such a transition is equivalent to the appearance of two new elementary excitations, with momenta $-\mathbf{q}$ (hole) and $\mathbf{q}+\mathbf{p}-\mathbf{p}'$. The law of conservation of energy in this process is expressed by the delta function in (21.17), in which $\omega+\mu$ acts as the initial energy of the quasi-particle $\varepsilon(\mathbf{p})$:

$$\varepsilon(\mathbf{p}) = \varepsilon(\mathbf{p}')+[\varepsilon(\mathbf{q}+\mathbf{p}-\mathbf{p}')-\varepsilon(\mathbf{q})];$$

here it is sufficient to put $\varepsilon(\mathbf{p}) = p^2/2m$ in the first approximation. In accordance with the significance mentioned, the energy $\varepsilon(\mathbf{p})$ determined by this equation in fact corresponds to a quasi-particle outside the Fermi sphere $(\varepsilon > \mu)$.

Similarly, the second term in the braces in (21.17) results from processes in which a pair is generated by a hole. This term gives the damping of elementary excitations with $\varepsilon < \mu$. In the language of the diagram technique, the possibility of creation of a pair by a quasi-particle is indicated by the possibility of dividing the G function diagram into two parts by cutting through three continuous lines, one of which is in the opposite direction to the other two. In the diagrams (21.1c) and (21.1d), such cuts are to be made between the two broken lines.

The case of a slightly non-ideal gas is special (in comparison with the general case of any Fermi liquid) in that the quasi-particle spectrum in it is meaningful for all values of the momenta, and not only near the Fermi surface: the decay of the quasi-particles (im ε) is relatively small, because the "gaseousness parameter" ap_F is relatively small. Here, however, we shall give the final result of the calculations only for two limiting cases.

Near the Fermi surface $(|p-p_F| \ll p_F)$, we find

$$\text{re}\,\varepsilon = \mu+(p-p_F)p_F/m^*,$$

with μ from (6.14) and m^* from (6.17). For the decay of the quasi-particles, we have

$$\text{im}\,\varepsilon = -\frac{1}{\pi m}(p_F a)^2(p-p_F)^2\,\text{sgn}\,(p-p_F). \tag{21.19}$$

The proportionality of this expression to $(p-p_F)^2$ has an obvious origin: one factor $p-p_F$ is the width of the region in momentum space (a thin spherical shell) which contains the momentum of the quasi-particle after it has created a pair, and the other factor $p-p_F$ is the width of the layer in which the pair is created. These considerations, it may be noted, apply to any Fermi liquid, so that we always have im $\varepsilon \propto (p-p_F)^2$ near the Fermi surface.†

† At non-zero temperatures, the averaging of this quantity over the thermal distribution makes the decay proportional to T^2, as discussed in §1.

For large momenta $p \gg p_F$ (but still $pa \ll 1$), we have

$$\varepsilon = \left(\frac{p^2}{2m} + \frac{2p_F^2}{3\pi m} p_F a\right) - i\frac{p_F p}{3\pi m}(p_F a)^2. \tag{21.20}$$

In both cases the ratio im ε/re ε is small. The maximum value of this ratio is reached when $p \sim p_F$, but even then it is $\sim (p_F a)^2 \ll 1$.

Lastly, the value of the renormalization constant for the Green's function of a slightly non-ideal gas, calculated as

$$\frac{1}{Z} = 1 - \left[\frac{\partial \Sigma(\omega, \mathbf{p})}{\partial \omega}\right]_{\omega = 0,\ p = p_F}$$

is

$$Z = 1 - \frac{8 \log 2}{\pi^2}(p_F a)^2. \tag{21.21}$$

CHAPTER III

SUPERFLUIDITY

§ 22. Elementary excitations in a quantum Bose liquid

LET us now consider quantum liquids with an energy spectrum of a completely different type, which may be called a *Bose spectrum*.†

This spectrum has the property that the elementary excitations (which are absent in the ground state of the liquid) can appear and disappear singly. But the angular momentum of any quantum-mechanical system (in this case, the liquid) can change only by an integer. Hence the elementary excitations appearing singly must have integral angular momenta, and therefore obey Bose statistics. Any quantum liquid consisting of particles with integral spin (such as the liquid isotope He^4) must certainly have a spectrum of this type.

For comparison, it may be recalled that in a Fermi liquid, described in terms of the spectrum of elementary excitations, which are absent in the ground state (see the end of §1), these excitations can only appear and disappear in pairs. This is the reason why elementary excitations in that type of spectrum can have a half-integral spin.

In a quantum Bose liquid, elementary excitations with small momenta p (wavelength large compared with the distances between atoms) correspond to ordinary hydrodynamic sound waves, i.e. are phonons. This means that the energy of such quasi-particles is a linear function of their momentum:

$$\varepsilon = up, \tag{22.1}$$

where u is the velocity of sound in the liquid. The latter is given by the usual formula $u^2 = \partial P/\partial \varrho$, and there is no need to specify whether the derivative is taken at constant temperature T or at constant entropy S, since $S \to 0$ when $T \to 0$.‡

The number of elementary excitations in a Bose liquid tends to zero as $T \to 0$, and at low temperatures, when their density is sufficiently small, the

† The theory of such quantum liquids was worked out by L. D. Landau in 1940–1941, following P. L. Kapitza's discovery of the superfluidity of liquid helium. These discoveries formed the basis for the whole of the modern physics of quantum liquids.

‡ The concept of phonons has been defined in Part 1, §§71 and 72, for elementary excitations in solids. It must be emphasized that the momentum of an elementary excitation in a microscopically homogeneous system (a liquid) is the actual momentum, and not the quasi-momentum as in the periodic field of the crystal lattice in a solid.

quasi-particles may be regarded as not interacting with one another, i.e. as forming an ideal Bose gas. Hence the statistical-equilibrium distribution of elementary excitations in a Bose liquid is given by the Bose distribution formula (with zero chemical potential; cf. the last footnote to §1)

$$n(\mathbf{p}) = [e^{\varepsilon(p)/T} - 1]^{-1}. \tag{22.2}$$

With this distribution, and knowing the function $\varepsilon(p)$ for small p, we can calculate the thermodynamic quantities for the liquid at temperatures so close to absolute zero that practically all the elementary excitations present in the liquid have low energies, i.e. are phonons. The corresponding formulae can be written down immediately by using the expressions for the thermodynamic quantities in a solid at low temperatures (see Part 1, §64). The only difference is that, instead of the three possible directions of polarization of sound waves in a solid (one longitudinal and two transverse), in a liquid there is only one (longitudinal), and so all the expressions for the thermodynamic quantities are to be divided by 3. For example, the free energy of the liquid is

$$F = F_0 - V.\pi^2 T^4/90(\hbar u)^3, \tag{22.3}$$

where F_0 is the free energy at absolute zero. The energy of the liquid is

$$E = E_0 + V.\pi^2 T^4/30(\hbar u)^3, \tag{22.4}$$

and the specific heat

$$C = V.2\pi^2 T^3/15(\hbar u)^3, \tag{22.5}$$

proportional to the cube of the temperature.

The phonon dispersion relation (22.1) is valid only if the wavelength $\hbar/p$ of the quasi-particle is large compared with the interatomic distances. As the momentum increases, the curve of $\varepsilon(p)$ of course deviates from the linear form; its subsequent form depends on the particular law of interaction of the liquid molecules, and therefore cannot be determined in a general form.

In liquid helium, the dispersion relation of the elementary excitations has the form shown in Fig. 2: after an initial linear increase, the function $\varepsilon(p)$ reaches a maximum, then decreases and passes through a minimum at a certain momentum value p_0.† In thermal equilibrium, the majority of the elementary excitations in the liquid have energies near the minima of $\varepsilon(p)$, i.e. in the region of small ε (near $\varepsilon = 0$) and in the region of $\varepsilon(p_0)$. These regions are therefore particularly important. Near $p = p_0$, the function $\varepsilon(p)$ may be expanded in powers of $p - p_0$. There is no linear term, and we have as far as the second-order terms

$$\varepsilon = \Delta + (p - p_0)^2/2m^*, \tag{22.6}$$

† This form of the spectrum was first suggested by L. D. Landau (1947) from an analysis of experimental results regarding the thermodynamic quantities for liquid helium; it was later confirmed by neutron scattering experiments.

A qualitative theory of such spectra was given by R.P. Feynman (1954); see the footnote following (87.5).

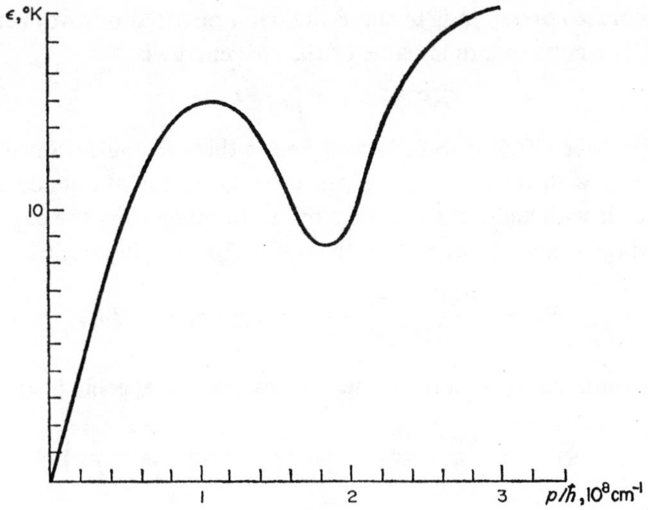

FIG. 2.

where $\Delta = \varepsilon(p_0)$ and m^* are constants. Quasi-particles of this type are called *rotons*. It must be emphasized, however, that both phonons and rotons are quasi-particles corresponding only to different parts of the same curve, and there is a continuous transition from one to the other.

The empirical values of the energy spectrum parameters for liquid helium (extrapolated to zero pressure and density $\varrho = 0.145$ g/cm³) are†

$$\left.\begin{aligned} u &= 2.4\times 10^4 \text{ cm/sec}, & \Delta &= 8.7°\text{K}, \\ p_0/\hbar &= 1.9\times 10^8 \text{ cm}^{-1}, & m^* &= 0.16m\,(\text{He}^4). \end{aligned}\right\} \tag{22.7}$$

Since the roton energy always includes the quantity Δ, which is large compared with T at temperatures sufficiently low for a "roton gas" to be considered, this gas may be described by the Boltzmann distribution instead of the Bose distribution. Accordingly, to calculate the roton part of the thermodynamic quantities for liquid helium we start from the formula for the free energy of a Boltzmann gas:

$$F = -NT\log\frac{eV}{N}\int e^{-\varepsilon/T}\,d\tau, \qquad d\tau = d^3p/(2\pi\hbar)^3;$$

see Part 1, §41. In this formula, N is to be taken as the number of rotons in the liquid, which is itself determined by the condition of thermodynamic equilibrium, i.e. by the condition of minimum free energy. Equating $\partial F/\partial N$ to zero, we find for the number of rotons

$$N_r = V\int e^{-\varepsilon/T}\,d\tau, \tag{22.8}$$

† The chemical potential of liquid helium at $T = 0$ is $\mu = -7.16°$K.

which of course corresponds to the Boltzmann distribution with zero chemical potential. The corresponding value of the free energy is

$$F_r = -VT \int e^{-\varepsilon/T}\, d\tau.$$

The expression (22.6) is to be substituted in these formulae. Since $p_0^2 \gg m^*T$, in integrating with respect to p we can take the factor p^2 outside the integral and replace it with sufficient accuracy by p_0^2. In integrating the exponential we can extend the range of integration from $-\infty$ to ∞. The result is

$$N_r = \frac{2(m^*T)^{1/2}\, p_0^2 V}{(2\pi)^{3/2}\, \hbar^3} e^{-\Delta/T}, \quad F_r = -TN_r. \tag{22.9}$$

Hence the roton contributions to the entropy and the specific heat are

$$S_r = N_r\left(\frac{3}{2}+\frac{\Delta}{T}\right), \quad C_r = N_r\left(\frac{3}{4}+\frac{\Delta}{T}+\frac{\Delta^2}{T^2}\right). \tag{22.10}$$

We see that the temperature dependence of the roton part of the thermodynamic quantities is essentially exponential. At sufficiently low temperatures (below about 0.8°K for liquid helium), the roton part is therefore less than the phonon part, while at high temperatures the position is reversed and the roton contribution is greater than that of the phonons.

§ 23. Superfluidity

A quantum liquid with an energy spectrum of the type described above possesses a remarkable property known as *superfluidity*: the property of flowing through narrow capillaries or slits without exhibiting viscosity. Let us first consider a liquid at absolute zero, at which temperature the liquid is in its ground state.

Let us consider a liquid flowing along a capillary at a constant velocity **v**. Because of the friction against the walls of the tube and the friction within the liquid itself, the presence of viscosity would have the effect that the kinetic energy of the liquid would be dissipated and the flow would gradually become slower.

It will be more convenient to discuss the flow in a coordinate system moving with the liquid. In such a system the liquid helium is at rest, and the walls of the capillary move with velocity $-\mathbf{v}$. When viscosity is present, the liquid at rest must also begin to move. It is physically evident that the entrainment of the liquid by the walls of the tube cannot initiate movement of the liquid as a whole. The motion must arise from a gradual excitation of internal motions, that is, from the appearance of elementary excitations in the liquid.

Let us suppose that a single elementary excitation appears in the liquid, with momentum $\mathbf{p}$ and energy $\varepsilon(p)$. Then the energy E_0 of the liquid (in the coordinate system in which it was originally at rest) is equal to the energy ε of the excitation, and its momentum $\mathbf{P}_0$ is equal to $\mathbf{p}$. Let us now return to the coordinate system in which the capillary is at rest. According to the familiar formulae of mechanics for the transformation of energy and momentum, we obtain for the energy E and momentum $\mathbf{P}$ of the liquid in this system

$$E = E_0 + \mathbf{P}_0 \cdot \mathbf{v} + \tfrac{1}{2}Mv^2, \qquad \mathbf{P} = \mathbf{P}_0 + M\mathbf{v}, \tag{23.1}$$

where M is the mass of the liquid. Substituting ε and $\mathbf{p}$ for E_0 and $\mathbf{P}_0$, we have

$$E = \varepsilon + \mathbf{p} \cdot \mathbf{v} + \tfrac{1}{2}Mv^2. \tag{23.2}$$

The term $\frac{1}{2}Mv^2$ is the original kinetic energy of the flowing liquid; the expression $\varepsilon + \mathbf{p} \cdot \mathbf{v}$ is the change in energy due to the appearance of the excitation. This change must be negative, since the energy of the moving liquid must decrease: $\varepsilon + \mathbf{p} \cdot \mathbf{v} < 0$.

For a given value of p, the quantity on the left-hand side of this inequality is a minimum when $\mathbf{p}$ and $\mathbf{v}$ are antiparallel; thus we must always have $\varepsilon - pv < 0$, or

$$v > \varepsilon/p. \tag{23.3}$$

This inequality must be satisfied for at least some values of the momentum p of the elementary excitation. Hence the final condition for the occurrence of excitations to be possible in the liquid as it moves along the capillary is obtained by finding the minimum of ε/p. Geometrically, the ratio ε/p is the slope of the line drawn from the origin (in the $p\varepsilon$-plane) to some point on the curve of $\varepsilon(p)$. Its minimum value is clearly given by the point at which the line from the origin is a tangent to the curve. If this minimum is not zero, then, for velocities of flow below a certain value, excitations cannot appear in the liquid. This means that the flow will not become slower, i.e. that the liquid exhibits the phenomenon of superfluidity.

The condition just derived for the presence of superfluidity is essentially equivalent to the requirement that the curve of $\varepsilon(p)$ should not touch the axis of abscissae at the origin (ignoring the unlikely possibility that it touches this axis at some other point). Thus any spectrum in which sufficiently small excitations are phonons will lead to superfluidity.

Let us now consider the same liquid at a temperature other than absolute zero (but close to it). In this case the liquid contains excitations, and is not in the ground state. The arguments given above remain valid, since they made no direct use of the fact that the liquid was originally in the ground state. The motion of the liquid relative to the walls of the tube when the above condition is satisfied still cannot cause any new elementary excitations to appear in it. It is, however, necessary to elucidate the effect of excitations already present in the liquid.

To do this, let us imagine that the "gas of quasi-particles" moves as a whole with respect to the liquid, with a translational velocity $\mathbf{v}$. The distribution function for the gas moving as a whole is obtained from the distribution function $n(\varepsilon)$ for the gas at rest by replacing the energy ε of a particle by $\varepsilon - \mathbf{p}.\mathbf{v}$, where $\mathbf{p}$ is the momentum of the particle. For an ordinary gas this is a direct consequence of Galileo's relativity principle, and is proved by a simple change of coordinates, but in the present case such arguments cannot be applied directly, since the quasi-particle gas is moving not in a vacuum but "through the liquid". Nevertheless, the statement remains valid, as can be seen from the following argument.

Let the gas of excitations be moving relative to the liquid with velocity $\mathbf{v}$. Let us take a coordinate system in which the gas is at rest as a whole, and the liquid is accordingly moving with velocity $-\mathbf{v}$ (system K). According to the transformation formula (23.1), the energy E of the liquid in the system K is related to the energy E_0 in a system K_0 where the liquid is at rest by

$$E = E_0 - \mathbf{P}_0 . \mathbf{v} + \tfrac{1}{2} M v^2.$$

Let an elementary excitation of energy $\varepsilon(p)$ in K_0 arise in the liquid. Then the additional energy of the liquid in K is $\varepsilon - \mathbf{p}.\mathbf{v}$, and this proves the statement.†

Thus the total momentum of the quasi-particle gas per unit volume is

$$\mathbf{P} = \int \mathbf{p} n(\varepsilon - \mathbf{p} . \mathbf{v})\, d\tau.$$

Let us assume that the velocity $\mathbf{v}$ is small, and expand the integrand in powers of $\mathbf{p} . \mathbf{v}$. The zero-order term gives zero on integration over the directions of the vector $\mathbf{p}$, leaving

$$\mathbf{P} = -\int \mathbf{p}(\mathbf{p} . \mathbf{v}) \frac{dn(\varepsilon)}{d\varepsilon}\, d\tau,$$

or, on averaging over the directions of $\mathbf{p}$,

$$\mathbf{P} = \frac{1}{3}\mathbf{v} \int \left(-\frac{dn}{d\varepsilon}\right) p^2\, d\tau. \qquad (23.4)$$

First of all, we see that the motion of the quasi-particle gas is accompanied by a transfer of mass: the effective mass per unit volume of the gas is determined by the proportionality coefficient between the momentum $\mathbf{P}$ and the velocity $\mathbf{v}$ in (23.4). On the other hand, in the flow of a liquid along a capillary (say) there is nothing to prevent the quasi-particles from colliding with the walls of the tube and exchanging momentum with them. In consequence the excitation gas will be slowed down, like any ordinary gas flowing along a capillary.

Thus we have the following fundamental result. At non-zero temperatures, part of the mass of the liquid will behave as a normal viscous liquid which

† For quasi-particles in a Bose liquid, $n(\varepsilon)$ is the distribution (22.2). It should be noted that the superfluidity condition $v < \varepsilon/p$ is precisely the condition for $n(\varepsilon - \mathbf{p}.\mathbf{v})$ to be positive and finite for all energies.

"sticks" as it moves along the walls of the vessel; the remaining part of the mass will behave as a superfluid without viscosity. Here it is very important that there is no friction between these two parts of the mass of the liquid as they pass "through one another", that is, there is no transfer of momentum from one part to the other. For the existence of such motion of one part of the mass of the liquid relative to the other has been derived by considering the statistical equilibrium in a uniformly moving excitation gas. But if any relative motion can occur in a state of thermal equilibrium, it is not accompanied by friction.

It should be emphasized that the treatment of the liquid as a "mixture" of normal and superfluid "parts" is simply a form of words convenient for the description of the phenomena in a quantum liquid. Like any description of quantum effects in classical terms, it is not entirely adequate. It does not at all mean that the liquid can actually be separated into two parts. In reality we should say that in a quantum Bose liquid there can exist simultaneously two motions, each of which has a corresponding "effective mass" such that the sum of these two masses is equal to the actual total mass of the liquid. One of these motions is "normal", i.e. has the same properties as that of an ordinary viscous liquid; the other is "superfluid". The two motions occur without transfer of momentum from one to the other.

Thus, in the hydrodynamic sense the density of a Bose liquid can be written as a sum $\varrho = \varrho_n + \varrho_s$ of normal and superfluid parts, each corresponding to a hydrodynamic velocity $\mathbf{v}_n$ or $\mathbf{v}_s$. An important property of superfluid motion is that it is a potential flow:

$$\operatorname{curl} \mathbf{v}_s = 0. \tag{23.5}$$

This property is the macroscopic expression of the fact that the elementary excitations with long wavelength (i. e. with small momentum) are sound quanta (phonons). Hence the macroscopic hydrodynamics of superfluid motion must not allow other than acoustic vibrations,† as is ensured by the condition (23.5); the proof of this condition will be considered in §26.‡

When $T = 0$, the normal part of the density $\varrho_n = 0$; the liquid can have only superfluid motion. For non-zero temperatures, ϱ_n is given by (23.4):

$$\varrho_n = \frac{1}{3} \int \left(-\frac{dn}{d\varepsilon} \right) p^2 \, d\tau. \tag{23.6}$$

To calculate the phonon contribution to ϱ_n, we put in (23.6) $\varepsilon = up$:

$$(\varrho_n)_{ph} = -\frac{1}{3u} \int_0^\infty \frac{dn}{dp} p^2 \frac{4\pi p^2 \, dp}{(2\pi\hbar)^3},$$

† The liquid is assumed infinite. When there is a free surface, surface capillary waves are also possible, and lead to a definite temperature dependence of the surface tension; see Problem 1.

‡ A detailed account of the hydrodynamics of a superfluid is given in *FM*, Chapter XVI.

and obtain, on integration by parts,

$$(\varrho_n)_{ph} = \frac{4}{3u}\int_0^\infty np\,\frac{4\pi p^2\,dp}{(2\pi\hbar)^3} = \frac{4}{3u^2}\int \varepsilon n\,d\tau.$$

The remaining integral here is just the energy of the phonon gas per unit volume; taking this from (22.4), we have finally

$$\begin{aligned}(\varrho_n)_{ph} &= 4E_{ph}/3u^2V \\ &= 2\pi^2 T^4/45\hbar^3u^5.\end{aligned} \qquad (23.7)$$

To calculate the roton contribution to ϱ_n we note that, since rotons can be described by a Boltzmann distribution, for them $dn/d\varepsilon = -n/T$, and from (23.6)

$$(\varrho_n)_r = \frac{1}{3T}\int p^2 n\,d\tau = \frac{\overline{p^2}}{3T}\frac{N_r}{V}.$$

Since $\overline{p^2} = p_0^2$ with sufficient accuracy, we find, taking N_r from (22.9),

$$(\varrho_n)_r = \frac{p_0^2 N_r}{3TV} = \frac{2(m^*)^{1/2}p_0^4}{3(2\pi)^{3/2}T^{1/2}\hbar^3}e^{-\Delta/T}. \qquad (23.8)$$

At very low temperatures, the phonon contribution to ϱ_n is large compared with the roton contribution. They become comparable at about 0.6°K, and at higher temperatures the roton contribution predominates.

As the temperature increases, an increasing fraction of the mass of the liquid becomes normal. At the point where $\varrho_n = \varrho$, the property of superfluidity disappears entirely. This is called the *λ-point* of the liquid, and is a phase transition point of the second kind.† The quantitative formulae (23.7) and (23.8) are, of course, inapplicable near the λ-point, where the quasi-particle concentration becomes large, so that even the concept of quasi-particles is largely meaningless.

We may also consider the behaviour of the atoms of substances dissolved in liquid helium; the concentration of the impurity is assumed to be so small that its atoms may be regarded as not interacting with one another (L. D. Landau and I. Ya. Pomeranchuk 1948).

The presence of an extraneous atom in the liquid gives rise to a new branch of the energy spectrum corresponding to the motion of this atom through the liquid; of course, owing to the strong interaction of the impurity atom with the atoms of the liquid, this motion is really a collective effect in which the liquid atoms also take part. A resultant conserved momentum **p** may be ascribed to this motion. Thus quasi-particles of a new type appear in the liquid,

† Liquid helium is called *helium II* at temperatures below this point. The λ-points form a curve in the phase diagram in the PT-plane. This curve intersects the liquid–vapour equilibrium curve at 2.19°K.

whose number is equal to the number of impurity atoms, and whose energy $\varepsilon_{\text{imp}}(p)$ is a definite function of the momentum. In thermal equilibrium, the energy of these quasi-particles is concentrated near the lowest minimum of the function $\varepsilon_{\text{imp}}(p)$. In practice, we are concerned with the He^3 isotope impurity, and empirical results show that this minimum is at $p = 0$; near that point, the quasi-particle energy is

$$\varepsilon_{\text{imp}}(p) = p^2/2m^*_{\text{imp}}, \tag{23.9}$$

with the effective mass m^*_{imp} equal to 2.8 times the mass of the He^3 atom.

Impurity quasi-particles interact with phonons and rotons when they collide with these, and therefore belong to the normal part of the liquid. Because of their low concentration, their thermal distribution is of the Boltzmann type, and their contribution to ϱ_n, determined from (23.6), is

$$(\varrho_n)_{\text{imp}} = \frac{N_{\text{imp}}}{V}\frac{\overline{p^2}}{3T} = \frac{N_{\text{imp}}}{V} m^*_{\text{imp}}, \tag{23.10}$$

where N_{imp}/V is the number of impurity atoms per unit volume.

PROBLEMS

PROBLEM 1. Find the limiting temperature dependence of the surface tension coefficient α of liquid helium near absolute zero (K. R. Atkins 1953).

SOLUTION. The coefficient α is the free energy per unit area of the liquid surface; see Part 1, (154.6). It is calculated from Part 1, (64.1), in which the frequencies ω_α now relate to surface vibrations. In the two-dimensional case, the change from summation to integration (over the wave vectors of the vibrations) is effected by including a factor $d^2k/(2\pi)^2$ or $2\pi k\,dk/(2\pi)^2$. Integration by parts gives

$$\begin{aligned}\alpha &= \alpha_0 + T\int \log\left(1-e^{-\hbar\omega/T}\right) k\,dk/2\pi \\ &= \alpha_0 - \frac{\hbar}{4\pi}\int \frac{k^2\,d\omega}{e^{\hbar\omega/T}-1},\end{aligned}$$

where α_0 is the surface tension at $T = 0$. At sufficiently low temperatures, only vibrations with low frequencies (i.e. long wavelengths) are important. Such vibrations are hydrodynamic capillary waves, for which $\omega^2 = \alpha k^3/\varrho \approx \alpha_0 k^3/\varrho$ (where ϱ is the density of the liquid). Hence

$$\alpha = \alpha_0 - \frac{\hbar}{4\pi}\left(\frac{\varrho}{\alpha_0}\right)^{2/3}\int_0^\infty \frac{\omega^{4/3}\,d\omega}{e^{\hbar\omega/T}-1};$$

since the integral converges rapidly, the upper limit may be replaced by infinity. The calculation of the integral (see the note in Part 1, §58) gives

$$\begin{aligned}\alpha &= \alpha_0 - \frac{T^{7/3}\varrho^{2/3}}{4\pi\hbar^{4/3}\alpha_0^{2/3}}\Gamma(7/3)\,\zeta(7/3) \\ &= \alpha_0 - 0.13\,T^{7/3}\varrho^{2/3}/\hbar^{4/3}\alpha_0^{2/3}.\end{aligned}$$

This applies to liquid He^4 at temperatures so low that the whole mass of the liquid may be regarded as superfluid.†

† In a Fermi liquid (liquid He^3) capillary waves of the type considered (like volume waves of ordinary sound) do not exist, since the viscosity increases without limit as $T \to 0$.

PROBLEM 2. Find the dispersion relation $\varepsilon_{\text{imp}}(\mathbf{p})$ for impurity particles in a moving superfluid if its form $\varepsilon^{(0)}_{\text{imp}}(p)$ in a liquid at rest is known (J. Bardeen, G. Baym and D. Pines 1967).

SOLUTION. After the addition to the liquid at rest ($T = 0$) of an impurity atom with mass m and momentum $\mathbf{p}_0$, the energy and momentum of the liquid, in the coordinate system in which it was originally at rest, are $E_0 = \varepsilon^{(0)}_{\text{imp}}(p_0)$, $\mathbf{P}_0 = \mathbf{p}_0$. In coordinates such that the liquid is moving with velocity $\mathbf{v}$, we have from (23.1)

$$E = \varepsilon^{(0)}_{\text{imp}}(p_0)+\mathbf{p}_0\cdot\mathbf{v}+\tfrac{1}{2}(M+m)\,v^2, \qquad \mathbf{P} = \mathbf{p}_0+(M+m)\,\mathbf{v}.$$

Hence we see that the changes of energy and momentum of the moving liquid when an impurity atom is added to it are

$$\varepsilon_{\text{imp}} = \varepsilon^{(0)}_{\text{imp}}(p_0)+\mathbf{p}_0\cdot\mathbf{v}+\tfrac{1}{2}mv^2, \qquad \mathbf{p} = \mathbf{p}_0+m\mathbf{v}.$$

Expressing ε_{imp} in terms of $\mathbf{p}$, we find

$$\varepsilon_{\text{imp}}(\mathbf{p}) = \varepsilon^{(0)}_{\text{imp}}(\mathbf{p}-m\mathbf{v})+\mathbf{p}\cdot\mathbf{v}-\tfrac{1}{2}mv^2.$$

For small $\mathbf{v}$, as far as the first-order terms, with a spectrum $\varepsilon^{(0)}_{\text{imp}}(p)$ of the form (23.9), we have

$$\varepsilon_{\text{imp}}(\mathbf{p}) = \frac{p^2}{2m^*_{\text{imp}}}+\mathbf{v}\cdot\mathbf{p}\left(1-\frac{m}{m^*_{\text{imp}}}\right).$$

§ 24. Phonons in a liquid

When we go from the classical picture of sound waves to the quantum concept of phonons, the hydrodynamic quantities (density, velocity of the liquid, etc.) are replaced by operators that can be expressed in terms of the phonon annihilation and creation operators $\hat{c}_{\mathbf{k}}$, $\hat{c}^+_{\mathbf{k}}$. We shall derive such expressions.

First, we recall that, in the classical description of a sound wave, the density of the liquid undergoes small oscillations whose frequencies and wave vectors are related by $\omega = uk$. The velocity $\mathbf{v}$ of the liquid is a quantity of the same order of smallness as the variable part $\varrho' = \varrho-\varrho_0$ of the density (where ϱ_0 is the equilibrium value of the density). The motion of the liquid in the wave is a potential flow, i.e. it can be described by a scalar velocity potential ϕ which determines the velocity according to

$$\mathbf{v} = \nabla\phi. \tag{24.1}$$

The velocity and the density are related by the equation of continuity $\partial\varrho'/\partial t = -\operatorname{div}(\varrho\mathbf{v}) \approx -\varrho_0 \operatorname{div}\mathbf{v}$, or

$$\partial\varrho'/\partial t = -\varrho_0\triangle\phi. \tag{24.2}$$

The energy of the liquid in the sound wave is given by the integral

$$E = \int \left(\tfrac{1}{2}\varrho_0\mathbf{v}^2+u^2\varrho'^2/2\varrho_0\right) d^3x. \tag{24.3}$$

The first term in the integrand is the kinetic energy density, and the second the internal energy density, of the liquid; both are quadratic in the small quantities $\mathbf{v}$ and ϱ'.

The subsequent quantization procedure could be carried out in an exactly similar way to that for phonons in solid crystals (see Part 1, §72). We shall take a somewhat different route, however, which illustrates some instructive points of methodology. Let us first consider the liquid density and velocity operators expressed in terms of microscopic variables, the coordinates of the particles.

In the classical theory, the density ϱ and the mass flow density $\mathbf{j}$ of the liquid can be written as sums

$$\varrho(\mathbf{r}) = \sum_a m_a \delta(\mathbf{r}_a - \mathbf{r}), \quad \mathbf{j}(\mathbf{r}) = \sum_a \mathbf{p}_a \delta(\mathbf{r}_a - \mathbf{r}),$$

taken over all the particles, where $\mathbf{r}_a$ and $\mathbf{p}_a$ are the position vectors and momenta of the particles. The integrals of these functions over any volume give the total mass and total momentum of the liquid in that volume. When we go to quantum theory, these functions are replaced by the corresponding operators. The density operator has the same form:

$$\hat{\varrho}(\mathbf{r}) = \sum_a m_a \delta(\mathbf{r}_a - \mathbf{r}); \tag{24.4}$$

the current density operator is

$$\hat{\mathbf{j}}(\mathbf{r}) = \tfrac{1}{2} \sum_a \{\hat{\mathbf{p}}_a \delta(\mathbf{r}_a - \mathbf{r}) + \delta(\mathbf{r}_a - \mathbf{r})\, \hat{\mathbf{p}}_a\}, \tag{24.5}$$

where $\hat{\mathbf{p}}_a = -i\hbar\nabla_a$ is the momentum operator of the particle.†

Let us find the commutation rule for the operators $\hat{\mathbf{j}}(\mathbf{r})$ and $\hat{\varrho}(\mathbf{r}')$ taken at points $\mathbf{r}$ and $\mathbf{r}'$: for brevity, we may consider just one term in the sums (24.4) and (24.5), since the operators corresponding to different particles commute. In the expansion of the commutator, the operators of the form $\delta(\mathbf{r}_1 - \mathbf{r})\nabla_1 \delta(\mathbf{r}_1 - \mathbf{r}')$ are transformed as follows:

$$\delta(\mathbf{r}_1 - \mathbf{r})\nabla_1 \delta(\mathbf{r}_1 - \mathbf{r}') = \delta(\mathbf{r}_1 - \mathbf{r})(\nabla \delta(\mathbf{r} - \mathbf{r}')) + \delta(\mathbf{r}_1 - \mathbf{r})\, \delta(\mathbf{r}_1 - \mathbf{r}')\nabla_1,$$

where in the first term $(\nabla\delta(\mathbf{r} - \mathbf{r}'))$ denotes simply the gradient of the delta function; because of the presence of the factor $\delta(\mathbf{r}_1 - \mathbf{r})$, we can replace $(\nabla_1 \delta(\mathbf{r}_1 - \mathbf{r}'))$ by $(\nabla\delta(\mathbf{r} - \mathbf{r}'))$ in that term. The result is

$$\hat{\mathbf{j}}(\mathbf{r})\, \hat{\varrho}(\mathbf{r}') - \hat{\varrho}(\mathbf{r}')\, \hat{\mathbf{j}}(\mathbf{r}) = -i\hbar\hat{\varrho}(\nabla\delta(\mathbf{r} - \mathbf{r}')). \tag{24.6}$$

Now, instead of $\hat{\mathbf{j}}$, we use the liquid velocity operator $\hat{\mathbf{v}}$, defined by

$$\hat{\mathbf{j}} = \tfrac{1}{2}(\hat{\varrho}\hat{\mathbf{v}} + \hat{\mathbf{v}}\hat{\varrho}).$$

† For simplicity, let the system consist of only one particle. Averaging the operator $\hat{\varrho}(\mathbf{r}) = m\delta(\mathbf{r}_1 - \mathbf{r})$ over the state with wave function $\psi(\mathbf{r}_1)$ gives $\int \psi^*(\mathbf{r}_1)\hat{\varrho}\psi(\mathbf{r}_1) d^3x_1 = m\,|\psi(\mathbf{r})|^2$, as it should. Similarly, averaging the operator $\hat{\mathbf{j}}(\mathbf{r})$ gives the correct expression for the current density,

$$(\hbar/2i)\{\psi^*(\mathbf{r})\,\nabla\psi(\mathbf{r}) - \psi(\mathbf{r})\,\nabla\psi^*(\mathbf{r})\}.$$

The commutation rule for the operators $\hat{\varrho}$ and $\hat{\mathbf{v}}$ is determined by the requirement that the expression (24.6) is obtained for the commutator of $\hat{\varrho}$ and $\hat{\mathbf{j}}$. It is easily verified that for this to be so we must put

$$\hat{\mathbf{v}}(\mathbf{r})\,\hat{\varrho}(\mathbf{r}') - \hat{\varrho}(\mathbf{r}')\,\hat{\mathbf{v}}(\mathbf{r}) = -i\hbar(\nabla\delta(\mathbf{r}-\mathbf{r}')),$$

using the obvious commutativity of the operators $\hat{\varrho}(\mathbf{r})$ and $\hat{\varrho}(\mathbf{r}')$. Lastly, putting $\hat{\mathbf{v}}(\mathbf{r}) = \nabla\hat{\phi}(\mathbf{r})$, we find the commutation rule for the density and velocity potential operators:

$$\hat{\phi}(\mathbf{r})\,\hat{\varrho}'(\mathbf{r}') - \hat{\varrho}'(\mathbf{r}')\,\hat{\phi}(\mathbf{r}) = -i\hbar\delta(\mathbf{r}-\mathbf{r}'); \tag{24.7}$$

here we must of course replace $\hat{\varrho}$ by the operator $\hat{\varrho}' = \hat{\varrho} - \varrho_0$ of the variable part of the density. The rule (24.7) is analogous to that for the particle coordinate and momentum operators; in this sense, ϱ' and ϕ here act as canonically conjugate generalized "coordinates" and "momenta".

Having used the expressions (24.4) and (24.5) to establish the rule (24.7), we can now write the operators $\hat{\phi}$ and $\hat{\varrho}'$ in the second-quantization representation (i.e. express them in terms of the phonon annihilation and creation operators), with the requirement that they satisfy the rule (24.7). To do so, we write

$$\hat{\phi}(\mathbf{r}) = \frac{1}{\sqrt{V}} \sum_{\mathbf{k}} (A_{\mathbf{k}}\hat{c}_{\mathbf{k}}\, e^{i\mathbf{k}\cdot\mathbf{r}} + A_{\mathbf{k}}^{*}\hat{c}_{\mathbf{k}}^{+}\, e^{-i\mathbf{k}\cdot\mathbf{r}})$$

with coefficients $A_{\mathbf{k}}$ as yet undetermined; the summation is over all values of the wave vector that occur for a liquid with large but finite volume V.† The operators $\hat{c}_{\mathbf{k}}$ and $\hat{c}_{\mathbf{k}}^{+}$ satisfy the Bose commutation rules

$$\hat{c}_{\mathbf{k}}\hat{c}_{\mathbf{k}'}^{+} - \hat{c}_{\mathbf{k}'}^{+}\hat{c}_{\mathbf{k}} = \delta_{\mathbf{k}\mathbf{k}'}. \tag{24.8}$$

For subsequent reference, the non-zero matrix elements of these operators are

$$\langle n_{\mathbf{k}}-1\,|\,\hat{c}_{\mathbf{k}}\,|\,n_{\mathbf{k}}\rangle = \langle n_{\mathbf{k}}\,|\,\hat{c}_{\mathbf{k}}^{+}\,|\,n_{\mathbf{k}}-1\rangle = \sqrt{n_{\mathbf{k}}}, \tag{24.9}$$

where $n_{\mathbf{k}}$ are the occupation numbers of the phonon states.

We shall later need, however, not the Schrödinger operator $\hat{\phi}(\mathbf{r})$ but the Heisenberg operator $\hat{\phi}(t, \mathbf{r})$. This is obtained from $\hat{\phi}(\mathbf{r})$ by simply including the factors $\exp(\pm i\omega t)$ with frequencies $\omega = uk$ in each term of the sum

$$\hat{\phi}(t, \mathbf{r}) = \frac{1}{\sqrt{V}} \sum_{\mathbf{k}} (A_{\mathbf{k}}\hat{c}_{\mathbf{k}}\, e^{i(\mathbf{k}\cdot\mathbf{r}-kut)} + A_{\mathbf{k}}^{*}\hat{c}_{\mathbf{k}}^{+}\, e^{-i(\mathbf{k}\cdot\mathbf{r}-kut)});$$

† Unlike the ψ operators of particles, the operator of the real quantity ϕ is Hermitian and contains both phonon creation and phonon annihilation operators. This property (like the corresponding property of the field operators in quantum electrodynamics) is due to the non-conservation of the number of "particles" in the phonon field.

cf. the relevant comment for the ψ operators at the beginning of §9. The density operator $\hat{\varrho}'(t, \mathbf{r})$ must be related to $\hat{\phi}(t, \mathbf{r})$ by (24.2), and is therefore given by a similar sum with factors $iA_{\mathbf{k}}\varrho_0 k/u$ in place of $A_{\mathbf{k}}$. The factors $A_{\mathbf{k}}$ must then be determined so as to satisfy the commutation rule (24.7). This gives the following final expressions:

$$\left.\begin{aligned} \hat{\phi}(t, \mathbf{r}) &= \sum_{\mathbf{k}} \left(\frac{\hbar u}{2V\varrho_0 k}\right)^{1/2} (\hat{c}_{\mathbf{k}}\, e^{i(\mathbf{k}\cdot\mathbf{r}-ukt)} + \hat{c}_{\mathbf{k}}^{+}\, e^{-i(\mathbf{k}\cdot\mathbf{r}-ukt)}), \\ \hat{\varrho}'(t, \mathbf{r}) &= \sum_{\mathbf{k}} i\left(\frac{\varrho_0 \hbar k}{2Vu}\right)^{1/2} (\hat{c}_{\mathbf{k}}\, e^{i(\mathbf{k}\cdot\mathbf{r}-ukt)} - \hat{c}_{\mathbf{k}}^{+}\, e^{-i(\mathbf{k}\cdot\mathbf{r}-ukt)}). \end{aligned}\right\} \quad (24.10)$$

For, on substituting these expressions on the left of (24.7) and using (24.8), we obtain the required delta function:

$$\begin{aligned} &-i\hbar \frac{1}{V} \sum_{\mathbf{k}} (\hat{c}_{\mathbf{k}}\hat{c}_{\mathbf{k}}^{+} - \hat{c}_{\mathbf{k}}^{+}\hat{c}_{\mathbf{k}})\, e^{i\mathbf{k}\cdot(\mathbf{r}-\mathbf{r}')} \\ &= -\frac{i\hbar}{V} \sum_{\mathbf{k}} e^{i\mathbf{k}\cdot(\mathbf{r}-\mathbf{r}')} \to -\frac{i\hbar}{V} \int e^{i\mathbf{k}\cdot(\mathbf{r}-\mathbf{r}')} \frac{V\, d^3k}{(2\pi)^3} = -i\hbar\delta(\mathbf{r}-\mathbf{r}'). \end{aligned}$$

It is also easy to see that the Hamiltonian of the liquid, obtained by substituting $\hat{\mathbf{v}} = \nabla\hat{\phi}$ and $\hat{\varrho}'$ in place of $\mathbf{v}$ and ϱ' in the integral (24.3), has the form

$$\hat{H} = \sum_{\mathbf{k}} u\hbar k(\hat{c}_{\mathbf{k}}^{+}\hat{c}_{\mathbf{k}} + \tfrac{1}{2}),$$

as it should; its eigenvalues are $\Sigma u\hbar k(n_{\mathbf{k}} + \frac{1}{2})$, in accordance with the concept of phonons having energies $\varepsilon = u\hbar k$.

The expression (24.3) for the energy of a liquid in a sound wave consists of the first (after the zero-order) terms in an expansion of the exact expression

$$E = \int \left[\tfrac{1}{2}\varrho \mathbf{v}^2 + \varrho e(\varrho)\right] d^3x,$$

where $e(\varrho)$ is the internal energy of the liquid per unit mass. This integral, with $\mathbf{v}$ and ϱ replaced by the operators $\hat{\mathbf{v}} = \nabla\hat{\phi}$ and $\hat{\varrho} = \varrho_0 + \hat{\varrho}'$ with $\hat{\phi}$ and $\hat{\varrho}'$ from (24.10), acts as the exact Hamiltonian of the liquid:

$$\hat{H} = \int \left[\tfrac{1}{2}\hat{\mathbf{v}}\cdot\varrho\hat{\mathbf{v}} + \hat{\varrho}e(\hat{\varrho})\right] d^3x; \qquad (24.11)$$

the kinetic energy operator is written in the symmetrized form $\frac{1}{2}\hat{\mathbf{v}}\cdot\varrho\hat{\mathbf{v}}$, so as to be Hermitian. Here it is important that ϱ and ϕ are canonically conjugate "generalized coordinates and momenta" in terms of which the Hamiltonian must be expressed. This is seen from the fact that the commutation rule (24.7) satisfied by the operators (24.10) is exact; the smallness of the oscillations is nowhere used in deriving it.

The terms of higher (third, etc.) degree in the expansion of this Hamiltonian represent the anharmonicity of the sound vibrations, or in terms of the phonon picture describe the interaction of phonons. They have non-zero matrix elements for transitions with simultaneous change of several phonon occupation numbers, and thus act as a perturbation causing various phonon scattering and decay processes. The matrix elements of the operators $\hat{c}_{\mathbf{k}}$ and $\hat{c}_{\mathbf{k}}^{+}$ have, of course, the previous form (24.9), since (as always in perturbation theory) the representation used is one in which the unperturbed Hamiltonian is diagonal. The terms of the third and fourth orders are

$$\hat{H}^{(3)} = \int \left[\frac{1}{2}\hat{\mathbf{v}}\cdot\hat{\varrho}'\hat{\mathbf{v}} + \left(\frac{d}{d\varrho_0}\frac{u^2}{\varrho_0}\right)\frac{\hat{\varrho}'^3}{6}\right] d^3x, \tag{24.12}$$

$$\hat{H}^{(4)} = \frac{1}{24}\left(\frac{d^2}{d\varrho_0^2}\frac{u^2}{\varrho_0}\right)\int \hat{\varrho}'^4\, d^3x. \tag{24.13}$$

§ 25. A degenerate almost ideal Bose gas

The fundamental properties of the Bose-type energy spectrum are clear from the model of a slightly non-ideal Bose gas at almost zero temperature. This model will be considered in the present section in the same way as in §6 for a Fermi gas.† The whole of the discussion in §6 relating to the general characteristics of models of a degenerate almost ideal gas applies here also. In particular, the condition of being only slightly non-ideal (the gaseousness parameter $a(N/V)^{1/3} \ll 1$, where a is the scattering length) can again be put in the form of the condition (6.1) that the particle momentum be small: $pa/\hbar \ll 1$.‡

The Hamiltonian of the system of bosons (assumed spinless) interacting in pairs differs from (6.6) only by the absence of the spin suffixes:

$$\hat{H} = \sum \frac{p^2}{2m}\hat{a}_{\mathbf{p}}^{+}\hat{a}_{\mathbf{p}} + \frac{1}{2}\sum \langle \mathbf{p}_1'\mathbf{p}_2' | U | \mathbf{p}_1\mathbf{p}_2\rangle \hat{a}_{\mathbf{p}_1'}^{+}\hat{a}_{\mathbf{p}_2'}^{+}\hat{a}_{\mathbf{p}_2}\hat{a}_{\mathbf{p}_1}, \tag{25.1}$$

with summation over all the momenta appearing as suffixes. The particle annihilation and creation operators now obey the commutation rules

$$\hat{a}_{\mathbf{p}}\hat{a}_{\mathbf{p}}^{+} - \hat{a}_{\mathbf{p}}^{+}\hat{a}_{\mathbf{p}} = 1.$$

As in §6, we again make the assumption that the momenta are small, and replace all the matrix elements in (25.1) by their values for zero momenta; then

$$\hat{H} = \sum \frac{p^2}{2m}\hat{a}_{\mathbf{p}}^{+}\hat{a}_{\mathbf{p}} + \frac{U_0}{2V}\sum \hat{a}_{\mathbf{p}_1'}^{+}\hat{a}_{\mathbf{p}_2'}^{+}\hat{a}_{\mathbf{p}_2}\hat{a}_{\mathbf{p}_1}. \tag{25.2}$$

† The method given below is due to N. N. Bogolyubov (1947). His application of it to the Bose gas was the first consistent microscopic derivation of the energy spectrum of "quantum liquids".

‡ We shall see below that, in a degenerate Bose gas, the majority of the particles (outside the "condensate") have momenta $p \sim \hbar\sqrt{(aN/V)}$, for which this inequality is indeed satisfied.

The starting-point for the application of perturbation theory to this Hamiltonian is the following remark. In the ground state of an ideal Bose gas, all particles are in the *condensate*, i.e. the state of zero energy; the occupation numbers $N_{\mathbf{p}=0} \equiv N_0 = N$, $N_{\mathbf{p}} = 0$ for $\mathbf{p} \neq 0$ (see Part 1, §62). In an almost ideal gas, in the ground state and in weakly excited states, the numbers $N_{\mathbf{p}}$ are not zero, but they are very small in comparison with the macroscopically large number N_0. The fact that the quantity $\hat{a}_0^+\hat{a}_0 = N_0 \approx N$ is very large in comparison with unity means that the expression

$$\hat{a}_0\hat{a}_0^+ - \hat{a}_0^+\hat{a}_0 = 1$$

is small compared with $\hat{a}_0$ and $\hat{a}_0^+$ themselves, which may therefore be regarded as ordinary numbers (equal to $\sqrt{N_0}$), their non-commutativity being neglected.

The application of perturbation theory now signifies formally the expansion of the fourfold sum in (25.2) in powers of the small quantities $\hat{a}_{\mathbf{p}}$, $\hat{a}_{\mathbf{p}}^+$ ($\mathbf{p} \neq 0$). The zero-order term in the expansion is

$$\hat{a}_0^+\hat{a}_0^+\hat{a}_0\hat{a}_0 = a_0^4. \tag{25.3}$$

The first-order terms are zero (since they cannot satisfy the law of conservation of momentum). The second-order terms are

$$a_0^2 \sum_{\mathbf{p} \neq 0} (\hat{a}_{\mathbf{p}}\hat{a}_{-\mathbf{p}} + \hat{a}_{\mathbf{p}}^+\hat{a}_{-\mathbf{p}}^+ + 4\hat{a}_{\mathbf{p}}^+\hat{a}_{\mathbf{p}}). \tag{25.4}$$

Taking only the second-order terms, we can replace $a_0^2 = N_0$ in (25.4) by the total number of particles N. In (25.3), the more accurate relation

$$a_0^2 + \sum_{\mathbf{p} \neq 0} \hat{a}_{\mathbf{p}}^+\hat{a}_{\mathbf{p}} = N$$

must be used. The sum of (25.3) and (25.4) is then

$$N^2 + N \sum_{\mathbf{p} \neq 0} (\hat{a}_{\mathbf{p}}\hat{a}_{-\mathbf{p}} + \hat{a}_{\mathbf{p}}^+\hat{a}_{-\mathbf{p}}^+ + 2\hat{a}_{\mathbf{p}}^+\hat{a}_{\mathbf{p}}),$$

and on substitution in (25.2) we get the following expression for the Hamiltonian:

$$\hat{H} = \frac{N^2}{2V} U_0 + \sum_{\mathbf{p}} \frac{p^2}{2m} \hat{a}_{\mathbf{p}}^+\hat{a}_{\mathbf{p}} + \frac{N}{2V} U_0 \sum_{\mathbf{p} \neq 0} (\hat{a}_{\mathbf{p}}\hat{a}_{-\mathbf{p}} + \hat{a}_{\mathbf{p}}^+\hat{a}_{-\mathbf{p}}^+ + 2\hat{a}_{\mathbf{p}}^+\hat{a}_{\mathbf{p}}). \tag{25.5}$$

The first term in this expression gives, in the first approximation, the energy E_0 of the ground state of the gas, and its derivative with respect to N the chemical potential μ at $T = 0$:

$$E_0 = N^2U_0/2V, \quad \mu = NU_0/V. \tag{25.6}$$

The remaining terms in (25.5) give the correction to E_0 and the spectrum of weakly excited states of the gas.

The integral U_0 in (25.5) has still to be expressed in terms of a real physical quantity, the scattering length a. In the second-order terms, this can be done directly from the formula (6.2): $U_0 = 4\pi\hbar^2 a/m$. In the first term, however, the more exact formula (6.5) is needed, which takes account of the second Born approximation in the scattering amplitude. Here we are considering the collision of two particles in the condensate, and accordingly in the sum in (6.5) we must put $\mathbf{p}_1 = \mathbf{p}_2 = 0$, $\mathbf{p}_1' = -\mathbf{p}_2' \equiv \mathbf{p}$, so that

$$U_0 = \frac{4\pi\hbar^2 a}{m}\left(1+\frac{4\pi\hbar^2 a}{V}\sum_{\mathbf{p}\neq 0}\frac{1}{p^2}\right).$$

Substitution in (25.5) gives for the Hamiltonian

$$\hat{H} = \frac{2\pi\hbar^2 a}{m}\frac{N^2}{V}\left(1+\frac{4\pi\hbar^2 a}{V}\sum_{\mathbf{p}\neq 0}\frac{1}{p^2}\right) + \frac{2\pi\hbar^2 a}{m}\frac{N}{V}\sum_{\mathbf{p}\neq 0}(\hat{a}_{\mathbf{p}}\hat{a}_{-\mathbf{p}}+\hat{a}_{\mathbf{p}}^{+}\hat{a}_{-\mathbf{p}}^{+}+2\hat{a}_{\mathbf{p}}^{+}\hat{a}_{\mathbf{p}})+\sum_{\mathbf{p}}\frac{p^2}{2m}\hat{a}_{\mathbf{p}}^{+}\hat{a}_{\mathbf{p}}. \tag{25.7}$$

To determine the energy levels, we must bring the Hamiltonian to diagonal form; this is done by a suitable linear transformation of the operators $\hat{a}_{\mathbf{p}}$, $\hat{a}_{\mathbf{p}}^{+}$. With new operators $\hat{b}_{\mathbf{p}}$ and $\hat{b}_{\mathbf{p}}^{+}$ defined by

$$\hat{a}_{\mathbf{p}} = u_{\mathbf{p}}\hat{b}_{\mathbf{p}}+v_{\mathbf{p}}\hat{b}_{-\mathbf{p}}^{+}, \quad \hat{a}_{\mathbf{p}}^{+} = u_{\mathbf{p}}\hat{b}_{\mathbf{p}}^{+}+v_{\mathbf{p}}\hat{b}_{-\mathbf{p}},$$

and the requirement that they satisfy the same commutation relations

$$\hat{b}_{\mathbf{p}}\hat{b}_{\mathbf{p}'}-\hat{b}_{\mathbf{p}'}\hat{b}_{\mathbf{p}} = 0, \quad \hat{b}_{\mathbf{p}}\hat{b}_{\mathbf{p}'}^{+}-\hat{b}_{\mathbf{p}'}^{+}\hat{b}_{\mathbf{p}} = \delta_{\mathbf{p}\mathbf{p}'}$$

as the $\hat{a}_{\mathbf{p}}$ and $\hat{a}_{\mathbf{p}}^{+}$ (it is easily seen that for this, we must have $u_{\mathbf{p}}^2-v_{\mathbf{p}}^2 = 1$), we can write the linear transformation as

$$\hat{a}_{\mathbf{p}} = \frac{\hat{b}_{\mathbf{p}}+L_{\mathbf{p}}\hat{b}_{-\mathbf{p}}^{+}}{\sqrt{(1-L_{\mathbf{p}}^2)}}, \quad \hat{a}_{\mathbf{p}}^{+} = \frac{\hat{b}_{\mathbf{p}}^{+}+L_{\mathbf{p}}\hat{b}_{-\mathbf{p}}}{\sqrt{(1-L_{\mathbf{p}}^2)}}. \tag{25.8}$$

The quantity $L_{\mathbf{p}}$ is to be defined so as to eliminate from the Hamiltonian the non-diagonal terms $\hat{b}_{\mathbf{p}}\hat{b}_{-\mathbf{p}}$ and $\hat{b}_{\mathbf{p}}^{+}\hat{b}_{-\mathbf{p}}^{+}$. A simple calculation gives

$$L_{\mathbf{p}} = \frac{1}{mu^2}\left\{\varepsilon(p)-\frac{p^2}{2m}-mu^2\right\}, \tag{25.9}$$

with the notation

$$\varepsilon(p) = [u^2p^2+(p^2/2m)^2]^{1/2}, \tag{25.10}$$

$$u = (4\pi\hbar^2 aN/m^2V)^{1/2}. \tag{25.11}$$

The Hamiltonian is then

$$\hat{H} = E_0+\sum_{\mathbf{p}\neq 0}\varepsilon(p)\,\hat{b}_{\mathbf{p}}^{+}\hat{b}_{\mathbf{p}}, \tag{25.12}$$

where

$$E_0 = \frac{1}{2}Nmu^2 + \frac{1}{2}\sum_{\mathbf{p}\neq 0}\left\{\varepsilon(p) - \frac{p^2}{2m} - mu^2 + \frac{m^3u^4}{p^2}\right\}. \qquad (25.13)$$

The form of the Hamiltonian (25.12) and the Bose commutation relations for the operators $\hat{b}_{\mathbf{p}}$, $\hat{b}_{\mathbf{p}}^+$ enable us to conclude that $\hat{b}_{\mathbf{p}}^+$ and $\hat{b}_{\mathbf{p}}$ are creation and annihilation operators for quasi-particles with energy $\varepsilon(p)$ which obey Bose statistics. The eigenvalues of the diagonal operator $\hat{b}_{\mathbf{p}}^+\hat{b}_{\mathbf{p}}$ represent the numbers $n_{\mathbf{p}}$ of quasi-particles with momentum $\mathbf{p}$, and formula (25.10) gives the dependence of their energy on the momentum. (The quasi-particle occupation numbers are again denoted by $n_{\mathbf{p}}$ to avoid confusion with the actual gas particle occupation numbers $N_{\mathbf{p}}$.) This completely determines the energy spectrum of weakly excited states of the gas in question.

The quantity E_0 is the energy of the ground state of the gas. Replacing the summation over the discrete values of $\mathbf{p}$ (in the volume V) by integration over $V\,d^3p/(2\pi\hbar)^3$ and completing the calculations, we get the expression

$$E_0 = \frac{2\pi\hbar^2 aN^2}{mV}\left[1 + \frac{128}{15}\sqrt{\frac{a^3N}{\pi V}}\right] \qquad (25.14)$$

(T. D. Lee and C. N. Yang 1957). The chemical potential of the gas (at $T = 0$) is correspondingly

$$\mu = \frac{\partial E_0}{\partial N} = \frac{4\pi\hbar^2 aN}{mV}\left[1 + \frac{32}{3}\sqrt{\frac{a^3N}{\pi V}}\right]. \qquad (25.15)$$

These formulae give the first two terms in an expansion in powers of $(a^3N/V)^{1/2}$. Even the next term, however, could not be obtained by the above method. It must contain the volume as V^{-2}, and a quantity of that order depends on triple collisions as well as on pair collisions.

For large momenta ($p \gg mu$) the quasi-particle energy (25.10) tends to $p^2/2m$, i.e. to the kinetic energy of an individual gas particle.

For small momenta ($p \ll mu$) we have $\varepsilon \approx up$. It is easy to see that the coefficient u is the same as the velocity of sound in the gas, so that this expression corresponds to phonons in accordance with the general theorems in §22. At $T = 0$, the free energy is equal to E_0; taking the leading term in the expansion of the latter, we find the pressure

$$P = -\partial E/\partial V = 2\pi\hbar^2 aN^2/mV^2.$$

The velocity of sound is $u = \sqrt{(\partial P/\partial \varrho)}$, where $\varrho = mN/V$ is the gas density; it is the same as (25.11).

In the model of a Bose gas here considered, the scattering length a must necessarily be positive (for a repulsive interaction between the particles). This is seen formally from the fact that imaginary terms would occur in the above formulae for the energy if $a < 0$. The thermodynamic significance of the condition $a > 0$ is that it is necessary to satisfy the inequality $(\partial P/\partial V)_T < 0$ in this model of a Bose gas.

The statistical distribution of elementary excitations (the mean values $\bar{n}_{\mathbf{p}}$ of their occupation numbers) at a non-zero temperature is given simply by the Bose distribution formula (22.2). The momentum distribution $\bar{N}_{\mathbf{p}}$ of the actual gas particles can be calculated by averaging the operator $\hat{a}_{\mathbf{p}}^{+}\hat{a}_{\mathbf{p}}$. Using (25.8) and the fact that the products $\hat{b}_{-\mathbf{p}}\hat{b}_{\mathbf{p}}$ and $\hat{b}_{\mathbf{p}}^{+}\hat{b}_{-\mathbf{p}}^{+}$ have zero diagonal matrix elements, we get

$$\bar{N}_{\mathbf{p}} = [\bar{n}_{\mathbf{p}} + L_{\mathbf{p}}^2(\bar{n}_{\mathbf{p}}+1)]/(1-L_{\mathbf{p}}^2). \tag{25.16}$$

This expression is, of course, valid only if $p \neq 0$. The number of particles with zero momentum is

$$\bar{N}_0 = N - \sum_{\mathbf{p}\neq 0} \bar{N}_{\mathbf{p}} = N - \frac{V}{(2\pi\hbar)^3}\int \bar{N}_{\mathbf{p}}\, d^3p. \tag{25.17}$$

In particular, at absolute zero all the $n_{\mathbf{p}} = 0$, and with (25.9) we obtain from (25.16) the distribution function in the form†

$$N_{\mathbf{p}} = \frac{m^2u^4}{2\varepsilon(p)\{\varepsilon(p)+p^2/2m+mu^2\}}; \tag{25.18}$$

when $T = 0$, the mean values of $N_{\mathbf{p}}$ are the same as the exact values, and the bar over the letter is therefore omitted. The non-idealness of the Bose gas naturally causes the presence of particles with non-zero momentum even at absolute zero; the integration in (25.17) with $N_{\mathbf{p}}$ from (25.18) is elementary, and gives

$$N_0 = N\left[1 - \frac{8}{3}\sqrt{\frac{Na^3}{\pi V}}\right]. \tag{25.19}$$

Lastly, a comment on the spectrum derived here. For small p, the derivative $d^2\varepsilon/dp^2 > 0$, i.e. the curve of $\varepsilon(p)$ turns upwards from the initial tangent $\varepsilon = up$. In such a case (see §34) there is an instability of the spectrum because of the possibility of spontaneous disintegration of the quasi-particles (phonons). The corresponding level width is, however, small (proportional to p^5 when p is small) and does not affect the expressions derived in the approximations considered above.

§ 26. The wave function of the condensate

As already mentioned in §23, the appearance or disappearance of superfluidity in liquid helium takes place by a phase transition of the second kind. Such a transition always involves some qualitative change in the properties of

† The maximum number of particles with a given momentum magnitude ($\sim p^2N_{\mathbf{p}}$) occurs for $p/\hbar \sim \sqrt{(aN/V)}$, where the change takes place from one limiting expression for $\varepsilon(p)$ to the other. This has already been mentioned in the second footnote to §25.

the body. At the λ-point of liquid helium, this change may be described macroscopically as the appearance or disappearance of the superfluid component of the liquid. From the more profound microscopic viewpoint it is a matter of certain properties of the momentum distribution of the (actual) liquid particles: in a superfluid, a finite fraction of the [illegible] macroscopically large number of them) have exactly zero mom[illegible]rticles form the *Bose–Einstein condensate*, or simply the conde[illegible]tum space. In an ideal Bose gas at $T = 0$, all its particles are i[illegible] (see Part 1, §62); in an almost ideal gas, almost all the partic[illegible]densate. In the general case of a Bose liquid with strong intera[illegible]e particles, the fraction of particles that are in the condensate at $T = 0$ is not close to unity.

We shall show how the property of Bose–Einstein condensation is formulated in terms of ψ operators. For an ideal Bose gas (a system of non-interacting bosons), the Heisenberg ψ operator is written explicitly as†

$$\hat{\Psi}(t, r) = \frac{1}{\sqrt{V}} \sum_{\mathbf{p}} \hat{a}_{\mathbf{p}} \exp\left\{\frac{i}{\hbar}\mathbf{p}\cdot\mathbf{r} - \frac{i}{\hbar}\frac{p^2}{2m}t\right\}. \tag{26.1}$$

As explained in §25, we may ignore the non-commutativity of the operators $\hat{a}_0$ and $\hat{a}_0^+$, regarding them as classical quantities. In other words, part of the ψ operator (26.1) is an ordinary number, which we denote by Ξ:

$$\hat{\Xi} = \hat{a}_0/\sqrt{V}. \tag{26.2}$$

To formulate this property of the ψ operators in the general case of an arbitrary Bose liquid, we note that, since the condensate contains a macroscopically large number of particles, changing this number by 1 does not essentially affect the state of the system; we may say that the result of adding (or removing) one particle in the condensate is to convert a state of a system of N particles into the "same" state of a system of $N\pm1$ particles.‡ In particular, the ground state remains the ground state. Let $\hat{\Xi}$ and $\hat{\Xi}^+$ denote the part of the ψ operators that changes the number of particles in the condensate by 1; then, by definition,

$$\hat{\Xi}\,|m, N+1\rangle = \Xi\,|m, N\rangle,$$
$$\hat{\Xi}^+\,|m, N\rangle = \Xi^*\,|m, N+1\rangle,$$

where the symbols $|m, N\rangle$ and $|m, N+1\rangle$ denote two "like" states differing only as regards the number of particles in the system, and Ξ is a complex number. These statements are rigorously valid in the limit $N \to \infty$. Hence the

† Cf. (9.3). We assume the gas particles spinless, and so the spin suffix is omitted. In (26.1) we have also used the fact that for an ideal Bose gas at $T = 0$ the chemical potential $\mu = 0$, and so the term $-\mu t/\hbar$ in the exponents is omitted.

‡ The addition or removal of the particle is to be regarded as occurring with infinite slowness. This prevents excitation of the system by the variable field.

definition of Ξ is to be written

$$\left.\begin{aligned}\lim_{N\to\infty}\langle m,N|\hat{\Xi}|m,N+1\rangle &= \Xi,\\ \lim_{N\to\infty}\langle m,N+1|\hat{\Xi}^{+}|m,N\rangle &= \Xi^{*};\end{aligned}\right\}\tag{26.3}$$

the limit is taken for a given finite value of the liquid density N/V.

If the ψ operators are written as

$$\hat{\Psi} = \hat{\Xi}+\hat{\Psi}', \quad \hat{\Psi}^{+} = \hat{\Xi}^{+}+\hat{\Psi}'^{+},\tag{26.4}$$

their remaining part ("above the condensate") converts the state $|m, N\rangle$ into states orthogonal to it, i.e. the matrix elements†

$$\left.\begin{aligned}\lim_{N\to\infty}\langle m,N|\hat{\Psi}'|m,N+1\rangle &= 0,\\ \lim_{N\to\infty}\langle m,N+1|\hat{\Psi}'^{+}|m,N\rangle &= 0.\end{aligned}\right\}\tag{26.5}$$

In the limit $N\to\infty$, the difference between the states $|m, N\rangle$ and $|m, N+1\rangle$ disappears entirely, and in this sense Ξ becomes the mean value of the operator $\hat{\Psi}$ for that state. It must be emphasized that the finiteness of the limiting value is a characteristic of systems containing a condensate.

The equations (26.3) complete the "operator" properties of $\hat{\Xi}$ and $\hat{\Xi}^{+}$, and they may be regarded as commuting with $\hat{\Psi}'$ and $\hat{\Psi}'^{+}$. In particular, the operators $\hat{\Xi}$ and $\hat{\Xi}^{+}$ will be replaced by Ξ and Ξ^{*} (i.e. will behave as classical quantities) in any averaging with respect to the ground state. We must emphasize again that (because the number of particles in the condensate is macroscopic) this approximation involves neglecting only quantities with relative order of smallness $1/N$.‡

If the time dependence of the wave functions is determined by the Hamiltonian $\hat{H}'=\hat{H}-\mu\hat{N}$, then Ξ is independent of time: the matrix element $\langle m,N|\Xi|m, N+1\rangle$ is proportional to

$$\exp\left\{-\frac{it}{\hbar}[E(N+1)-E(N)-(N+1)\mu+N\mu]\right\},$$

and the exponent tends to zero, since (to within a quantity $\sim 1/N$) $E(N+1)-$ $-E(N) = \mu$.

In a homogeneous liquid at rest, Ξ is independent also of the coordinates and is simply (with the appropriate choice of the phase of the complex quantity)

$$\Xi = \sqrt{n_0},\tag{26.6}$$

† To avoid misunderstanding, it may be mentioned again that these equations refer only to transitions between "like" states.

‡ In particular, to this accuracy we must regard as equal the matrix elements of the operators $\hat{\Psi}'$ for transitions between states differing by the same (small) number of particles in the system.

where n_0 is the number of condensate particles per unit volume of the liquid: $\hat{\Xi}^+\hat{\Xi}$ is the operator of the particle number density in the condensate, and the mean value of this operator is just n_0.

The existence of the condensate brings about a qualitative difference in the properties of the density matrix for particles in a Bose liquid in comparison with the density matrix in an ordinary liquid. In an arbitrary state of a homogeneous Bose liquid, the density matrix is given by

$$N\varrho(\mathbf{r}_1, \mathbf{r}_2) = \langle m, N \,|\, \hat{\Psi}^+(t, \mathbf{r}_2)\hat{\Psi}(t, \mathbf{r}_1) \,|\, m, N\rangle, \tag{26.7}$$

and this function depends only on the difference $\mathbf{r} = \mathbf{r}_1 - \mathbf{r}_2$; cf. (7.13). Substituting here the ψ operators in the form (26.4) and using the properties (26.3) and (26.5), we get

$$N\varrho(\mathbf{r}_1, \mathbf{r}_2) = n_0 + N\varrho'(\mathbf{r}_1, \mathbf{r}_2). \tag{26.8}$$

The density matrix ϱ' "above the condensate" tends to zero as $|\mathbf{r}_1 - \mathbf{r}_2| \to \infty$; the density matrix ϱ tends to the finite limit n_0/N. This expresses the existence of "long-range order" in a superfluid, which is not present in ordinary liquids; in these, we always have $\varrho \to 0$ as $|\mathbf{r}_1 - \mathbf{r}_2| \to \infty$. It is this symmetry property that distinguishes the superfluid and non-superfluid phases (V. L. Ginzburg and L. D. Landau 1950).

The Fourier component of the density matrix determines the momentum distribution of the liquid particles by

$$N(\mathbf{p}) = N \int \varrho(\mathbf{r})\, e^{-i\mathbf{p}\cdot\mathbf{r}}\, d^3x; \tag{26.9}$$

cf. (7.20). Substituting ϱ from (26.8), we obtain

$$N(\mathbf{p}) = (2\pi)^3 n_0 \delta(\mathbf{p}) + N \int \varrho'(\mathbf{r})\, e^{-i\mathbf{p}\cdot\mathbf{r}}\, d^3x. \tag{26.10}$$

The delta function term corresponds to the finite probability for the particle to have exactly zero momentum.

If superfluid motion takes place in the liquid, or if it is in non-uniform and non-stationary external conditions (which, however, vary considerably only over distances large in comparison with interatomic distances), the Bose–Einstein condensation again occurs, but we cannot now assert that it will occur in the state with $p = 0$. The quantity Ξ, again defined by (26.3), will now be a function of coordinates and time, representing the particle wave function in the condensate state. It is normalized by the condition $|\Xi|^2 = n_0$, and can therefore be expressed as

$$\Xi(t, \mathbf{r}) = \sqrt{[n_0(t, \mathbf{r})]}\, e^{-i\Phi(t, \mathbf{r})}. \tag{26.11}$$

Since there is a macroscopically large number of particles in the condensate state, the wave function of this state becomes a classical macroscopic quantity.†

† Just as the field strength of an electromagnetic wave becomes a classical quantity for large photon occupation numbers in every state (cf. *RQT*, §5).

Thus there is a new characteristic of macroscopic states in a superfluid, including states of thermodynamic equilibrium.

The current density calculated from the wave function (26.11) is

$$\mathbf{j}_{\text{cond}} = \frac{i\hbar}{2m}(\Xi\nabla\Xi^* - \Xi^*\nabla\Xi)$$

$$= \frac{\hbar}{m} n_0 \nabla\Phi,$$

where m is the mass of a liquid particle. This has the significance of the macroscopic current density of condensate particles, and may be [illegible]uated to $n_0\mathbf{v}_s$, where $\mathbf{v}_s$ is the macroscopic velocity of that motion [illegible]arison of the two expressions, we find

$$\mathbf{v}_s = (\hbar/m)\nabla\Phi. \tag{26.12}$$

Since the motion can occur in a state of thermody[illegible]n (characterized by the quantity Ξ), it is non-dissipative, and (2[illegible] therefore determines the velocity of the superfluid motion. We thus arrive at the property of such motion already mentioned in §23: it is a potential flow. The velocity potential ϕ is equal (apart from a constant factor) to the phase of the condensate wave function:

$$\phi = (\hbar/m)\Phi. \tag{26.13}$$

To avoid misunderstanding, however, we should emphasize that, although the condensate velocity is the same as the velocity of the superfluid component of the liquid (and although the condensate and the superfluid component appear simultaneously at the λ-point), the densities mn_0 of the condensate and ϱ_s of the superfluid component are not at all the same. The identity of these two quantities would be impossible to justify, and its incorrectness is also evident from the fact that at absolute zero the whole mass of the liquid is superfluid, whereas not all its particles are in the condensate.‡

§ 27. Temperature dependence of the condensate density

The particle number density in the condensate is greatest at $T = 0$, and decreases with rising temperature. The limiting form of its temperature dependence as $T \to 0$ can be found by considering the fluctuations of a macroscopic quantity, the condensate wave function Ξ (R. A. Ferrell, N. Menyhárd, H. Schmidt, F. Schwabl and P. Szépfalusy 1968).

First, we recall that Ξ is a classical quantity which corresponds to the operator $\hat{\Psi}$ in the quantum-mechanical formalism. Hence, to calculate the fluctuations, we ought in principle to use that operator. However, near absolute

‡ In practice, the density of the condensate in liquid helium seems to be only a small fraction of the total density of the liquid.

zero, long-wavelength oscillations play the main part in the fluctuation spectrum of a macroscopic quantity. These oscillations in the liquid are sound waves described by the macroscopic equations of hydrodynamics, and it is therefore possible to construct an operator corresponding to Ξ by independent quantization of Ξ.

In the present case, for $\Xi = \sqrt{n_0} \exp i\Phi$, in the long-wavelength limit, the phase Φ fluctuates most strongly, and is directly related to the superfluid velocity potential by (26.13). Both ϕ and Φ, it may be recalled, are defined only to within additive constants. The uniquely defined quantity $\sqrt{n_0}$ can therefore be expressed only in terms of the derivatives of Φ, and so the Fourier components of its fluctuations will contain extra powers of the wave vector **k**, i.e. will be small when **k** is small.

The relation of the phase Φ to the potential ϕ allows Φ to be directly related to quantities characterizing the phonon distribution in the liquid. For this purpose, we regard ϕ, and therefore Φ, as second-quantized operators, expressing ϕ by (24.10) in terms of the phonon creation and annihilation operators:

$$\hat{\Phi} = \sum_{\mathbf{p}} \left(\frac{mu}{2Vnp}\right)^{1/2} (\hat{c}_{\mathbf{p}} e^{i\mathbf{p}.\mathbf{r}/\hbar} + \hat{c}_{\mathbf{p}}^{+} e^{-i\mathbf{p}.\mathbf{r}/\hbar}); \tag{27.1}$$

the unperturbed liquid density is written as $\varrho = nm$, where n is the particle number density, and the suffix 0 is omitted. According to the foregoing discussion, this means that the operator of the macroscopic quantity Ξ, i.e. the long-wave part of the operator $\hat{\Psi}$, can be expressed as

$$\hat{\Psi} = \sqrt{n_0} \exp i\hat{\Phi}, \tag{27.2}$$

where n_0 is the condensate particle density.

We first apply this formula to calculate the momentum distribution of particles "above the condensate" in a Bose liquid (for small momenta). In the single-particle density matrix $\varrho(\mathbf{r}_1, \mathbf{r}_2)$, for large distances $|\mathbf{r}_1 - \mathbf{r}_2|$, we can use the long-wave expression (27.2) for the ψ operator:

$$N\varrho(\mathbf{r}_1, \mathbf{r}_2) = \langle \hat{\Psi}^{+}(\mathbf{r}_2) \hat{\Psi}(\mathbf{r}_1) \rangle \approx n_0 \langle e^{-i\hat{\Phi}^{+}(\mathbf{r}_2)} e^{i\hat{\Phi}(\mathbf{r}_1)} \rangle, \tag{27.3}$$

where the mean value is taken with respect to the state of the liquid at a given temperature. Since the fluctuations are small, this expression is to be expanded in powers of $\hat{\Phi}$, retaining only the first non-vanishing (the quadratic) terms. Since $\hat{\Phi}^{+} = \hat{\Phi}$, we obtain

$$N\varrho(\mathbf{r}_1, \mathbf{r}_2) = n_0 - n_0 \langle \hat{\Phi}^2(\mathbf{r}) \rangle + n_0 \langle \hat{\Phi}(\mathbf{r}_2) \hat{\Phi}(\mathbf{r}_1) \rangle. \tag{27.4}$$

The third term tends to zero as $|\mathbf{r}_2 - \mathbf{r}_1| \to \infty$, and gives the required above-condensate part of the density matrix; the second term is independent of $\mathbf{r}$ in a homogeneous liquid, and gives a correction to the condensate density that will be calculated below by a somewhat different method. Using (27.1), we can write the above-condensate part as

$$N\varrho'(\mathbf{r}_1, \mathbf{r}_2) = \frac{n_0 mu}{2Vn} \sum_{\mathbf{p}} \frac{1}{p} \{\langle \hat{c}_{\mathbf{p}}^+ \hat{c}_{\mathbf{p}} \rangle e^{-i\mathbf{p}\cdot(\mathbf{r}_1-\mathbf{r}_2)/\hbar} + \langle \hat{c}_{\mathbf{p}} \hat{c}_{\mathbf{p}}^+ \rangle e^{i\mathbf{p}\cdot(\mathbf{r}_1-\mathbf{r}_2)/\hbar}\}$$

$$= \frac{n_0 mu}{Vn} \sum_{\mathbf{p}} \frac{1}{p} \left(n_{\mathbf{p}} + \frac{1}{2}\right) e^{i\mathbf{p}\cdot(\mathbf{r}_1-\mathbf{r}_2)/\hbar},$$

where

$$n_{\mathbf{p}} = [e^{pu/T} - 1]^{-1}.$$

Changing from summation to integration, we have

$$N\varrho'(\mathbf{r}_1, \mathbf{r}_2) = \frac{n_0 mu}{n} \int \frac{n_{\mathbf{p}} + \frac{1}{2}}{p} e^{i\mathbf{p}\cdot(\mathbf{r}_1-\mathbf{r}_2)} \frac{d^3p}{(2\pi\hbar)^3}. \tag{27.5}$$

This expression applies, of course, only to the contribution from small p ($\hbar/p$ large compared with interatomic distances). The integrand in (27.5) immediately gives the particle momentum distribution

$$N(\mathbf{p}) = \frac{n_0 mu}{np} \left(n_{\mathbf{p}} + \frac{1}{2}\right). \tag{27.6}$$

When $T = 0$, this becomes

$$N(\mathbf{p}) = n_0 mu/2np \tag{27.7}$$

(J. Gavoret and P. Nozières 1964); when $T \neq 0$ and $up \ll T$,

$$N(\mathbf{p}) = n_0 mT/np^2 \tag{27.8}$$

(P. C. Hohenberg and P. C. Martin 1965).

We can now determine the temperature dependence of the condensate density. By definition,

$$n_0(T) = n - \int N(\mathbf{p})\, d^3p/(2\pi\hbar)^3. \tag{27.9}$$

If we substitute (27.6) here directly, the integral diverges because of the zero-point vibrations. This is related to the invalidity of (27.6) for large p, and means only that we cannot calculate in such a way the value of the condensate density at $T = 0$, which must here be regarded as a given quantity. To find the required temperature dependence, we must subtract from $n_0(T)$ its value at $T = 0$; the integral is then convergent. The result is

$$\frac{n_0(T) - n_0(0)}{n_0(0)} = -\frac{mu}{n} \int \frac{n_{\mathbf{p}}}{p} \frac{d^3p}{(2\pi\hbar)^3}$$

$$= -\frac{mT^2}{2\pi^2 nu\hbar^3} \int_0^\infty \frac{x\,dx}{e^x - 1} = -\frac{mT^2}{12nu\hbar^3}. \tag{27.10}$$

In the calculation we have neglected the temperature dependence of the total density of the liquid; this is legitimate, since the thermal expansion of the liquid (due to the excitation of phonons) is proportional to a higher power of the temperature, namely T^4 (cf. Part 1, §67).†

Finally, we may make some remarks on the methodologically important subject of a two-dimensional Bose liquid. In this case, the temperature-dependent part of the integral (27.9) diverges logarithmically for small **p**, where the formula for $N(\mathbf{p})$ should have been correct. This means that in the two-dimensional case the basic assumption is incorrect, namely that there exists a condensate at non-zero temperatures. In the two-dimensional case, the condensate can exist only at $T = 0$.‡ The position here is analogous to that of two-dimensional crystals (see Part 1, §137). As with the latter the fluctuations of the atomic displacements smooth out the lattice, so the phase fluctuations eliminate the condensate. The formal analogy between the two systems is that in both cases the energy depends on quantities that can appear in it only as derivatives. In the first case these are the atomic displacement vectors, which cannot themselves appear in the energy, because the latter is invariant under displacements of the system as a whole. In the second case it is the phase of the condensate wave function, which cannot itself appear in the energy, because it is not uniquely determined. The dependence of the energy on only the gradients of these quantities is the ultimate reason for the divergence of the fluctuations.

Next, we have seen in Part 1, §138, that the weak (logarithmic) divergence of the fluctuations causes in a two-dimensional crystal a slow (power-function) decrease of the correlation function in the system. Similarly, in a two-dimensional Bose system the density matrix (27.3) decreases as $|\mathbf{r}_1-\mathbf{r}_2| \to \infty$ according to a power law, and does not tend to a constant limit as in the presence of the condensate.§ Such a system thereby differs qualitatively from an ordinary liquid, and so, in the two-dimensional case also, there can be a phase transition of the second kind between the ordinary liquid with an exponential decrease of $\varrho(\mathbf{r}_1, \mathbf{r}_2)$ and a liquid with a power-law decrease.

§ 28. Behaviour of the superfluid density near the λ-point

As already mentioned in §23, with increasing temperature the fraction ϱ_s/ϱ of the superfluid density in a Bose liquid decreases, becoming zero at the λ-point of the liquid, a phase transition point of the second kind. The temperature T_λ of this point is a function of the pressure P; the equation $T = T_\lambda(P)$ defines the curve of λ-points in the phase diagram in the PT-plane.

In the general theory of phase transitions of the second kind, the change in state of the body is described by the behaviour of the order parameter, which

† The formulae obtained, which are valid for any Bose liquid, are of course in agreement with those of §25 for a slightly non-ideal Bose gas. In the comparison, it must be noted that for such a gas $n_0 \approx n$, and the condition for p to be small is $p \ll mu \sim \hbar(an)^{1/2}$.

‡ These statements relate also to a two-dimensional ideal Bose gas.

§ See J. W. Kane and L. Kadanoff, *Physical Review* **155**, 80, 1967.

characterizes its symmetry properties. For the λ-transition of a Bose liquid, the condensate wave function Ξ acts as such a parameter, and describes, as explained in §26, the "long-range order" in the liquid. The fact that Ξ is complex means that the order parameter has two components, and the effective Hamiltonian of the system (see Part 1, §147) depends only on $|\Xi|^2$, i.e. is invariant under the transformation $\Xi \to e^{i\alpha}\Xi$ for any real α.

The empirical results concerning the λ-transition in liquid helium seem to indicate that there is no region in which the Landau theory of phase transitions is valid: the condition in Part 1 (146.15) is not satisfied anywhere in the neighbourhood of the λ-point (i. e. anywhere in the region $|T-T_\lambda| \ll T_\lambda$). Hence, to describe the properties of this transition, we must use the fluctuation theory of phase transitions of the second kind, which makes it possible to relate the temperature dependences of various quantities.

The temperature dependence of the order parameter (and therefore of the condensate density n_0) as $T \to T_\lambda$ is given by the critical index β (see Part 1, §148):

$$|\Xi| = \sqrt{n_0} \propto (T_\lambda - T)^\beta. \tag{28.1}$$

A more interesting question, however, is that of the behaviour of the superfluid density ϱ_s. To calculate it, let us consider a liquid in which the phase Φ of the condensate wave function varies slowly in space. This means that there is in the liquid a macroscopic superfluid motion with the velocity (26.12) and accordingly with kinetic energy (per unit volume of the liquid)

$$\tfrac{1}{2}\varrho_s v_s^2 = \varrho_s(\hbar^2/2m^2)(\nabla\Phi)^2. \tag{28.2}$$

This expression may also be applied to the long-wavelength fluctuations of the order parameter. According to the hypothesis of scale invariance, the only parameter of length that determines the fluctuation picture near the transition point is the correlation radius r_c of the fluctuations. This therefore determines the order of magnitude of the distances at which the fluctuational change of the phase Φ is of the order of unity; hence the mean square of the fluctuational velocity varies with temperature according to

$$\overline{v_s^2} \propto 1/r_c^2 \propto (T_\lambda - T)^{2\nu}, \tag{28.3}$$

where ν is the critical index of the correlation radius. On the other hand, since it is the long-wavelength fluctuations that govern the singularity of the thermodynamic quantities at the transition point, we may naturally assume that near this point the fluctuational kinetic energy (28.2) varies with temperature in the same way as the singular part of the thermodynamic potential of the liquid, i.e. as $(T_\lambda - T)^{2-\alpha}$ (where α is the critical index of the specific heat C_p). Thus we find

$$\varrho_s\overline{v_s^2} \propto \varrho_s(T_\lambda - T)^{2\nu} \propto (T_\lambda - T)^{2-\alpha},$$

whence $\varrho_s \propto (T_\lambda - T)^{2-\alpha-2\nu}$. Lastly, with the relation $3\nu = 2-\alpha$ (which follows from the hypothesis of scale invariance; see Part 1, §149), we have

$$\varrho_s \propto (T_\lambda - T)^{(2-\alpha)/3}. \qquad (28.4)$$

This establishes the relation between the temperature dependences of ϱ_s and the specific heat near the λ-point (B. D. Josephson 1966).†

§ 29. Quantized vortex filaments

An ordinary liquid enclosed in a cylindrical vessel rotating about its axis is carried along by friction against the vessel walls, and is ultimately caused to rotate as a whole together with the vessel. In a superfluid, only the normal component is brought into rotation; the superfluid component remains at rest, in accordance with the fact that this component cannot rotate as a whole, since this would make the superfluid motion no longer a potential flow.‡

For sufficiently large rates of rotation, however, such a state becomes thermodynamically unfavourable. The condition of thermodynamic equilibrium is that the quantity

$$E_{\text{rot}} = E - \mathbf{M}.\mathbf{\Omega} \qquad (29.1)$$

is a minimum; this is the energy in a rotating coordinate frame, with E and $\mathbf{M}$ the energy and angular momentum of the system in a fixed coordinate frame (see Part 1, §26). The term $-\mathbf{M}.\mathbf{\Omega}$ in (29.1) causes (for sufficiently large $\mathbf{\Omega}$) the state with $\mathbf{M}.\mathbf{\Omega} > 0$ to be thermodynamically more favourable than that with $\mathbf{M} = 0$.

Thus, as the rate of rotation of the vessel increases, superfluid motion must eventually occur. The apparent contradiction between this statement and the condition for superfluid motion to be a potential flow is removed by assuming that the potential flow is lost only at certain lines of singularity in the liquid, known as *vortex filaments* or *vortex lines*.§ The liquid executes a motion about these lines which may be called *potential rotation*, since curl $\mathbf{v}_s = 0$ throughout the volume outside the lines.

The vortex filaments in a liquid have a thickness of atomic dimensions, and macroscopically they must be regarded as being of infinitesimal thickness. Their existence does not contradict the expression (26.12) for the velocity, since the latter assumes that $\mathbf{v}_s$ varies sufficiently slowly in space, whereas it

† The indices α and ζ for liquid helium are very small, and so we have with high accuracy $\beta \approx 1/3$, and $\varrho_s \propto n_0 \propto (T_\lambda - T)^{2/3}$.

‡ When the liquid rotates as a whole, the velocity $\mathbf{v} = \mathbf{\Omega} \times \mathbf{r}$, where $\mathbf{\Omega}$ is the angular velocity and the position vector $\mathbf{r}$ is drawn from some point on the axis. Then curl $\mathbf{v} = 2\mathbf{\Omega} \neq 0$.

§ This assumption was proposed by L. Onsager (1949) and further developed by R. P. Feynman (1955).

‖ This statement does not apply, however, to the immediate nieghbourhood of the λ-point; there, the thickness of a vortex filament is of the order of the correlation radius of the fluctuations.

varies with arbitrarily great rapidity near a vortex filament; see (29.3) below. It also does not contradict the proof in §23 that superfluid motion is a potential flow, which made use of properties of a Bose liquid energy spectrum, since a vortex filament is associated with a particular macroscopically large energy (see (29.8) below), and the state of a liquid containing a filament cannot be regarded as weakly excited.

Let us first consider vortex filaments from a purely kinematic standpoint, as lines of singularity in the velocity distribution for potential flow of the liquid. Each vortex filament has a particular value ($2\pi\varkappa$, say) of the velocity circulation along a closed contour round the filament:

$$\oint \mathbf{v}_s \cdot d\mathbf{l} = 2\pi\varkappa. \tag{29.2}$$

This value is independent of the choice of the contour of integration: if C_1 and C_2 are two contours enclosing the vortex filament, the difference between the circulations along them is, by Stokes's theorem, equal to the flux of the vector curl $\mathbf{v}_s$ through a surface spanning C_1 and C_2; since this surface does not meet the vortex filament, curl $\mathbf{v}_s = 0$ at all points on it, and the integral is zero. Hence it follows also that a vortex filament cannot terminate: either it is closed or it ends at the boundary of the liquid (or, in an infinite liquid, has both ends at infinity), since the existence of a free end of a vortex filament would imply that there could be a surface spanning the contour C but nowhere meeting the filament, and so the integral on the left of (29.2) would be zero.

The condition (29.2) enables us to determine the velocity distribution in a liquid moving round a vortex filament. In the simplest case of a straight filament in an infinite liquid, the streamlines are circles in planes perpendicular to the filament, with centres lying on the filament. The circulation along such a curve is $2\pi r v_s$, so that

$$v_s = \varkappa / r, \tag{29.3}$$

where r is the distance from the filament. We may note that in potential rotation the velocity decreases away from the axis of rotation (the vortex filament), in contrast to rigid rotation, where the velocity increases in proportion to r.

For a vortex filament of any shape, the velocity distribution is given by

$$\mathbf{v}_s = \tfrac{1}{2}\varkappa \int d\mathbf{l} \times \mathbf{R}/R^3, \tag{29.4}$$

where the integration is along the filament, and $\mathbf{R}$ is the radius vector from $d\mathbf{l}$ to the point where the velocity is observed.† At distances from the filament

† This expression may be written down immediately by analogy with the familiar Biot–Savart formula for the magnetic field of line currents. The formal equivalence of the two problems is evident from a comparison of the velocity circulation (29.2) with the circulation of the magnetic field $\mathbf{H}$ round the line current J:

$$\oint \mathbf{H} \cdot d\mathbf{l} = 4\pi J/c.$$

One problem is obtained from the other by substituting $\mathbf{v}_s$ for $\mathbf{H}$ and $\frac{1}{2}\varkappa$ for J/c.

that are small compared with its radius of curvature, formula (29.4) of course reduces approximately to (29.3).

As already mentioned, formulae (29.2)–(29.4) result simply from the fact that the motion of the liquid is a potential flow. The quantum nature of vortex filaments in a superfluid is shown by the fact that the constant $\varkappa$ can only have values in a certain discrete series. Using (26.12) for the velocity $\mathbf{v}_s$ expressed in terms of the phase Φ of the condensate wave function, we find as its circulation

$$\oint \mathbf{v}_s \cdot d\mathbf{l} = (\hbar/m)\, \Delta\Phi, \tag{29.5}$$

where $\Delta\Phi$ is the change of phase on traversing the contour. Since the wave function is single-valued, its change of phase on returning to the original point must be an integral multiple of 2π, and so

$$\varkappa = n\hbar/m, \tag{29.6}$$

where n is an integer. We shall see below that in fact only vortex filaments with the lowest possible circulation ($n = 1$) are thermodynamically stable. We shall therefore put

$$\varkappa = \hbar/m. \tag{29.7}$$

Let us now determine the critical rate of rotation of the cylindrical vessel at which a vortex filament first appears. It is evident from symmetry that this filament will be along the axis of the vessel. The change in the energy of the liquid due to the appearance of the vortex filament in it is

$$\Delta E = \int \tfrac{1}{2} \varrho_s v_s^2\, dV = \tfrac{1}{2} \varrho_s L \int v_s^2 . 2\pi r\, dr = L\varrho_s \pi \varkappa^2 \int dr/r,$$

where L is the length of the vessel. The integration with respect to r is to be taken between the radius R of the vessel and some value $r \sim a$ of the order of atomic distances, at which the macroscopic treatment ceases to be meaningful; because the integral is logarithmically divergent, its value does not depend greatly on the precise choice of a. Thus

$$\Delta E = L\pi\varrho_s(\hbar^2/m^2) \log (R/a); \tag{29.8}$$

this expression is said to have logarithmic accuracy, i.e. not only the ratio R/a but also its logarithm is large.† The angular momentum of the rotating liquid is

$$M = \int \varrho_s v_s r\, dV = \varrho_s \varkappa \int dV = L\pi R^2(\hbar/m)\, \varrho_s. \tag{29.9}$$

† The motion round the vortex filament is in general accompanied by a change in the density of the liquid. The neglect of this change in the calculation given here is justified by the fact that the main contribution to the energy (29.8) comes (because of the logarithmic divergence of the integral) from large distances r, at which the density change is small. For the same reason, we may neglect the contribution to ΔE from the change in the internal energy of the liquid.

The occurrence of the vortex filament is thermodynamically favourable if $\Delta E_{rot} = \Delta E - M\Omega < 0$, i.e. if

$$\Omega > \Omega_{cr} = (\hbar/mR^2)\log(R/a). \tag{29.10}$$

The above arguments also indicate the reason why vortex filaments with $n > 1$ in (29.6) are thermodynamically unstable: when $n = 1$ is replaced by a value $n > 1$, the energy ΔE is increased by a factor of n^2, and M by a factor n, which must increase ΔE_{rot}.

When the rate of rotation of the cylindrical vessel increases further beyond the critical value (29.10), new vortex filaments appear, and when $\Omega \gg \Omega_{cr}$ their number is very large. Their distribution over the cross-section of the vessel tends to a uniform one, and in the limit they simulate the rotation of the superfluid part of the liquid as a rigid body.† The number of vortex filaments for a given (large) value of Ω is easily determined by the condition that the velocity circulation along a contour enclosing a large number of filaments should have a value corresponding to rotation of the liquid as a whole. If such a contour encloses unit area in the plane perpendicular to the axis of rotation, then

$$\oint \mathbf{v}_s \cdot d\mathbf{l} = \nu . 2\pi\varkappa = 2\pi\nu\hbar/m,$$

where ν is the distribution density of the vortex filaments over the cross-section of the vessel. On the other hand, when the liquid rotates as a whole, curl $\mathbf{v}_s = 2\mathbf{\Omega}$, and this circulation is 2Ω. Equating the two expressions, we find

$$\nu = m\Omega/\pi\hbar. \tag{29.11}$$

The occurrence of vortex filaments to some extent eliminates the property of superfluidity. The elementary excitations that form the normal component of the liquid are then scattered by the filaments, transferring to these (and thus to the superfluid component of the liquid) a part of their momentum. This consequently implies the presence of friction between the two components of the liquid.

Vortex filaments in general move about in space with the flow of the liquid. When $T = 0$ and the liquid is entirely superfluid, each element $d\mathbf{l}$ of the filament moves with the velocity $\mathbf{v}_s$ of the liquid at the position of that element. At non-zero temperatures, the frictional force on the filament causes it to have a velocity relative to the superfluid component.

Vortex filaments formed by rotation are straight. The flow of a liquid through capillaries, slits, etc., may be accompanied by the formation of closed filaments or

† This is easily seen by noting that, since the number of filaments increases in proportion to Ω (see (29.11) below), the second term in $\Delta E_{rot} = \Delta E - M\Omega$ increases as Ω^2, but the first term increases as Ω, and may therefore be neglected when $\Omega \gg \Omega_{cr}$. Then the minimization of ΔE_{rot} is equivalent to the maximization of M, which occurs when the liquid rotates as a rigid body.

vortex rings. These eliminate the superfluidity in flow at velocities above a certain critical value. The actual values of these *critical velocities* depend on the specific conditions of the flow; they are much less than the value above which the condition (23.3) is violated.

Unlike straight vortex filaments, which can remain stationary in a liquid that is at rest (far from them), vortex rings move relative to the liquid. The displacement velocity of each line element is the value of $\mathbf{v}_s$ which results (according to (29.4)) at its position from the action of all the rest of the filament; for curved filaments this is not in general zero. Consequently, vortex rings have as a whole not only definite energies but also definite momenta, and in this sense are a special type of elementary excitations.

PROBLEMS

PROBLEM 1. Find the velocity and momentum of a circular vortex ring.

SOLUTION. Each element of the ring moves with the velocity $\mathbf{v}_s$ at a given point, and from the symmetry of a circular ring this velocity is the same at every point of it. It is therefore sufficient to determine the velocity $\mathbf{v}_s$ at any one point P of the ring due to the rest of the ring. The elements $d\mathbf{l}$ of the ring and the radius vectors $\mathbf{R}$ from $d\mathbf{l}$ to the point P are in the plane of the ring; hence the velocity at the point P, given by (29.4), is perpendicular to the plane of the ring, as a result of which the ring moves without change of shape or size.

Let us define the position of the element dl by the angle ϑ (Fig. 3). Then

$$dl = R_0\, d\vartheta, \qquad R = 2R_0 \sin \tfrac{1}{2}\vartheta, \qquad |\, d\mathbf{l}\times\mathbf{R}\,| = R \sin \tfrac{1}{2}\vartheta . dl,$$

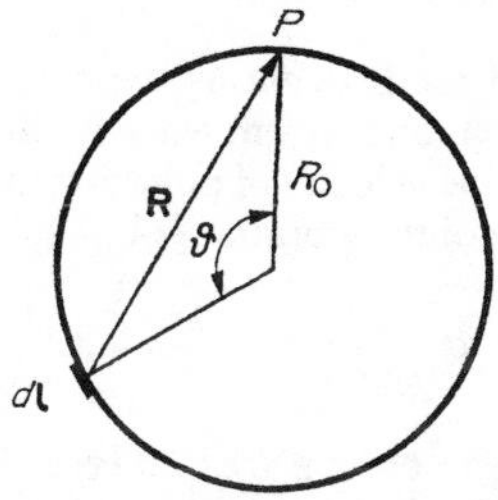

FIG. 3.

where R_0 is the radius of the ring, and we find from (29.4) for the ring velocity v

$$v = \frac{\varkappa}{8R_0} . 2 \int_0^{\pi} \frac{d\vartheta}{\sin \frac{1}{2}\vartheta} .$$

This integral, however, is logarithmically divergent at the lower limit, and must be cut off at a value $\vartheta \sim a/R_0$ corresponding to atomic distances ($\sim a$) of the element dl from the point P. The integral is determined, with logarithmic accuracy, by the range of values $a/R_0 \ll \vartheta \ll \pi$, and is

$$\int_{-a/R_0}^{-1} \frac{2\, d\vartheta}{\vartheta} = 2 \log \frac{R_0}{a} ,$$

so that

$$v = (\varkappa/2R_0)\log(R_0/a)$$
$$= (\hbar/2mR_0)\log(R_0/a). \qquad (1)$$

With the same logarithmic accuracy, the energy of the vortex ring is

$$\varepsilon = 2\pi^2 R_0 \varrho_s(\hbar^2/m^2)\log(R_0/a), \qquad (2)$$

which is (29.8) with R_0 and $2\pi R_0$ in place of R and L. The energy ε is related to the velocity v by $d\varepsilon/dp = v$, where p is the momentum of the ring. Hence

$$dp = d\varepsilon/v$$
$$= 4\pi^2\varrho_s(\hbar/m)\, R_0 dR_0$$

(with logarithmic accuracy, the large logarithm is to be regarded as constant in the differentiation), and so

$$p = 2\pi^2\varrho_s(\hbar/m)\, R_0^2. \qquad (3)$$

Formulae (2) and (3) determine the function $\varepsilon(p)$ for vortex rings in parametric form (with R_0 as parameter).

It may be noted that, because of the logarithmic nature of the integration that leads to formula (1), this formula (with some changes of notation) remains valid also for the velocity $\mathbf{v}$ with which any given element moves in a curved vortex ring of any shape:

$$\mathbf{v} = (\varkappa/2R_0)\,\mathbf{b}\log(\lambda/a). \qquad (4)$$

Here $\mathbf{b}$ is a unit vector perpendicular to the tangent plane at the given point on the filament (the binormal vector), R_0 is the radius of curvature at that point, and λ is the characteristic distance over which the curvature of the filament varies.

PROBLEM 2. Find the dispersion relation for small vibrations of a straight vortex filament (W. Thomson 1880).

SOLUTION. We take the line of the filament as the z-axis, and let $\mathbf{r} = (x, y)$ be a vector giving the displacement of points on the filament when it vibrates; $\mathbf{r}$ is a function of z and the time t, of the form $\exp[i(kz-\omega t)]$. The velocity of points on the filament is given by formula (4), with λ here taken as the wavelength of the vibrations ($\lambda \sim 1/k$):

$$\mathbf{v} = d\mathbf{r}/dt = -i\omega\mathbf{r} = \frac{1}{2}\varkappa\log\frac{1}{ak}\frac{\mathbf{b}}{R_0}.$$

The binormal vector $\mathbf{b} = \mathbf{t}\times\mathbf{n}$, where $\mathbf{t}$ and $\mathbf{n}$ are unit vectors along the tangent and the principal normal to the curve. According to a well-known formula of differential geometry, $d^2\mathbf{r}/dl^2 = \mathbf{n}/R_0$, where l is the length measured along the curve. For small vibrations, the filament is only slightly curved, and we can therefore take $l \approx z$ and $\mathbf{t} = \mathbf{t}_z$ (a unit vector along the z-axis); then

$$\mathbf{b}/R_0 \approx \mathbf{t}_z\times d^2\mathbf{r}/dz^2 = -k^2\mathbf{t}_z\times\mathbf{r}.$$

The equation of motion of the filament is then

$$-i\omega\mathbf{r} = -\tfrac{1}{2}\varkappa k^2\mathbf{t}_z\times\mathbf{r}\log(1/ak).$$

On expansion, this gives two linear homogeneous equations for x and y; equating the determinant to zero, we get the required relation between ω and k:

$$\omega = \tfrac{1}{2}\varkappa k^2\log(1/ak).$$

§ 30. A vortex filament in an almost ideal Bose gas

As already mentioned, the thickness of a vortex filament in a liquid is comparable with atomic distances. An exception in this respect, however, occurs in the case of an almost ideal Bose gas. Here the "core" of the vortex filament, where the properties of the medium are significantly altered, has (as we shall see below) a macroscopic thickness, and its structure may be macroscopically described (V. L. Ginzburg and L. P. Pitaevskiĭ 1958, L. P. Pitaevskiĭ 1961, E. P. Gross 1961).

Let us consider a slightly non-ideal gas at absolute zero. In such a gas, almost all the particles are in the condensate state. In terms of ψ operators, this means that the "above-condensate" part of the operator ($\hat{\Psi}'$) is small in comparison with its mean value, i.e. in comparison with the condensate wave function Ξ. If we neglect this small part completely, Ξ will satisfy the same "Schrödinger equation" (7.8) as the complete operator $\hat{\Psi}$. If only pair interactions are taken into account, this equation is (for spinless particles)

$$i\hbar \frac{\partial}{\partial t} \Xi(t, \mathbf{r}) = -\left(\frac{\hbar^2}{2m} \triangle + \mu\right) \Xi(t, \mathbf{r}) + \Xi(t, \mathbf{r}) \int |\Xi(t, \mathbf{r}')|^2 U(\mathbf{r}-\mathbf{r}')\, d^3x'. \tag{30.1}$$

Regarding the function $\Xi(t, \mathbf{r}')$ as varying only slightly over atomic distances, we can replace it by $\Xi(t, \mathbf{r})$ and take it outside the integral, which then becomes $\int U(r) d^3x \equiv U_0$. Substituting also $\mu = nU_0$ (see (25.6); n is the unperturbed value of the particle number density in the gas), we get

$$i\hbar\, \partial\Xi/\partial t = -(\hbar^2/2m) \triangle\Xi + U_0\{\Xi\,|\Xi|^2 - n\Xi\}. \tag{30.2}$$

In a stationary state, Ξ is independent of the time. A straight vortex filament corresponds to a solution having the form

$$\Xi = \sqrt{n} e^{i\phi} f(r/r_0), \quad r_0 = \hbar/\sqrt{(2mU_0 n)}, \tag{30.3}$$

where r and ϕ are the distance from the axis of the filament and the polar angle round the axis. The phase of this function corresponds to the value (29.7) of the circulation. The squared modulus $|\Xi|^2$ is the particle number density in the condensate; in the approximation considered, it is the same as the total density of the gas. When $r \to \infty$, this density must tend to the fixed value n, and therefore f must tend to unity.

With the dimensionless variable $\xi = r/r_0$, we find for the function $f(\xi)$ the equation

$$\frac{1}{\xi}\frac{d}{d\xi}\left(\xi \frac{df}{d\xi}\right) - \frac{f}{\xi^2} + f - f^3 = 0. \tag{30.4}$$

Figure 4 shows the solution obtained by numerical integration of (30.4). When $\xi \to 0$ it tends to zero as ξ; when $\xi \to \infty$ it tends to unity as $1 - 1/2\xi^2$.

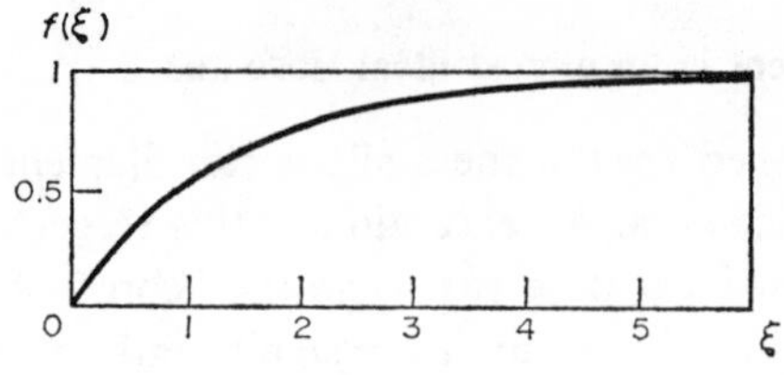

FIG. 4.

The parameter r_0 determines the order of magnitude of the filament "core" radius. Using the scattering length a instead of U_0, with $U_0 = 4\pi\hbar^2 a/m$ (6.2), we find

$$r_0 \sim n^{-1/3}\eta^{-1/2} \gg n^{-1/3},$$

where $\eta = an^{1/3}$ is the gaseousness parameter. This radius is therefore in fact large in comparison with interatomic distances if the gaseousness parameter is sufficiently small.

PROBLEM

Find the spectrum of elementary excitations in an almost ideal Bose gas, regarding it as the dispersion relation for small oscillations of the condensate wave function.

SOLUTION. We consider small oscillations of Ξ about a constant mean value $\sqrt{n}$:

$$\Xi = \sqrt{n} + Ae^{i(\mathbf{k}\cdot\mathbf{r}-\omega t)} + B^* e^{-i(\mathbf{k}\cdot\mathbf{r}-\omega t)},$$

where A and B^* are small complex amplitudes. Substituting this expression in equation (30.2), linearizing, and separating terms with different exponential factors, we find a set of two equations:

$$\hbar\omega A = (p^2/2m)\, A + nU_0(A+B),$$
$$-\hbar\omega B = (p^2/2m)\, B + nU_0(A+B),$$

with $p = \hbar k$. Hence, equating the determinant to zero, we have

$$(\hbar\omega)^2 = (p^2/2m)^2 + (p^2/m)\, nU_0,$$

in agreement with (25.10).

§ 31. Green's functions in a Bose liquid†

The mathematical formalism of Green's functions in a Bose liquid is very similar in its structure to that for a Fermi system. Without reiterating all the arguments, we shall give here first of all the basic definitions and formulae, stressing the differences due either to the different particle statistics or to the presence of the condensate.‡ As in the preceding sections of this chapter, the particles in the liquid are assumed spinless.

† In §§31–33 and 35, the units used have $\hbar = 1$.

‡ The mathematical technique of Green's functions was first applied to Bose systems with condensate by S. T. Belyaev (1958).

In determining the Green's function for a Bose liquid, we must separate the condensate part of the Heisenberg ψ operators by putting them in the form (26.4). The Green's function is found from the above-condensate part of the operators according to the formula

$$G(X_1, X_2) = -i\langle \mathrm{T}\hat{\Psi}'(X_1)\hat{\Psi}'^{+}(X_2)\rangle, \tag{31.1}$$

where the angle brackets $\langle \ldots \rangle$ again denote averaging with respect to the ground state of the system, and T denotes the chronological product. However, in contrast to the fermion case, the interchange of ψ operators to put them in the necessary order need not be accompanied by a change of sign of the product, so that, unlike (7.10),

$$iG(X_1, X_2) = \begin{cases} \langle \hat{\Psi}'(X_1)\hat{\Psi}'^{+}(X_2)\rangle, & t_1 > t_2, \\ \langle \hat{\Psi}'^{+}(X_2)\hat{\Psi}'(X_1)\rangle, & t_1 < t_2. \end{cases} \tag{31.2}$$

A similar mean value to (31.1) but with the complete ψ operators instead of the above-condensate ones would give

$$-i\langle \mathrm{T}\hat{\Psi}(X_1)\hat{\Psi}^{+}(X_2)\rangle = -in_0 + G(X_1, X_2), \tag{31.3}$$

where n_0 is the particle number density in the condensate.† In a homogeneous liquid, the function G depends, of course, only on the difference $X = X_1 - X_2$.

The above-condensate density matrix ϱ' is expressed in terms of the Green's function by

$$N\varrho'(\mathbf{r}_1, \mathbf{r}_2) = iG(t_1, \mathbf{r}_1; t_1+0, \mathbf{r}_2) = iG(t = -0, \mathbf{r}); \tag{31.4}$$

it will be noted that the sign is opposite to that of (7.18). In particular, for $\mathbf{r}_1 = \mathbf{r}_2$ we get from this the total above-condensate particle number density,

$$\frac{N}{V} - n_0 = iG(t = -0, \mathbf{r} = 0); \tag{31.5}$$

cf. (7.19).

The change to the momentum representation is made by the same formulae (7.21), (7.22). The normalization of $G(\omega, \mathbf{p})$ is expressed by

$$\frac{N}{V} = n_0 + i \lim_{t \to -0} \int G(\omega, \mathbf{p})\, e^{-i\omega t} \frac{d\omega\, d^3p}{(2\pi)^4}; \tag{31.6}$$

cf. (7.24).

† As with Fermi systems, we shall consider states of a Bose system for a given value of the chemical potential μ (rather than of the number N). Accordingly, the difference $\hat{H}' = \hat{H} - \mu\hat{N}$ (7.1) acts as the Hamiltonian of the system. The condensate part of the ψ operator is then independent of time.

For the Green's function of a Bose system in the momentum representation we can derive an expansion similar to the one obtained in §8 for a Fermi system. Exactly analogous calculations lead first to the formula

$$G(\omega, \mathbf{p}) = (2\pi)^3 \sum_m \left\{ \frac{A_m \delta(\mathbf{p}-\mathbf{P}_m)}{\omega + E_0(N) - E_m(N+1) + \mu + i0} - \frac{B_m \delta(\mathbf{p}+\mathbf{P}_m)}{\omega - E_0(N) + E_m(N-1) + \mu - i0} \right\}, \tag{31.7}$$

where

$$A_m = |\langle 0 | \hat{\psi}'(0) | m \rangle|^2, \quad B_m = |\langle m | \hat{\psi}'(0) | 0 \rangle|^2,$$

$\hat{\psi}'(\mathbf{r})$ being the Schrödinger above-condensate operator.† To bring this expansion to the final form, we note that the excitation energies $\varepsilon_m(N)$ in a Bose system are determined as the (always positive) differences between the energies of the excited states of the system and the energy of its ground state for a constant particle number N. Since $E_0(N) + \mu \approx E_0(N+1)$, we therefore find that

$$E_m(N+1) - E_0(N) - \mu \approx E_m(N+1) - E_0(N+1) = \varepsilon_m(N+1) > 0,$$
$$E_m(N-1) - E_0(N) + \mu \approx E_m(N-1) - E_0(N-1) = \varepsilon_m(N-1) > 0.$$

But the addition or removal of one particle changes the properties of the system only in the terms of relative order $\sim 1/N$; for a macroscopic system these terms are negligible, and so the excitation energies $\varepsilon_m(N \pm 1)$ are to be regarded as coinciding with each other and with $\varepsilon_m(N)$. Thus we have finally

$$G(\omega, \mathbf{p}) = (2\pi)^3 \sum_m \left\{ \frac{A_m \delta_m(\mathbf{p}-\mathbf{P}_m)}{\omega - \varepsilon_m + i0} - \frac{B_m \delta_m(\mathbf{p}+\mathbf{P}_m)}{\omega + \varepsilon_m - i0} \right\}. \tag{31.8}$$

By the same method as in deriving (8.14), we easily find from this that for Bose systems the imaginary part of the Green's function is always negative:

$$\operatorname{im} G(\omega, \mathbf{p}) < 0. \tag{31.9}$$

The asymptotic form of the Green's function for $\omega \to \infty$ remains the same as for Fermi systems:

$$G(\omega, \mathbf{p}) \to 1/\omega \quad \text{as} \quad |\omega| \to \infty; \tag{31.10}$$

cf. (8.15). In deriving this, we must use the commutation rule

$$\hat{\Psi}(t, \mathbf{r}_1) \hat{\Psi}^+(t, \mathbf{r}_2) - \hat{\Psi}^+(t, \mathbf{r}_2) \hat{\Psi}(t, \mathbf{r}_1) = \delta(\mathbf{r}_1 - \mathbf{r}_2),$$

in which the commutator of the operators $\hat{\Psi}$ and $\hat{\Psi}^+$ now replaces the anticommutator.‡

† Formula (31.7) corresponds to (8.7). The factor $\frac{1}{2}$ is absent here, because the particles are spinless. It should be noted that the sign of the second term in (31.7) is the opposite of that in (8.7).

‡ The fact that the condensate part of the ψ operators is separated in the definition of G is here unimportant: the constant term $-in_0$ in (31.3) corresponds in the momentum representation to the delta function $\delta(\omega)\,\delta(\mathbf{p})$, which does not affect (31.10).

Next, arguments similar to those in §8 lead to the fundamental result that the poles of the Green's function determine the spectrum of elementary excitations

$$G^{-1}(\varepsilon, \mathbf{p}) = 0, \tag{31.11}$$

and only the positive roots of this equation are to be taken; the subtraction of μ from ε is here unnecessary, in contrast to (8.16).

Near its pole, the Green's function has the form

$$G(\omega, \mathbf{p}) \approx Z_{\pm}/[\omega \mp \varepsilon(\mathbf{p})], \qquad Z_+ > 0, \quad Z_- < 0; \tag{31.12}$$

the sign of the residue at the pole is the same as that of ω, as follows from the fact that the coefficients A_m and B_m in (31.8) are positive. The magnitude of the residue is subject to no conditions such as (10.4) for example, for Fermi systems. Using the expression (31.12), we can easily verify (as in §8) that the inequality (31.9) automatically makes the quasi-particle damping coefficients positive, i.e. gives the necessary sign $-\text{im}\, \varepsilon > 0$, when the values of ε move into the complex domain.

The possible passage of above-condensate particles into the condensate and back has the result that, in the mathematical formalism of Green's functions for Bose systems, as well as the function (31.1), the following functions automatically appear, as we shall see in §33:

$$iF(X_1, X_2) = \langle N-2 | \mathrm{T} \hat{\Psi}'(X_1) \hat{\Psi}'(X_2) | N \rangle, \tag{31.13}$$

$$\begin{aligned} iF^+(X_1, X_2) &= \langle N | \mathrm{T} \hat{\Psi}'^+(X_1) \hat{\Psi}'^+(X_2) | N-2 \rangle \\ &= \langle N+2 | \mathrm{T} \hat{\Psi}'^+(X_1) \hat{\Psi}'^+(X_2) | N \rangle, \end{aligned} \tag{31.14}$$

where the matrix element is taken for transitions with change in the total number of particles in the system, and $|N\rangle$ denotes the ground state of the system with N particles; the last equation in (31.14) is valid to within quantities $\sim 1/N$ (cf. the fourth footnote to §26). The functions F and F^+ thus defined are called *anomalous Green's functions*. We shall show that in a homogeneous liquid at rest they are equal.

Like the function G, the functions F and F^+ for a homogeneous liquid depend only on the difference $X = X_1 - X_2$.† Since interchanging X_1 and X_2 changes only the order of the operators in the product, which is in any case governed by the chronological operator, we have

$$F(X) = F(-X). \tag{31.15}$$

† The fact that the function F is independent of the sum of times $t_1 + t_2$ arises because the term $-\mu\hat{N}$ is included in the definition of the Hamiltonian $\hat{H}' = \hat{H} - \mu\hat{N}$. This excludes from the difference of energy eigenvalues of systems with different numbers of particles the term

$$E(N+2) - E(N) \approx 2\partial E/\partial N = 2\mu,$$

and correspondingly excludes the factor $\exp[-i\mu(t_1+t_2)]$ from the matrix elements of the operator $\hat{\Psi}_1'\hat{\Psi}_2'$.

Hence it follows, of course, that in the momentum representation also F is an even function of its argument:

$$F(P) = F(-P). \tag{31.16}$$

Next, a relation between F and F^+ results from the following property of the Heisenberg ψ operator of a liquid at rest:†

$$\hat{\Psi}^+(t, \mathbf{r}) = \tilde{\hat{\Psi}}(-t, -\mathbf{r}). \tag{31.17}$$

Taking, say, $t_2 > t_1$, we thus have

$$\begin{aligned} iF^+(X_1, X_2) &\doteq \langle N+2 | \hat{\Psi}'^+(X_2) \hat{\Psi}'^+(X_1) | N \rangle \\ &= \langle N | \tilde{\hat{\Psi}}'^+(X_1) \tilde{\hat{\Psi}}'^+(X_2) | N+2 \rangle \\ &= \langle N | \hat{\Psi}'(-X_1) \hat{\Psi}'(-X_2) | N+2 \rangle \\ &= iF(-X_1, -X_2), \end{aligned}$$

or $F^+(X) = F(-X)$. Using (31.15), we then obtain the required relation

$$F^+(X) = F(X). \tag{31.18}$$

Expressing $F(X)$ in terms of the matrix elements of the ψ operators, we can derive for $F(\omega, \mathbf{p})$ an expansion similar to (31.8), and thus determine the poles of the function, but we shall not pause to do so here, merely mentioning that the poles of $F(\omega, \mathbf{p})$ coincide with those of $G(\omega, \mathbf{p})$.

To conclude this section, let us calculate the Green's function $G^{(0)}$ of an ideal Bose gas. First of all, since in the ground state of such a gas all particles are in the condensate, the above-condensate particle annihilation operator $\hat{\Psi}'$ acting on the wave function of the ground state gives zero. Hence the function $G^{(0)}(t, \mathbf{r})$ is non-zero only for $t = t_1 - t_2 > 0$ (when, according to (31.2), the creation operator $\hat{\Psi}'^+$ acts first).

† This property may be proved as follows. All non-zero matrix elements of the operators $\hat{a}_\mathbf{p}$ and $\hat{a}_\mathbf{p}^+$ can be defined as real quantities; see *QM*, (64.7), (64.8). In this sense the operators are real, i.e. $\hat{a}_\mathbf{p}^+ \equiv \tilde{\hat{a}}_\mathbf{p}^* = \tilde{\hat{a}}_\mathbf{p}$. ($\tilde{\hat{a}}$ denotes the transposed operator; cf. *QM*, §3.) The Schrödinger ψ operator

$$\hat{\psi}(\mathbf{r}) = V^{-1/2} \sum_\mathbf{p} \hat{a}_\mathbf{p} e^{i\mathbf{p}\cdot\mathbf{r}}$$

therefore has the property $\hat{\psi}^+(\mathbf{r}) = \tilde{\hat{\psi}}(-\mathbf{r})$. Hence in turn we have the equation (31.17) for the Heisenberg operator

$$\hat{\Psi}(t, \mathbf{r}) = \exp(i\hat{H}t)\, \hat{\psi}(\mathbf{r}) \exp(-i\hat{H}t),$$

as is easily seen by noting that (for a system without spin interactions) the Hamiltonian $\hat{H}$ is real (so that $\hat{H}^+ = \tilde{\hat{H}}$) and unchanged by inversion. We must emphasize, however, that if the Hamiltonian is real, there can be no macroscopic superfluid motion in the liquid. For a Bose system with condensate, the Hamiltonian depends on a macroscopic parameter, the condensate wave function Ξ. In a moving liquid, this parameter is complex, and therefore the Hamiltonian also is complex (but, of course, Hermitian).

Although the chemical potential $\mu = 0$ for an ideal gas, we shall not assert this here, regarding μ as a free parameter not specified beforehand; such a procedure is necessary with a view to the subsequent application of $G^{(0)}$ in the diagram technique for an arbitrary liquid, where μ acts as such a parameter. Accordingly, the operator $\hat{\Psi}^{(0)\prime}(t, \mathbf{r})$ is written as

$$\hat{\Psi}^{(0)\prime}(t, \mathbf{r}) = \frac{1}{\sqrt{V}} \sum_{\mathbf{p} \neq 0} \hat{a}_{\mathbf{p}} \exp\left[i\left(\mathbf{p}\cdot\mathbf{r} - \frac{p^2}{2m}t + \mu t\right)\right], \tag{31.19}$$

differing from (26.1) by the term $i\mu t$ in the exponents. When this expression is substituted in the definition of $G^{(0)}$, in accordance with (31.2), we note that on averaging (i.e. taking the diagonal matrix element) we can obtain a non-zero result only from the products $\hat{a}_{\mathbf{p}}\hat{a}_{\mathbf{p}}^{+}$ and $\hat{a}_{\mathbf{p}}^{+}\hat{a}_{\mathbf{p}}$. But, since in the ground state of the gas the occupation numbers of all particle states with $\mathbf{p} \neq 0$ are zero, we have

$$\langle \hat{a}_{\mathbf{p}}^{+}\hat{a}_{\mathbf{p}} \rangle = 0, \quad \langle \hat{a}_{\mathbf{p}}\hat{a}_{\mathbf{p}}^{+} \rangle = 1.$$

Now changing in the usual way from summation over $\mathbf{p}$ to integration, we get

$$G^{(0)}(t, \mathbf{r}) = \begin{cases} -i\int \exp\left[-i\frac{p^2}{2m} + i\mu t + i\mathbf{p}\cdot\mathbf{r}\right]\frac{d^3p}{(2\pi)^3} & \text{for} \quad t > 0, \\ 0 & \text{for} \quad t < 0. \end{cases} \tag{31.20}$$

Hence the Green's function in the momentum representation is

$$G^{(0)}(\omega, \mathbf{p}) = -i\int_0^\infty \exp\left(-i\frac{p^2}{2m}t + i\mu t + i\omega t\right) dt.$$

The integration is effected by means of the formula

$$\int_0^\infty e^{i\alpha t}\, dt = \frac{i}{\alpha + i0}, \tag{31.21}$$

derived by including in the integrand a factor $e^{-\lambda t}$ with $\lambda > 0$ and then taking the limit as $\lambda \to 0$. Finally we have

$$G^{(0)}(\omega, \mathbf{p}) = \left[\omega - \frac{p^2}{2m} + \mu + i0\right]^{-1}. \tag{31.22}$$

For an ideal gas the function $F^{(0)}(X) = 0$, as is evident from the definition (31.13), in which both operators annihilate above-condensate particles. In the momentum representation also, therefore,

$$F^{(0)}(\omega, \mathbf{p}) = 0. \tag{31.23}$$

This equation expresses the fact that particles appear above the condensate (at $T = 0$) only as a result of interaction.

PROBLEM

Find the Green's function of a phonon field, defined as

$$D(X_1, X_2) \equiv D(X_1 - X_2) = -i\langle \mathrm{T}\, \hat{\varrho}'(X_1)\, \hat{\varrho}'(X_2)\rangle, \qquad (1)$$

where the angle brackets denote averaging with respect to the ground state of the field, and $\hat{\varrho}'$ is the density operator from (24.10); the chronological product is expanded by the rule (31.2).

SOLUTION. When substituting (24.10) in the definition (1) we note that, since in the ground state all the phonon state occupation numbers are zero, only the mean values $\langle \hat{c}_{\mathbf{k}} \hat{c}_{\mathbf{k}}^{+}\rangle = 1$ are other than zero. Then, changing from summation over **k** to integration, we obtain

$$D(t, \mathbf{r}) = \int \frac{\varrho k}{2iu}\, e^{i(\mathbf{k}\cdot\mathbf{r} \mp ukt)} \frac{d^3k}{(2\pi)^3}$$

where the minus and plus signs in the exponent refer to $t > 0$ and $t < 0$ respectively; in the integral for $t < 0$, we have renamed the variable of integration, $\mathbf{k} \to -\mathbf{k}$. The integrand (without the factor $e^{i\mathbf{k}\cdot\mathbf{r}}$) is the Fourier component of the function $D(t, \mathbf{r})$ with respect to the coordinates. Expanding with respect to time also, we find the Green's function in the momentum representation.

$$D(\omega, \mathbf{k}) = \frac{\varrho k}{2iu} \left\{ \int_0^{\infty} e^{i(\omega - uk)t}\, dt + \int_{-\infty}^{0} e^{i(\omega + uk)t}\, dt \right\}.$$

The integration is carried out by means of (31.21);

$$D(\omega, \mathbf{k}) = \frac{\varrho k}{2u} \left[\frac{1}{\omega - uk + i0} - \frac{1}{\omega + uk - i0} \right] = \frac{\varrho k^2}{\omega^2 - u^2k^2 + i0}.$$

§ 32. The diagram technique for a Bose liquid

The diagram technique for the calculation of Green's functions in a Bose system can be set up similarly to those for Fermi systems in §§12 and 13. We shall again formulate the rules of this technique for systems with a pair interaction between particles, described by the operator

$$\hat{V}(t) = \tfrac{1}{2} \int \hat{\Psi}^{+}(t, \mathbf{r}_1)\, \hat{\Psi}^{+}(t, \mathbf{r}_2)\, U(\mathbf{r}_1 - \mathbf{r}_2)\, \hat{\Psi}(t, \mathbf{r}_2)\, \hat{\Psi}(t, \mathbf{r}_1)\, d^3x_1\, d^3x_2. \qquad (32.1)$$

The chief distinctive feature of Bose liquids with condensate is that all the Heisenberg ψ operators must be put in the form $\hat{\Psi} = \hat{\Psi}' + \Xi$, where $\hat{\Psi}'$ is the above-condensate part and Ξ the condensate wave function, which for a liquid at rest is simply the real number $\sqrt{n_0}$.† After this substitution, the operator (32.1) separates into a series of terms containing from four to no operators $\hat{\Psi}'$ (together with the corresponding additional number of factors $\sqrt{n_0}$).

The whole discussion in §12 concerning the change to the interaction representation remains valid, and the subsequent expansion of the expressions obtained is carried out by means of Wick's theorem, except that the inter-

† We must emphasize that, since this quantity arises from the separation into parts of the exact (Heisenberg) ψ operator, n_0 is the exact value of the condensate density in the liquid (at $T = 0$).

change of ψ operators in the product being averaged does not now involve a change of sign. The difference in the form of the terms into which the operator (32.1) separates leads, however, to new elements in the Feynman diagrams. These will be described in their final form in the momentum representation.

At each vertex of the diagram we again have three lines meeting: a broken line associated with the factor $-iU(Q)$, with 4-momentum $Q = (q_0, \mathbf{q})$, and two particle lines, one ingoing and one outgoing. Here we must distinguish condensate and above-condensate particles. The continuous lines will now correspond to above-condensate particles, and such a line (with 4-momentum $P = (\omega, \mathbf{p})$) is again associated with a factor $iG^{(0)}(P)$. The lines of condensate particles will be drawn as wavy lines; these have an assigned 4-momentum $P = 0$ and an associated factor $\sqrt{n_0}$.† Thus four kinds of vertices arise:

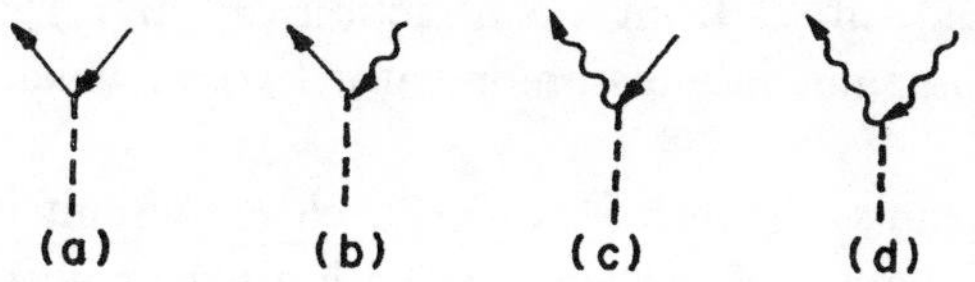

(32.2)

Vertices with one or two wavy lines are said to be *incomplete*. At each vertex there must be "conservation of 4-momentum"; in vertices (b) and (c), therefore, the 4-momentum of the broken line is equal to that of the continuous line, and in vertex (d) it is zero. The wavy lines are always external lines of the diagram, i.e. are joined to it at only one end, the other end remaining free.

Each diagram that occurs in the definition of the Green's function $G(P)$ has two continuous external lines with 4-momenta P (ingoing and outgoing), and may also have some (even) number of external wavy lines; the total numbers of ingoing and outgoing external lines are equal in every diagram (this expresses the conservation of the total number of particles, condensate and above-condensate, in the system). As for a Fermi system, and for the same reason (see §13), only those diagrams are admissible which do not separate into two (or more) disconnected parts. Unlike the case of Fermi systems, however, the diagrams in iG all have the same sign, i.e. rule 3 in §13 is eliminated.

Each broken line in the diagram has a complete or incomplete vertex at its two ends. These, however, cannot be two vertices of the type (32.2d): having no continuous end, such a figure cannot be attached to a Green's function diagram. They also cannot be vertices of the types (32.2d and c) or (32.2d and b): when there are three wavy external lines, the conservation of 4-momentum at the vertices would mean that the 4-momentum of the fourth external line would also be zero in such a figure, and we should have a figure with four condensate (wavy) external lines.

† More precisely, a factor Ξ is to be associated with a wavy line coming to a vertex, and a factor Ξ^* with one leaving; since Ξ is real, these factors are actually the same.

A considerable number of diagrams in each order of perturbation theory, constructed by the above rules, are identically zero, however. This is due to the absence of above-condensate particles in the ground state of an ideal Bose gas, as is particularly clear if we trace back to the origin of the diagrams in the coordinate representation: all contractions of the form $\langle \hat{\Psi}'^{+}\hat{\Psi}' \rangle$, in which the above-condensate particle annihilation operator is to the right and acts first on the ground state, are zero; this leaves only contractions of the form $\langle \hat{\Psi}'\hat{\Psi}'^{+} \rangle$.†

Diagrams with a closed continuous line are zero: such a line arises from a contraction $\langle \hat{\Psi}'^{+}(t, \mathbf{r})\, \hat{\Psi}'(t, \mathbf{r}) \rangle$, which is the above-condensate particle density. Diagrams containing a continuous line closed by a broken line

are zero: such a line arises from a contraction $\langle \hat{\Psi}'^{+}(t, \mathbf{r}_2)\, \hat{\Psi}'(t, \mathbf{r}_1) \rangle$ of two ψ operators within the same interaction operator $\hat{V}(t)$, in which $\hat{\Psi}'^{+}$ is to the left of $\hat{\Psi}'$.

Lastly, all diagrams are zero in which a closed circuit is formed by any sequence of continuous and broken lines with all the continuous lines in the same direction. Such a circuit can be represented as follows, with the time arguments of the ψ operators shown at the end-points of the lines:

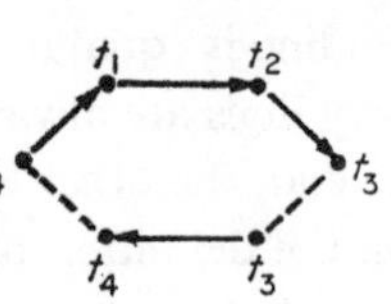

The arguments at the ends of each broken line are the same.‡ Those of the functions $G^{(0)}$ corresponding to the continuous lines are equal to the differences $t_2 - t_1$, $t_3 - t_2$, $t_4 - t_3$, $t_1 - t_4$; for each closed circuit their sum is zero, so that at least one of them is negative and the corresponding function $G^{(0)}$ is zero.

The above rules relate also to the diagrams which determine the anomalous Green's function, the only difference being that both the continuous external lines must be outgoing (for F) or both ingoing (for F^{+}). Accordingly, in these diagrams the numbers of ingoing and outgoing wavy lines are no longer equal, but the total number of outgoing lines remains equal to the total number of ingoing ones. The 4-momentum P is assigned to one of the continuous external lines, and $-P$ to the other, where P is the argument of the required function $F(P)$ or $F^{+}(P)$;§ the sum of the 4-momenta of these two lines must be zero, by the "law of conservation of 4-momentum" applied to the whole diagram.

† For a similar reason, some diagrams were zero for two-particle scattering in vacuum; cf. §16.

‡ In the space–time representation of the diagrams, a factor $iU(X_1 - X_2)$, which contains the delta function $\delta(t_1 - t_2)$, corresponds to a broken line between points 1 and 2.

§ Since F is an even function of its argument, the choice of sign for P is here unimportant.

The Green's functions calculated by the diagram technique contain two parameters: the chemical potential μ and the condensate density n_0; these parameters have also to be related to the liquid density $n = N/V$. One relation between these three quantities is given by formula (31.6), which follows immediately from the definition of the Green's function. As a second relation, we use the equation (33.11) derived below, which expresses μ explicitly in terms of the concepts of the diagram technique.

§ 33. Self-energy functions

Let us examine more closely the structure of the diagrams for Green's functions, using the concept of the *self-energy function* in the same way as was done in §14 for Fermi systems: by considering the set of all diagrams (with two continuous external lines) that cannot be cut into two parts by dividing just one continuous line. In contrast to §14, however, there are now various possibilities as regards the direction of the external lines in the diagrams: as well as diagrams with one ingoing and one outgoing line, there are those with two ingoing, or two outgoing, lines. Accordingly, there are self-energy parts of three kinds:

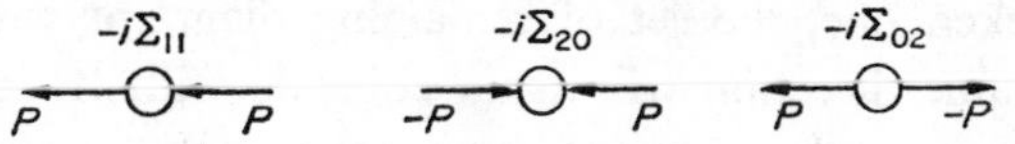

(33.1)

(in this notation, the two suffixes to Σ denote respectively the numbers of ingoing and outgoing continuous external lines). As well as the continuous external lines, the self-energy diagrams in general also have wavy (condensate) free ends. These are included in the definition of the self-energy function, which is represented here by a circle. We shall see later that the functions $\Sigma_{02}(P)$ and $\Sigma_{20}(P)$ are in fact the same:

$$\Sigma_{02}(P) = \Sigma_{20}(P). \tag{33.2}$$

We may also note at this point that, since P and $-P$ occur symmetrically in the definition of these functions, they are even functions:

$$\Sigma_{02}(P) = \Sigma_{02}(-P). \tag{33.3}$$

As an illustration, the following are all the non-zero diagrams of the functions Σ_{11} and Σ_{02} in the first two orders of perturbation theory:

(33.4)

(33.5)

Let us now establish the equations giving the exact functions G and F in terms of the self-energy functions.

In terms of perturbation theory, the difference $G(P)-G^{(0)}(P)$ is expressed by a sum of an infinite number of chain diagrams

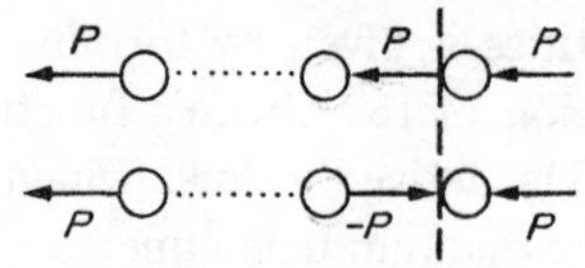

consisting of various numbers of circles joined in all possible ways by forward and backward (relative to the two outermost) arrows. Similarly, the exact function F (the function $F^{(0)} \equiv 0$) is represented by a sum of chain diagrams in which the two outermost arrows have opposite directions:

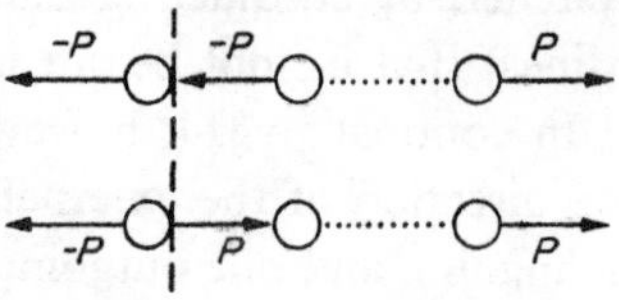

If the end member (circle and arrow) is detached from each chain, as shown by the vertical broken line, the set of remaining diagrams with the outermost arrows in the same direction will again coincide with the exact function G, and the set of those with the outermost arrows in the opposite direction will coincide with the exact F.

We shall introduce the graphical notation for these functions, of thick arrows in one or both directions

 (33.6)

Then the foregoing assertions can be written as graphical equations consisting of skeleton diagrams:

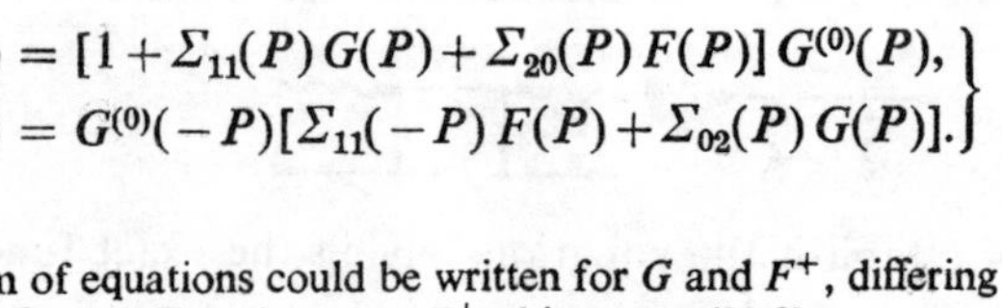 (33.7)

Cf. the analogous equation (14.4). In analytical form, these equations give†

$$\left.\begin{aligned} G(P) &= [1+\Sigma_{11}(P)\,G(P)+\Sigma_{20}(P)\,F(P)]\,G^{(0)}(P), \\ F(P) &= G^{(0)}(-P)[\Sigma_{11}(-P)\,F(P)+\Sigma_{02}(P)\,G(P)]. \end{aligned}\right\} \tag{33.8}$$

† A similar system of equations could be written for G and F^+, differing from (33.8) only by the interchange of Σ_{02} and Σ_{20}. Since $F = F^+$, this proves (33.2).

Solving these equations for G and F and substituting (31.22) for $G^{(0)}$, we obtain the required formulae

$$G(P) = \frac{1}{D}\left[\omega+\frac{p^2}{2m}-\mu+\Sigma_{11}(-P)\right], \quad F(P) = -\frac{1}{D}\Sigma_{02}(P), \quad (33.9)$$

where

$$D = [\Sigma_{02}(P)]^2-\left[\Sigma_{11}(P)-\omega-i0+\frac{p^2}{2m}-\mu\right]\left[\Sigma_{11}(-P)+\omega-i0+\frac{p^2}{2m}-\mu\right]. \quad (33.10)$$

It must be emphasized that these relations do not depend on the internal structure of the self-energy functions, and therefore are not connected with the assumption of pair interactions between particles; they are thus valid for any Bose liquid.

The energy of elementary excitations in the liquid, as a function of the momentum **p**, is determined by the poles of G and F as functions of the variable ω. For small **p**, these excitations are phonons, and their energy tends to zero with **p**. Hence the function (33.10) must vanish when $\mathbf{p} = 0$ and $\omega = 0$. From this we find the equation

$$[\Sigma_{11}(0)-\mu]^2 = \Sigma_{02}^2(0).$$

As an equation for μ, it has two roots, of which we must choose

$$\mu = \Sigma_{11}(0)-\Sigma_{02}(0). \quad (33.11)$$

For, in the long-wave limit, the ψ operator is given by (27.2), and its above-condensate part $\hat{\Psi}' = \hat{\Psi}-\sqrt{n_0} \approx i\sqrt{n_0}\hat{\Phi}$, so that $\hat{\Psi}'^+ = -\hat{\Psi}'$ and $F \approx -G$; the latter equation is satisfied with the choice (33.11), when the numerators in (33.9) (in the limit $P \to 0$) differ only in sign. The equation (33.11) is the second relation (see the end of §32), which, together with (31.6), enables us to express the parameters μ and n_0 in terms of the density n of the liquid.

The subsequent expansion of (33.10) in series in ω and **p** determines the form of the Green's function for small values of the arguments. Here we must take into account that the scalar functions Σ_{11} and Σ_{02} are expanded in powers of $\mathbf{p}^2$, and the expansion of Σ_{02}, an even function of all its arguments, contains only even powers of ω also. Putting (33.10) in the form

$$D = \left\{\omega-\frac{1}{2}[\Sigma_{11}(P)-\Sigma_{11}(-P)]\right\}^2 - \left\{\frac{p^2}{2m}-\mu+\frac{1}{2}[\Sigma_{11}(P)+\Sigma_{11}(-P)]\right\}^2+\Sigma_{02}^2(P),$$

we can immediately conclude that the first non-vanishing terms in the expansion have the form $D = \text{constant}\times(\omega^2-u^2p^2+i0)$, where u is a constant, which is clearly the velocity of sound in the liquid. Noting also that, from (33.11), the numerators in (33.9) for $\omega \to 0$ and $\mathbf{p} \to 0$ differ only in sign, we find

$$G = -F = \frac{\text{constant}}{\omega^2 - u^2p^2 + i0}.$$

The contour rule is determined by comparing with (31.8).

The value of the constant in the numerator can be determined by calculating from this Green's function the particle momentum distribution $N(\mathbf{p})$ (for small $\mathbf{p}$) and comparing it with the known distribution (27.7). The integral

$$N(\mathbf{p}) = i \lim_{t \to -0} \int_{-\infty}^{\infty} G(\omega, \mathbf{p})\, e^{-i\omega t} \frac{d\omega}{2\pi}$$

(cf. (7.23)) is calculated by closing the contour of integration by an infinite semicircle in the upper half-plane (cf. the note at the end of §7), and accordingly is determined by the residue at the pole $\omega = -up + i0$. The result is $N(\mathbf{p}) = \text{constant}/2up$, and by comparison with (27.7) the constant is found to be $n_0 mu^2/n$. Thus we have finally the following expression for the Green's functions with small ω and $\mathbf{p}$:

$$G = -F = n_0 mu^2/n(\omega^2 - u^2p^2 + i0). \tag{33.12}$$

This function coincides (apart from a normalization coefficient) with the Green's function of the phonon field (see §31, Problem)—an entirely reasonable result, since for small ω and $\mathbf{p}$ the elementary excitations in a Bose liquid are phonons.

Lastly, let us show the application of the above formulae to the model (§25) of an almost ideal Bose gas with pair interaction between particles. In first-order perturbation theory, Σ_{11} and Σ_{02} are determined by the first two diagrams (33.4) and the first diagram (33.5). Expanding these in analytical form, we find

$$\Sigma_{11} = n_0[U_0 + U(\mathbf{p})], \quad \Sigma_{02} = n_0 U(\mathbf{p}).$$

With the same accuracy, the condensate density n_0 in these formulae may be replaced by the total gas density n. As mentioned in §25, the gas particle momenta in this model may be regarded as small, and accordingly the Fourier components $U(\mathbf{p})$ may be replaced by their value U_0 at $\mathbf{p} = 0$. Then

$$\Sigma_{11} = 2nU_0, \quad \Sigma_{02} = nU_0. \tag{33.13}$$

Substitution of these expressions in (33.11) gives $\mu = nU_0$, in agreement with (25.6). Substitution in (33.9) and (33.10) leads to the following formulae for the Green's functions:

$$\left.\begin{aligned} G(\omega, \mathbf{p}) &= \frac{\omega + p^2/2m + nU_0}{\omega^2 - \varepsilon^2(p) + i0}, \\ F(\omega, \mathbf{p}) &= \frac{-nU_0}{\omega^2 - \varepsilon^2(p) + i0}, \end{aligned}\right\} \tag{33.14}$$

where

$$\varepsilon(p) = \left[\left(\frac{p^2}{2m}\right)^2 + \frac{p^2}{m} nU_0\right]^{1/2}.$$

From the form of the denominators in these functions, it is clear that $\varepsilon(p)$ is the energy of the elementary excitations, in agreement with the result (25.10), (25.11) obtained previously by a different method.

§ 34. Disintegration of quasi-particles

The finite lifetime (decay) of a quasi-particle in a quantum liquid may be due either to collisions with other quasi-particles or to spontaneous disintegration into two or more new quasi-particles. As the temperature $T \to 0$, the first cause of decay disappears, since the collision probability tends to zero with the quasi-particle number density, and the decay is then due only to the disintegration of quasi-particles.

Let us consider the disintegration of a quasi-particle (with momentum $\mathbf{p}$) into two. If $\mathbf{q}$ is the momentum of one of the resulting quasi-particles, that of the other is $\mathbf{p}-\mathbf{q}$, and the law of conservation of energy gives the condition

$$\varepsilon(p) = \varepsilon(q) + \varepsilon(|\mathbf{p}-\mathbf{q}|). \tag{34.1}$$

It can happen that in some range of values of p this equation is not satisfied for any $\mathbf{q}$; the quasi-particles in such a range do not decay at all (if, of course, disintegration into a larger number of quasi-particles is also impossible). As p varies, decay begins at the value $p = p_c$ (disintegration threshold) for which equation (34.1) first has solutions.

First of all, it should be noted that at the point $p = p_c$ the right-hand side of (34.1) has an extremum as a function of $\mathbf{q}$. For let the extremum value of the sum $\varepsilon(q)+\varepsilon(|\mathbf{p}-\mathbf{q}|)$ for a given p be $E(p)$; we shall take the particular case where this is a minimum. Then, in the equation

$$\varepsilon(p) - E(p) = \varepsilon(q) + \varepsilon(|\mathbf{p}-\mathbf{q}|) - E(p),$$

the right-hand side is non-negative. The equation therefore certainly has no roots for values of p such that $\varepsilon(p) - E(p) < 0$; a root appears only at the point $p = p_c$ for which $\varepsilon(p_c) = E(p_c)$.

Putting equation (34.1) in the symmetrical form

$$\varepsilon(p) = \varepsilon(q_1) + \varepsilon(q_2), \quad \mathbf{q}_1 + \mathbf{q}_2 = \mathbf{p},$$

we find that the condition for an extremum of its right-hand side may be written $\partial\varepsilon/\partial\mathbf{q}_1 = \partial\varepsilon/\partial\mathbf{q}_2$, or

$$\mathbf{v}_1 = \mathbf{v}_2, \tag{34.2}$$

i.e. the two quasi-particles formed at the threshold point have equal velocities. Here we may distinguish various cases (L. P. Pitaevskiĭ 1959).

(a) The quasi-particle velocity in the Bose liquid is zero for momentum $p = p_0$ corresponding to the roton minimum on the curve in Fig. 2 (§22). Hence, if $\mathbf{v}_1 = \mathbf{v}_2 = 0$, the quasi-particle disintegrates at the threshold into two rotons with momenta p_0 and energies Δ. Accordingly, the energy of the disintegrating quasi-particle is $\varepsilon(p_c) = 2\Delta$ and its momentum p_c is related to p_0 by the condition $\mathbf{p}_c = \mathbf{p}_{01}+\mathbf{p}_{02}$, i.e. $2p_0 \cos\theta = p_c$, where 2θ is the angle between the directions of the two rotons. Hence it follows that we must always have

$$p_c < 2p_0. \tag{34.3}$$

(b) If the velocities are $\mathbf{v}_1 = \mathbf{v}_2 \neq 0$, and the corresponding momenta $\mathbf{q}_1$ and $\mathbf{q}_2$ are finite, the disintegration at the threshold yields two quasi-particles with collinear (parallel or antiparallel) momenta.†

(c) If the velocities $\mathbf{v}_1$ and $\mathbf{v}_2$ are non-zero but one of the momenta ($\mathbf{q}_1$, say) tends to zero near the threshold, the corresponding quasi-particle is a phonon and the velocity $v_1 = u$. We then have a threshold beyond which the creation of a phonon by the quasi-particle becomes possible. At the threshold itself, the phonon energy is zero, and the quasi-particle velocity reaches that of sound (equal to $v_1 = v_2 = u$).

(d) There is one special case, in which there is a disintegration of one phonon into two, the threshold being at the start of the spectrum, $p = 0$. Such a disintegration is, however, possible only for one sign of the curvature of the initial (phonon) part of the spectrum: we must have $d^2\varepsilon(p)/dp^2 > 0$, i.e. the curve of $\varepsilon(p)$ must turn upwards from the initial tangent $\varepsilon = up$. This is easily seen by representing this part of the spectrum as

$$\varepsilon(p) \approx up+\alpha p^3, \tag{34.4}$$

which includes both the linear term and the next term in the expansion in powers of the small momentum.‡ The equation of conservation of energy (34.1) then gives

$$u(p-q-|\mathbf{p}-\mathbf{q}|) = -\alpha(p^3-q^3-|\mathbf{p}-\mathbf{q}|^3).$$

Near the threshold, the phonon is emitted at a small angle θ to the direction of the initial quasi-particle momentum $\mathbf{p}$; on the left-hand side we have

$$p-q-|\mathbf{p}-\mathbf{q}| \approx -\frac{pq}{p-q}(1-\cos\theta), \tag{34.5}$$

† Because of the isotropy of the liquid, the directions of the quasi-particle momentum $\mathbf{p}$ and velocity $\mathbf{v} = \partial\varepsilon/\partial\mathbf{p}$ are collinear, but they may be in either the same or opposite directions.

‡ The dispersion relation for acoustic vibrations gives the squared frequency ω^2 as a function of the wave vector. Accordingly, the squared phonon energy $\varepsilon^2(p)$ has a regular expansion in powers of the momentum $\mathbf{p}$; the expansion begins with a term in $\mathbf{p}^2$, and continues in powers of $\mathbf{p}^2$ because of the isotropy of the liquid. The expansion of $\varepsilon(p)$ itself therefore contains odd powers of p.

and on the right-hand side it is sufficient to put $|\mathbf{p}-\mathbf{q}| \approx p-q$. Then

$$1-\cos\theta = 3\alpha(p-q)^2. \tag{34.6}$$

Hence it follows that necessarily $\alpha > 0$.

We shall see later (§35) that in cases (a) and (b) the function $\varepsilon(p)$ cannot be continued beyond the threshold, which is thus the end of the spectrum. In cases (c) and (d) the disintegration of a quasi-particle with emission of a long-wavelength phonon causes a slight decay which may be determined by means of perturbation theory.†

Let us calculate the decay of a phonon due to its disintegration into two phonons (case (d)). The matrix elements for this process come from the third-order terms in the Hamiltonian, given by (24.12). For a transition from the initial (i) state with one phonon $\mathbf{p}$ to the final (f) state with phonons $\mathbf{q}_1$ and $\mathbf{q}_2$, the matrix element of the perturbation operator is

$$V_{fi} = \delta(\mathbf{p}-\mathbf{q}_1-\mathbf{q}_2)\frac{3!(2\pi\hbar)^3}{2(2V)^{3/2}}\left(\frac{u}{\varrho}pq_1q_2\right)^{1/2}\left\{1+\frac{\varrho^2}{3u^2}\frac{d}{d\varrho}\frac{u^2}{\varrho}\right\}; \tag{34.7}$$

the suffix 0 of the unperturbed density ϱ_0 is omitted. The factor $(pq_1q_2)^{1/2}$ should be noted; its smallness (in this disintegration of a long-wavelength phonon) ensures the applicability of perturbation theory.‡

The differential disintegration probability per unit time is given by

$$dw = \frac{2\pi}{\hbar}|V_{fi}|^2\,\delta(E_f-E_i)\frac{V^2\,d^3q_1\,d^3q_2}{(2\pi\hbar)^6};$$

see *QM*, (43.1). When (34.7) is substituted, a squared delta function appears, which is to be interpreted as§

$$[\delta(\mathbf{p}-\mathbf{q}_1-\mathbf{q}_2)]^2 = \frac{V}{(2\pi\hbar)^3}\delta(\mathbf{p}-\mathbf{q}_1-\mathbf{q}_2). \tag{34.8}$$

The remaining delta function is removed by integration over d^3q_2; putting also $E_i = up$, $E_f = u(q_1+q_2)$, we obtain

$$w = \frac{1}{2}\left\{1+\frac{\varrho^2}{3u^2}\frac{d}{d\varrho}\frac{u^2}{\varrho}\right\}^2\frac{9\pi}{4\hbar p}\int pq_1(p-q_1)\,\delta(p-q_1-|\mathbf{p}-\mathbf{q}_1|)\frac{d^3q_1}{(2\pi\hbar)^3};$$

† Which of these cases can actually occur in practice depends on the specific form of the quasi-particle spectrum $\varepsilon(p)$. Empirical results for liquid He^4 indicate the presence (at pressures below 15 atm) of a short initial section of the phonon spectrum in which is an instability of type (d). The spectrum in liquid helium terminates at a point of type (a).

‡ To calculate the matrix element (34.7), we must take into account that each of the phonon operators $\hat{c}_\mathbf{p}$ and $\hat{c}_\mathbf{p}^+$ can be taken from any of three factors $\hat{\varrho}'$ and $\hat{\mathbf{v}}$; this is the reason for the factor 3!. The delta function in (34.7) arises from the integration of the factor $\exp[i(\mathbf{p}-\mathbf{q}_1-\mathbf{q}_2)\cdot\mathbf{r}/\hbar]$. Lastly, we have used the fact that the directions of $\mathbf{p}$, $\mathbf{q}_1$ and $\mathbf{q}_2$ are almost the same.

§ The function $\delta(\mathbf{k})$ arises from the integral $\int e^{i\mathbf{k}\cdot\mathbf{r}}d^3x/(2\pi)^3$. If the other similar integral is calculated at $\mathbf{k} = 0$ (because one delta function is already present), and the integration is taken over a finite volume V, we get $V/(2\pi)^3$, as is expressed by formula (34.8).

with independent integration over d^3q_1 and d^3q_2, the result must be halved, because of the identity of the two phonons. Finally, expressing the argument of the delta function in the form (34.5) and carrying out the integration over $d^3q_1 = 2\pi q_1^2\, dq_1 d\cos\theta$ (in the range $q_1 \leqslant p$), we find the total disintegration probability

$$w = \frac{3p^5}{320\pi\varrho\hbar^4}\left\{1+\frac{\varrho^2}{3u^2}\frac{d}{d\varrho}\frac{u^2}{\varrho}\right\}^2. \tag{34.9}$$

The phonon decay coefficient $\gamma \equiv -\operatorname{im}\varepsilon = \frac{1}{2}\hbar w$. In particular, for an almost ideal gas, according to (25.11), $u^2/\varrho \approx 4\pi\hbar^2 a/m^3$ is independent of the density. In this case

$$\gamma = 3p^5/640\pi\hbar^3\varrho \tag{34.10}$$

(S. T. Belyaev 1958).

For phonon emission by a quasi-particle near a threshold of type (c), the form of the perturbation operator is established by considering the change in the quasi-particle energy in the sound wave. This change consists of two parts:

$$\delta\varepsilon(\mathbf{p}) = \frac{\partial\varepsilon}{\partial\varrho}\varrho' + \mathbf{v}\cdot\mathbf{p}.$$

The first term is due to the change in the density of the liquid, on which the quasi-particle energy depends as a parameter. The second term, in which $\mathbf{v}$ is the liquid velocity in the sound wave, is the change in the energy of the quasi-particle because of the macroscopic motion of the liquid; since the wavelength of the phonon emitted (near the threshold) is large compared with the wavelength of the quasi-particle, we may suppose that the latter is in a uniform flow of liquid, and the change in its energy is then determined as shown at the beginning of §23. The perturbation operator is found from $\delta\varepsilon$ by replacing $\mathbf{v} = \nabla\phi$ and ϱ' by the second-quantized operators (24.10), and $\mathbf{p}$ by the quasi-particle momentum operator $\hat{\mathbf{p}} = -i\hbar\nabla$:

$$\hat{V} = \frac{\partial\varepsilon}{\partial\varrho}\hat{\varrho}' + \frac{1}{2}(\hat{\mathbf{v}}\cdot\hat{\mathbf{p}}+\hat{\mathbf{p}}\cdot\hat{\mathbf{v}});$$

in the second term, the product has been symmetrized in order to bring it to Hermitian form. The phonon emission probability is then calculated as previously for phonon disintegration (see Problem).

PROBLEM

Determine the probability of phonon emission by a quasi-particle whose momentum p is close to the threshold value p_c at which the quasi-particle velocity reaches that of sound.

SOLUTION. The matrix element of the operator (34.11) is taken for the creation of one phonon (with momentum $\mathbf{q}$) and the simultaneous transition of the quasi-particle between states (plane waves) with momenta **p** and **p'**. Near the threshold, the phonon momentum $q \ll p_c$, and the direction of **q** is almost the same as that of **p**.† We then find

$$V_{fi} = -i(2\pi\hbar)^3\, \delta(\mathbf{p}-\mathbf{q}_1-\mathbf{q}_2)\, \frac{A}{V^{3/2}} \left(\frac{qu}{2\varrho}\right)^{1/2},$$

where

$$A = p_c + \left[\frac{\varrho}{u}\frac{\partial \varepsilon}{\partial \varrho}\right]_{p=p_c}.$$

Hence the differential phonon emission probability is

$$dw = \frac{\pi q u}{\hbar \varrho} A^2 \delta[\varepsilon(p) - \varepsilon(|\mathbf{p}-\mathbf{q}|) - uq] \frac{d^3q}{(2\pi\hbar)^3};$$

the delta function of momenta is removed by the integration over d^3p'. Writing the argument of the delta function in the approximate form $-uq(1-\cos\theta)$ and integrating over d^3q, we obtain

$$w = 2A^2(p-p_c)^3/3\pi\varrho\hbar^4.$$

§ 35. Properties of the spectrum near its termination point

In this section we shall consider the properties of the spectrum of a Bose liquid near the decay thresholds (thresholds of disintegration) of elementary excitations into two quasi-particles, neither of which is a phonon (cases (a) and (b) in §34).‡ In contrast to phonon-creating disintegrations, these cases do not allow the application of perturbation theory, and in order to investigate them it is necessary to elucidate the nature of the singularities of the Green's functions of the liquid at the threshold points. On the other hand, since we are interested only in these singularities, we can to a large extent schematize and thus simplify the calculations. In particular, we need not distinguish between the functions G and F (since their analytical properties are the same), and can proceed as if there were only one type of Green's function; taking account of the difference between G and F would simply produce some terms with analogous analytical properties in the equations, which would not affect the results.

The fact that the relevant singularity of the Green's function is related to the disintegration (decay) of a quasi-particle into two others means, in terms of the diagram technique, that it arises from diagrams of the type

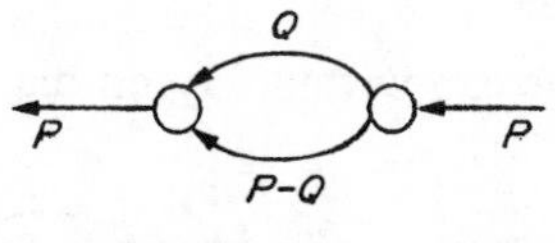

(35.1)

† We are taking the particular case where the phonon is emitted in that direction (and not the opposite one). For this to be so, $\varepsilon(p)$ near the threshold must have the form

$$\varepsilon(p) \approx \varepsilon(p_c) + (p-p_c)\,u + \alpha(p-p_c)^2,$$

with a plus sign in the linear term. From the law of conservation of energy, we easily see that phonon emission is then possible if $\alpha > 0$, and occurs when $p > p_c$; the momentum of the emitted phonon takes values in the range $0 \leqslant q \leqslant 2(p-p_c)$.

‡ The results in this section are due to L. P. Pitaevskiĭ (1959).

which may be cut across two continuous lines, i.e. which contain two-particle intermediate states. In these diagrams, there is integration with respect to the intermediate 4-momentum $Q = (q_0, \mathbf{q})$, and a decisive role (as regards the occurrence of the singularity) is played by the range of values of Q and $P-Q$ with which quasi-particle decay products are formed near the threshold. The basic proposition in the theory given below is that this range of 4-momentum values is not special as regards the Green's function $G(Q)$, which has there the usual pole form

$$G(Q) \equiv G(q_0, \mathbf{q}) \propto [q_0 - \varepsilon(q) + i0]^{-1}, \tag{35.2}$$

where the function $\varepsilon(q)$ is the energy of the quasi-particles formed and has no singularity. The only physically distinctive feature of this range is that within it the quasi-particle can "stick" to another quasi-particle, but this process is impossible at zero temperature, because of the absence of real excitations. The only special region as regards the Green's function is the range of P values (external lines in the diagrams (35.1)) near the decay threshold of the original quasi-particle.

The two joining lines in the diagram (35.1) correspond to factors $G(Q)G(P-Q)$, and there is integration with respect to Q. Here, since only a small range of Q values is important, the remaining factors in the diagram may be taken as constant in the integration and equal to their values at the threshold $Q = Q_c$.† Thus the diagram includes a factor expressed by the integral

$$\Pi(P) = \frac{i}{(2\pi)^4} \int \frac{d^4Q}{[q_0 - \varepsilon(q) + i0][\omega - q_0 - \varepsilon(|\mathbf{p}-\mathbf{q}|) + i0]},$$

where $P = (\omega, \mathbf{p})$. The integration with respect to q_0 is carried out by closing the contour of integration with an infinite semicircle in one half-plane of the complex variable q_0, and gives

$$\Pi(P) = \frac{1}{(2\pi)^3} \int \frac{d^3q}{\omega - \varepsilon(q) - \varepsilon(|\mathbf{p}-\mathbf{q}|) + i0}. \tag{35.3}$$

We shall return later to the study of this integral; first, we must express in terms of it the required exact function $G(P)$, summing for this purpose all diagrams of the form (35.1).

For the function $G(P)$ we can write a Dyson diagram equation:

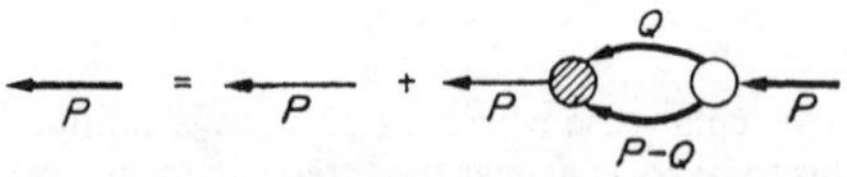

(35.4)

† This statement needs to be made more precise. Since the factors $G(Q)G(P-Q)$ are independent of the angle ϕ which defines the position of the $(\mathbf{p}, \mathbf{q})$ plane. Hence the integration with respect to ϕ amounts to averaging the rest of the integrand with respect to ϕ, and then d^4Q can be taken as $2\pi q^2 dq_0 dq\, d\cos\theta$. In this integration over d^4Q, only a small range is important. This remark applies also to corresponding stages in the following calculations.

Here the thick lines represent the exact function iG, and the thin lines the "non-singular" part of this function, determined by the set of diagrams that cannot be divided across two lines. The second term on the right of (35.4) represents the set of diagrams of the form (35.1). The white circle stands for the exact "three-point" vertex function, which we denote by $\Gamma(Q, P-Q, P)$; the shaded circle is its non-singular part, which excludes diagrams that can be divided across two continuous lines.† As explained previously, the integration over d^4Q leads to the presence of a factor $\Pi(P)$, and the remaining factors in the diagram are replaced by their values at $Q = Q_c$. Thus equation (35.4) signifies that

$$G(P) = a(P)+b(P)\,G(P)\,\Gamma_c(P)\,\Pi(P), \tag{35.5}$$

where $\Gamma_c(P) = \Gamma(Q_c, P-Q_c, P)$, and $a(P)$, $b(P)$ are some functions regular near the threshold $P = P_c$.

In (35.5) there are two singular functions G and Γ_c and to express them in terms of Π a further equation is therefore needed. This is found by noting that the exact vertex function Γ is represented by a "ladder" series:

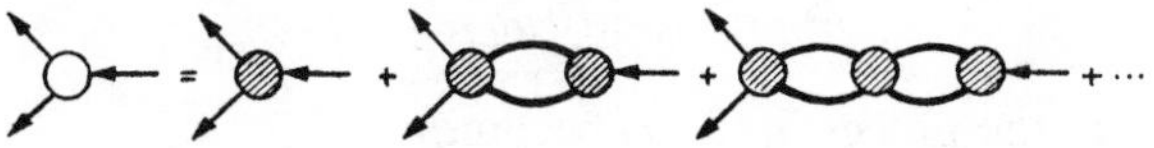

similar to (17.3) for a four-point vertex function. Summation of this gives the equation

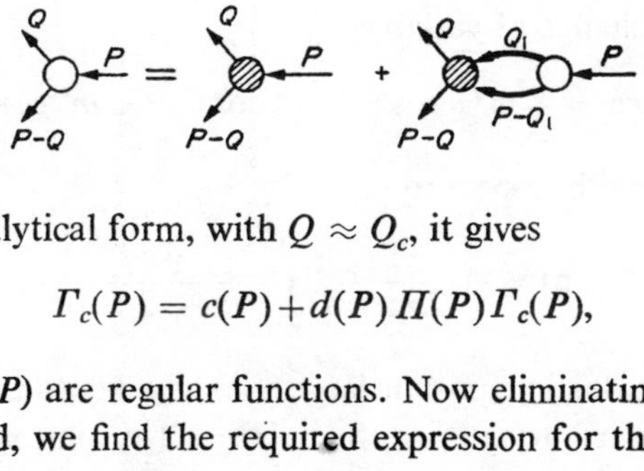

(cf. (17.4)); in analytical form, with $Q \approx Q_c$, it gives

$$\Gamma_c(P) = c(P)+d(P)\,\Pi(P)\,\Gamma_c(P),$$

where $c(P)$ and $d(P)$ are regular functions. Now eliminating Γ_c from the two equations obtained, we find the required expression for the Green's function in terms of Π:

$$G^{-1}(P) = \frac{A(P)\,\Pi(P)}{1+B(P)\,\Pi(P)}+C(P), \tag{35.6}$$

where A, B and C are functions that are likewise regular (near $P = P_c$).

The subsequent calculations are different for the various types of quasi-particle disintegration.

† The situation here is analogous to Dyson's equation in quantum electrodynamics (see *RQT*, §104): as there, the whole required set of diagrams is obtained by applying corrections to only one of the vertex functions.

(a) *Threshold for disintegration into two rotons*

In this case, the energy $\varepsilon(q)$ of the particles formed near the threshold is given by (22.6), and the integral (35.3) becomes

$$\Pi(\omega, q) = \int \left\{\omega - 2\Delta - \frac{1}{2m^*}[(q-p_0)^2 + (|\mathbf{p}-\mathbf{q}|-p_0)^2]\right\}^{-1} \frac{d^3q}{(2\pi)^3}. \quad (35.7)$$

For the integration, we use new variables q'_z, q'_ϱ, according to the definitions

$$q_x = (p_0 \sin\theta + q'_\varrho)\cos\phi,$$
$$q_y = (p_0 \sin\theta + q'_\varrho)\sin\phi,$$
$$q_z = p_0 \cos\theta + q'_z,$$

the z-axis being in the direction of $\mathbf{p}$, and the angle θ defined by the equation $2p_0 \cos\theta = p$. Near the threshold, q'_z and q'_ϱ are small, and we have with the necessary accuracy

$$q \approx p_0 + q'_\varrho \sin\theta + q'_z \cos\theta,$$
$$|\mathbf{p}-\mathbf{q}| \approx p_0 + q'_\varrho \sin\theta - q'_z \cos\theta,$$
$$d^3q \approx p_0 \sin\theta \, dq'_\varrho \, dq'_z \, d\phi.$$

The expression in the braces in (35.7) becomes

$$\left\{\omega - 2\Delta - \frac{1}{m^*}(q'^2_\varrho \sin^2\theta + q'^2_z \cos^2\theta)\right\}$$

and after a further change of variables

$$q'_\varrho \sin\theta = \sqrt{m^*}\,\varrho\cos\psi, \quad q'_z \cos\theta = \sqrt{m^*}\,\varrho\sin\psi$$

we find, integrating with respect to ψ,

$$\Pi(\omega, \mathbf{p}) = -\frac{m^* p_0}{2\pi\cos\theta}\int \frac{\varrho\, d\varrho}{-\omega + 2\Delta + \varrho^2}.$$

The divergence of this integral for large ϱ is due only to the approximations made, and is not important; cutting off the integral at some value $\varrho^2 \gg |2\Delta - \omega|$ gives a contribution only to the regular part of Π. The singular part of this function, with which we are concerned, arises from the range near the lower limit of integration, and is found to be

$$\Pi \propto \log\frac{1}{2\Delta - \omega}. \quad (35.8)$$

For small values of $2\Delta - \omega$, this logarithm is large; substituting (35.8) in (35.6) and expanding in inverse powers of the logarithm, we obtain

$$G^{-1}(\omega, p) = b + c\left(\log\frac{a}{2\Delta - \omega}\right)^{-1},$$

where a, b and c are further regular functions of ω and $\mathbf{p}$. At the threshold ($p = p_c$), the energy of the disintegrating quasi-particle is 2Δ. Since the energy of quasi-particles is determined by the zeros of G^{-1}, this means that $G^{-1}(2\Delta, p_c) = 0$, and for this we must have $b(2\Delta, p_c) = 0$. The regular function $b(\omega, p)$ is expanded in integral powers of the differences $p-p_c$ and $\omega-2\Delta$; replacing also the regular functions $a(\omega, p)$ and $c(\omega, p)$ by their values at the threshold, we arrive at the following expression for the Green's function in the region near the threshold:

$$G^{-1}(\omega, \mathbf{p}) = \beta\left[p-p_c+\alpha\left(\log\frac{a}{2\Delta-\omega}\right)^{-1}\right], \qquad (35.9)$$

where a, α and β are constants.

Equating this expression to zero, we find the spectrum $\varepsilon(p)$ near the threshold. If the range in which disintegration is impossible lies at $p < p_c$, $\varepsilon < 2\Delta$, the constants α and a must be positive, and the equation $G^{-1} = 0$ has the non-decaying solution

$$\varepsilon = 2\Delta - a\exp\left(-\frac{\alpha}{p_c-p}\right). \qquad (35.10)$$

We see that the spectrum reaches the threshold with a horizontal tangent of infinite order. In the range $p > p_c$, however, the equation $G^{-1} = 0$ has no solutions, real or complex, with $\varepsilon \approx 2\Delta$ for $p \approx p_c$. In this sense the spectrum does not continue beyond the threshold, but terminates there.†

(b) *Threshold for disintegration into two quasi-particles with parallel momenta.*

Since at the threshold, with $p = p_c$, the expression $\varepsilon(q)+\varepsilon(|\mathbf{p}-\mathbf{q}|)$ as a function of $\mathbf{q}$ must have a minimum, its form near the threshold is

$$\varepsilon(q)+\varepsilon(|\mathbf{p}-\mathbf{q}|) = \varepsilon_c+v_c(p-p_c)+\alpha(\mathbf{q}-\mathbf{q}_0)^2+\beta[(\mathbf{q}-\mathbf{q}_0)\cdot\mathbf{p}_c]^2, \qquad (35.11)$$

where α and β are constants, v_c is the velocity of each of the quasi-particles formed by disintegration at the threshold, and $\mathbf{q}_0$ is the momentum of one of them. Substituting (35.11) in (35.3) and using new variables of integration defined by

$$\boldsymbol{\rho} = \mathbf{q}-\mathbf{q}_0, \qquad \boldsymbol{\rho}\cdot\mathbf{p}_c = \varrho p_c\cos\psi,$$

we obtain

$$\Pi(\omega, \mathbf{p}) = \frac{1}{(2\pi)^2}\int\frac{\varrho^2\,d\varrho\,d\cos\psi}{\varepsilon-\varepsilon_c-v_c(p-p_c)-\alpha\varrho^2-\beta\varrho^2p_c^2\cos^2\psi}.$$

This integral has a square-root singularity at the threshold:

$$\Pi \propto [v_c(p-p_c)-(\varepsilon-\varepsilon_c)]^{1/2}. \qquad (35.12)$$

† As already mentioned in the third footnote to §34, the spectrum in liquid helium in fact ends at a point of this type; the curve in Fig. 2 approximates to the straight line $\varepsilon = 2\Delta$, with a horizontal tangent.

Substituting this in (35.6), we find the Green's function near the threshold:

$$G^{-1}(\omega, \mathbf{p}) = A(\omega, \mathbf{p}) + B(\omega, \mathbf{p})[v_c(p-p_c)-(\omega-\varepsilon_c)]^{1/2}.$$

Since $G^{-1}(\varepsilon_c, \mathbf{p}_c) = 0$, and A and B are regular functions, we can expand the latter in powers of $p-p_c$ and $\omega-\varepsilon_c$, and finally obtain

$$G^{-1} \propto [v_c(p-p_c)-(\omega-\varepsilon_c)]^{1/2} + [a(p-p_c)+b(\omega-\varepsilon_c)], \tag{35.13}$$

where a and b are constants.

The form of the spectrum is determined by the equation $G^{-1}(\varepsilon, \mathbf{p}) = 0$. We seek its solution in the form $\varepsilon-\varepsilon_c = v_c(p-p_c)+\text{constant}\times(p-p_c)^2$; if this exists for $p < p_c$, we must have $a+bv_c > 0$, and then

$$\varepsilon = \varepsilon_c + v_c(p-p_c)-(a+bv_c)^2(p-p_c)^2. \tag{35.14}$$

With the same condition at $p > p_c$, the equation $G^{-1} = 0$ has no solution with $\varepsilon \approx \varepsilon_c$ for $p \approx p_c$. Thus in this case also the spectrum terminates at the threshold.

CHAPTER IV

GREEN'S FUNCTIONS AT NON-ZERO TEMPERATURES

§ 36. Green's functions at non-zero temperatures†

THE definition of the Green's function of a macroscopic system at non-zero temperatures differs from that at zero temperature only in that the averaging with respect to the ground state of a closed system is replaced by an averaging over the Gibbs distribution: the symbol $\langle \ldots \rangle$ now denotes

$$\langle \ldots \rangle = \sum_n w_n \langle n | \ldots | n \rangle, \qquad w_n = \exp\left(\frac{\Omega - E_n'}{T}\right), \tag{36.1}$$

where the summation is over all states of the system (distinguished both by the energy E_n and by the particle number N_n), $E_n' = E_n - \mu N_n$, and $\langle n | \ldots | n \rangle$ is the diagonal matrix element with respect to the nth state. The mean values thus defined are functions of the thermodynamic variables T, μ and V.

In the study of the analytical properties of Green's functions at non-zero temperatures (L. D. Landau 1958) it is convenient to use what are called retarded and advanced Green's functions, whose analytical properties are simpler.‡ We shall take the particular case of Fermi systems.

The *retarded Green's function* is defined by

$$iG^R_{\alpha\beta}(X_1, X_2) = \begin{cases} \langle \hat{\Psi}_\alpha(X_1)\hat{\Psi}^+_\beta(X_2) + \hat{\Psi}^+_\beta(X_2)\hat{\Psi}_\alpha(X_1)\rangle, & t_1 > t_2, \\ 0, & t_1 < t_2. \end{cases} \tag{36.2}$$

For a microscopically homogeneous non-ferromagnetic system, in the absence of an external field, this function (like the ordinary $G_{\alpha\beta}$) reduces to a scalar function depending only on the difference $X = X_1 - X_2$:

$$G^R_{\alpha\beta}(X_1, X_2) = \delta_{\alpha\beta} G^R(X), \quad G^R = \tfrac{1}{2} G^R_{\alpha\alpha}. \tag{36.3}$$

The change to the momentum representation is made in the usual way. But, since $G^R(t, \mathbf{r}) = 0$ for $t < 0$, in the definition

$$G^R(\omega, \mathbf{p}) = \int \int_0^\infty e^{i(\omega t - \mathbf{p}\cdot\mathbf{r})} G^R(t, \mathbf{r})\, dt\, d^3x \tag{36.4}$$

† In §§36–38, the units used are such that $\hbar = 1$.

‡ These functions are customarily denoted by indices R and A respectively.

the integration with respect to t is actually taken only from 0 to ∞. The displacement of the variable ω into the upper half-plane simply improves the convergence of such an integral. Hence the integral (36.4) defines an analytic function without singularities in the upper half-plane of ω.† In the lower half-plane, where the function G^R is defined by analytical continuation, it has poles (see below).

We can obtain for G^R an expansion similar to (8.7) for G at $T = 0$. Expanding the matrix element $\langle n | \dots | n\rangle$ of the product of ψ operators by the matrix multiplication rule and expressing the matrix elements in the form (8.4), we have

$$iG^R(t, \mathbf{r}) = \tfrac{1}{2} \sum_{n,m} w_n \{ e^{-i(\omega_{mn}t - \mathbf{k}_{mn}\cdot\mathbf{r})} \langle n | \hat{\psi}_\alpha(0) | m\rangle \langle m | \hat{\psi}^+_\alpha(0) | n\rangle$$
$$+ e^{i(\omega_{mn}t - \mathbf{k}_{mn}\cdot\mathbf{r})} \langle n | \hat{\psi}^+_\alpha(0) | m\rangle \langle m | \hat{\psi}_\alpha(0) | n\rangle \},$$

where $\omega_{mn} = E'_m - E'_n$, $\mathbf{k}_{mn} = \mathbf{P}_m - \mathbf{P}_n$. The summation over n and m has slightly different meanings for the two terms in the braces: in the first term, the numbers of particles in the states n and m are related by $N_m = N_n + 1$, and in the second term by $N_m = N_n - 1$. In order to eliminate this difference, we interchange the suffixes m and n in the second sum. Noting also that

$$\langle n | \hat{\psi}_\alpha(0) | m\rangle \langle m | \hat{\psi}^+_\alpha(0) | n\rangle = |\langle n | \hat{\psi}_\alpha(0) | m\rangle|^2 \equiv A_{mn},$$

we can write the whole expression as

$$iG^R(t, \mathbf{r}) = \tfrac{1}{2} \sum_{m,n} w_n e^{-i(\omega_{mn}t - \mathbf{k}_{mn}\cdot\mathbf{r})} A_{mn}(1 + e^{-\omega_{mn}/T}), \qquad t > 0. \qquad (36.5)$$

Lastly, in calculating the integral (36.4) we replace ω (as in §8) by $\omega + i0$, obtaining finally

$$G^R(\omega, \mathbf{p}) = \tfrac{1}{2}(2\pi)^3 \sum_{m,n} w_n \frac{A_{mn}\delta(\mathbf{p} - \mathbf{k}_{mn})}{\omega - \omega_{mn} + i0} (1 + e^{-\omega_{mn}/T}). \qquad (36.6)$$

It should be noted that all the poles of this expression lie (in accordance with the above analysis) below the real axis, in the lower half-plane of ω.

The latter property is sufficient to establish a certain relation between the real and imaginary parts of the function, called the *Kramers–Kronig relation* or the *dispersion relation*:

$$\operatorname{re} G^R(\omega, \mathbf{p}) = \frac{1}{\pi} \mathrm{P} \int_{-\infty}^{\infty} \frac{\operatorname{im} G^R(u, \mathbf{p})}{u - \omega} du; \qquad (36.7)$$

compare the similar relation for $\alpha(\omega)$ in Part 1, §123. The validity of this can also be verified directly, by separating the real and imaginary parts in (36.6)

† Compare the analogous discussion for the function $\alpha(\omega)$ in Part 1, §123. The correspondence of the analytical properties of the functions G^R and α is, of course, not due to chance. From Part 1, (126.8), α is expressed similarly in terms of a certain operator commutator.

through the use of (8.11). It may also be noted that the latter formula allows (36.7) to be rewritten as

$$G^R(\omega, \mathbf{p}) = \frac{1}{\pi} \int_{-\infty}^{\infty} \frac{\varrho(u, \mathbf{p})}{u-\omega-i0} du, \tag{36.8}$$

where

$$\varrho(u, \mathbf{p}) = -4\pi^4 \sum_{m,n} w_n A_{mn} \delta(u-\omega_{mn})\, \delta(\mathbf{p}-\mathbf{k}_{mn})(1+e^{-\omega_{mn}/T}).$$

For real ω, $\varrho = \text{im}\, G^R$.

The representation (36.8) acquires a deeper significance if we take the "macroscopic limit" $V \to \infty$ (for a given ratio N/V). In this limit, the poles ω_{mn} coalesce, and $\varrho(u)$ becomes non-zero for all u, and is not just a sum of delta functions at discrete points. Then formula (36.8) determines $G^R(\omega)$ directly in the upper half-plane of ω and on the real axis. To determine $G^R(\omega)$ in the lower half-plane of ω, it is necessary to make an analytical continuation of the integral, and this requires the contour of integration to be deformed in such a way that it always passes below the point $u = \omega$. Here $G^R(\omega)$ may have singularities in the lower half-plane (at a finite distance from the real axis), and the contour of integration is then "pinched" between the pole $u = \omega$ and the singularity of the numerator.

The *advanced Green's function* is defined similarly by

$$iG^A_{\alpha\beta}(X_1, X_2) = \begin{cases} 0, & t_1 > t_2, \\ -\langle \hat{\Psi}_\alpha(X_1)\, \hat{\Psi}^+_\beta(X_2) + \hat{\Psi}^+_\beta(X_2)\, \hat{\Psi}_\alpha(X_1)\rangle, & t_1 < t_2. \end{cases} \tag{36.9}$$

The function $G^A(\omega, \mathbf{p})$ in the momentum representation is an analytical function of the variable ω, without singularities in the lower half-plane. Its expansion differs from (36.6) by a change of sign of $i0$ in the denominators. This means that on the real axis $G^A(\omega) = G^{R*}(\omega)$, and throughout the ω-plane

$$G^A(\omega^*) = G^{R*}(\omega). \tag{36.10}$$

As $\omega \to \infty$, G^R and G^A tend to zero in the same way as G:

$$G^R, G^A \to 1/\omega \quad \text{as} \quad |\omega| \to \infty. \tag{36.11}$$

The coefficient unity in this asymptotic expression is determined (see the derivation of (8.15)) by the discontinuity of the function at $t_2 = t_1$, which is independent of temperature and is the same for all three functions G^R, G^A, G, as is clear from their definitions.

To establish the relation between the functions G^R and G^A thus defined and the ordinary Green's function

$$iG_{\alpha\beta}(X_1, X_2) = \langle \mathrm{T}\, \hat{\Psi}_\alpha(X_1)\, \hat{\Psi}^+_\beta(X_2)\rangle, \tag{36.12}$$

we obtain for the latter an expansion analogous to (36.5). Calculations exactly similar to those above give the result†

$$G(\omega, \mathbf{p}) = -\frac{1}{2}(2\pi)^3 \sum_{m,n} w_n A_{mn} \delta(\mathbf{p}-\mathbf{k}_{mn}) \times$$
$$\times \left\{ \frac{1}{\omega_{mn}-\omega}(1+e^{-\omega_{mn}/T}) + i\pi\delta(\omega-\omega_{mn})(1-e^{-\omega_{mn}/T}) \right\}. \quad (36.13)$$

Comparison of (36.13) and (36.6) shows that

$$\left.\begin{matrix} G^R(\omega, \mathbf{p}) \\ G^A(\omega, \mathbf{p}) \end{matrix}\right\} = \text{re}\, G(\omega, \mathbf{p}) \pm i \coth(\omega/2T)\, \text{im}\, G(\omega, \mathbf{p}). \quad (36.14)$$

The same expression (36.13) also shows that

$$\text{sgn im}\, G(\omega, \mathbf{p}) = -\text{sgn}\, \omega. \quad (36.15)$$

It should be noted that G, unlike G^R and G^A, is not an analytic function of ω.

As $T \to 0$, $\coth(\omega/2T) \to \text{sgn}\, \omega$, it follows from (36.14) that on the real axis

$$G = \begin{cases} G^R, & \omega > 0, \\ G^A, & \omega < 0. \end{cases} \quad (36.16)$$

Thus the function $G(\omega)$ for $T = 0$ is equal, on the two halves of the real axis of ω, to the limits (as $|\text{im}\, \omega| \to 0$) of two different analytic functions: $G^R(\omega)$ on the right half and $G^A(\omega)$ on the left half.

It is easy to write down expressions for G^R and G^A in an ideal Fermi gas. We need only observe that they satisfy the same equation (9.6), the derivation of which used only the magnitude of the discontinuity of the function at $t_1 = t_2$. The method of passing round the pole is known from the fact that the pole must be below the real axis for $G^{(0)R}$ and above it for $G^{(0)A}$. Hence

$$G^{(0)R,A}(\omega, \mathbf{p}) = \left[\omega - \frac{p^2}{2m} + \mu \pm i0\right]^{-1}, \quad (36.17)$$

which is valid at both zero and non-zero temperatures. For the function $G^{(0)}$, we find, according to (36.14),

$$G^{(0)}(\omega, \mathbf{p}) = \mathrm{P}\frac{1}{\omega - p^2/2m + \mu} - i\pi \tanh\frac{\omega}{2T} \cdot \delta\left(\omega - \frac{p^2}{2m} + \mu\right). \quad (36.18)$$

As $T \to 0$, we return to formula (9.7), which differs from (36.17) in that $\pm i0$ is replaced by $i0 .\, \text{sgn}\, \omega$.

† In changing to the momentum representation, the integral with respect to t is divided into two parts, from $-\infty$ to 0 and from 0 to ∞; in one of these, the summation suffixes m and n are interchanged.

The corresponding formulae for a Bose system are as follows. The retarded and advanced Green's functions are defined by

$$\left.\begin{aligned} iG^R(X_1, X_2) &= \begin{cases} \langle \hat{\Psi}(X_1)\,\hat{\Psi}^+(X_2) - \hat{\Psi}^+(X_2)\,\hat{\Psi}(X_1)\rangle, & t_1 > t_2, \\ 0, & t_1 < t_2; \end{cases} \\ iG^A(X_1, X_2) &= \begin{cases} 0, & t_1 > t_2, \\ -\langle \hat{\Psi}(X_1)\,\hat{\Psi}^+(X_2) - \hat{\Psi}^+(X_2)\,\hat{\Psi}(X_1)\rangle, & t_1 < t_2. \end{cases} \end{aligned}\right\} \tag{36.19}$$

For temperatures above the λ-point, these definitions involve complete ψ operators; below the λ-point, they relate to above-condensate operators. Instead of (36.6) we now have

$$G^R(\omega, \mathbf{p}) = (2\pi)^3 \sum_{m,n} w_n \frac{A_{mn}\delta(\mathbf{p}-\mathbf{k}_{mn})}{\omega-\omega_{mn}+i0}(1-e^{-\omega_{mn}/T}). \tag{36.20}$$

This function is related to G by

$$G^R(\omega, \mathbf{p}) = \operatorname{re} G(\omega, \mathbf{p}) + i \tanh(\omega/2T).\operatorname{im} G(\omega, \mathbf{p}). \tag{36.21}$$

On the real axis,

$$\operatorname{im} G(\omega, \mathbf{p}) < 0; \tag{36.22}$$

G is defined, according to (31.1), with averaging over the Gibbs distribution instead of averaging with respect to the ground state. For an ideal Bose gas, the function G^R is given by the same formula (36.17), and G is

$$G^{(0)}(\omega, \mathbf{p}) = \mathrm{P}\frac{1}{\omega-p^2/2m+\mu} - i\pi \coth\frac{\omega}{2T}\,\delta\left(\omega-\frac{p^2}{2m}+\mu\right). \tag{36.23}$$

The physical significance of the Green's functions at non-zero temperatures is essentially the same as at $T = 0$. The formulae relating G to the particle momentum distribution (7.23) and to the density matrix (7.18), (31.4), remain valid, of course.

The basic propositions which assert the coincidence of the poles of the Green's function with the energy of the elementary excitations also remain valid (but, since G itself is not analytic, it is here more convenient to refer to the poles of the analytic function G^R in the lower half-plane of ω, or to those of G^A in the upper half-plane). This statement again (as in §8) follows from the expansion (36.6). Although different terms of this expansion now contain the transition frequencies ω_{mn} between any two states of the system, there still remain (after taking the macroscopic limit) poles corresponding only to transitions from the ground state to states with one elementary excitation. Transitions between two excited states do not produce a pole in the macroscopic one-particle Green's function, for the same reason that no pole results from transitions from the ground state to states with more than one quasi-particle (see §8): the energy difference of such states is not uniquely determined by their momentum difference.

We must also emphasize that at non-zero temperatures the lifetime of quasi-particles is governed not only by their intrinsic instability but also by their collisions with one another. The decay from both causes must be weak if the concept of quasi-particles is to continue to be meaningful.

§ 37. Temperature Green's functions

To construct diagram techniques for calculating the Green's function at non-zero temperatures, it would be necessary to change from the Heisenberg representation of the ψ operators to the interaction representation, as in §12. This would again lead to an expression differing from (12.12) only in that the averaging is not with respect to the ground state. This is, however, a very important difference: the averaging of the operator $\hat{S}^{-1}$ can no longer be separated from that of the other factors as was done in going from (12.12) to (12.14). The reason is that a state other than the ground state is not converted into itself by the operator $\hat{S}^{-1}$, but into some superposition of excited states having the same energy (which includes the results of all possible mutual scattering processes of quasi-particles). This causes a considerable complication of the diagram techniques, and new terms arise from contractions involving also ψ operators from $\hat{S}^{-1}$.

We can, however, alter the definition of the Green's function in such a way that such complications do not occur. The mathematical formalism based on this definition, which was developed by T. Matsubara (1955), is especially suitable for calculating the thermodynamic quantities of a macroscopic system.

We define *Matsubara ψ operators* by†

$$\left.\begin{aligned}\hat{\Psi}^M_\alpha(\tau, \mathbf{r}) &= e^{\tau\hat{H}'}\,\hat{\psi}_\alpha(\mathbf{r})\,e^{-\tau\hat{H}'},\\ \hat{\bar{\Psi}}^M_\alpha(\tau, \mathbf{r}) &= e^{\tau\hat{H}'}\,\hat{\psi}^+_\alpha(\mathbf{r})\,e^{-\tau\hat{H}'},\end{aligned}\right\} \tag{37.1}$$

where τ is an auxiliary real variable; these operators formally differ from the Heisenberg ones in that the real variable t in the latter is replaced by the imaginary variable $-i\tau$.‡ A similar change ($\hat{\Psi} \to \hat{\Psi}^M$, $\hat{\Psi}^+ \to \hat{\bar{\Psi}}^M$, $i\partial/\partial t \to -\partial/\partial\tau$), for example in (7.8), gives the equations satisfied by the operators (37.1). With these operators, a new Green's function $\mathcal{G}$ is defined similarly to the definition of the ordinary Green's function G in terms of the Heisenberg ψ operators:

$$\mathcal{G}_{\alpha\beta}(\tau_1, \mathbf{r}_1; \tau_2, \mathbf{r}_2) = -\langle \mathrm{T}_\tau\, \hat{\Psi}^M_\alpha(\tau_1, \mathbf{r}_1)\, \hat{\bar{\Psi}}^M_\beta(\tau_2, \mathbf{r}_2)\rangle, \tag{37.2}$$

† In this section we shall write the formulae simultaneously for Fermi systems and for Bose systems (above the λ-point). When there are alternative signs, the upper signs correspond to Fermi systems and the lower ones to Bose systems. The spin indices are to be omitted for Bose systems.

‡ It must be emphasized that, because of this change, the operator $\hat{\bar{\Psi}}^M$ is not the same as $\hat{\Psi}^{M+}$.

where T_τ is the "τ-chronological operator", which places the operators from right to left in order of increasing τ (with change of sign when operators are interchanged for Fermi systems); the brackets $\langle\ldots\rangle$ denote averaging over the Gibbs distribution. This averaging may be written explicitly if the definition (37.2) is expressed as

$$\mathcal{G}_{\alpha\beta} = -\operatorname{tr}\{\hat{w}\,\mathrm{T}_\tau \hat{\Psi}^M_\alpha(\tau_1, \mathbf{r}_1)\,\hat{\bar{\Psi}}^M_\beta(\tau_2, \mathbf{r}_2)\}, \qquad \hat{w} = \exp\left(\frac{\Omega - \hat{H}'}{T}\right), \tag{37.3}$$

where tr denotes the sum of all the diagonal matrix elements. This is called a *temperature Green's function*, in contrast to the "ordinary" function G, called in this connection a *time Green's function*.

Like $G_{\alpha\beta}$, $\mathcal{G}_{\alpha\beta}$ for a non-ferromagnetic system in the absence of an external magnetic field reduces to a scalar: $\mathcal{G}_{\alpha\beta} = \mathcal{G}\delta_{\alpha\beta}$. For a spatially homogeneous system, its dependence on $\mathbf{r}_1$ and $\mathbf{r}_2$ again reduces to a dependence on the difference $\mathbf{r} = \mathbf{r}_1 - \mathbf{r}_2$.

It is also easily seen that, by its definition (37.3), $\mathcal{G}$ depends only on the difference $\tau = \tau_1 - \tau_2$. For example, let $\tau_1 < \tau_2$; then†

$$\mathcal{G} = \pm\frac{1}{(2)} e^{\Omega/T} \operatorname{tr}\{e^{-\hat{H}'/T} e^{\tau_2\hat{H}'}\,\hat{\psi}_\alpha(\mathbf{r}_2)\,e^{-(\tau_2+\tau_1)\hat{H}'}\,\hat{\psi}_\alpha(\mathbf{r}_1)\,e^{-\tau_1\hat{H}'}\},$$

or, with a cyclic interchange of factors in the trace,

$$\mathcal{G} = \pm\frac{1}{(2)} e^{\Omega/T} \operatorname{tr}\{e^{-(1/T+\tau)\hat{H}'}\,\hat{\psi}^+_\alpha(\mathbf{r}_2)\,e^{\tau\hat{H}'}\,\hat{\psi}_\alpha(\mathbf{r}_1)\}, \qquad \tau < 0, \tag{37.4}$$

which makes evident the truth of the statement.

The variable τ in practice takes values only in a finite range

$$-1/T \leqslant \tau \leqslant 1/T. \tag{37.5}$$

The values of $\mathcal{G}(\tau)$ for $\tau < 0$ and $\tau > 0$ are related in a simple manner. When $\tau = \tau_1 - \tau_2 > 0$, we find, similarly to the derivation of (37.4),

$$\begin{aligned}\mathcal{G} &= -\frac{1}{(2)} e^{\Omega/T} \operatorname{tr}\{e^{-(1/T-\tau)\hat{H}'}\,\hat{\psi}_\alpha(\mathbf{r}_1)\,e^{-\tau\hat{H}'}\,\hat{\psi}^+_\alpha(\mathbf{r}_2)\}\\ &= -\frac{1}{(2)} e^{\Omega/T} \operatorname{tr}\{e^{-\tau\hat{H}'}\,\hat{\psi}^+_\alpha(\mathbf{r}_2)\,e^{-(1/T-\tau)\hat{H}'}\,\hat{\psi}_\alpha(\mathbf{r}_1)\}, \qquad \tau > 0,\end{aligned}$$

and comparison with (37.4) gives

$$\mathcal{G}(\tau) = \mp\mathcal{G}(\tau + 1/T), \qquad \tau < 0; \tag{37.6}$$

from (37.5), the argument of the function on the right is positive when $\tau < 0$.

† The factor 2 in parentheses applies to Fermi systems; it is to be replaced by 1 for Bose systems.

Let us now expand $\mathfrak{G}(\tau, \mathbf{r})$ as a Fourier integral with respect to the coordinates and a Fourier series in τ (over the range (37.5)):†

$$\mathfrak{G}(\tau, \mathbf{r}) = T \sum_{s=-\infty}^{\infty} \int e^{i(\mathbf{p}\cdot\mathbf{r}-\zeta_s\tau)} \mathfrak{G}(\zeta_s, \mathbf{p}) \frac{d^3p}{(2\pi)^3}; \tag{37.7}$$

for Fermi systems

$$\zeta_s = (2s+1)\pi T, \tag{37.8a}$$

and for Bose systems

$$\zeta_s = 2s\pi T, \tag{37.8b}$$

$s = 0, \pm 1, \pm 2, \ldots$; the condition (37.6) is then automatically satisfied. The inverse transformation to (37.7) is

$$\mathfrak{G}(\zeta_s, \mathbf{p}) = \int_0^{1/T} \int e^{-i(\mathbf{p}\cdot\mathbf{r}-\zeta_s\tau)} \mathfrak{G}(\tau, \mathbf{r})\, d^3x\, d\tau; \tag{37.9}$$

the integral over the range $-1/T \leqslant \tau \leqslant 1/T$ is converted into one from 0 to $1/T$, using (37.6) and (37.8).

Calculations similar to those in §36 enable us to express $\mathfrak{G}(\zeta_s, \mathbf{p})$ in terms of the matrix elements of the Schrödinger ψ operators, with the result

$$\mathfrak{G}(\zeta_s, \mathbf{p}) = \frac{(2\pi)^3}{(2)} \sum_{m,n} w_{mn} \frac{A_{mn}\delta(\mathbf{p}-\mathbf{k}_{mn})}{i\zeta_s - \omega_{mn}} (1 \pm e^{-\omega_{mn}/T}). \tag{37.10}$$

Hence we see, first of all, that

$$\mathfrak{G}(-\zeta_s, \mathbf{p}) = \mathfrak{G}^*(\zeta_s, \mathbf{p}). \tag{37.11}$$

Next, comparing (37.10) with the expansions (36.6) and (36.20) for G^R, we find that

$$\mathfrak{G}(\zeta_s, \mathbf{p}) = G^R(i\zeta_s, \mathbf{p}), \qquad \zeta_s > 0. \tag{37.12}$$

The condition $\zeta_s > 0$ is due to the fact that the expressions (36.6) and (36.20) are immediately valid only in the upper half-plane of ω, as explained in §36. Similarly, we find that $\mathfrak{G}(\zeta_s, \mathbf{p}) = G^A(i\zeta_s, \mathbf{p})$, $\zeta_s < 0$. Thus the temperature Green's function in Fourier components is the same as the retarded or advanced Green's function at discrete points on the imaginary ω-axis. In particular, this result leads at once to an expression for the temperature Green's function in an ideal gas: replacing ω by $i\zeta_s$, we find from (36.17)

$$\mathfrak{G}^{(0)}(\zeta_s, \mathbf{p}) = \left[i\zeta_s - \frac{p^2}{2m} + \mu\right]^{-1}. \tag{37.13}$$

In the next section, the diagram technique for calculating the function $\mathfrak{G}(\zeta_s, \mathbf{p})$ will be described. To determine $G^R(\omega, \mathbf{p})$ (and therefore, in particular,

† This device is due to A. A. Abrikosov, L. P. Gor'kov and I. E. Dzyaloshinskiĭ (1959) and E. S. Fradkin (1959).

to determine the energy spectrum of the system), we must construct an analytic function equal to $\mathcal{G}(\zeta_s, \mathbf{p})$ at the points $\omega = i\zeta_s$ and having no singularity in the upper half-plane of ω. This procedure is unique if we add the requirement that $G^R(\omega, \mathbf{p}) \to 0$ as $|\omega| \to \infty$; see (36.11). Nevertheless, in specific cases such an analytical continuation may involve some difficulties. It is, however, unnecessary in calculating the thermodynamic quantities.

For example, to calculate the potential Ω, we can start from the expression for the density matrix averaged over the Gibbs distribution,

$$N\varrho_{\alpha\beta}(\mathbf{r}_1, \mathbf{r}_2) = \pm \mathcal{G}_{\alpha\beta}(\tau_1, \mathbf{r}_1; \tau_1+0, \mathbf{r}_2), \tag{37.14}$$

which is evident from the definition (37.2); cf. (7.17). Putting $\mathbf{r}_2 = \mathbf{r}_1$ and summing over $\alpha = \beta$, we find as the density of the system

$$\frac{N}{V} = \pm T \sum_{s=-\infty}^{\infty} \left[\int \mathcal{G}(\zeta_s, \mathbf{p})\, e^{-i\zeta_s\tau} \frac{d^3p}{(2\pi)^3} \right]_{\tau \to -0}. \tag{37.15}$$

This expression determines N as a function of μ, T and V, and $\Omega(\mu, T, V)$ is then calculated by integrating the equation $N = -\partial\Omega/\partial\mu$.

§ 38. The diagram technique for temperature Green's functions

The diagram technique for calculating the temperature Green's function $\mathcal{G}$ is established in a similar way to that in §§12 and 13 for the time function G. The fact that the definition of the Matsubara ψ operators (37.1) differs from that of the Heisenberg operators only in the formal replacement of it by τ enables us to make considerable use of direct analogy.

First of all, let us define the Matsubara operators in the interaction representation; they differ from (37.1) in that the exact Hamiltonian $\hat{H}'$ is replaced by the free-particle Hamiltonian $\hat{H}_0'$:

$$\hat{\Psi}^M_{0\alpha}(\tau, \mathbf{r}) = \exp(\tau\hat{H}_0')\, \hat{\psi}_\alpha(\mathbf{r}) \exp(-\tau\hat{H}_0'). \tag{38.1}$$

The relation between the operators $\hat{\Psi}^M_{0\alpha}$ and $\hat{\Psi}^M_\alpha$ is given by the Matsubara S-matrix, constructed similarly to (12.8):

$$\hat{\sigma}(\tau_2, \tau_1) = \mathrm{T}_\tau \exp\left\{ -\int_{\tau_1}^{\tau_2} \hat{V}_0(\tau)\, d\tau \right\}, \tag{38.2}$$

where

$$\hat{V}_0(\tau) = \exp(\tau\hat{H}_0')\, \hat{V} \exp(-\tau\hat{H}_0') \tag{38.3}$$

is the interaction operator in that representation. But, whereas in §12 the relation between $\hat{\Psi}$ and $\hat{\Psi}_0$ was established with the initial condition that the interaction was "switched on" at $t = -\infty$, the "initial" condition must now be that $\hat{\Psi}^M$ and $\hat{\Psi}^M_0$ are the same at $\tau = 0$. Accordingly, we have instead of (12.11)

$$\hat{\Psi}^M_\alpha(\tau) = \hat{\sigma}^{-1}(\tau, 0)\, \hat{\Psi}^M_{0\alpha}(\tau)\, \hat{\sigma}(\tau, 0). \tag{38.4}$$

We substitute this expression in the definition (37.3) of the Green's function; taking the particular case $\tau_1 > \tau_2$, we find

$$\mathcal{G}_{\alpha\beta}(\tau_1, \tau_2) = -\operatorname{tr}\{\hat{w}\hat{\sigma}^{-1}(\tau_1, 0)\,\hat{\Psi}^M_{0\alpha}(\tau_1)\,\hat{\sigma}(\tau_1, 0)\,\hat{\sigma}^{-1}(\tau_2, 0)\,\hat{\bar{\Psi}}^M_{0\beta}(\tau_2)\,\hat{\sigma}(\tau_2, 0)\};$$

the arguments $\mathbf{r}_1$ and $\mathbf{r}_2$ are omitted, for brevity. Noting that, when $\tau_1 > \tau_2 > \tau_3$,

$$\hat{\sigma}(\tau_1, \tau_3) = \hat{\sigma}(\tau_1, \tau_2)\,\hat{\sigma}(\tau_2, \tau_3),$$
$$\hat{\sigma}(\tau_2, \tau_1)\,\hat{\sigma}^{-1}(\tau_3, \tau_1) = \hat{\sigma}(\tau_2, \tau_3),$$

we obtain

$$\mathcal{G}_{\alpha\beta}(\tau_1, \tau_2) = -\operatorname{tr}\left\{\hat{w}\hat{\sigma}^{-1}\left(\frac{1}{T}, 0\right)\left[\hat{\sigma}\left(\frac{1}{T}, \tau_1\right)\hat{\Psi}^M_{0\alpha}(\tau_1)\,\hat{\sigma}(\tau_1, \tau_2)\,\hat{\bar{\Psi}}^M_{0\beta}(\tau_2)\,\hat{\sigma}(\tau_2, 0)\right]\right\}.$$

The factors in the square brackets are already in order of increasing τ from right to left. We can therefore write

$$\mathcal{G}_{\alpha\beta}(\tau_1, \tau_2) = -\operatorname{tr}\{\hat{w}\hat{\sigma}^{-1}[T_\tau\,\hat{\Psi}^M_{0\alpha}(\tau_1)\,\hat{\bar{\Psi}}^M_{0\beta}(\tau_2)\,\hat{\sigma}]\}, \tag{38.5}$$

where

$$\hat{\sigma} \equiv \hat{\sigma}(1/T, 0).$$

It is easily verified that in this form the expression remains valid for $\tau_1 < \tau_2$ also.

In contrast to (12.12), equation (38.5) contains an extra (Gibbs) factor, and the averaging is over states of a system of interacting particles. We shall show that these two differences cancel out, and a complete analogy with (12.14) exists. To do so, we use the formula

$$e^{-\tau\hat{H}'} = e^{-\tau\hat{H}_0'}\,\hat{\sigma}(\tau, 0), \tag{38.6}$$

which is obtained by substituting (38.1) in (38.4) and then comparing the resulting expression with the definition of $\hat{\Psi}^M$ (37.1). By means of (38.6) we can substitute in (38.5)

$$e^{-\tau\hat{H}'/T}\,\hat{\sigma}^{-1}(1/T, 0) = e^{-\hat{H}_0'/T}.$$

The factor $e^{\Omega/T}$ is taken outside the trace, moved from the numerator to the denominator, and put in the form

$$e^{-\Omega/T} = \operatorname{tr} e^{-\hat{H}'/T} = \operatorname{tr} e^{-\hat{H}_0'/T}\,\hat{\sigma}(1/T, 0).$$

Lastly, multiplying the numerator and denominator by $\exp(\Omega_0/T)$, where Ω_0 is the thermodynamic potential of an ideal gas for the same values of μ, T and V, we find

$$\mathcal{G}_{\alpha\beta}(\tau_1, \tau_2) = -\frac{1}{\langle\hat{\sigma}\rangle_0}\langle T_\tau\,\hat{\Psi}^M_{0\alpha}(\tau_1)\,\hat{\bar{\Psi}}^M_{0\beta}(\tau_2)\,\hat{\sigma}\rangle_0, \tag{38.7}$$

where the averaging is with respect to the states of a system of non-interacting particles:

$$\langle \ldots \rangle_0 = \mathrm{tr}\,\{\hat{w}_0 \ldots\}.$$

There is an evident analogy with (12.14).

To change to the diagrams of perturbation theory, as in §13, we expand (38.7) in powers of the interaction operator $\hat{V}_0(\tau)$. For a system with pair interaction between particles, this operator differs from (13.2) only in that the Heisenberg operators $\hat{\Psi}_0$, $\hat{\Psi}_0^+$ are replaced by Matsubara operators $\hat{\Psi}^M$, $\hat{\bar{\Psi}}^M$. The mean values of the products of ψ operators are again expanded by Wick's theorem (i.e. by taking all possible ways of contracting pairs of operators); the validity of this theorem in the macroscopic limit is proved in this case by the same arguments as in §13.

The rules of the diagram technique thus obtained are entirely analogous to the rules derived in §13 for $T = 0$. The graphical form of the diagrams is exactly the same. There is only a slight change in the rules for analytical reading of the diagrams.

In the coordinate representation, each continuous line from point 2 to point 1 is associated with a factor $-\mathcal{G}^{(0)}_{\alpha\beta}(\tau_1, \mathbf{r}_1; \tau_2, \mathbf{r}_2)$ (with a minus sign). Each broken line joining points 1 and 2 corresponds to a factor $-U(\mathbf{r}_1-\mathbf{r}_2) \times \times \delta(\tau_1-\tau_2)$. For all variables τ and $\mathbf{r}$ of internal points in the diagram, there is integration over d^3x through all space, and over $d\tau$ from 0 to $1/T$.

In changing to the momentum representation, we must expand all functions $\mathcal{G}^{(0)}$ in the form (37.7). After the integration with respect to all the internal variables $\mathbf{r}$, a delta function appears at each vertex of the diagram, expressing the law of conservation of momentum ($\Sigma\mathbf{p} = 0$). There is also at each vertex an integral of the form

$$T \int_0^{1/T} \exp\{-i\tau(\zeta_{s_1} + \zeta_{s_2} + \zeta_{s_3})\}\, d\tau.$$

This integral is (from (37.8)) zero unless $\Sigma\zeta_s = 0$, and in this case is equal to unity. Thus the law of conservation of discrete frequencies is also satisfied at each vertex. Each continuous line is now associated with a factor $-\mathcal{G}^{(0)}_{\alpha\beta}(\zeta_s, \mathbf{p})$; a closed continuous line again has a factor $n^{(0)}(\mu, T)$, the ideal-gas density for given μ and T. For each broken line there is a factor $-U(\mathbf{q})$. There is integration and summation over all momenta and frequencies that remain undetermined by the conservation laws at all vertices, in the form

$$T \sum_{s=-\infty}^{\infty} \int \frac{d^3p}{(2\pi)^3} \cdots.$$

The coefficient of the whole diagram in $-\mathcal{G}_{\alpha\beta}$ is $(-1)^L$ for Fermi systems, where L is the number of closed sequences of continuous lines in the diagram. For Bose systems, the coefficient is unity.

In these techniques also, of course, as when $T = 0$, we can make a partial summation and define various "blocks" in the diagram. In particular, we can determine the vertex part, which is expressed in terms of the two-particle Green's function. This vertex part is related to $\mathcal{G}$ by a Dyson equation analogous to (15.14). We shall not write out the corresponding formulae, whose derivation is entirely similar to that in the diagram technique for $T = 0$.

When we make the transition to the case $T = 0$, the sums over s in the Matsubara diagrams become integrals over ζ, and the Matsubara techniques become very reminiscent of the ordinary ones described in Chapter II. There is a difference, however, in that for real ζ the Matsubara functions are the same as the values of G^R and G^A on the corresponding halves of the imaginary axis; see (37.11), (37.12). In changing to the ordinary technique for $T = 0$, we must also rotate the contour of integration until it becomes the real ω-axis.

CHAPTER V

SUPERCONDUCTIVITY

§ 39. A superfluid Fermi gas. The energy spectrum

THE whole of the Landau theory given in Chapter I applies only to one class of Fermi liquids—those whose energy spectrum is not such as to lead to superfluidity. This is not the only possible type of spectrum for a quantum Fermi liquid, and we shall now go on to consider Fermi systems with spectra of a different kind. The origin of such spectra and their basic properties can be most clearly perceived from a simple model which allows a complete theoretical analysis: a degenerate almost ideal Fermi gas with attraction between the particles.†

A slightly non-ideal Fermi gas with repulsion between the particles has been discussed in §6. At first sight, the calculations given there are equally valid whether there is repulsion or attraction, i.e. whether the scattering length a is positive or negative. In fact, however, for the case of attraction ($a < 0$) the ground state of the system thus found is unstable with respect to a certain rearrangement that alters its character and lowers its energy.

The physical nature of this instability consists in a tendency of the particles to "pair" by forming bound states of pairs of particles lying near the Fermi surface in $\mathbf{p}$-space and having equal and opposite momenta and antiparallel spins—the *Cooper effect* (L. N. Cooper 1957). It is noteworthy that this effect occurs in a Fermi gas, however weak the attraction between the particles.

Because of this effect, the set of operators $\hat{a}_{\mathbf{p}\alpha}$, $\hat{a}^+_{\mathbf{p}\alpha}$ used in the problem of a Fermi gas with repulsion, corresponding to free states of individual particles of the gas, cannot now serve as a correct initial approximation in perturbation theory.‡ Instead, we must use from the start new operators, which we shall seek in the form of linear combinations

$$\left.\begin{aligned} \hat{b}_{\mathbf{p}-} &= u_p\hat{a}_{\mathbf{p}-} + v_p\hat{a}^+_{-\mathbf{p},+}, \\ \hat{b}_{\mathbf{p}+} &= u_p\hat{a}_{\mathbf{p}+} - v_p\hat{a}^+_{-\mathbf{p},-}, \end{aligned}\right\} \tag{39.1}$$

of the operators of particles with opposite momenta and spins; the suffixes +

† This problem is the basis of the theory of superconductivity due to J. Bardeen, L. N. Cooper and J. R. Schrieffer (1957). The method of solution given below is by N. N. Bogolyubov (1958).

‡ An indication of the inapplicability of perturbation theory (in the form used in §6) to pairs of particles with spin components $\pm\frac{1}{2}$ and momenta $\mathbf{p}_2 \approx -\mathbf{p}_1$ is already given by the presence of the singularity at $\vartheta = \pi$ of the expression given by this theory for the quasi-particle interaction function (6.16); this singularity exists only with antiparallel spins, corresponding to the eigenvalue -3 of the operator $\boldsymbol{\sigma}_1 . \boldsymbol{\sigma}_2$.

and $-$ refer to the two values of the spin component. Because of the isotropy of the gas, the coefficients u_p and v_p can depend only on the absolute magnitude of the momentum $\mathbf{p}$. In order that these new operators should correspond to the creation and annihilation of quasi-particles, they must obey similar Fermi commutation rules to those of the old operators:

$$\hat{b}_{\mathbf{p}\alpha}\hat{b}^{+}_{\mathbf{p}\alpha}+\hat{b}^{+}_{\mathbf{p}\alpha}\hat{b}_{\mathbf{p}\alpha} = 1, \tag{39.2}$$

and all other pairs of operators anticommute (the suffix α labels the two values of the spin components). For this to be so, the transformation coefficients must be such that

$$u_p^2+v_p^2 = 1; \tag{39.3}$$

u_p and v_p may be made real by a suitable choice of the phase factor. The inverse transformation to (39.1) is

$$\left.\begin{aligned}\hat{a}_{\mathbf{p}+} &= u_p\hat{b}_{\mathbf{p}+}+v_p\hat{b}^{+}_{-\mathbf{p},-},\\ \hat{a}_{\mathbf{p}-} &= u_p\hat{b}_{\mathbf{p}-}-v_p\hat{b}^{+}_{-\mathbf{p},+}.\end{aligned}\right\} \tag{39.4}$$

For the same reasons (the predominant role of the interaction between pairs of particles with opposite momenta and spins), we shall retain in the second sum in the Hamiltonian (6.7) only terms in which $\mathbf{p}_1=-\mathbf{p}_2\equiv\mathbf{p}$, $\mathbf{p}_1'=-\mathbf{p}_2'\equiv\mathbf{p}'$:

$$\hat{H} = \sum_{\mathbf{p},\alpha}\frac{p^2}{2m}\hat{a}^{+}_{\mathbf{p}\alpha}\hat{a}_{\mathbf{p}\alpha}-\frac{g}{V}\sum_{\mathbf{p},\mathbf{p}'}\hat{a}^{+}_{\mathbf{p}'+}\hat{a}^{+}_{-\mathbf{p}',-}\hat{a}_{-\mathbf{p},-}\hat{a}_{\mathbf{p}+}, \tag{39.5}$$

again with the "coupling constant" $g = 4\pi\hbar^2|a|/m$ (the scattering length $a < 0$).

In subsequent calculations, it will again be convenient to use the customary procedure for avoiding the need to take explicit account of the constancy of the number of particles in the system: as a new Hamiltonian, we use the difference $\hat{H}' = \hat{H}-\mu\hat{N}$, where

$$\hat{N} = \sum_{\mathbf{p},\alpha}\hat{a}^{+}_{\mathbf{p}\alpha}\hat{a}_{\mathbf{p}\alpha}$$

is the particle number operator; the chemical potential is then determined, in principle, by the condition that the mean value $\bar{N}$ is equal to the given number of particles in the system.

We shall also use the notation

$$\eta_p = p^2/2m-\mu. \tag{39.6}$$

Since $\mu \approx p_F^2/2m$, we have near the Fermi surface

$$\eta_F = v_F(p-p_F), \tag{39.7}$$

where $v_F = p_F/m$. Subtracting $\mu\hat{N}$ from (39.5), we can thus write the initial Hamiltonian as

$$\hat{H}' = \sum_{\mathbf{p},\alpha}\eta_p\hat{a}^{+}_{\mathbf{p}\alpha}\hat{a}_{\mathbf{p}\alpha}-\frac{g}{V}\sum_{\mathbf{p},\mathbf{p}'}\hat{a}^{+}_{\mathbf{p}'+}\hat{a}^{+}_{-\mathbf{p}',-}\hat{a}_{-\mathbf{p},-}\hat{a}_{\mathbf{p}+}. \tag{39.8}$$

Here we make the transformation (39.4). Using the relations (39.2) and (39.3) and the possibility of replacing the summation suffix **p** by $-\mathbf{p}$, we obtain

$$\hat{H}' = 2\sum_{\mathbf{p}} \eta_p v_p^2 + \sum_{\mathbf{p}} \eta_p(u_p^2 - v_p^2)(\hat{b}_{\mathbf{p}+}^+ \hat{b}_{\mathbf{p}+} + \hat{b}_{\mathbf{p}-}^+ \hat{b}_{\mathbf{p}-})$$

$$+ 2\sum_{\mathbf{p}} \eta_p u_p v_p (\hat{b}_{\mathbf{p}+}^+ \hat{b}_{-\mathbf{p},-}^+ + \hat{b}_{-\mathbf{p},-} \hat{b}_{\mathbf{p}+}) - \frac{g}{V} \sum_{\mathbf{p},\mathbf{p}'} \hat{B}_{\mathbf{p}'}^+ \hat{B}_{\mathbf{p}}, \tag{39.9}$$

$$\hat{B}_{\mathbf{p}} = u_p^2 \hat{b}_{-\mathbf{p},-} \hat{b}_{\mathbf{p}+} - v_p^2 \hat{b}_{\mathbf{p}+}^+ \hat{b}_{-\mathbf{p},-}^+ + v_p u_p (\hat{b}_{-\mathbf{p},-} \hat{b}_{-\mathbf{p},-}^+ - \hat{b}_{\mathbf{p}+}^+ \hat{b}_{\mathbf{p}+}).$$

The coefficients u_p and v_p are now chosen from the condition that the energy E of the system be a minimum for a given entropy. The entropy is given by the combinatorial expression

$$S = -\sum_{\mathbf{p},\alpha} [n_{\mathbf{p}\alpha} \log n_{\mathbf{p}\alpha} + (1 - n_{\mathbf{p}\alpha}) \log (1 - n_{\mathbf{p}\alpha})].$$

The condition stated is therefore equivalent to minimizing the energy for given quasi-particle occupation numbers $n_{\mathbf{p}\alpha}$.

In the Hamiltonian (39.9) the diagonal matrix elements are zero except for terms containing the products

$$\hat{b}_{\mathbf{p}\alpha}^+ \hat{b}_{\mathbf{p}\alpha} = n_{\mathbf{p}\alpha}, \quad \hat{b}_{\mathbf{p}\alpha} \hat{b}_{\mathbf{p}\alpha}^+ = 1 - n_{\mathbf{p}\alpha}.$$

Hence

$$E = 2\sum_{\mathbf{p}} \eta_p v_p^2 + \sum_{\mathbf{p}} \eta_p (u_p^2 - v_p^2)(n_{\mathbf{p}+} + n_{\mathbf{p}-}) - \frac{g}{V} \Big[\sum_{\mathbf{p}} u_p v_p (1 - n_{\mathbf{p}+} - n_{\mathbf{p}-})\Big]^2. \tag{39.10}$$

Varying this expression with respect to the parameters u_p and using the relation (39.3), we find as the condition for a minimum

$$\frac{\delta E}{\delta u_p} = -\frac{2}{v_p}(1 - n_{\mathbf{p}+} - n_{\mathbf{p}-}) \Big[2\eta_p u_p v_p - \frac{g}{V}(u_p^2 - v_p^2) \sum_{\mathbf{p}'} u_{p'} v_{p'} (1 - n_{\mathbf{p}'+} - n_{\mathbf{p}'-})\Big] = 0.$$

Hence

$$2\eta_p u_p v_p = \Delta (u_p^2 - v_p^2), \tag{39.11}$$

where Δ denotes the sum

$$\Delta = \frac{g}{V} \sum_{\mathbf{p}} u_p v_p (1 - n_{\mathbf{p}+} - n_{\mathbf{p}-}). \tag{39.12}$$

From (39.11) and (39.3) we can express u_p and v_p in terms of η_p and Δ:

$$\left.\begin{matrix} u_p^2 \\ v_p^2 \end{matrix}\right\} = \frac{1}{2}\left(1 \pm \frac{\eta_p}{\sqrt{(\Delta^2 + \eta_p^2)}}\right). \tag{39.13}$$

Substituting these values in (39.12), we obtain an equation for Δ:

$$\frac{g}{2V} \sum_{\mathbf{p}} \frac{1 - n_{\mathbf{p}+} - n_{\mathbf{p}-}}{\sqrt{(\Delta^2 + \eta_p^2)}} = 1.$$

In equilibrium, the quasi-particle occupation numbers are independent of the spin direction and are given by the Fermi distribution formula (with zero chemical potential; cf. the last footnote to §1):

$$n_{\mathbf{p}+} = n_{\mathbf{p}-} \equiv n_{\mathbf{p}} = [e^{\varepsilon/T}+1]^{-1}. \tag{39.14}$$

Changing also from summation to integration over **p**-space, we can write this equation in the form

$$\frac{1}{2}g\int \frac{1-2n_{\mathbf{p}}}{\sqrt{(\Delta^2+\eta_p^2)}}\frac{d^3p}{(2\pi\hbar)^3} = 1. \tag{39.15}$$

Let us now analyse the relations obtained above. We shall see that Δ plays a basic role in the theory of spectra of the type under consideration. We shall first calculate its value Δ_0 for $T = 0$.

When $T = 0$ there are no quasi-particles, so that $n_{\mathbf{p}} = 0$ and equation (39.15) becomes

$$\frac{g}{2(2\pi\hbar)^3}\int \frac{4\pi p^2\,dp}{\sqrt{(\Delta_0^2+\eta_p^2)}} = 1. \tag{39.16}$$

We may note immediately that this equation certainly could not have a solution for Δ_0 if $g < 0$, i.e. in the case of repulsion, since the two sides would then have opposite signs.

The main contribution to the integral in (39.16) comes from the range of momenta where $\Delta_0 \ll v_F\,|p_F-p| \ll v_F p_F \sim \mu$, and the integral is logarithmic; the smallness of Δ_0 relative to μ is confirmed by the result. Cutting off the logarithmic integral at some $\eta = \tilde{\varepsilon} \sim \mu$, we have†

$$\int \frac{p^2\,dp}{[\Delta_0^2+v_F^2(p_F-p)^2]^{1/2}} \approx \frac{p_F^2}{v_F}\int \frac{d\eta}{(\Delta_0^2+\eta^2)^{1/2}} \approx \frac{2p_F^2}{v_F}.\log\frac{\tilde{\varepsilon}}{\Delta_0}.$$

Hence

$$(gmp_F/2\pi^2\hbar^3)\log(\tilde{\varepsilon}/\Delta_0) = 1, \tag{39.17}$$

or

$$\Delta_0 = \tilde{\varepsilon}\exp(-2\pi^2\hbar^3/gmp_F) = \tilde{\varepsilon}\exp(-\pi\hbar/2p_F\,|a|). \tag{39.18}$$

This expression may also be written

$$\Delta_0 = \tilde{\varepsilon}\exp(-2/gv_F), \tag{39.19}$$

† When $p \gg p_F$, $\eta_p \propto p^2$, and the integral (39.16) as written diverges as p. In reality, however, this divergence is spurious, and is eliminated by renormalizing the relation between the constant g (i.e. the scattering length a) and the interaction potential, as in §§6 and 25. A consistent performance of this quite complicated calculation allows us to determine also the proportionality factor between the cut-off parameter $\tilde{\varepsilon}$ and the chemical potential μ: $\tilde{\varepsilon} = (2/e)^{7/3}\mu = 0.49\mu$ (L. P. Gor'kov and T. K. Melik-Barkhudarov, *Soviet Physics JETP* **13,** 1018, 1961).

where $\nu_F = mp_F/\pi^2\hbar^3$ is the energy density of the number of states of a particle on the Fermi surface ($\nu\, d\varepsilon$ is the number of states in the range $d\varepsilon$).

The most interesting feature is the form of the energy spectrum of the system, i.e. the energy of the elementary excitations $\varepsilon_{\mathbf{p}+} = \varepsilon_{\mathbf{p}-} \equiv \varepsilon(\mathbf{p})$. We can find this from the change of energy of the whole system when the quasi-particle occupation numbers change, i.e. by varying E from (39.10) with respect to $n_{\mathbf{p}\alpha}$. Since the values of u_p and v_p have already been taken from the condition that the derivatives of E with respect to them are zero, the variation of E with respect to $n_{\mathbf{p}\alpha}$ can be carried out with constant u_p and v_p. Then

$$\varepsilon = (\delta E/\delta n_{\mathbf{p}\alpha})_{u_p,\, v_p}.$$

The calculation of the derivative, using (39.11)–(39.13), leads to the simple result

$$\varepsilon(p) = \sqrt{(\Delta^2 + \eta_p^2)}. \tag{39.20}$$

We see that the quasi-particle energy cannot be less than the value Δ, which is reached when $p = p_F$. In other words, the excited states of the system are separated from the ground state by an energy gap. The quasi-particles, having half-integral spin, must appear in pairs. In this sense we may say that the gap is 2Δ. Since $p_F|a|/\hbar \ll 1$, Δ_0 is exponentially small with regard to μ. Moreover, the expression (39.18) cannot be expanded in powers of the small parameter, the coupling constant g; the latter occurs in the denominator of the exponent, and so $g = 0$ is an essential singularity of $\Delta_0(g)$.

The spectrum (39.20) satisfies the superfluidity condition established in §23: the minimum value of ε/p is not zero. Thus a Fermi gas with attraction between the particles must have the property of superfluidity.†

Figure 5 compares the dispersion relations of quasi-particles in a superfluid Fermi system (upper curve) and in a normal one. In the latter, the dispersion

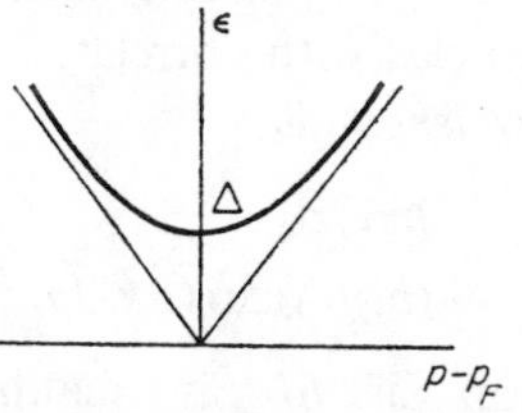

FIG. 5.

† Note, however, that the Landau condition has different meanings for the Bose and Fermi spectra. Fot the Bose spectrum, violation of the condition would lead to unlimited growth of excitations and there could not be an equilibrium motion of the normal part relative to the superfluid part. (This is shown by the fact that the Bose distribution function is negative; see the first footnote to § 23.) An unlimited growth of Fermi excitations is prevented by the Pauli principle, and the presence of a Fermi branch not satisfying the Landau condition need not imply the absence of superfluidity, only that the normal part is present at $T = 0$.

relation is represented by the two straight lines $\varepsilon = v_F|p-p_F|$, in accordance with the treatment mentioned at the end of §1.

The magnitude of the gap Δ depends on the temperature, i.e. the form of the spectrum itself depends on the statistical distribution of quasi-particles—a situation analogous to that of a normal Fermi liquid. Since the quasi-particle occupation numbers increase (tending to unity) with increasing temperature, it is evident from (39.15) that Δ decreases, and becomes zero at some finite temperature T_c, at which the system passes from the superfluid to the normal state. This point is a phase transition of the second kind, like the λ-transition in a Bose superfluid.

The presence of the energy gap in the spectrum of a degenerate Fermi gas is a manifestation of the "pairing"effect mentioned at the beginning of this section. The quantity 2Δ may be regarded as the binding energy of the Cooper pair, which would have to be expended in order to break it up.

The Hamiltonian (39.5) takes account (as already noted in §6) of the interaction only between pairs of particles in the singlet *s*-state: the orbital angular momentum of the relative motion of the particles is zero, and their spins are antiparallel. The pairs, having zero total spin, behave as Bose objects and may accumulate in any numbers at the level (of their motion as a whole) with the least energy, namely that for which the total momentum is zero. In this intuitive treatment, the phenomenon is entirely analogous to the accumulation of particles in a state with zero energy (Bose–Einstein condensation) in a Bose gas; in this case the condensate is the ensemble of paired particles.

The concept of bound pairs must not, of course, be taken very literally. It would be more precise to speak of a correlation between the states of a pair of particles in **p**-space, leading to a finite probability of the particles' having zero total momentum. The spread δp of the momentum values in the correlation range corresponds to an energy of the order of Δ, i.e. $\delta p \sim \Delta/v_F$. The corresponding length $\xi \sim \hbar/\delta p \sim \hbar v_F/\Delta$ determines the order of magnitude of the distances between particles with correlated momenta. When $T = 0$ this length, called the *coherence length*, is

$$\begin{aligned}\xi_0 &\sim \hbar v_F/\Delta_0 \\ &\sim (\hbar/p_F)\exp(\pi\hbar/2p_F|a|).\end{aligned} \tag{39.21}$$

Since, in a degenerate Fermi gas, $\hbar/p_F$ is equal in order of magnitude to the interatomic distances, we see that ξ_0 is very large in comparison with these. This shows particularly clearly the conventionality of the concept of bound pairs.

The origin of the Cooper effect is closely connected with the existence of the Fermi surface which bounds (in **p**-space) a finite region of occupied states at $T = 0$; an important point is that the energy density of the number of states on this surface is not zero. The relationship is evident in formula (39.19) for the gap Δ_0, which becomes zero as $v_F \to 0$.

§ 40. A superfluid Fermi gas. Thermodynamic properties

We shall begin the study of the thermodynamic properties of a superfluid Fermi gas by calculating the temperature dependence of the energy gap. Rewriting (39.15) as

$$-1+\frac{1}{2}g\int\frac{d^3p}{\varepsilon(2\pi\hbar)^3}=g\int\frac{n_p\,d^3p}{\varepsilon(2\pi\hbar)^3},$$

we note that the integral on the left differs from that for $T = 0$ only in that Δ_0 is replaced by Δ. Hence, using (39.17), we see that the left-hand side is $(gp_Fm/2\pi^2\hbar^3)\log(\Delta_0/\Delta)$. On the right we substitute n_p from (39.14) and change to integration over $dp = d\eta/v_F$:

$$\log\frac{\Delta_0}{\Delta}=\int\limits_{-\infty}^{\infty}\frac{d\eta}{\varepsilon(e^{\varepsilon/T}+1)}\equiv 2I(\Delta/T), \tag{40.1}$$

where

$$I(u)=\int\limits_0^{\infty}\frac{dx}{\sqrt{(x^2+u^2)}[\exp\sqrt{(x^2+u^2)}+1]};$$

on account of the rapid convergence of the integral, the limits of integration can be extended to $\pm\infty$.

At low temperatures ($T \ll \Delta_0$) the integral is easily† calculated to give

$$\Delta=\Delta_0[1-\sqrt{(2\pi T/\Delta_0)}\,e^{-\Delta_0/T}]. \tag{40.2}$$

Near the transition point, Δ is small, and the leading terms in the expansion of the integral $I(\Delta/T)$ give‡

$$\log(\Delta_0/\Delta)=\log(\pi T/\gamma\Delta)+7\zeta(3)\,\Delta^2/8\pi^2T^2. \tag{40.3}$$

Hence, first of all, we see that Δ is zero at a temperature

† For large u, the first term in the expansion of $I(u)$ in powers of $1/u$ is

$$I(u)\approx\int\limits_0^{\infty}\frac{dx}{u}\exp\left[-u\left(1+\frac{x^2}{2u^2}\right)\right]$$
$$=(\pi/2u)^{1/2}\,e^{-u}.$$

‡ To expand the integral $I(u)$ when $u\to 0$, we add to and subtract from it the integral

$$I_1=\frac{1}{2}\int\limits_0^{\infty}\left(\frac{1}{\sqrt{(x^2+u^2)}}-\frac{1}{x}\tanh\frac{1}{2}x\right)dx.$$

Then $I = I_1+I_2$, where

$$I_2=\frac{1}{2}\int\limits_0^{\infty}\left(\frac{1}{x}\tanh\frac{1}{2}x-\frac{1}{\sqrt{(x^2+u^2)}}\tanh\frac{1}{2}\sqrt{(x^2+u^2)}\right)dx.$$

In I_1, the first term in the integrand is integrated by elementary means, and the second is

$$T_c = \gamma\Delta_0/\pi = 0.57\Delta_0 \tag{40.4}$$

which is small in comparison with the degeneracy temperature $T_0 \sim \mu$. Then, in the first order in $T_c - T$, we obtain

$$\Delta = T_c\left[\frac{8\pi^2}{7\zeta(3)}\left(1-\frac{T}{T_c}\right)\right]^{1/2} = 3.06T_c\sqrt{\left(1-\frac{T}{T_c}\right)}. \tag{40.5}$$

It remains to calculate the thermodynamic quantities for the gas. Let us first consider the region of low temperatures.

To find the specific heat in this region, it is simplest to start from the formula

$$\delta E = \sum_{\mathbf{p}} \varepsilon(\delta n_{\mathbf{p}+} + \delta n_{\mathbf{p}-}) = 2\sum_{\mathbf{p}} \varepsilon\delta n_{\mathbf{p}}$$

for the change in the total energy when the quasi-particle occupation numbers vary. Dividing by δT and changing from summation to integration, we obtain the specific heat

$$C = V\frac{mp_F}{\pi^2\hbar^3}\int_{-\infty}^{\infty} \varepsilon\frac{\partial n}{\partial T}\,d\eta.$$

When $T \ll \Delta$, the quasi-particle distribution function $n \approx e^{-\varepsilon/T}$, and the quasi-particle energy $\varepsilon \approx \Delta_0 + \eta^2/2\Delta_0$; a simple integration gives

$$C = V\frac{\sqrt{2}mp_F\Delta_0^{5/2}}{\pi^{3/2}\hbar^3T^{3/2}}e^{-\Delta_0/T}. \tag{40.6}$$

---

integrated by parts, giving

$$2I_1 = -\log\frac{1}{2}u + \frac{1}{2}\int_0^{\infty}\frac{\log x}{\cosh^2\frac{1}{2}x}\,dx.$$

The integral is equal to $2\log(\pi/2\gamma)$, where $\log\gamma = C = 0.577$ is Euler's constant; thus $2I_1 = \log(\pi/\gamma u)$.

The integral I_2 is zero when $u = 0$. The first term of its expansion in powers of u^2 is

$$I_2 = -\frac{1}{4}u^2\int_0^{\infty}\frac{dx}{x}\left(\frac{1}{x}\tanh\frac{1}{2}x\right).$$

Substituting the expansion

$$\tanh\tfrac{1}{2}x = 4x\sum_{n=0}^{\infty}[\pi^2(2n+1)^2+x^2]^{-1},$$

derived as in the second footnote to §42, we obtain

$$2I_2 = 4u^2\sum_{n=0}^{\infty}\int_0^{\infty}\frac{dx}{[(2n+1)^2\pi^2+x^2]^2} = \frac{u^2}{\pi^2}\sum_{n=0}^{\infty}(2n+1)^{-3} = u^2\frac{7\zeta(3)}{8\pi^2}.$$

Thus, as $T \to 0$, the specific heat decreases exponentially—a direct consequence of the presence of the gap in the energy spectrum.

In subsequent calculations it is convenient to start from the thermodynamic potential Ω, since the whole analysis is for a given chemical potential of the system, not for a given number of particles in it.† We use the formula

$$(\partial\Omega/\partial\lambda)_{T,V,\mu} = \langle\partial\hat{H}/\partial\lambda\rangle, \tag{40.7}$$

where λ is any parameter characterizing the system (cf. Part 1, (11.4), (15.11)); in this case we take as the parameter the coupling constant g, which appears in the second term in the Hamiltonian (39.8). The mean value of this term is given by the last term in (39.10), which by (39.12) is $-V\Delta^2 g \propto g$. Hence

$$\partial\Omega/\partial g = -V\Delta^2/g^2.$$

As $g \to 0$, the energy gap Δ tends to zero. Hence, integrating this equation with respect to g from 0 to g, we find the difference between the thermodynamic potential Ω in the superfluid state and the value it would have in the normal state ($\Delta = 0$) at the same temperature:‡

$$\Omega_s - \Omega_n = -V\int_0^g \frac{\Delta^2}{g^2}\,dg. \tag{40.8}$$

According to the general theorem of small increments (Part 1, (24.16)), the correction (40.8), when expressed in terms of the appropriate variables, is the same for all the thermodynamic potentials.

At absolute zero $\Delta = \Delta_0$, and from (39.18)

$$d\Delta_0/dg = 2\pi^2\hbar^3\,\Delta_0/mp_F g^2.$$

Changing in (40.8) from integration over dg to that over $d\Delta_0$, we find the following expression for the difference between the ground-state energies of the superfluid and normal systems:

$$E_s - E_n = -V\frac{mp_F}{4\pi^2\hbar^3}\,\Delta_0^2. \tag{40.9}$$

† The chemical potential of the gas itself is not to be confused with the zero chemical potential of the quasi-particle gas.

‡ A comment is necessary here in connection with the approximations made throughout. When $g = 0$, no interaction between particles remains in the Hamiltonian (39.8), and one might suppose that we then have an ideal Fermi gas, not a "normal" non-ideal gas. In reality, however, approximations have already been made in the Hamiltonian (39.8), after which one cannot speak of calculating the absolute value of the energy. Interaction terms (which are not important in finding the form of the spectrum and the difference $\Omega_s - \Omega_n$) have been omitted whose contribution to the energy is large compared with the exponentially small quantity (40.8); this is the contribution proportional to Ng given by (6.13).

The negative sign indicates that, as mentioned at the beginning of this section, the "normal" ground state is unstable when there is attraction between the gas particles. The difference (40.9) per particle is $\sim \Delta^2/\mu$.

Let us now take the opposite case, $T \to T_c$. Differentiating (40.3) with respect to g, we find

$$\frac{7\zeta(3)}{4\pi^2 T^2}\,\Delta\, d\Delta = \frac{d\Delta_0}{\Delta_0} = \frac{2\pi^2\hbar^3}{mp_F}\,\frac{dg}{g^2}\,.$$

From here we substitute dg/g^2 in (40.8), regarding it as the difference of free energies:

$$F_s - F_n = -V\frac{7\zeta(3)mp_F}{8\pi^4\hbar^3 T^2}\int_0^\Delta \Delta^3\, d\Delta$$

and finally obtain, using (40.5),

$$F_s - F_n = -V\frac{2mp_F T_c^2}{7\zeta(3)\,\hbar^3}\left(1-\frac{T}{T_c}\right)^2. \tag{40.10}$$

The difference of entropies is therefore

$$S_s - S_n = -V\frac{4mp_F T_c}{7\zeta(3)\,\hbar^3}\left(1-\frac{T}{T_c}\right).$$

As $T \to T_c$, the difference of specific heats tends to a finite limit,

$$C_s - C_n = V\frac{4mp_F T_c}{7\zeta(3)\,\hbar^3}, \tag{40.11}$$

i.e. there is a discontinuity at the transition point, with $C_s > C_n$. The specific heat of the normal state is given (in the first approximation) by the ideal-gas formula (see Part 1, (58.6)); expressed in terms of p_F, it is $C_n = Vmp_F T/3\hbar^3$. The ratio of specific heats at the transition point is therefore

$$\frac{C_s(T_c)}{C_n(T_c)} = \frac{12}{7\zeta(3)} + 1 = 2.43. \tag{40.12}$$

As regards its superfluidity, the gas is characterized by the division of its density ϱ into normal and superfluid parts. According to (23.6), the normal part of the density is

$$\varrho_n = -\frac{8\pi}{3(2\pi\hbar)^3}\int p^4 \frac{dn}{d\varepsilon}\, dp$$

$$\approx -\frac{p_F^4}{3\pi^2\hbar^3 v_F}\int_{-\infty}^{\infty} \frac{dn}{d\varepsilon}\, d\eta.$$

The total density of the gas is related to p_F by

$$\varrho = mN/V = 8\pi p_F^3 m/3(2\pi\hbar)^3.$$

Hence

$$\frac{\varrho_n}{\varrho} = -2\int_0^\infty \frac{dn}{d\varepsilon}\,d\eta. \tag{40.13}$$

This integral does not need to be calculated specially, since it can be reduced to the known function $\Delta(T)$. Differentiating (40.1) with respect to T and comparing the resulting integral with (40.13), we see that

$$\frac{\varrho}{\varrho_n} = 1 - \frac{\Delta}{T\Delta'}. \tag{40.14}$$

Substituting here the limiting expressions (40.2) and (40.5), we obtain

$$T \to 0: \quad \frac{\varrho_n}{\varrho} = \left(\frac{2\pi\Delta_0}{T}\right)^{1/2} e^{-\Delta_0/T}, \tag{40.15}$$

$$T \to T_c: \quad \frac{\varrho_s}{\varrho} = 2\left(1 - \frac{T}{T_c}\right). \tag{40.16}$$

Lastly, two comments are needed concerning the range of temperature in which the above formulae are valid.

As the transition point T_c is approached, processes of interaction of quasi-particles (not taken into account in the above theory) become important; they are responsible in this case for the occurrence of the singularities of the thermodynamic quantities that are characteristic of phase transitions of the second kind. Sufficiently close to such a point, the formulae derived above must become invalid. However, because of the presence of a small parameter (the coupling constant g) in the model considered, this happens only for extremely small values of $T_c - T$; we shall discuss this in more detail in §45.

As in a superfluid Bose liquid, sound can be propagated in the Fermi gas under consideration (unlike one with repulsion; cf. §4), with a velocity $u \sim p_F/m$ determined in the usual manner by the compressibility of the medium. This means that, as well as the Fermi-type excitation spectrum dealt with here, the spectrum of such a gas also contains a phonon (Bose) branch. The specific heat due to phonons is proportional to T^3 with a small coefficient, but as $T \to 0$ it must ultimately predominate over the exponentially decreasing specific heat (40.6).

§ 41. Green's functions in a superfluid Fermi gas

Let us now set up the mathematical formalism of Green's functions for application to superfluid Fermi systems.†

We have seen in §26 that, in terms of ψ operators, the Bose–Einstein condensation in a Bose system is expressed by the existence of non-zero limiting values (as the particle number $N \to \infty$) of the matrix elements between states that differ only in that N changes by unity. The physical significance of this statement is that the removal or addition of one condensate particle does not alter the state of a macroscopic system.

For a superfluid Fermi system, the same must be true of the condensate of Cooper pairs: the state of the system cannot be altered when the number of pairs in the condensate changes by unity. This is expressed mathematically by the presence of non-zero limiting values ($N \to \infty$) of the matrix elements for the product $\hat{\Psi}_\beta(X_2)\hat{\Psi}_\alpha(X_1)$, the two-particle annihilation operator, and its Hermitian conjugate, the pair creation operator $\hat{\Psi}_\alpha^+(X_1)\hat{\Psi}_\beta^+(X_2)$. These matrix elements relate the "like" states of systems, differing only by the removal or addition of one pair of particles:

$$\lim_{N\to\infty} \langle m, N | \hat{\Psi}_\beta(X_2)\hat{\Psi}_\alpha(X_1) | m, N+2\rangle$$
$$= \lim_{N\to\infty} \langle m, N+2 | \hat{\Psi}_\alpha^+(X_1)\hat{\Psi}_\beta^+(X_2) | m, N\rangle^* \neq 0. \tag{41.1}$$

We shall henceforward omit the symbol for taking the limit, and for brevity also the diagonal matrix suffix m which labels the "like" states of systems with different numbers of particles.

As with Bose systems (§31), the mathematical formalism of Green's functions for superfluid Fermi systems involves several different functions. Together with the ordinary Green's functions

$$iG_{\alpha\beta}(X_1, X_2) = \langle N | \mathrm{T}\hat{\Psi}_\alpha(X_1)\hat{\Psi}_\beta^+(X_2) | N\rangle \tag{41.2}$$

we need also the "anomalous" functions defined by

$$\begin{aligned} iF_{\alpha\beta}(X_1, X_2) &= \langle N | \mathrm{T}\hat{\Psi}_\alpha(X_1)\hat{\Psi}_\beta(X_2) | N+2\rangle, \\ iF_{\alpha\beta}^+(X_1, X_2) &= \langle N+2 | \mathrm{T}\hat{\Psi}_\alpha^+(X_1)\hat{\Psi}_\beta^+(X_2) | N\rangle. \end{aligned} \tag{41.3}$$

Since each of the functions $F_{\alpha\beta}$ and $F_{\alpha\beta}^+$ is composed of two equal operators,

$$F_{\alpha\beta}(X_1, X_2) = -F_{\beta\alpha}(X_2, X_1), \quad F_{\alpha\beta}^+(X_1, X_2) = -F_{\beta\alpha}^+(X_2, X_1). \tag{41.4}$$

The interchange of Fermi ψ operators with the factors in chronological order causes a change in the sign of the product.

† The technique described in this section is due to L. P. Gor'kov (1958).

According to the fundamental principles of statistical physics, the result of statistical averaging does not depend on whether it is with respect to the exact wave function of a stationary state of a closed system or by means of the Gibbs distribution. The only difference is that in the first case the result of the averaging is expressed in terms of E and N, the energy and number of particles in the body; in the second case, in terms of T and μ, the temperature and the chemical potential. The first method is the more convenient for the discussion that follows in this section.

In the model of a Fermi gas considered in §39, the bound pairs are in a singlet state. The spin dependence of the matrix elements of the creation and annihilation operators of such a pair reduces to a unit antisymmetric spinor:

$$g_{\alpha\beta} = \begin{pmatrix} 0 & 1 \\ -1 & 0 \end{pmatrix}. \tag{41.5}$$

The functions (41.3) may be written†

$$F_{\alpha\beta} = g_{\alpha\beta} F(X_1, X_2), \quad F^{+}_{\alpha\beta} = g_{\alpha\beta} F^{+}(X_1, X_2); \tag{41.6}$$

from (41.4), F and F^+ are symmetrical in X_1 and X_2. The spin dependence of the Green's function $G_{\alpha\beta}$ for a non-ferromagnetic system reduces to a unit matrix $\delta_{\alpha\beta}$:

$$G_{\alpha\beta} = \delta_{\alpha\beta} G.$$

In a homogeneous system macroscopically at rest, the Green's functions G, F and F^+ depend only on the differences of the coordinates of the points and the difference of times (see the sixth footnote to §31).

Just as the function $\Xi(X)$ defined in §26 had the sense of a wave function for particles in the condensate, so the function $iF(t, \mathbf{r}_1; t, \mathbf{r}_2)$ may be regarded as the wave function of particles bound in Cooper pairs in the condensate. Then the function

$$\Xi(X) = iF(X, X) \tag{41.7}$$

will be the wave function for the motion of these pairs as a whole. From the definitions (41.3) and (41.5) it is easily seen that then $F^+(X, X) = i\Xi^*(X)$. In a stationary system macroscopically at rest, the function $\Xi(X)$ reduces to a constant, which may be made real by a suitable choice of the phases of the ψ operators.

Let us now calculate the Green's functions thus defined for the model of a Fermi gas with weak attraction between the particles.

The Heisenberg ψ operator satisfies the equation (7.8). Because the range of the forces between particles in the gas considered is small, in the integral term

† Cf. the fifth footnote to §7. Whereas $G_{\alpha\beta}$ in its spin structure is a mixed spinor of rank two, $F_{\alpha\beta}$ and $F^{+}_{\alpha\beta}$ are contravariant and covariant spinors respectively.

in this equation we can take the factors $\hat{\Psi}(t, \mathbf{r}')$ at the point $\mathbf{r}' = \mathbf{r}$ and bring them outside the integral. The equation then becomes†

$$i\frac{\partial \hat{\Psi}_\alpha}{\partial t} = -\left(\frac{\triangle}{2m}+\mu\right)\hat{\Psi}_\alpha - g\hat{\Psi}^+_\gamma\hat{\Psi}_\gamma\hat{\Psi}_\alpha. \tag{41.8}$$

By taking the Hermitian conjugate of each term in this equation, we get a corresponding equation for the operator $\hat{\Psi}^+$:

$$i\frac{\partial \hat{\Psi}^+_\alpha}{\partial t} = \left(\frac{\triangle}{2m}+\mu\right)\hat{\Psi}^+_\alpha + g\hat{\Psi}^+_\alpha\hat{\Psi}^+_\gamma\hat{\Psi}_\gamma. \tag{41.9}$$

Substituting (41.8) in the derivative $\partial G_{\alpha\beta}/\partial t$ (9.5), we obtain the equation

$$\left(i\frac{\partial}{\partial t}+\frac{\triangle}{2m}+\mu\right)G_{\alpha\beta}(X-X') - ig\langle N|\mathrm{T}\hat{\Psi}^+_\gamma(X)\hat{\Psi}_\gamma(X)\hat{\Psi}_\alpha(X)\hat{\Psi}^+_\beta(X')|N\rangle = \delta_{\alpha\beta}\,\delta^{(4)}(X-X'); \tag{41.10}$$

cf. (15.12). The diagonal matrix element of the product of four ψ operators can be written out, by the matrix multiplication rule, as a sum of products of matrix elements of two pairs of operators. Of all such products, we keep only those containing matrix elements for transitions in which the change in the number of particles is $N \leftrightarrow N+2$, and omit all other terms:

$$\begin{aligned}\langle N|\mathrm{T}\hat{\Psi}^+_\gamma\hat{\Psi}_\gamma\hat{\Psi}_\alpha\hat{\Psi}^{+\prime}_\beta|N\rangle &\to \langle N|\mathrm{T}\hat{\Psi}_\gamma\hat{\Psi}_\alpha|N+2\rangle\langle N+2|\mathrm{T}\hat{\Psi}^+_\gamma\hat{\Psi}^{+\prime}_\beta|N\rangle \\ &= -F_{\gamma\alpha}(X,X)\,F^+_{\gamma\beta}(X,X') = -\delta_{\alpha\beta}F(0)\,F^+(X-X');\end{aligned} \tag{41.11}$$

the expressions (41.5) are used to derive the last formula. This term corresponds physically to the pairing of particles, and it has the same order of magnitude as the condensate density.

We must emphasize, however, that there is a fundamental difference from the approximations used for a slightly non-ideal Bose gas. In the latter, almost all particles are in the condensate at $T = 0$, and the number of above-condensate particles, which occur only because of the weak interaction of the particles, is relatively small. In the present case, on the other hand, the condensate itself is due to the weak interaction, and therefore contains only a small fraction of the particles. In other words, the terms omitted in making the substitution (41.11) are large, not small, compared with those retained. The latter, however, give rise to a qualitatively new effect, a change in the nature of the spectrum, whereas the former would be needed only to calculate the corrections (which are of no interest here) to the ground state of the system; cf. the last footnote to §40.

† As in §39, we use the notation g for the coupling constant, equal to $-U_0 = -\int U d^3x$. In §§41 and 42, we have put $\hbar = 1$.

After the substitution (41.11), equation (41.10) becomes

$$\left(i\frac{\partial}{\partial t}+\frac{\triangle}{2m}+\mu\right)G(X)+g\Xi F^{+}(X)=\delta^{(4)}(X); \qquad (41.12)$$

the argument $X-X'$ is replaced by X, and the constant $iF(0)$ is denoted by Ξ, in accordance with the definition (41.7). Here there are two unknown functions, $G(X)$ and $F^{+}(X)$, and another equation is therefore needed to calculate them.

It may be found by calculating the derivative

$$i\frac{\partial F^{+}_{\alpha\beta}(X-X')}{\partial t}=\left\langle N+2\left|\mathrm{T}\frac{\partial\hat{\Psi}^{+}_{\alpha}(X)}{\partial t}\hat{\Psi}^{+}_{\beta}(X')\right|N\right\rangle;$$

a delta-function term (similar to the second term in (9.5)) does not arise here, since the function $F^{+}_{\alpha\beta}(X-X')$, unlike $G_{\alpha\beta}(X-X')$, is continuous at $t=t'$.† Substituting (41.9) and again separating the condensate term as in (41.11), we obtain the equation

$$\left(i\frac{\partial}{\partial t}-\frac{\triangle}{2m}-\mu\right)F^{+}(X)+g\Xi^{*}G(X)=0. \qquad (41.13)$$

It contains the same two functions G and F^{+} as (41.12); the two equations are therefore sufficient to calculate these functions. To calculate F, a further equation would have to be derived in a similar way.

In these equations we can change to the momentum representation by using the Fourier components $G(P)$ and $F^{+}(P)$ in the customary way:

$$\left.\begin{aligned}(\omega-\eta_p)\,G(P)+g\Xi F^{+}(P)&=1,\\(\omega+\eta_p)\,F^{+}(P)+g\Xi^{*}G(P)&=0,\end{aligned}\right\} \qquad (41.14)$$

where $P=(\omega,\mathbf{p})$ and $\eta_p=p^2/2m-\mu$. Since $F^{+}(X)$ is an even function, so are its Fourier components: $F^{+}(P)=F^{+}(-P)$.

Eliminating F^{+} from the two equations, we find the equation for G

$$(\omega^2-\eta_p^2-\Delta^2)\,G(P)=\omega+\eta_p, \qquad (41.15)$$

with the notation

$$\Delta=g\,|\,\Xi\,|. \qquad (41.16)$$

The formal solution of (41.15) is

$$G(P)=\frac{\omega+\eta_p}{\omega^2-\varepsilon^2(p)}=\frac{u_p^2}{\omega-\varepsilon(p)}+\frac{v_p^2}{\omega+\varepsilon(p)}, \qquad (41.17)$$

where $\varepsilon(p)=\sqrt{(\Delta^2+\eta_p^2)}$ and u_p and v_p are given by (39.13). It is evident from this that the spectrum of elementary excitations, determined by the positive

† This is easily seen by calculating the discontinuity of $F^{+}_{\alpha\beta}$ in the same way as for $G_{\alpha\beta}$ in §9, and noting that the operators $\hat{\Psi}^{+}_{\alpha}(t,\mathbf{r})$ and $\hat{\Psi}^{+}_{\beta}(t,\mathbf{r}')$ anticommute.

pole of the Green's function, is given by $\varepsilon(p)$, and we recover the result (39.20). We also see that the energy gap Δ and the modulus of the condensate wave function for the motion of pairs as a whole are proportional to each other.

The expression (41.17) for $G(P)$ is, however, not yet complete, since the manner of passing round the poles has not been defined. That is, the imaginary part of G is not yet determined; it contains the delta function $\delta(\omega \pm \varepsilon)$, and therefore disappears on multiplication by $\omega^2 - \varepsilon^2$ in (41.15).

When $T = 0$, the rule for passing round the poles is established by direct comparison of (41.17) with the expansion (8.7): in terms with positive and negative poles, the variable is to be replaced by $\omega + i0$ and $\omega - i0$ respectively; then (41.17) becomes

$$G(\omega, \mathbf{p}) = \frac{u_p^2}{\omega - \varepsilon(p) + i0} + \frac{v_p^2}{\omega + \varepsilon(p) - i0}$$

$$= \frac{\omega + \eta_p}{(\omega - \varepsilon + i0)(\omega + \varepsilon - i0)}. \tag{41.18}$$

Now expressing F^+ by means of the second equation (41.14), we find

$$F^+(\omega, \mathbf{p}) = \frac{-g\Xi^*}{(\omega - \varepsilon + i0)(\omega + \varepsilon - i0)}. \tag{41.19}$$

But, by definition,

$$i\Xi^* \equiv F^+(X = 0) = \int \int_{-\infty}^{\infty} F^+(P) \frac{d\omega \, d^3p}{(2\pi)^4}. \tag{41.20}$$

Substituting (41.19), we integrate with respect to ω by closing the contour with an infinite semicircle in the upper half-plane, and so express the integral in terms of the residue at the pole $\omega = \varepsilon$. Then, after cancelling Ξ^*, we find equation (39.16) for Δ_0.

When $T \neq 0$, it is somewhat more complicated to determine the imaginary part of the Green's functions. To construct $G(\omega, \mathbf{p})$ with the correct analytical properties with respect to the variable ω, we first write down the retarded function $G^R(\omega, \mathbf{p})$; it must be analytic in the upper half-plane, and is therefore obtained from (41.17) by the substitution $\omega \to \omega + i0$. The imaginary part of this function is

$$\operatorname{im} G^R = -\pi[u_p^2\delta(\omega - \varepsilon) + v_p^2\delta(\omega + \varepsilon)].$$

The imaginary part of the required function G is found from this by means of (36.14), which gives

$$\operatorname{im} G(\omega, \mathbf{p}) = \tanh(\omega/2T) \operatorname{im} G^R(\omega, \mathbf{p})$$
$$= -(1 - 2n_{\mathbf{p}})\pi[u_p^2\delta(\omega - \varepsilon) - v_p^2\delta(\omega + \varepsilon)],$$

where $n_{\mathbf{p}}$ is the Fermi distribution function (39.14); by using this formula, we change from averaging with respect to a given stationary state of the system to averaging over the Gibbs distribution. The function G with this imaginary part may be written

$$G(\omega, \mathbf{p}) = \frac{u_p^2}{\omega-\varepsilon+i0} + \frac{v_p^2}{\omega+\varepsilon-i0} + 2\pi i n_\mathbf{p}[u_p^2\delta(\omega-\varepsilon) - v_p^2\delta(\omega+\varepsilon)]. \quad (41.21)$$

We now find for the function $F^+(\omega, \mathbf{p})$

$$F^+(\omega, \mathbf{p}) = [F^+(\omega, \mathbf{p})]_{T=0} - \frac{i\pi g \Xi n_\mathbf{p}}{\varepsilon}[\delta(\omega-\varepsilon)+\delta(\omega+\varepsilon)], \quad (41.22)$$

where the first term is the function (41.19), relating to $T = 0$. Substituting this expression in (41.20) and carrying out the integration, we return to equation (39.15) for $\Delta(T)$.

Equations (41.14) can be put in diagram form, similarly to the representation of equations (33.7) for a superfluid Bose system. The functions G, F and F^+ are represented by the same graphical elements (33.6)—one-way and two-way arrows. The two equations (41.14) are written

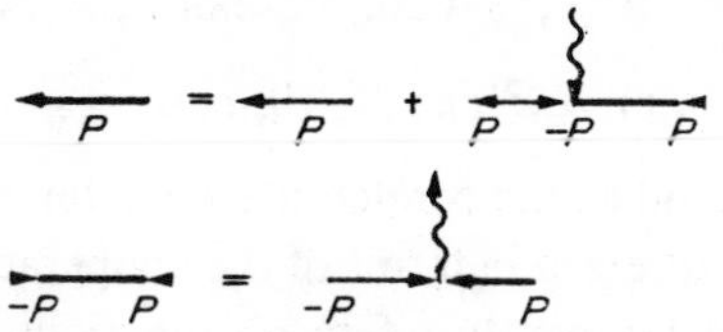

(41.23)

A thin arrow corresponds to a factor $iG^{(0)}(P)$, where $G^{(0)}(P)$ is the Green's function of an ideal Fermi gas. The wavy lines entering and leaving a vertex correspond to factors $ig\Xi$ and $-ig\Xi^*$ respectively. Comparing (41.23) and (33.7), we see that these latter factors correspond to the self-energy functions $i\Sigma_{02}$ and $i\Sigma_{20}$ respectively, i.e. are first approximations to these quantities. The new elements (two-way arrows, wavy lines) are the only special features of the diagram technique for superfluid Fermi systems; unlike the case of Bose systems, "triple" vertices do not appear. The diagram technique is therefore much simpler here and closer to the "ordinary" kind than for superfluid Bose systems.

§ 42. Temperature Green's functions in a superfluid Fermi gas

In §41 we have determined the energy spectrum of a superfluid Fermi gas by using the ordinary time Green functions. However, in order to solve more complex problems (in particular, to investigate the properties of the system in external fields), it is more convenient to use the mathematical formalism of temperature Green's functions (A. A. Abrikosov and L. P. Gor'kov 1958).

The temperature function $\mathcal{G}_{\alpha\beta}$ is defined by the same formula (37.3) as for a normal Fermi gas. The temperature functions $\mathcal{F}_{\alpha\beta}$ and $\overline{\mathcal{F}}_{\alpha\beta}$ (corresponding to the time functions $F_{\alpha\beta}$ and $F^+_{\alpha\beta}$) will be defined by the analogous formulae

$$\left.\begin{aligned}\mathcal{F}_{\alpha\beta}(\tau_1, \mathbf{r}_1; \tau_2, \mathbf{r}_2) &= \sum_m \langle m, N \,|\, \hat{w} T_\tau \hat{\Psi}^M_{\alpha 1} \hat{\Psi}^M_{\beta 2} \,|\, m, N+2\rangle, \\ \overline{\mathcal{F}}_{\alpha\beta}(\tau_1, \mathbf{r}_1; \tau_2, \mathbf{r}_2) &= \sum_m \langle m, N+2 \,|\, \hat{w} T_\tau \hat{\overline{\Psi}}{}^M_{\alpha 1} \hat{\overline{\Psi}}{}^M_{\beta 2} \,|\, m, N\rangle.\end{aligned}\right\} \quad (42.1)$$

The spin dependence of these functions is separated (as in (41.6)) in the form of factors $g_{\alpha\beta}$:†

$$\mathcal{F}_{\alpha\beta} = g_{\alpha\beta}\mathcal{F}, \quad \overline{\mathcal{F}}_{\alpha\beta} = -g_{\alpha\beta}\overline{\mathcal{F}}. \quad (42.2)$$

Like $\mathcal{G}$, the functions $\mathcal{F}$ and $\overline{\mathcal{F}}$ depend only on the difference $\tau = \tau_1 - \tau_2$, and satisfy the relations (37.6) with the upper sign:

$$\mathcal{F}(\tau) = -\mathcal{F}(\tau + 1/T), \quad \overline{\mathcal{F}}(\tau) = -\overline{\mathcal{F}}(\tau + 1/T). \quad (42.3)$$

The Fourier series in τ for these functions therefore contain only odd "frequencies" (37.8a): $\zeta_s = (2s+1)\pi T$.

The Matsubara ψ operators for $\tau = 0$ are the same as the Heisenberg operators for $t = 0$:

$$\hat{\Psi}^M(\tau = 0, \mathbf{r}) = \hat{\Psi}(t = 0, \mathbf{r}).$$

Comparing the definitions of $\mathcal{F}, \overline{\mathcal{F}}$ with those of F, F^+, we thus find that

$$\mathcal{F}(0, \mathbf{r}; 0, \mathbf{r}) = \Xi(\mathbf{r}), \quad \overline{\mathcal{F}}(0, \mathbf{r}; 0, \mathbf{r}) = \Xi^*(\mathbf{r}), \quad (42.4)$$

where Ξ is to be understood as the condensate wave function averaged over the Gibbs distribution, i.e. expressed in terms of the temperature of the system.

We shall show how the temperature Green's functions may be used to obtain again the energy spectrum of a superfluid Fermi gas at non-zero temperatures.

The equations for the temperature functions $\mathcal{G}, \mathcal{F}, \overline{\mathcal{F}}$ are derived in an exactly analogous way to equations (41.12) and (41.13); differentiation with respect to τ replaces that with respect to t, and equations (41.8) and (41.9) are replaced by others which differ by the substitution of τ for it. As in (41.11), we separate from the mean value of the product of four Matsubara ψ operators the terms containing matrix elements for transitions in which the number of particles changes by 2. The resulting equations are

$$\left.\begin{aligned}\left(-\frac{\partial}{\partial\tau} + \frac{\triangle}{2m} + \mu\right)\mathcal{G}(\tau, \mathbf{r}; \tau', \mathbf{r}') + g\Xi\overline{\mathcal{F}}(\tau, \mathbf{r}; \tau', \mathbf{r}') &= \delta(\tau - \tau')\,\delta(\mathbf{r} - \mathbf{r}'), \\ \left(\frac{\partial}{\partial\tau} + \frac{\triangle}{2m} + \mu\right)\overline{\mathcal{F}}(\tau, \mathbf{r}; \tau', \mathbf{r}') - g\Xi^*\mathcal{G}(\tau, \mathbf{r}; \tau', \mathbf{r}') &= 0.\end{aligned}\right\} \quad (42.5)$$

After the change to Fourier components, these equations become

$$\left.\begin{aligned}(i\zeta_s - \eta_p)\,\mathcal{G}(\zeta_s, \mathbf{p}) + g\Xi\overline{\mathcal{F}}(\zeta_s, \mathbf{p}) &= 1, \\ -(i\zeta_s + \eta_p)\,\overline{\mathcal{F}}(\zeta_s, \mathbf{p}) - g\Xi^*\mathcal{G}(\zeta_s, \mathbf{p}) &= 0.\end{aligned}\right\} \quad (42.6)$$

† The different signs in the definitions of $\mathcal{F}$ and $\overline{\mathcal{F}}$ (in contrast to the same signs in (41.5)) are appropriate because the factor i in (41.3) does not appear in (42.1).

The solutions are

$$\mathcal{G}(\zeta_s, \mathbf{p}) = -\frac{i\zeta_s + \eta_p}{\zeta_s^2 + \varepsilon^2}, \tag{42.7}$$

$$\overline{\mathcal{F}}(\zeta_s, \mathbf{p}) = g\Xi^*/(\zeta_s^2 + \varepsilon^2) = F^+(i\zeta_s, \mathbf{p}), \tag{42.8}$$

where again $\varepsilon^2 = \Delta^2 + \eta_p^2$, $\Delta = g\Xi$; this solution is uniquely defined, and contains no delta functions, unlike G and F^+.

The condition which determines the energy gap in the spectrum is now obtained from the equation

$$\Xi^* = \overline{\mathcal{F}}(\tau = 0, \mathbf{r} = 0) = T \sum_{s=-\infty}^{\infty} \int \overline{\mathcal{F}}(\zeta_s, \mathbf{p}) \frac{d^3p}{(2\pi)^3},$$

or, after the substitution of (42.8),

$$\frac{gT}{(2\pi)^3} \sum_{s=-\infty}^{\infty} \int \frac{d^3p}{\zeta_s^2 + \varepsilon^2(p)} = 1. \tag{42.9}$$

The summation with respect to s is given by the formula†

$$\sum_{s=-\infty}^{\infty} [(2s+1)^2\pi^2 + a^2]^{-1} = \frac{1}{2a} \tanh \frac{1}{2} a \tag{42.10}$$

and leads to

$$\frac{1}{2} g \int \frac{1}{\varepsilon} \tanh \frac{\varepsilon}{2T} \frac{d^3p}{(2\pi)^3} = 1, \tag{42.11}$$

in agreement with (39.15).

§ 43. Superconductivity in metals

The phenomenon of superconductivity in metals is a superfluidity of the electron Fermi liquid in them, similar to that of the degenerate Fermi gas considered in the preceding sections. Of course, in many important respects the electron liquid and the Fermi gas are quite different physical systems. The basic physical aspects of the energy spectrum are, however, the same for both. Let us examine qualitatively which features of the above model can be applied to electrons in metals, and to what extent.

† This may be derived by writing

$$\frac{1}{(2s+1)^2\pi^2 + a^2} = \frac{1}{2a}\left[\frac{1}{a + i\pi(2s+1)} + \frac{1}{a - i\pi(2s+1)}\right]$$

$$= \frac{1}{2a} \int_0^\infty e^{-ax}[e^{-i\pi(2s+1)x} + e^{i\pi(2s+1)x}]\, dx$$

and summing the geometrical progression before integrating.

An important property of a metal is the anisotropy of its electron energy spectrum, in contrast to the isotropy of the spectrum for the Fermi gas considered above. This does not, however, prevent the occurrence of the Cooper effect, which depends only on the existence of a sharply defined Fermi surface (of whatever shape) and a finite density for the number of states on that surface. It is also necessary that electrons with opposite momenta and spins should have the same energy, i.e. should both be on the Fermi surface. This condition automatically follows from symmetry under time reversal. We may say that the electrons are paired in states that are obtained from each other by time reversal.

Next, there is the question of the sign of the interaction of the electrons in a metal. In a very simplified way, we may say that this interaction is made up of the Coulomb repulsion, screened at interatomic distances, and the interaction via the lattice. The latter is describable as resulting from the exchange of virtual phonons, and is attractive (§64). If this interaction preponderates, the metal will be a superconductor at sufficiently low temperatures.

It is important to note that the interaction by phonon exchange involves only electrons in a comparatively thin shell of **p**-space near the Fermi surface, whose thickness ($\sim \hbar\omega_D$, where ω_D is the Debye frequency of the crystal) is small in comparison with the electron chemical potential μ. Hence, if we describe the superconductivity by a model of a slightly non-ideal Fermi gas, the cut-off parameter $\tilde{\varepsilon}$ in (39.19) is to be taken as

$$\tilde{\varepsilon} \sim \hbar\omega_D \tag{43.1}$$

instead of $\tilde{\varepsilon} \sim \mu$.

As to the assumption regarding the weakness of the interaction, we in fact have for all actual superconductors

$$T_c \ll \hbar\omega_D \ll \mu. \tag{43.2}$$

The assumption made in §39, however, embodies something further, namely that the coupling constant g is small, and therefore that the dimensionless exponent in (39.19) is large. In the present case, this condition is expressed as

$$\log (\hbar\omega_D/T_c) \gg 1; \tag{43.3}$$

not only the ratio $\hbar\omega_D/T_c$ but also its logarithm must be large. In practice, this condition is considerably less well satisfied.‡

When all the actual differences between the electron liquid in a metal and the model of a slightly non-ideal Fermi gas are taken into account, the theory of superconductivity becomes very complicated. It is, however, found that even a simple theory based on this model gives in many respects a good description

† This, incidentally, eliminates the problem of the divergence of the integral (39.16) for large momenta (cf. the last footnote to §39).

‡ The ratio $\hbar\omega_D/T_c$ varies between about 10 for lead and 300 for aluminium and cadmium.

of the properties of superconductors, both qualitatively and quantitatively. As already mentioned, this theory is due to Bardeen, Cooper and Schrieffer; the model of a Fermi gas with weak attraction between the particles is therefore known as the *BCS model.*

§ 44. The superconductivity current

The two types of motion in an electrically neutral superfluid (liquid helium) correspond, in a superconducting metal, to two types of electric current that can simultaneously flow in it. The *superconductivity current* transfers no heat and involves no dissipation of energy; it can exist in a system in thermodynamic equilibrium. The *normal current* is associated with the evolution of Joule heat. We shall denote the two current densities by $\mathbf{j}_s$ and $\mathbf{j}_n$; the total current density $\mathbf{j} = \mathbf{j}_s + \mathbf{j}_n$.

Several important conclusions about the properties of the superconductivity current can be drawn regardless of any particular model, simply from the existence of a new macroscopic quantity, the condensate wave function $\Xi(t, \mathbf{r})$.

As in §26, we use the phase Φ of this function:

$$\Xi(t, \mathbf{r}) = |\Xi| e^{i\Phi}. \tag{44.1}$$

Just as, in liquid helium, the gradient of Φ determines the velocity $\mathbf{v}_s$ of the superfluid flow by (26.12), so in a superconductor the gradient of the phase determines the observable quantity, the superconductivity current density. Because of the anisotropy of the metal, the direction of $\mathbf{j}_s$ does not in general coincide with that of $\nabla\Phi$, and the components of these vectors are related by a tensor of rank two. To avoid inessential complications, however, we shall here consider only a metal crystal having cubic symmetry.

The tensor then reduces to a scalar, and there is simple proportionality between $\mathbf{j}_s$ and $\nabla\Phi$, which may be written

$$\mathbf{j}_s = (e\hbar/2m)\, n_s \nabla\Phi. \tag{44.2}$$

Here, by definition, $e = -|e|$ is the electron charge and m its (actual) mass. The quantity n_s thus defined, a function of temperature, is called the *number density of superconducting electrons*, and acts here as an analogue of the density of the superfluid component in liquid helium. It must be emphasized that this is not the same as the density of the condensate of Cooper pairs, just as in liquid helium ϱ_s is not the density of condensate atoms.†

† The coefficient in (44.2) is written in such a way that in a free superfluid Fermi gas (BCS model) mn_s is equal to ϱ_s as calculated in §40. The latter is defined so that the current $\mathbf{j}_s$ must be expressed as $en_s\mathbf{v}_s$, where $\mathbf{v}_s$ is the velocity of superfluid motion. In turn, $\mathbf{v}_s$ is related to the phase gradient by $\mathbf{v}_s = (\hbar/2m)\nabla\Phi$; twice the mass occurs here (instead of m as in (26.12)) because the condensate consists of paired particles.

Formula (44.2), like (26.12) for liquid helium, presupposes that the phase varies sufficiently slowly in space. However, in a Bose liquid Φ had to vary only slightly over interatomic distances, but the condition here is considerably stronger. The characteristic dimension for a superfluid Fermi liquid is the coherence length $\xi_0 \sim \hbar v_F/\Delta_0$, and the phase Φ must vary only slightly over this distance, which is large in comparison with interatomic distances. We must emphasize that this is a constant (not temperature-dependent) length parameter ξ_0. A rigorous justification of the above condition will be given later (see the end of §51).

The relation between $\mathbf{j}_s$ and Φ becomes more complicated if the superconductor is in an external magnetic field. We shall consider here the case of a field constant in time. The necessary changes in formula (44.2) can be ascertained from the condition that the theory is gauge-invariant.

This condition states that all observable physical quantities must remain unchanged by a gauge transformation of the vector potential of the magnetic field:

$$\mathbf{A} \rightarrow \mathbf{A} + \nabla\chi(\mathbf{r}), \tag{44.3}$$

where $\chi(\mathbf{r})$ is an arbitrary function of the coordinates. The ψ operators are transformed in the same way as the wave functions:

$$\hat{\Psi} \rightarrow \hat{\Psi} \exp(ie\chi/\hbar c), \quad \hat{\Psi}^+ \rightarrow \hat{\Psi}^+ \exp(-ie\chi/\hbar c), \tag{44.4}$$

where e is the charge of the particles described by the ψ operator; see *QM*, (111.9).† The Green's functions $G(X, X')$ and $F(X, X')$, as matrix elements of the products $\hat{\Psi}\hat{\Psi}'^+$ or $\hat{\Psi}\hat{\Psi}'$, are transformed according to

$$\left.\begin{aligned} G(X, X') &\rightarrow \exp\left\{\frac{ie}{\hbar c}[\chi(\mathbf{r}) - \chi(\mathbf{r}')]\right\} G(X, X'), \\ F(X, X') &\rightarrow \exp\left\{\frac{ie}{\hbar c}[\chi(\mathbf{r}) + \chi(\mathbf{r}')]\right\} F(X, X'). \end{aligned}\right\} \tag{44.5}$$

Here

$$\Xi = iF(X, X) \rightarrow \exp(2ie\chi/\hbar c)\,\Xi,$$

i.e. the phase of the condensate wave function

$$\Phi \rightarrow \Phi + (2e/\hbar c)\,\chi(\mathbf{r}). \tag{44.6}$$

The relation (44.2) is not invariant under such a phase transformation. To obtain the required invariance, this relation must include a further term containing the vector potential of the magnetic field:

$$\mathbf{j}_s = \frac{e\hbar}{2m} n_s \left(\nabla\Phi - \frac{2e}{\hbar c}\mathbf{A}\right). \tag{44.7}$$

† Since the ψ operators appear in the second-quantized Hamiltonian (7.7) as pairs $\hat{\Psi}(X)$ and $\hat{\Psi}^+(X)$, it is transformed by the changes (44.3), (44.4) in the same way as the ordinary Hamiltonian for a similar transformation of ordinary (not operator) wave functions. A transformation in the form (44.3), (44.4) has in fact already been used in §19.

The doubling of the charge in this term corresponds to the pairing of electrons in the superconductor.

This expression is now sufficient to account for the fundamental macroscopic property of a superconductor; the displacement away from it of a magnetic field (the *Meissner effect*).†

Let us consider a homogeneous superconductor in a magnetic field that is weak compared with the critical field H_c at which the superconductivity is lost. This condition excludes any significant influence of the magnetic field on the value of n_s. Let the body be in a state of thermodynamic equilibrium, so that there is no normal current and $\mathbf{j}_s = \mathbf{j}$.‡ Taking the curl of both sides of (44.7) and noting that curl $\mathbf{A} = \mathbf{B}$, the magnetic induction in the body, we get the *London equation*

$$\operatorname{curl} \mathbf{j} = -(e^2 n_s/mc)\,\mathbf{B} \tag{44.8}$$

(F. and H. London 1935).§

This equation is specific to superconductors. We shall also make use of the general Maxwell's equations

$$\operatorname{curl} \mathbf{B} = (4\pi/c)\,\mathbf{j}, \tag{44.9}$$

$$\operatorname{div} \mathbf{B} = 0. \tag{44.10}$$

Substituting $\mathbf{j}$ from (44.9) in (44.8) and noting that, from (44.10), curl curl $\mathbf{B} = -\triangle\mathbf{B}$, we obtain an equation for the magnetic field in a superconductor:

$$\triangle\mathbf{B} = \mathbf{B}/\delta^2, \tag{44.11}$$

where

$$\delta^2 = mc^2/4\pi e^2 n_s. \tag{44.12}$$

We can use (44.11) to find the field distribution near the surface (assumed plane) of a superconductor. The surface is taken as the yz-plane, with the x-axis into the body. In these conditions, the field distribution depends only on one coordinate, x, and (44.10) gives $dB_x/dx = 0$; then, from (44.11), we necessarily have $B_x = 0$. Equation (44.11) now becomes $d^2\mathbf{B}/dx^2 = \mathbf{B}/\delta^2$, whence

$$\mathbf{B}(x) = \mathfrak{H} e^{-x/\delta}, \tag{44.13}$$

where the vector $\mathfrak{H}$ is parallel to the surface.

We see that the magnetic field decreases exponentially into the superconductor, penetrating only to distances $\sim \delta$. This distance is macroscopic, but small compared with the usual dimensions of solid objects ($\delta \sim 10^{-6}$–10^{-5} cm), and

† See *ECM*, Chapter VI, for the phenomenological electrodynamics of superconductors.

‡ This will be assumed throughout the rest of Chapter V, and $\mathbf{j}$ will therefore everywhere denote the superconductivity current density.

§ The derivation of (44.8) given here is due to L. D. Landau (1941).

so the field actually penetrates only into a thin surface layer. The distance δ is called the *London penetration depth* of the field. We must emphasize that it is directly measurable and has an entirely definite meaning, unlike the conventional significance of the parameter n_s.

The above derivation requires an important proviso, however. The original formula (44.7) is valid only if all quantities vary sufficiently slowly in space: the characteristic distances over which they vary considerably must be large compared with the coherence length ξ_0.† In the present case, this means that we must have

$$\delta \gg \xi_0. \tag{44.14}$$

This requirement, of course, does not affect the proof that the field is displaced from the superconductor: to suppose that the field is not displaced would lead to a logical contradiction, since it would then certainly vary slowly and equation (44.11) would be valid, but the specific equation (44.11) and the resulting law of field decay (44.13) are valid only if (44.14) is satisfied.

A superconductor in which the inequality $\delta \gg \xi_0$ is satisfied is called a *London superconductor*; the opposite case, with $\delta \ll \xi_0$, is called the *Pippard case* (the field decay in the superconductor is then of the kind to be discussed in §52). As $T \to T_c$, the superconducting electron density $n_s \to 0$, so that $\delta \to \infty$. Thus we always have the London case sufficiently close to the transition point. As $T \to 0$, however, the relation between δ and ξ_0 depends on the specific properties of the metal.‡

Lastly, let us consider a further consequence of equation (44.7) that is independent of the relation between δ and ξ_0. As we know from the macroscopic electrodynamics of superconductors, if there is a magnetic flux linking a superconducting torus, it remains constant regardless of any changes in state of the body (if these do not destroy its superconductivity). Here we assume that the torus has a diameter and thickness large compared with the coherence length and the field penetration depth. We shall show that the magnetic flux "frozen" in the aperture of the torus can only be an integral multiple of a certain "flux quantum" (F. London 1954).

Within the body (beyond the range of penetration of the field) the current density $\mathbf{j} = 0$; the vector potential, however, is not zero, but only its curl, i.e. the magnetic induction $\mathbf{B}$. We take any closed contour C embracing the aperture of the torus and passing through the torus far from the surface, so that the condition for equation (44.7) is satisfied, namely the slowness of the spatial variation of the phase Φ and the potential $\mathbf{A}$. The circulation of the vector $\mathbf{A}$

† The induction $\mathbf{B}$ itself is the true microscopic strength of the magnetic field, averaged over physically infinitesimal volume elements that are large only in comparison with the lattice constant.

‡ The London case occurs at all temperatures, for example, in pure transition metals and in some intermetallic compounds. The Pippard case occurs (far from T_c) in pure non-transition metals.

along the contour C is equal to the flux of the magnetic induction through a surface spanning the contour, i.e. the flux ϕ through the aperture of the torus:

$$\oint \mathbf{A} \cdot d\mathbf{l} = \int \operatorname{curl} \mathbf{A} \cdot d\mathbf{f} = \int \mathbf{B} \cdot d\mathbf{f} = \phi.$$

On the other hand, equating (44.7) to zero and integrating it along the contour, we have

$$\oint \mathbf{A} \cdot d\mathbf{l} = \frac{\hbar c}{2e} \oint \nabla \Phi \cdot d\mathbf{l} = \frac{\hbar c}{2e} \delta \Phi,$$

where $\delta\Phi$ is the change of phase of the wave function on passing round the contour. Since this function must be one-valued, it follows that the phase change can only be an integral multiple of 2π. Thus we have the result

$$\phi = n\phi_0, \quad \phi_0 = \pi\hbar c/|e| = 2\times 10^{-7} \text{ G.cm}^2, \tag{44.15}$$

where n is an integer. The quantity ϕ_0 is the *quantum of magnetic flux*.

The quantization of the magnetic flux has another aspect: it causes the values of the total current J that can flow along a superconducting ring (in the absence of an external magnetic field) to be discrete. This current J creates a magnetic flux through the ring equal to LJ/c, where L is the self-inductance. Equating this to $n\phi_0$, we find as the possible values of the current

$$J = c\phi_0 n/L = \pi\hbar c^2 n/|e|L. \tag{44.16}$$

In contrast to the magnetic flux quantum, the "quantum of total current", like the self-inductance L, depends on the shape and size of the ring.

PROBLEM

Determine the magnetic moment of a superconducting sphere with radius $R \ll \delta$ in a magnetic field, in the London case.

SOLUTION. When $R \ll \delta$, the magnetic field within the sphere may be regarded as constant and equal to the external field $\mathfrak{H}$. If the vector potential is taken in the form $\mathbf{A} = \frac{1}{2}\mathfrak{H}\times\mathbf{r}$, we can put simply

$$\mathbf{j} = -(n_s e^2/mc)\,\mathbf{A},$$

i.e. take $\Phi = 0$ in (44.7): the boundary condition for the normal component of the current to be zero on the surface of the sphere ($\mathbf{n}.\mathbf{j} = 0$) is then automatically satisfied. The magnetic moment is calculated as the integral

$$\mathbf{M} = \frac{1}{2c}\int \mathbf{r}\times\mathbf{j}\, dV$$

over the volume of the sphere, and is

$$\mathbf{M} = -R^5\,\mathfrak{H}/30\delta^2.$$

§ 45. The Ginzburg–Landau equations

The complete theory of the behaviour of a superconductor in a magnetic field is very complex. However, the position is considerably simpler in the temperature range near the transition point. Here it is possible to set up a system of relatively simple equations, valid in both weak and strong fields.†

In the general Landau theory of phase transitions of the second kind, the difference between the "unsymmetrical" and the "symmetrical" phase is described by the order parameter, which is zero at the transition point (see Part 1, §142). For a superconducting phase, the natural order parameter is the condensate wave function Ξ. To avoid complications that are in principle unnecessary, we shall assume that the metal crystal has cubic symmetry; as already mentioned in §44, the superconducting state is then characterized by a scalar quantity n_s, the superconducting electron density. A more convenient choice as the order parameter in this case is a quantity ψ that is proportional to Ξ but is normalized by the condition $|\psi|^2 = \frac{1}{2}n_s$. The phase of ψ is the same as that of Ξ:

$$\psi = \sqrt{(\tfrac{1}{2}n_s)}\, e^{i\Phi}. \tag{45.1}$$

The superconductivity current density (44.2), expressed in terms of ψ, is

$$\mathbf{j}_s = \frac{e\hbar}{m}|\psi|^2 \nabla\Phi = -\frac{ie\hbar}{2m}(\psi^*\nabla\psi - \psi\nabla\psi^*). \tag{45.2}$$

The starting-point of the theory is the expression for the free energy of the superconductor as a functional of $\psi(\mathbf{r})$. In accordance with the general ideas of the Landau theory, this is found by expanding the free energy density in powers of the small (near the transition point) order parameter ψ and its derivatives with respect to the coordinates. As a first step, let us consider a superconductor in the absence of a magnetic field.

In accordance with its significance as a quantity proportional to the Green's function $F(X, X) \equiv -i\Xi(X)$, the order parameter ψ is not unique: since $F(X, X)$ is constructed from two operators $\hat{\Psi}$, an arbitrary change of phase of these operators, $\hat{\Psi} \to \hat{\Psi}e^{i\alpha/2}$, causes a change of phase of the function F by α. Physical quantities, of course, must not be affected by this arbitrariness, i.e. must be invariant under a transformation of the complex order parameter $\psi \to \psi e^{i\alpha}$. This excludes odd powers of ψ in the expansion of the free energy.

The specific form of this expansion is established by the same considerations as in the general theory of phase transitions of the second kind (see Part 1, §146). Without repeating the arguments, we can write down the following ex-

† The theory given below is due to V. L. Ginzburg and L. D. Landau (1950). It is noteworthy that this theory was constructed phenomenologically, before the microscopic theory of superconductivity.

pansion of the total free energy of a superconducting body:†

$$F = F_n + \int \left\{ \frac{\hbar^2}{4m} |\nabla\psi|^2 + a|\psi|^2 + \frac{1}{2} b |\psi|^4 \right\} dV. \qquad (45.3)$$

Here F_n is the free energy in the normal state (i.e. for $\psi = 0$); b is a positive coefficient depending only on the density of the substance (not on the temperature); a is a function of the temperature given by

$$a = \alpha(T - T_c), \qquad (45.4)$$

and is zero at the transition point; the coefficient $\alpha > 0$, in accordance with the fact that the superconducting phase corresponds to the range $T < T_c$; the coefficient of $|\nabla\psi|^2$ in (45.3) is chosen so that the expression (45.2) is obtained for the current (see below).‡ The fact that (45.3) contains only the first derivatives of ψ is the result of assuming sufficient slowness of spatial variation of ψ.

In a homogeneous superconductor, with no external field, the parameter ψ is independent of the coordinates. Then the expression (45.3) reduces to

$$F = F_n + aV|\psi|^2 + \tfrac{1}{2} bV |\psi|^4. \qquad (45.5)$$

The equilibrium value of $|\psi|^2$ (for $T < T_c$) is determined by the condition for this expression to be a minimum:

$$|\psi|^2 = -a/b = \alpha(T_c - T)b; \qquad (45.6)$$

the superconducting electron density, as a function of temperature, decreases linearly to zero at the transition point.

Substituting (45.6) back into (45.5), we find the difference in the free energies of the superconducting and normal states:

$$F_s - F_n = -V(\alpha^2/2b)(T_c - T)^2. \qquad (45.7)$$

From this, by differentiating with respect to the temperature, we can find the difference in the entropies, and then the discontinuity in the specific heat at the transition point:§

$$C_s - C_n = V\alpha^2 T_c/b. \qquad (45.8)$$

† We shall only mention again that this form of the gradient term depends on the assumption that the crystal has cubic symmetry. With lower symmetry, it would have a more general quadratic dependence on the derivatives $\partial\psi/\partial x_i$.

‡ This choice (including the identification of m with the actual mass of the electron) has, of course, no deep significance, and is conventional to the same extent as the definition of n_s in (44.2).

§ Comparison of (45.6) and (45.8) for $|\psi|^2 = \varrho_s/2m$ and for the discontinuity in the specific heat with (40.16) and (40.11) for the same quantities in the BCS model gives the values of the coefficients α and b in that model (L. P. Gor'kov 1959):

$$\alpha = 6\pi^2 T_c/7\zeta(3)\,\mu = 7.04 T_c/\mu, \quad b = \alpha T_c/n;$$

here we have used the relation between the particle number density $n = \varrho/m$, the chemical potential μ (at $T = 0$) and the limiting momentum for an ideal gas:

$$n = p_F^3/3\pi^2\hbar^3, \quad \mu = p_F^2/2m.$$

Near the transition point, the difference (45.7) is a small addition to the free energy. According to the theorem of small increments (Part 1, §15), the same quantity, expressed as a function of temperature and pressure, instead of temperature and volume, gives the difference in the thermodynamic potentials, $\Phi_s - \Phi_n$. On the other hand, according to a general formula in the thermodynamics of superconductors (see *ECM*, (43.7)), this difference is $-VH_c^2/8\pi$, where H_c is the critical field which destroys the superconductivity. Thus we find for this field the following temperature dependence near the transition point:†

$$H_c = (4\pi a^2/b)^{1/2} = (4\pi\alpha^2/b)^{1/2}(T_c - T). \tag{45.9}$$

When a magnetic field is present, the expression (45.3) for the free energy has to be modified in two ways. Firstly, the magnetic field energy density $\mathbf{B}^2/8\pi$ (where $\mathbf{B} = \operatorname{curl} \mathbf{A}$ is the magnetic induction in the body) has to be added to the integrand. Secondly, the gradient term has to be changed so as to satisfy the requirement of gauge invariance. In the previous section it has been shown that this condition makes it necessary to replace the gradient $\nabla\Phi$ of the condensate wave function phase by $\nabla\Phi - 2e\mathbf{A}/\hbar c$. In the present case, this means making the substitution

$$\nabla\psi = e^{i\Phi}\nabla|\psi| + i\psi\nabla\Phi \rightarrow \nabla\psi - (2ie/\hbar c)\mathbf{A}\psi.$$

Thus we have the following basic equation:

$$F = F_{n0} + \int \left\{ \frac{\mathbf{B}^2}{8\pi} + \frac{\hbar^2}{4m}\left| \left(\nabla - \frac{2ie}{\hbar c}\mathbf{A} \right)\psi \right|^2 + a|\psi|^2 + \frac{1}{2}b|\psi|^4 \right\} dV, \tag{45.10}$$

where F_{n0} is the free energy of the body in the normal state in the absence of the magnetic field. It must be emphasized that the coefficient $2ie/\hbar c$ in this expression is not arbitrary (in contrast to the above-mentioned conventional choice of the coefficient $\hbar^2/4m$). The doubling of the electron charge is due to the Cooper effect (L. P. Gor'kov 1959); this coefficient could not, of course, be found by purely phenomenological means.

The differential equations which determine the distribution of the wave function ψ and the magnetic field in a superconductor are now found by minimizing the free energy as a functional of the three independent functions ψ, ψ^* and $\mathbf{A}$.

The complex quantity ψ is a set of two real quantities, so that ψ and ψ^* must be regarded as independent functions in the variation. Varying the integral

† In the BCS model,

$$H_c = 2.44\,(mp_F/\hbar^3)^{1/2}(T_c - T) \quad \text{as} \quad T \rightarrow T_c.$$

In the same model at $T = 0$,

$$H_c = 0.99T_c\,(mp_F/\hbar^3)^{1/2},$$

as is found by equating $-VH_c^2/8\pi$ to the energy difference (40.9).

with respect to ψ^* and integrating by parts in the integral of the term $(\nabla\psi - 2ieA\psi/\hbar c)\nabla\delta\psi^*$, we find

$$\delta F = \int \left\{ -\frac{\hbar^2}{4m}\left(\nabla - \frac{2ie}{\hbar c}\mathbf{A}\right)^2 \psi + a\psi + b\,|\psi|^2\psi \right\} \delta\psi^*\, dV$$

$$+\frac{\hbar^2}{4m}\oint \left(\nabla\psi - \frac{2ie}{\hbar c}\mathbf{A}\psi\right)\delta\psi^*\, d\mathbf{f}; \tag{45.11}$$

the second integral is taken over the surface of the body. Putting $\delta F = 0$, we obtain as the condition for the volume integral to be zero for arbitrary $\delta\psi^*$

$$\frac{1}{4m}\left(-i\hbar\nabla - \frac{2e}{c}\mathbf{A}\right)^2 \psi + a\psi + b\,|\psi|^2\psi = 0; \tag{45.12}$$

varying the integral with respect to ψ gives the complex conjugate equation, and therefore nothing new.

Similarly, varying the integral with respect to **A** gives Maxwell's equation

$$\operatorname{curl}\mathbf{B} = (4\pi/c)\,\mathbf{j}, \tag{45.13}$$

and the current density is

$$\mathbf{j} = -\frac{ie\hbar}{2m}(\psi^*\nabla\psi - \psi\nabla\psi^*) - \frac{2e^2}{mc}|\psi|^2\mathbf{A}, \tag{45.14}$$

which agrees with (44.7); we have written $\mathbf{j}$ for $\mathbf{j}_s$, since in thermodynamic equilibrium there is no normal current. From (45.13) we have the equation of continuity $\operatorname{div}\mathbf{j} = 0$, which may also be obtained by direct differentiation of (45.14), using (45.12).

Equations (45.12)–(45.14) form the complete set of *Ginzburg–Landau equations.*

The boundary conditions on these equations are found from the condition that the surface integrals in the variation δF are zero. Thus we get from (45.11) the boundary condition

$$\mathbf{n}\cdot\left(-i\hbar\,\nabla\psi - \frac{2e}{c}\mathbf{A}\psi\right) = 0, \tag{45.15}$$

where $\mathbf{n}$ is the normal vector at the surface of the body. As a result of this condition, the normal component of the current (45.14) is also zero, as it should be: $\mathbf{n}.\mathbf{j} = 0$.†

† With the boundary condition (45.15), ψ itself is not zero, as the wave function apparently ought to be at the boundary of the body. This is because ψ actually falls to zero only at distances $\sim \xi_0$ from the surface, but such distances are regarded as negligible in the Ginzburg–Landau theory.

The condition (45.15) has been derived here essentially for a superconductor–vacuum boundary. It remains valid for a boundary with an insulator, but it is not correct for an inter-

The boundary conditions for the field are as follows. From equation (45.13), since $\mathbf{j}$ is finite in all space (up to the surface of the body), the tangential component $\mathbf{B}_t$ of the induction is continuous. The equation div $\mathbf{B} = 0$ shows that the normal component B_n of the induction is continuous. Thus the boundary conditions require the continuity of the whole vector $\mathbf{B}$.

In a weak magnetic field, we can neglect the influence of the field on $|\psi|^2$, and take the latter to have the value (45.6) at all points in the body. Then the substitution of (45.14) in (45.13), followed by taking the curl of both sides, gives the London equation (44.11), with penetration depth

$$\delta = \left[\frac{mc^2 b}{8\pi e^2 |a|}\right]^{1/2} = \left[\frac{mc^2 b}{8\pi e^2 \alpha(T_c - T)}\right]^{1/2}. \tag{45.16}$$

The Ginzburg–Landau equations contain another characteristic length besides this: the correlation radius of the fluctuations of the order parameter ψ (in the absence of the field), which we denote by $\xi(T)$. From the formulae of fluctuation theory (see Part 1, §146), this radius is expressed in terms of the coefficients in the free energy (45.3) by

$$\begin{aligned} \xi(T) &= \hbar/2(m|a|)^{1/2} \\ &= \hbar/2(m\alpha)^{1/2}(T_c - T)^{1/2}. \end{aligned} \tag{45.17}$$

The characteristic lengths (45.16) and (45.17) determine the order of magnitude of the distances over which there is a significant change in the order parameter ψ and the magnetic field, as described by the Ginzburg–Landau equations. The length δ is in general characteristic of the magnetic field, and $\xi(T)$ of the distribution of ψ. Both these lengths must be large in comparison with the "dimensions of the pair" ξ_0, in order to satisfy the assumption that all quantities vary sufficiently slowly in space. Since both lengths increase as the transition point is approached (in proportion to $(T_c - T)^{-1/2}$), this condition is in general satisfied near the transition point (see below).

In the theory given here, the *Ginzburg–Landau parameter* is important; it is defined as the constant (temperature-independent) ratio of the two lengths:

$$\varkappa = \delta(T)/\xi(T) = mcb^{1/2}/(2\pi)^{1/2}|e|\hbar. \tag{45.18}$$

---

face between different metals (one superconducting, the other normal), since it does not take into account the partial penetration of superconducting electrons into the normal metal. In this case, (45.15) is replaced by a more general condition compatible with $\mathbf{n}.\mathbf{j} = 0$:

$$\mathbf{n}.\left(-i\hbar\nabla\psi - \frac{2e}{c}\mathbf{A}\psi\right) = i\psi/\lambda, \tag{45.15a}$$

where λ is a real constant (with the dimensions of length); however, an estimate of this constant would need a more detailed microscopic investigation.

In order of magnitude, $\varkappa \sim \delta_0/\xi_0$, where ξ_0 is the coherence length (39.21), and δ_0 is the London penetration depth at absolute zero. There is also a formula

$$\varkappa = 2\sqrt{2}(|e|/\hbar c)\, H_c(T)\, \delta^2(T), \tag{45.19}$$

obtained from (45.9) and (45.16), expressing $\varkappa$ directly in terms of observable quantities.

Having established the form of the equations, let us now consider their range of applicability.

At low temperatures, this range is in any case limited by the condition $T_c - T \ll T_c$, enabling the order parameter to be regarded as small, which is thus fundamental to the expansion that has been obtained for the free energy. The same condition ensures that $\xi(T) \gg \xi_0$, but it is not strong enough to ensure that $\delta(T) \gg \xi_0$ in superconductors for which the parameter $\varkappa$ is small;† in such cases, the inequality $\delta \gg \xi_0$ gives the condition

$$T_c - T \ll \varkappa^2 T_c. \tag{45.20}$$

As $T \to T_c$, the validity of the equations is limited only by the general condition for the validity of the Landau theory of phase transitions, relating to the occurrence of fluctuations in the order parameter. In the present case, however, this condition is extremely weak: it is expressed in terms of the coefficients in the expansion (45.3) by the inequality

$$T_c - T \gg b^2 T_c^2 / \alpha(\hbar^2/m)^3$$

(see Part 1, (146.15)). For instance, an estimate of the expression on the right, using the values of b and α in the BCS model, gives

$$(T_c - T)/T_c \gg (T_c/\mu)^4. \tag{45.21}$$

Since the ratio $T_c/\mu \sim 10^{-3}$–10^{-4} is very small, we can regard this condition as satisfied almost up to the transition point itself. The fluctuation region for the transition of the second kind between the superconducting and normal phases practically disappears.

PROBLEM

Find the critical magnetic field (parallel to the film plane) which destroys the superconductivity for a plane film with thickness $d \ll \xi, \delta$ (V. L. Ginzburg and L. D. Landau 1950).‡

SOLUTION. We take the median plane of the film as the xz-plane, with the x-axis in the direction of the field. In equation (45.13) for the field $B \equiv B_z(y)$ (which varies along the y-axis normal to the film), we can take ψ = constant. Then the first term in the expression (45.14) for

† As examples, the values of $\varkappa$ for some pure metals are: aluminium 0.01, tin 0.13, mercury 0.16, lead 0.23.

‡ See §47 for the corresponding problem of a small sphere.

the current is zero, and taking the curl of (45.13) gives $B''=\theta^2B/\delta^2$, where $\theta=\psi/\psi_0$, $\psi_0^2=|a|/b$. The solution of this equation, symmetrical in y, is

$$B(y) = \mathfrak{H}\frac{\cosh(y\theta/\delta)}{\cosh(d\theta/2\delta)} \approx \mathfrak{H}\left[1+\frac{y^2-(\tfrac{1}{2}d)^2}{2\delta^2}\theta^2\right],$$

where $\mathfrak{H}$ is the external field. This corresponds to the current distribution

$$j = j_z = -cB'/4\pi \approx -c\theta^2\,\mathfrak{H}y/4\pi\delta^2.$$

In equation (45.12), however, we cannot completely neglect the dependence of ψ on y: the small derivative $\partial^2\psi/\partial y^2$ is here multiplied by $\hbar^2/m|a| \sim \xi^2$, and thus acquires the large (from the condition $d \ll \xi$) coefficient $(\xi/d)^2$. We *can* neglect the potential $A = A_z(y)$ in this equation, which here leads to terms of a higher order of smallness in d/ξ. In order to avoid the need to consider the dependence of ψ on y, we average equation (45.12) over the film thickness; the derivatives with respect to y then disappear, because of the boundary condition $\partial\psi/\partial y = 0$ at the surface of the film. Noting also that

$$-\frac{\partial^2\psi}{\partial z^2} \approx \left(\frac{mj}{|e|\,\hbar\,|\psi|^2}\right)^2\psi$$

because of the z-dependence of the phase of ψ (and the relation between its gradient and the current), we find, after cancelling ψ,

$$\frac{m\overline{j^2}}{4e^2|\psi|^4}-|a|+b|\psi|^2 = 0,$$

where

$$\overline{j^2} = \frac{1}{d}\int_{-d/2}^{d/2} j^2\,dy = \frac{c^2d^2\theta^4\mathfrak{H}^2}{3(8\pi)^2\delta^4}.$$

Using also (45.9) and (45.16), we arrive at the equation

$$\frac{1}{24}\left(\frac{\mathfrak{H}d}{H_c\delta}\right)^2 = 1-\frac{|\psi|^2}{\psi_0^2},$$

which determines ψ for a film in a magnetic field. The critical field H_c^f for the film is that for which $\psi = 0$. It is related to the critical field H_c for a bulk superconductor by

$$H_c^f = \sqrt{(24)}\,H_c\delta/d.$$

In the conditions considered, the removal of superconductivity by the field takes place through a phase transition of the second kind: ψ tends to zero continuously as $\mathfrak{H}$ increases. This is entirely reasonable, since for $d \ll \delta$ the field actually penetrates into the superconducting film and there is no cause for a transition of the first kind, which would consist in a sudden penetration of the field into the body.

§ 46. Surface tension at the boundary of superconducting and normal phases

The Ginzburg–Landau equations allow, in particular, the calculation of the surface tension at the boundary of superconducting (s) and normal (n) phases (in the same sample) in terms of bulk characteristics of the material (V. L. Ginzburg and L. D. Landau 1950). Such boundaries exist in metallic samples that are in the "intermediate" state in a magnetic field. Since the only difference between the two phases is that ψ is zero in one but not in the other, the transition between them is continuous over a certain layer and is described by the Ginzburg–Landau equations with boundary conditions established only at large distances on either side of this layer.

Let us consider a plane interface between n and s phases in a metal, taking the interface as the yz-plane and the x-axis into the s phase; the distribution of all quantities in both phases depends only on the coordinate x. The vector potential of the field, the choice of which is not yet uniquely specified, will now be subjected to the gauge in which div $\mathbf{A} = 0$; in the present problem this gives $dA_x/dx = 0$, whence we see that it is possible to take $A_x = 0$. It is evident from symmetry that the vector $\mathbf{A}$ is everywhere in one plane; let this be the xy-plane, so that $A_y \equiv A$; then the induction vector is in the xz-plane, with

$$B \equiv B_z = A' \tag{46.1}$$

(the prime denoting differentiation with respect to x).

Next, we rewrite (45.13) in the form usual in macroscopic electrodynamics, curl $\mathbf{H} = 0$, with the field strength $\mathbf{H}$ given by†

$$\mathbf{H} = \mathbf{B} - 4\pi\mathbf{M}, \qquad c \operatorname{curl} \mathbf{M} = \mathbf{j}.$$

From this equation, it follows in the present case that $H =$ constant. Far from the interface, within the normal phase, the induction and the field are the same and equal to the critical field: $B = H = H_c$ (we neglect the magnetic susceptibility of the normal phase). Hence $H \equiv H_z = H_c$ in all space.

Neglecting the change in the density of the material in the superconducting phase transition, we shall regard the density (and the temperature) as constant throughout the body.‡ Let f denote the free energy per unit volume (in contrast to F, the free energy of the whole body). At constant temperature and density, and with surface effects neglected, the differential df is

$$df = \mathbf{H} \cdot d\mathbf{B}/4\pi; \tag{46.2}$$

cf. *ECM*, §30. Hence we see that the additional requirement of constant $\mathbf{B}$ would lead in these conditions to constancy of

$$\tilde{f} = f - \mathbf{H} \cdot \mathbf{B}/4\pi. \tag{46.3}$$

Thus the whole contribution to the integral $\tilde{F} = \int \tilde{f}\, dV$ from the variable part of $\tilde{F}$ is due only to the presence of the interface. Taking this contribution per

† To avoid misunderstanding, we may mention that the comment in *ECM* §41 about the unsuitability of using $\mathbf{H}$ referred to the electrodynamics of superconductors, where the range of penetration of the magnetic field was regarded as infinitesimally short. The Ginzburg–Landau equations, however, are applied to the structure of precisely this region.

‡ Strictly speaking, in phase equilibrium the chemical potential, not the density, is constant throughout the system. When taking account of the change in the density it would therefore be necessary to consider the thermodynamic potential Ω, not the free energy.

unit area of the interface, we can therefore calculate the surface tension coefficient as the integral

$$\alpha_{ns} = \int_{-\infty}^{\infty} (\tilde{f} - \tilde{f}_n)\, dx, \tag{46.4}$$

where the constant $\tilde{f}_n$ is the value of $\tilde{f}$ far from the interface, for example within the normal phase.

For the normal phase, the free energy $f_n = f_{n0} + B^2/8\pi = f_{n0} + H_c^2/8\pi$, so that

$$\begin{aligned}\tilde{f}_n &= f_n - H_c^2/4\pi \\ &= f_{n0} - H_c^2/8\pi \\ &= f_{n0} - a^2/2b,\end{aligned}$$

where the last expression is found by means of (45.9). The quantity $\tilde{f}$ at any point is expressed in terms of the free energy density f by

$$\tilde{f} = f - H_c B/4\pi.$$

Now, using (45.10), we reach the following formula for the surface tension:

$$\alpha_{ns} = \int_{-\infty}^{\infty} \left\{ \frac{B^2}{8\pi} + \frac{\hbar^2}{4m}\left(|\psi'|^2 + \frac{4e^2}{\hbar^2 c^2} A^2 |\psi|^2 + a|\psi|^2 \right) + \frac{1}{2} b |\psi|^4 - \frac{H_c B}{4\pi} + \frac{a^2}{2b} \right\} dx. \tag{46.5}$$

The integrand vanishes, as it should, both within the normal phase ($x \to -\infty$), where $\psi = 0$ and $B = H_c$, and within the superconducting phase ($x \to \infty$), where $|\psi|^2 = -a/b$, $B = 0$.

It should be noted that, in the integrand of (46.5), the term $i\mathbf{A}.\nabla\psi$ does not appear, since $A_x = 0$. The corresponding term also does not appear in (45.12), and so the equation remaining has real coefficients; its solution may therefore be taken real, as will be assumed below. The first term in the current density (45.14) disappears, leaving

$$\mathbf{j} = -(2e^2/mc)\, \psi^2 \mathbf{A}. \tag{46.6}$$

We shall use instead of the variable x and the functions $A(x)$ and $\psi(x)$ the dimensionless quantities

$$\bar{x} = x/\delta, \quad \bar{\psi} = \psi\sqrt{(b/|a|)}, \quad \bar{A} = A/H_c\delta, \quad \bar{B} = d\bar{A}/d\bar{x} = B/H_c. \tag{46.7}$$

In the rest of this section, only these quantities will be used, and the bars over the letters will be omitted, for brevity. In these variables, equation (45.12) becomes

$$\psi'' = \varkappa^2[(\tfrac{1}{2}A^2 - 1)\psi + \psi^3]. \tag{46.8}$$

Equation (45.13) with **j** from (46.6) is

$$A'' = A\psi^2. \tag{46.9}$$

The boundary conditions on these equations in the problem considered (corresponding to the n and s phases as $x \to -\infty$ and $x \to \infty$) are

$$\left.\begin{array}{lll} \psi = 0, & B = A' = 1 & \text{at} \quad x = -\infty, \\ \psi = 1, & A' = 0 & \text{at} \quad x = \infty. \end{array}\right\} \tag{46.10}$$

It is easily verified that equations (46.8), (46.9) have the first integral

$$(2/\varkappa^2)\,\psi'^2 + (2 - A^2)\,\psi^2 - \psi^4 + A'^2 = \text{constant} = 1, \tag{46.11}$$

the value of the constant being determined from the boundary conditions.†

Lastly, the expression (46.5) becomes

$$\alpha_{ns} = \frac{\delta H_c^2}{8\pi} \int_{-\infty}^{\infty} \left[\frac{2}{\varkappa^2}\psi'^2 + (A^2 - 2)\,\psi^2 + \psi^4 + (A' - 1)^2\right] dx$$

$$= \frac{\delta H_c^2}{4\pi} \int_{-\infty}^{\infty} \left[\frac{2}{\varkappa^2}\psi'^2 + A'(A' - 1)\right] dx, \tag{46.12}$$

with the second equation obtained by taking ψ^4 from (46.11).

Let us now examine the above equations, and take first the case $\varkappa \ll 1$ (which usually occurs in superconducting pure metals). This inequality signifies that $\delta(T) \ll \xi(T)$, i.e. the magnetic field varies considerably over a distance small compared with the characteristic distance of variation of the function $\psi(x)$.

Figure 6 shows diagrammatically the distribution pattern for the field and ψ in this case. Where the field is large we have $\psi \approx 0$; then the field falls abruptly and $\psi(x)$ begins to change slowly (over distances $\sim 1/\varkappa$) in the absence of the field. Putting $A = 0$ in (46.11), we find the equation

$$\psi' = \frac{\varkappa}{\sqrt{2}}(1 - \psi^2),$$

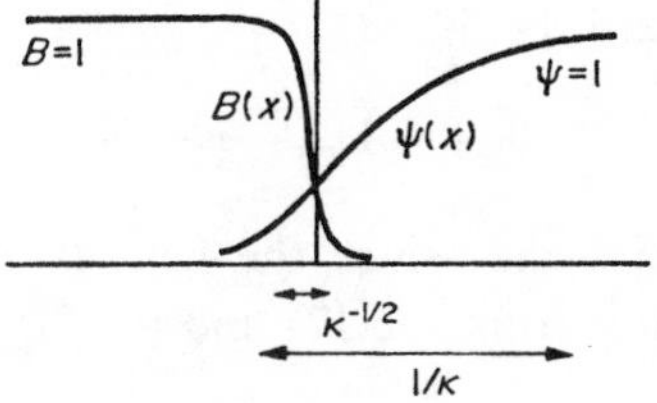

FIG. 6.

† From the conditions (46.10) it necessarily follows that $\psi' = 0$ at $x = \pm\infty$, and from the same conditions and (46.9), that $A'' = 0$ and $A = 0$ at $x = \infty$; the definite value of $A(\infty)$ is the result of taking ψ real.

which is to be solved with the condition $\psi = 0$ at $x = 0$, taken somewhere in the region of decreasing field. This solution is

$$\psi = \tanh \varkappa x/\sqrt{2}, \tag{46.13}$$

and a calculation of the integral (46.12) with this function (and $A = 0$) gives

$$\alpha_{ns} = \frac{H_c^2 \delta}{3\sqrt{2}\pi\varkappa} = \frac{H_c^2}{8\pi} \cdot \frac{1.9\delta}{\varkappa}. \tag{46.14}$$

The error in this value is due to neglecting here the contribution to the integral from the region where the field decreases. To estimate the width δ_1 of this region,† we note, first, that $1/\delta_1^2 \sim \psi^2$ from (46.9); second, that formula (46.13) must remain valid in order of magnitude even at the boundary of the region $x \sim \delta_1$, whence $\psi \sim \varkappa\delta_1$. From these two relations we find $\delta_1 \sim 1/\sqrt{\varkappa}$. The contribution to the surface tension from this region is $\sim H_c^2\delta/\sqrt{\varkappa}$, i.e. is small in comparison with (46.14) only in the ratio $\sim \sqrt{\varkappa}$ (so that the accuracy of (46.14) is fairly low).

When the parameter $\varkappa$ increases, the surface tension coefficient passes through zero and becomes negative. This is evident from the fact that the inequality $\alpha_{ns} < 0$ is always satisfied for sufficiently large $\varkappa$: the characteristic distances of variation of $\psi(x)$ in this problem cannot be less than those for $A(x)$, since any change in A causes a change in ψ; hence, for large $\varkappa$, the term $\psi'^2/\varkappa^2$ in the integrand in (46.12) may be neglected, and the integrand is negative since $0 < A' < 1$ (i.e. $0 < B < H_c$ in ordinary units). We shall show that α_{ns} is zero at

$$\varkappa = 1/\sqrt{2}. \tag{46.15}$$

To do so, we rewrite the expression for α_{ns} as

$$\alpha_{ns} = \frac{H_c^2\delta}{8\pi} \int_{-\infty}^{\infty} [(A'-1)^2 - \psi^4]\, dx, \tag{46.16}$$

which is obtained from the first integral (46.12) by integrating the term ψ'^2 by parts and then substituting ψ'' from (46.8). The integral is certainly zero if the integrand is identically zero, i.e. if

$$A' - 1 = -\psi^2; \tag{46.17}$$

the opposite sign cannot occur, since the field $B = A'$ must decrease with increasing x. Eliminating ψ from (46.17) and (46.9), we find

$$A'' = A(1 - A'), \tag{46.18}$$

† We must emphasize that δ_1 is not the same as the field penetration depth in a superconductor adjoining a vacuum. In the latter case, $\psi \sim 1$ in the field penetration region, whereas in penetration from the n phase the field decreases in a region where ψ is small.

the solution of which (with the boundary conditions $A' = 1$ at $x = -\infty$ and $A = 0$ at $x = \infty$) determines the field distribution; by virtue of (46.17), the boundary conditions (46.10) for ψ are then satisfied automatically. We need not actually solve (46.18), but simply verify that for $\varkappa^2 = \frac{1}{2}$ the equation (46.8), which has not yet been used, or equivalently its first integral (46.11), must necessarily be satisfied. Substituting (46.17) in (46.9), we obtain $\psi' = -\frac{1}{2}A\psi$; this value of ψ', with A' from (46.17), in fact satisfies equation (46.11) identically with $\varkappa^2 = \frac{1}{2}$.

PROBLEM

For a superconductor with $\varkappa \ll 1$, find the first field correction to the penetration depth in weak fields.

SOLUTION. We take the surface of the superconductor as the yz-plane, with the z-axis in the direction of the external field $\mathfrak{H}$, and the x-axis into the body. The distribution of the field and ψ in the superconductor is given by equations (46.8), (46.9), which are to be solved with the boundary conditions

$$\psi' = 0, \quad B = A' = \mathfrak{H} \quad \text{at} \quad x = 0,$$
$$\psi' = 1, \quad A = 0 \qquad \text{at} \quad x = \infty;$$

the first of these is (45.15). We seek the solution in the form

$$\psi = 1+\psi_1(x), \quad A = -\mathfrak{H}\, e^{-x}+A_1(x),$$

where ψ_1 and A_1 are small corrections to the solution at $\varkappa = 0$, which corresponds to the London decay of the field (44.13). The correction ψ_1 is given by the equation

$$\psi_1'' = 2\varkappa^2\psi_1+\tfrac{1}{2}\varkappa^2\mathfrak{H}^2\, e^{-2x},$$

whence, with the boundary conditions,

$$\psi_1 = \frac{1}{8}\varkappa^2\mathfrak{H}^2\, e^{-2x} - \frac{1}{4\sqrt{2}}\varkappa\mathfrak{H}^2\, e^{-\sqrt{2}\varkappa x}. \tag{1}$$

For A_1 we can now write the equation

$$A_1'' = A_1 - 2\mathfrak{H}\, e^{-x}\psi_1,$$

and substitute for ψ_1 only the second term in (1), which is of the first order in $\varkappa$. Using the boundary condition ($A_1' = 0$ when $\varkappa = 0$) and neglecting where possible the higher-order terms with respect to $\varkappa$ in the coefficients, we find

$$A_1 = -\tfrac{1}{8}\mathfrak{H}^3[(1+\sqrt{2}\varkappa)\, e^{-x} - e^{-(1+\sqrt{2}\varkappa)x}]. \tag{2}$$

This gives the corrections to the field decay within the superconductor. The effective penetration depth δ_{eff} is, by definition, such that

$$\mathfrak{H}\delta_{\text{eff}} = \int_0^\infty B(x)\, dx = -A(0) = \mathfrak{H} - A_1(0).$$

Returning to ordinary units, we find from (2)

$$\delta_{\text{eff}} = \delta\left[1+\frac{\varkappa}{4\sqrt{2}}\left(\frac{\mathfrak{H}}{H_c}\right)^2\right].$$

§ 47. The two types of superconductor

The sign of the surface tension α_{ns} has a considerable influence on the properties of superconductors. This forms the basis for the division of all superconductors into two classes: those of the first kind, with $\alpha_{ns} > 0$, and those of the second kind, with $\alpha_{ns} < 0$. Since the sign of α_{ns} is governed by the value of the Ginzburg–Landau parameter $\varkappa$, values $\varkappa < 1/\sqrt{2}$ correspond (near T_c) to the former and $\varkappa > 1/\sqrt{2}$ to the latter.†

Let us consider a solid cylindrical superconductor in a longitudinal external magnetic field $\mathfrak{H}$. If the superconductor is of the first kind, it undergoes a phase transition of the first kind as the field increases to a critical value H_c. The role of the surface tension is then (as in any phase transition of the first kind) to impede the formation of the first nuclei of the new phase and thus make possible a metastable continuance of the s phase at fields somewhat above H_c.

If the superconductor is of the second kind, however, the occurrence of "inclusions" of the n phase may be thermodynamically favourable even before H_c is reached; the increase of volume energy is compensated by the negative surface energy of such a nucleus. The lower limit of fields for which this is possible is usually denoted by H_{c1} and called the *lower critical field*. Similarly, starting from a metal in the normal state in a high external field, we reach a value $H_{c2} > H_c$, the *upper critical field*, below which the occurrence of "inclusions" of the s phase is thermodynamically favourable, again because of the advantage from the negative energy of the boundaries. Thus, over a certain range of fields, $H_{c1} < \mathfrak{H} < H_{c2}$, the superconductor is in a *mixed state*.‡ Its properties in this state gradually change from purely superconducting at

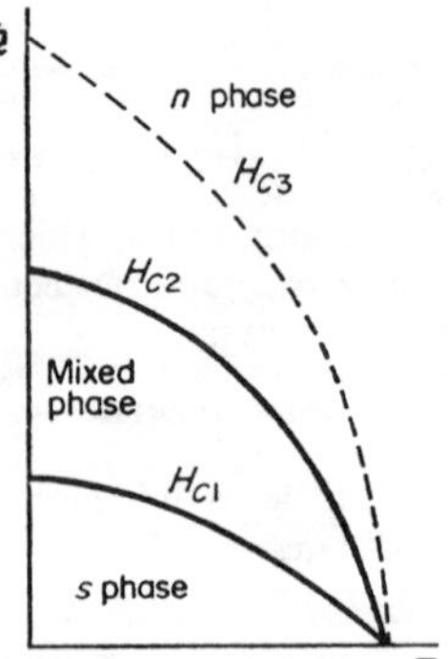

FIG. 7.

† The first kind includes superconducting pure metallic elements; the second kind includes superconducting alloys. The hypothesis that $\varkappa > 1/\sqrt{2}$ in alloys was first put forward by L. D. Landau.

‡ Not to be confused with the intermediate state of a superconductor of the first kind, which results from certain configurations of the sample in the external magnetic field.

H_{c1} to purely normal at H_{c2}; at the same time, the magnetic field gradually penetrates into it. The value of H_c determined only by the relation between the bulk energies of the *n* and *s* phases has no special significance in this case.

The two critical fields depend, of course, on the temperature, and become zero at $T = T_c$. This gives a phase diagram as shown in Fig. 7 for superconductors of the second kind. The broken curve in Fig. 7 is explained below.

The upper critical field can be determined (in the Ginzburg–Landau theory) even without previously ascertaining the structure of the mixed state. We need only observe that in fields somewhat below H_{c2} a nucleus of the *s* phase must have a small value of the order parameter ψ; it is evident that $\psi \to 0$ when $\mathfrak{H}$ tends to H_{c2}. Hence the state of these nuclei can be described by the Ginzburg–Landau equations linearized with respect to ψ. Omitting the non-linear term in (45.12), we obtain the equation

$$\frac{1}{4m}\left(-i\hbar\nabla - \frac{2e}{c}\mathbf{A}\right)^2 \psi = |a|\,\psi, \tag{47.1}$$

where $\mathbf{A}$ is to be taken as the vector potential of the uniform field $\mathfrak{H}$ at $\psi = 0$, when the body is in the normal state and the external field penetrates it completely.

But (47.1) is in form just the Schrödinger's equation for a particle of mass $2m$ and charge $2e$ in a magnetic field, with $|a|$ as the energy level. The boundary conditions also agree: $\psi = 0$ at infinity. It is known (see *QM*, §112) that the minimum energy of a particle moving in a uniform magnetic field is $E_0 = \frac{1}{2}\hbar\omega_H$, where $\omega_H = 2|e|\mathfrak{H}/2mc$; this is the energy value at which the continuous energy spectrum begins. The analogy between the two problems therefore shows that the *s*-phase nuclei described by (47.1) can exist only if

$$|a| > |e|\,\hbar\mathfrak{H}/2mc,$$

so that the critical field $H_{c2} = 2mc|a|/|e|\hbar$. By means of (45.9), (45.17) and (45.18) this formula may be written

$$H_{c2} = \sqrt{2}\varkappa H_c \tag{47.2}$$

(A. A. Abrikosov 1952).

The solution of equation (47.1), with the boundary condition $\psi = 0$ at infinity, corresponds to the formation of an *s*-phase nucleus within the sample, far from its surface. We shall show that the presence of the surface favours nucleation, and that nuclei may thus be formed in a thin surface layer even if $\mathfrak{H} > H_{c2}$ (P. G. de Gennes and D. Saint-James 1963).

The solution of equation (47.1), describing an *s*-phase nucleus near the surface of the body (assumed plane), must satisfy on the surface the boundary condition $\partial\psi/\partial x = 0$, where x is the coordinate along the normal to the surface (the condition (45.15) with $A_x = 0$). To establish the required quantum-mechanical analogy, we recall that the above-mentioned problem of the motion of a particle in a uniform magnetic field is in turn equivalent to that of motion in a one-dimensional parabolic potential well

$$U = \tfrac{1}{2}\,2m\omega_H^2(x-x_0)^2,$$

where x_0 is a constant corresponding to the "centre of the orbit" (see *QM*, §112). Let us now consider a double well consisting of two equal parabolic wells lying symmetrically relative to the plane $x = 0$ (Fig. 8). The ground state of a particle in such a field corresponds to a wave function $\psi(x)$ that is

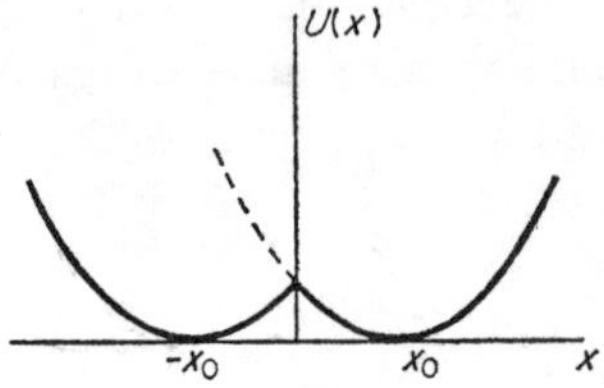

FIG. 8.

even in x and has no zeros; such a function automatically satisfies the condition $\psi' = 0$ at $x = 0$. The ground level of a particle in the double well is, however, below that in the single well;† when applied to the nucleation problem, this proves the above assertion about their easier formation near the surface.

Numerical calculation of the level in a double well gives the result that its minimum value (as a function of the parameter x_0) is $0.59E_0$. Repeating the arguments that led to (47.2), we find that the upper limit of fields in which surface nuclei of the s phase occur is $H_{c3} = H_{c2}/0.59$, i.e.

$$H_{c3} = 1.7H_{c2} = 2.4\varkappa H_c. \tag{47.3}$$

Thus, in the range of fields between H_{c2} and H_{c3}, there arises the phenomenon of surface superconductivity; the boundary of this region is shown by the broken curve in Fig. 7. The thickness of the superconducting layer at the surface of the normal phase is of the order of $\xi(T)$. This estimate is easily derived from the same quantum-mechanical analogy: the wave function of a particle in a potential well (at the level E_0) is concentrated in the region $x \sim \hbar/\sqrt{(mE_0)}$. The corresponding dimension of the nucleus is obtained by replacing E_0 by $|a|$ and is, according to (45.17), $\xi(T)$.

The above discussion relates entirely to superconductors of the second kind, but the critical fields H_{c2} and H_{c3} defined here may have a certain physical significance for those of the first kind also.

If $\varkappa$ is in the range $1/\sqrt{2} = 0.71 > \varkappa > 0.59/\sqrt{2} = 0.42$, then $H_{c2} < H_c$ but $H_{c3} > H_c$. Although no mixed phase occurs in this case, there is surface superconductivity in the range of fields between H_c and H_{c3}.

† This is because the potential energy in the half-space $x < 0$ is less than it would be for a single well shown by the broken curve in Fig. 8; see, for example, *QM*, §50, Problem 3.

Finally, in the sense of the above derivation, the value of H_{c2} (47.2) determines (for any $\varkappa$) the upper limit of fields in which the formation of s-phase nuclei with arbitrarily small ψ is possible. Hence, in a superconductor of the first kind (where $H_{c2} < H_c$) in fields $\mathfrak{H} < H_{c2}$, the thermodynamically unfavourable normal phase is absolutely unstable; but in the range $H_{c2} < \mathfrak{H} < H_c$ the normal phase can exist as a metastable phase. In this range, the phase transition of the first kind from the n to the s phase can only occur by the formation of s-phase nuclei with finite values of ψ, which is opposed by the positive surface tension at their boundaries (V. L. Ginzburg 1956).

PROBLEM

Determine the critical field for a superconducting sphere of small radius $R \ll \delta$ (V. L. Ginzburg 1958).

SOLUTION. In this case (as in a thin film; see §45, Problem) the superconductivity is destroyed by a phase transition of the second kind. The critical field for a sphere may be found as that below which the n phase ceases to be stable with respect to the formation of s-phase nuclei. As in the text, this amounts to finding the lowest eigenvalue of the Schrödinger's equation (47.1). With the condition $R \ll \delta$, this may be sought by means of perturbation theory with respect to the external field, the unperturbed wave function being ψ=constant (the nucleus occupies the whole volume of the sphere). The eigenvalue is then simply the mean value of the perturbation operator $(2e\mathbf{A}/c)^2/4m$ (the mean value of the operator $(ie\hbar/mc)\,\mathbf{A}.\nabla$ for ψ=constant is zero). The vector potential of the uniform field must here be taken as $\mathbf{A}=\frac{1}{2}\mathfrak{H}\times\mathbf{r}$; with this gauge, the solution ψ = constant satisfies on the surface of the sphere the boundary condition (45.15), which reduces to $\mathbf{n}.\mathbf{A} = 0$. The result of the averaging is

$$E_0 = \frac{e^2}{4mc^2}\,\frac{2}{3}\,\mathfrak{H}^2\overline{r^2} = \frac{e^2\mathfrak{H}^2R^2}{10mc^2}.$$

The critical field is found, as in the text, from the condition $E_0 = |a|$, which gives

$$H_c^{\mathrm{sph}} = \sqrt{20}H_c\delta/R.$$

The legitimacy of using perturbation theory is confirmed by the fact that the value found for E_0 (at $\mathfrak{H} = H_c^{\mathrm{sph}}$), with the condition $R \ll \delta$, is in fact small compared with the next eigenvalue, which would correspond to a wave function varying within the sphere and would be of the order of $\hbar^2/mR^2$.

§ 48. The structure of the mixed state

We shall again consider (as in §47) a cylindrical sample of a superconductor of the second kind in a longitudinal magnetic field $\mathfrak{H}$, and ascertain the structure of the mixed state of the body in fields slightly exceeding the lower critical field H_{c1}.†

In this case there are nuclei of the normal phase in the main superconducting phase. To attain the maximum thermodynamic favourability they must have (with negative surface tension) the largest possible surface. The structure expected is therefore one in which the n-phase nuclei are filaments parallel to the

† The results in this section and in the Problems are due to A. A. Abrikosov (1957).

field. The magnetic field that penetrates into the body and the annular superconductivity currents surrounding these *vortex filaments* are concentrated near the filaments.

As the external field approaches H_{c1}, the number of such filaments in the body decreases and the distance between them increases. When this distance is sufficiently great, the arguments given at the end of §44 become applicable to the individual vortex filaments, whereby the total magnetic flux concentrated near a filament must be an integral multiple of the flux quantum $\phi_0 = \pi\hbar c/|e|$; we shall see later that filaments with the lowest possible flux, ϕ_0 itself, are thermodynamically favourable. The fact that ϕ_0 is not zero is what sets a limit to the further fragmentation of the n-phase nuclei.

When the external field, increasing from low values, reaches H_{c1}, one vortex filament appears in the cylinder. We can write down the thermodynamic condition that determines this point without for the present investigating the structure of the filament itself, but merely using the fact that it is associated with some (positive) energy; this energy per unit length of the filament will be denoted by ε (and calculated below).

It is evident that, in a cylindrical body in a longitudinal external field, the induction $\mathbf{B}$ also will be everywhere parallel to the axis of the cylinder. The same is true of the macroscopic field $\mathbf{H} = \mathbf{B} - 4\pi\mathbf{M}$ defined in §46. The equation $\operatorname{curl} \mathbf{H} = 0$ then shows that $\mathbf{H}$ is constant over the cross-section (and therefore throughout the volume) of the cylinder; because of the boundary condition that the tangential component of $\mathbf{H}$ is continuous, this constant value must be equal to the external field: $\mathbf{H} = \mathfrak{H}$. Thus we have to consider the thermodynamic equilibrium of the body for given volume, temperature, and field strength $\mathbf{H}$. The condition for such an equilibrium is for $\tilde{F}$, the thermodynamic potential with respect to these variables, to be a minimum (see *ECM*, §30). Let $\tilde{F}_s$ be this potential for a superconducting cylinder; since $\mathbf{B} = 0$ in the superconducting phase, $\tilde{F}_s$ is the same as the free energy F_s. Then the potential $\tilde{F}$ for a cylinder with one vortex filament will be

$$\begin{aligned}\tilde{F} &= \tilde{F}_s + L\varepsilon - \int \mathbf{H}\cdot\mathbf{B}\, dV/4\pi \\ &= F_s + L\varepsilon - \mathfrak{H}\int B\, dV/4\pi.\end{aligned}$$

The term $L\varepsilon$ is the free energy of the filament (L being the length of the filament, which is equal to that of the cylinder), and the last term is the difference between the potential $\tilde{F}$ and the free energy F. Since the induction $\mathbf{B}$ is entirely concentrated near the vortex filament in the body, we have $\int B\, dV = L\phi_0$, where ϕ_0 is the flux of the induction through the cross-section of the filament. Thus

$$\tilde{F} = \tilde{F}_s + L\varepsilon - L\phi_0\mathfrak{H}/4\pi. \tag{48.1}$$

The occurrence of vortex filaments becomes thermodynamically favourable when the quantity added to $\tilde{F}_s$ is negative. Equating it to zero, we thus have as the critical value of the external field

$$H_{c1} = 4\pi\varepsilon/\phi_0. \tag{48.2}$$

Let us now consider the structure of a single vortex filament. We shall take only the important case where

$$\varkappa \gg 1, \tag{48.3}$$

i. e. $\delta \gg \xi$. The length ξ determines the order of magnitude of the radius of the "core" of the filament, in which $|\psi|^2$ varies from zero (corresponding to the normal state on the filament axis) to the finite value corresponding to the main s phase; at large distances r from the filament axis, $|\psi|^2$ remains constant.† The induction $B(r)$ varies much more slowly, decaying only at distances $r \sim \sim \delta \gg \xi$. Thus essentially the whole of the magnetic flux passes through the region outside the core, where $|\psi|^2 =$ constant (Fig. 9).

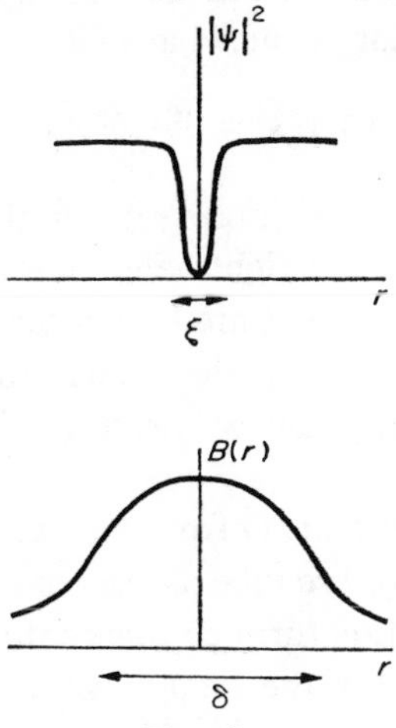

FIG. 9.

The latter fact enables us to use the London equation (whose validity, let us recall, does not depend on the temperature's being close to T_c) in order to find the field distribution. To put it in the appropriate form, we first rewrite the relation (44.7) between the superconductivity current density and the phase of the wave function:

$$\mathbf{A} + \delta^2 \operatorname{curl} \mathbf{B} = \phi_0 \nabla\Phi/2\pi, \tag{48.4}$$

using the penetration depth δ and expressing $\mathbf{j}$ in terms of the induction by $\mathbf{j} = c \operatorname{curl} \mathbf{B}/4\pi$. The London approximation corresponds to the assumption that δ is constant. We integrate (48.4) along a closed contour C that embraces the filament and passes at distances $r \gg \xi$ from its axis. Transforming the integral

† In this section, r will denote a cylindrical coordinate, the distance from the axis.

of **A** by Stokes's theorem into an integral over a surface spanning the contour C, we find

$$\int \mathbf{B}\,.\,d\mathbf{f} + \delta^2 \oint \operatorname{curl} \mathbf{B}\,.\,d\mathbf{l} = \phi_0, \tag{48.5}$$

and a similar transformation of the second integral gives

$$\int (\mathbf{B} + \delta^2 \operatorname{curl} \operatorname{curl} \mathbf{B})\,.\,d\mathbf{f} = \phi_0; \tag{48.6}$$

on the right we have written the lowest possible (non-zero) value, corresponding to a phase increase of only 2π. If the contour C passes at distances $r \gg \delta$ from the filament, where the field and the currents may be regarded as zero, the second integral in (48.5) may be omitted, and we see that ϕ_0 is equal to the total flux of the induction concentrated around the isolated vortex filament. The filament axis itself is a line singularity, a passage around which alters the phase of the wave function.

Since the equation (48.6) must be satisfied for any contour C that satisfies the conditions stated, it shows that we must have

$$\mathbf{B} + \delta^2 \operatorname{curl} \operatorname{curl} \mathbf{B} = \mathbf{B} - \delta^2 \triangle \mathbf{B} = \phi_0 \delta(\mathbf{r}), \tag{48.7}$$

where $\mathbf{r}$ is the two-dimensional position vector in the plane of cross-section of the vortex filament. Writing the right-hand side of this equation as a delta function signifies that distances $\sim \xi$ are here regarded as zero. In all space except the line $\mathbf{r} = 0$, equation (48.7) is the same as the London equation (44.11), but to describe the vortex filament we need a solution with a singularity at $r = 0$.

The field distribution at distances r from the axis in the range $\delta \gg r \gg \xi$ can be found directly from (48.5). We take as the contour C a circle of radius r in this range. The flux of induction through this contour, the first term on the left of (48.5), is only a small part of the total magnetic flux, in the ratio $\sim (r/\delta)^2$; it will be neglected. In the second term, $d\mathbf{l}$ is an element of length of the circle; since the vector $\mathbf{B}$ is along the z-axis (in cylindrical polar coordinates with the axis along the filament) and depends only on r, we have

$$\mathbf{l}\,.\,(\nabla \times \mathbf{B}) = (\mathbf{l} \times \nabla)\,.\,\mathbf{B} = -\partial B_z/\partial r = -dB/dr,$$

where $\mathbf{l}$ is a unit tangent vector to the circle. Thus we obtain the equation

$$\mathbf{l}\,.\operatorname{curl} \mathbf{B} = -dB/dr = \phi_0/2\pi r\delta^2, \tag{48.8}$$

whence

$$B(r) = \frac{\phi_0}{2\pi\delta^2} \log \frac{\delta}{r}, \qquad \xi \ll r \ll \delta. \tag{48.9}$$

Because of the logarithmic dependence, the upper limit of integration (at which we should have $B \approx 0$) may be taken equal to the upper limit of the range of distances r under consideration.

To continue the distribution found above into the range $r \gtrsim \delta$, we use equation (48.7), which is valid for all $r \gg \xi$. Expanding the Laplacian operator in cylindrical polar coordinates (with $B = B_z(r)$), we can rewrite the equation (for $r \neq 0$) as

$$B'' + B'/r + B/\delta^2 = 0.$$

The solution of this equation that decreases as $r \to \infty$ is

$$B(r) = \text{constant} \times K_0(r/\delta),$$

where K_0 is the Macdonald function (the Hankel function with imaginary argument). The constant coefficient is determined by "joining" to the solution (48.9), using the known limiting form $K_0(z) \approx \log(2/z\gamma)$ for $z \ll 1$ ($\gamma = e^C = 1.78$). Thus we have finally

$$B(r) = \frac{\phi_0}{2\pi\delta^2} K_0(r/\delta), \qquad r \gg \xi. \tag{48.10}$$

By means of the known asymptotic expression $K_0(z) \approx (\pi/2z)^{1/2}e^{-z}$ for $z \to \infty$, we therefore find, in particular, the law of decrease of the field far from the axis of the filament:

$$B(r) = \frac{\phi_0}{(8\pi r\delta^3)^{1/2}} e^{-r/\delta}. \tag{48.11}$$

Attention may be drawn to the evident analogy between the properties of vortex filaments in superconductors and those in liquid helium (§29). In both cases, they are line singularities, a passage around which alters the phase of the condensate wave function. The circular paths of superfluid motion round vortex filaments in liquid helium correspond to circular currents in the superconductor; in the former, the velocity v_s of superfluid motion decreases as $1/r$, and in the latter the superconductivity current density

$$j = \frac{c}{4\pi} |\operatorname{curl} \mathbf{B}| = \frac{c\phi_0}{8\pi^2\delta^2 r} \tag{48.12}$$

decreases in the same way. This agreement is to be expected, since in both cases the relation is a direct consequence of the existence of the line singularity. But, whereas in liquid helium this relation $v_s(r)$ extends to all distances, in a superconductor the decrease of $j(r)$ becomes exponential for $r \gg \delta$. The difference is due to the charged state of the electron liquid: the motion of charged particles creates a magnetic field, which in turn screens the field; if the particle charge e is made to tend to zero, the penetration depth $\delta \to \infty$.

We can now calculate the free energy of a vortex filament. The contribution from the region of space outside the core ($r \gg \xi$), is given by the integrals

$$F_{\text{fil}} = \frac{1}{8\pi} \int \mathbf{B}^2 \, dV + \frac{\delta^2}{8\pi} \int (\operatorname{curl} \mathbf{B})^2 \, dV \tag{48.13}$$

taken over this region. For, on varying this expression with respect to **B** (at a given temperature i.e. given δ), we immediately obtain the London equation (48.7) (for $r \neq 0$).† The second integral in (48.13), which diverges logarithmically at both ends of the range $\delta \gg r \gg \xi$, is large in comparison with the first. Substituting here $|\text{curl}\,\mathbf{B}|$ from (48.8), we obtain for the energy per unit length of the filament

$$\varepsilon = (\phi_0/4\pi\delta)^2 \log(\delta/\xi). \qquad (48.14)$$

This expression has logarithmic accuracy, i.e. we assume not only $\delta/\xi \gg 1$ but also $\log(\delta/\xi) \gg 1$; to this accuracy we can neglect the contribution to ε from the core of the filament.

The result (48.14) allows, in particular, the proof of the statement made above, that the formation of vortex filaments with the lowest magnitude of the magnetic flux is thermodynamically favourable. Since the free energy of the filament is proportional to the square of the magnetic flux attached to the filament, the energy for a filament with flux $n\phi_0$ would contain a further factor n^2, and the break-up of such filament into n filaments with flux ϕ_0 would lead to an n-fold gain in energy.

Substituting (48.14) in (48.2), we find the lower critical field

$$H_{c1} = \frac{\phi_0}{4\pi\delta^2} \log(\delta/\xi). \qquad (48.15)$$

As $T \to T_c$, this expression may also be written, by means of (45.19), as‡

$$H_{c1} = H_c \frac{\log \varkappa}{\sqrt{2}\varkappa}. \qquad (48.16)$$

As the external field increases, so does the number of vortex filaments and therefore the penetration of the magnetic field into the superconductor. When the interaction between filaments is taken into account, thermodynamic equilibrium corresponds to a certain ordered configuration of the filaments, forming a two-dimensional lattice in the plane of cross-section of the cylinder.§ For any number density of filaments, the axis of each remains a line such that passage around it alters the phase of the wave function ψ by 2π. The mean value (over the cross-section of the cylinder) of the induction is

$$\bar{B} = \nu\phi_0, \qquad (48.17)$$

† The second term in (48.13), expressed as a function of the current **j**, is

$$2\pi c^2\delta^2 \int \mathbf{j}^2\, dV = \int \tfrac{1}{2}\varrho_s \mathbf{v}_s^2\, dV;$$

in the second expression we have also substituted $\delta^2 = mc^2/4\pi e^2 n_s$, and the density and velocity of the superfluid component according to $\mathbf{j} = e\varrho_s\mathbf{v}_s/m$; see the first footnote to §44. We see that this term can be regarded as the kinetic energy of the superconducting electrons.

‡ Since this formula has been derived on the assumption that $\log\varkappa \gg 1$, it cannot be used when $\varkappa \sim 1$. In particular, for $\varkappa = 1/\sqrt{2}$ the field H_{c1} (like H_{c2}) must be simply H_c.

§ The most favourable lattice seems to be that formed by equilateral triangles with vortex filaments at their vertices.

where ν is the number of filaments per unit area of the cross-section. For, if we integrate the relation (48.4) along the contour around the whole cross-section of the sample, we obtain equation (48.5) with $S\nu\phi_0$ on the right (S being the cross-sectional area); on the left-hand side, the first integral is the total flux $S\bar{B}$ of the induction, and the second represents an edge effect that is small in comparison with the first, in the ratio $\sim \delta/R$, and is therefore negligible (R is the linear size of the cross-section); here it is important, of course, that the field around the filaments decays at distances $\sim \delta$.

So long as the distances d between the filaments remain large in comparison with the correlation radius ξ, we can assert that the magnetic fields of the vortex filaments are simply additive: when $d \gg \xi$, we can still draw a contour embracing any number of vortex filaments in such a way as to pass everywhere far (at distances $\gg \xi$) from their cores. On such a contour, the condition for the London approximation is satisfied (δ is constant), and we therefore again arrive at an equation that differs from (48.7) only in that the delta function on the right is replaced by a sum of delta functions of the distances from each filament; since this equation is linear, the statement made is proved.

When the external field approaches H_{c2}, the distances between vortex filaments become comparable with ξ. This is clear also from the expression (47.2) for the critical field if it is written, by means of (45.9) and (45.16)–(45.18), in the form

$$H_{c2} = \phi_0/2\pi\xi^2; \tag{48.18}$$

it corresponds to a flux ϕ_0 concentrated on an area $\sim \xi^2$.

The disappearance of the superconductivity at $\mathfrak{H} = H_{c2}$ takes place as a phase transition of the second kind. In accordance with the general theory of such transitions, we can assert that the order parameter ψ as a function of the external field vanishes as $|\psi|^2 \propto H_{c2}-\mathfrak{H}$. On the other hand, the magnetization of the substance $M = (B-H)/4\pi$, a quantity independent of the choice of the phase of ψ, is itself proportional to $|\psi|^2$ in this range. Since at $\mathfrak{H} = H_{c2}$ we must also have $B = H_{c2}$, we thus obtain a linear relation between the induction B in a superconductor and the external field near the transition point:

$$B - H_{c2} \propto \mathfrak{H} - H_{c2}. \tag{48.19}$$

PROBLEMS

PROBLEM 1. Calculate the energy of interaction of two vortex filaments at a distance $d \gg \xi$ apart.

SOLUTION. We transform the expression (48.13) for the free energy of a system of two vortex filaments to a form in which the integrations are taken only near each separate filament. To do so, we write, using equation (48.7),

$$\begin{aligned}\mathbf{B}^2+\delta^2(\mathrm{curl}\,\mathbf{B})^2 &= \delta^2\{-\mathbf{B}.\mathrm{curl}\,\mathrm{curl}\,\mathbf{B}+(\mathrm{curl}\,\mathbf{B})^2\}\\ &= \delta^2\,\mathrm{div}\,(\mathbf{B}\times\mathrm{curl}\,\mathbf{B}).\end{aligned}$$

The volume in tegral is transformed to

$$F_{\text{fils}} = (\delta^2/8\pi) \int_{f_1+f_2} \mathbf{B}\times\operatorname{curl}\mathbf{B}.d\mathbf{f}, \tag{1}$$

taken over the cylindrical surfaces f_1 and f_2 (of small radius r_0: $\xi \ll r_0 \ll \delta$) embracing the cores of the filaments. When $d \gg \xi$, the filament fields are additive, i.e. $\mathbf{B} = \mathbf{B}_1+\mathbf{B}_2$. The filament interaction energy is then given by the part of the integral (1) that depends on both $\mathbf{B}_1$ and $\mathbf{B}_2$:

$$L\varepsilon_{12} = (\delta^2/8\pi)\{\int \mathbf{B}_2\times\operatorname{curl}\mathbf{B}_1.d\mathbf{f}_1+\int \mathbf{B}_1\times\operatorname{curl}\mathbf{B}_2.d\mathbf{f}_2\};$$

the integrals of the form $\int \mathbf{B}_2\times\operatorname{curl}\mathbf{B}_1\cdot d\mathbf{f}_2$ tend to zero with r_0. Using (48.8) and (48.10), we now find

$$\varepsilon_{12} = 2\frac{\delta^2}{8\pi}2\pi r_0\frac{\phi_0}{2\pi r_0\delta^2}B(d) = \frac{\phi_0^2}{8\pi^2\delta^2}K_0(d/\delta).$$

In particular, at distances $d \gg \delta$,

$$\varepsilon_{12} = \frac{\phi_0^2}{2^{7/2}\pi^{3/2}\delta^2}(\delta/d)^{1/2}e^{-d/\delta}. \tag{2}$$

PROBLEM 2. Determine the dependence of the mean (over the cross-section of a cylindrical sample) magnetic induction $\bar{B}$ on the external field $\mathfrak{H}$ in the mixed state when the vortex filaments are at distances $d \gg \delta$ apart, forming (in the cross-section of the sample) a lattice of equilateral triangles.

SOLUTION. The area of an equilateral triangle is $\sqrt{3}d^2/4$ (where d is the length of a side), and the number of filaments is half the number of triangles in the lattice (N triangles have $3N$ vertices, but each vertex in the lattice belongs to six triangles that meet there); hence $\nu = 2/\sqrt{3}d^2$.

The thermodynamic potential $\tilde{f}$ per unit volume of the body in the mixed state is

$$\tilde{f} = \tilde{f}_s - \frac{\phi_0}{4\pi}\nu(-H_{c1}+\mathfrak{H})+\frac{1}{2}\sum_{i,k}\varepsilon_{ik},$$

where the second term corresponds to the expression (48.1) with H_{c1} from (48.2); in the third term, ε_{12} is the energy of interaction of two filaments, and the summation is over all filaments passing through a unit area. Because of the exponential decrease of ε_{12} when $d \gg \delta$, it is sufficient to consider only pairs of neighbouring filaments. In the triangular lattice, each filament has six nearest neighbours, so that

$$\frac{1}{2}\sum_{i,k}\varepsilon_{ik} = 6.\frac{1}{2}\sum_{i}\varepsilon_{i1} = 3\nu\varepsilon_{12}(d).$$

Substituting ε_{12} from (2), Problem 1, we find

$$\tilde{f} = \tilde{f}_s + \frac{\phi_0}{2\sqrt{3}\pi\delta^2}\left[-\frac{\mathfrak{H}-H_{c1}}{a^2}+\frac{3\phi_0}{2\sqrt{(2\pi)}\delta^2}\frac{e^{-a}}{a^{5/2}}\right],$$

where $a = d/\delta$. The dependence of a on $\mathfrak{H}$ is determined by the condition for the function $\tilde{f}(a)$ to be a minimum, which gives

$$\mathfrak{H}-H_{c1} = \frac{3\phi_0}{4\sqrt{(2\pi)}\delta^2}\sqrt{a}\,e^{-a}; \tag{3}$$

he term of h igher order in $1/a \ll 1$ is omitted. This equation, together with $\bar{B} = \nu\phi_0$, i.e.

$$a = (2\phi_0/\sqrt{3}\delta^2\bar{B})^{1/2},$$

gives the required dependence $\bar{B}(\mathfrak{H})$. As $\mathfrak{H}\to H_{c1}$, the derivative $d\bar{B}/d\mathfrak{H}$ tends to infinity according to

$$\frac{d\bar{B}}{d\mathfrak{H}} \propto \frac{1}{\mathfrak{H}-H_{c1}}\log^{-3}\frac{1}{\mathfrak{H}-H_{c1}}.$$

§ 49. Diamagnetic susceptibility above the transition point

It has been mentioned at the end of §45 that the range of temperatures T_c in which the fluctuations of the order parameter ψ become large is extremely narrow in superconductors. Outside this range, the fluctuation corrections to the thermodynamic quantities are in general very small. They may, however, be important as regards the magnetic susceptibility of a metal above the transition point: the occurrence, because of fluctuations, of even a relatively small number of superconducting electrons may give a contribution to the susceptibility that exceeds the ordinarily very small susceptibility of the normal metal far from the transition point.†

Let us consider a metal in a weak ($\mathfrak{H} \ll H_c$) external magnetic field at a temperature above but close to T_c. The equilibrium value of the order parameter is here $\psi = 0$, and to calculate its fluctuations we can use the free energy from the Ginzburg–Landau theory. In the expression (45.10), since the fluctuations are small, we need retain only the terms quadratic in ψ, omitting the term in $|\psi|^4$ and taking $\mathbf{A}$ to be the vector potential of the uniform field $\mathfrak{H}$. The fluctuations of the induction $\mathbf{B}$ due to those of ψ are quadratic in ψ (since the current density $\mathbf{j}$ is quadratic). Hence, in the term $\mathbf{B}^2/8\pi$, we can take $\mathbf{B}$ to be the mean (thermodynamic) value of the induction, and neglect its fluctuations. Thus the change in the total free energy of the metal in a fluctuation is given by the following functional of ψ:‡

$$\Delta F[\psi] = \int \left\{ \frac{1}{4m} \left| \left(-i\hbar\nabla - \frac{2e}{c}\mathbf{A} \right) \psi \right|^2 + a|\psi|^2 \right\} dV. \tag{49.1}$$

To calculate the fluctuational contribution ΔF to the free energy, we must regard the functional (49.1) as the "effective Hamiltonian", which determines ΔF from the formula

$$\exp(-\Delta F/T) = \int \exp(-\Delta F[\psi]/T)\, D\psi. \tag{49.2}$$

where the (functional) integration is taken over all distributions $\psi(\mathbf{r})$; see Part 1, §147. In practice, it is carried out by expanding ψ in terms of some complete set of eigenfunctions and integrating over the infinite number of coefficients in this expansion. For a homogeneous system (without external field), the expansion is made simply with respect to plane waves (see, for example, the calculation of the fluctuational correction to the specific heat in Part 1, §147, Problem).

† This effect was pointed out by V. V. Shmidt (1966).

‡ To avoid misunderstanding, we should mention that the magnetic field is not, with regard to the superconductor, the "external field" h in the sense in which it was defined in Part 1, §144. The latter would have to appear in the free energy as a term $-h(\psi+\psi^*)$, which in the present case is certainly impossible because such a term is not invariant with respect to the choice of the phase of ψ.

In the present case, the expansion is to be made with respect to the eigenfunctions of the "Schrödinger's equation"

$$\frac{1}{4m}\left(-i\hbar\nabla-\frac{2e}{c}\mathbf{A}\right)^2\psi = E\psi, \tag{49.3}$$

corresponding to the Hamiltonian (49.1). It has been mentioned in §47 that this equation is formally identical with Schrödinger's equation for the motion of a particle (with mass $2m$ and charge $2e$) in a uniform magnetic field. Its eigenfunctions are labelled by one discrete (n) and two continuous (p_x, p_z) quantum numbers, the eigenvalues depending only on n and p_z (the z-axis is in the direction of $\mathfrak{H}$) and being given by

$$E(n+\tfrac{1}{2}, p_z) = (n+\tfrac{1}{2})\,|e|\,\hbar\mathfrak{H}/mc+p_z^2/4m; \tag{49.4}$$

the number of different eigenfunctions for a given n with p_z in the range dp_z and any possible p_x is

$$[V.\, 2\,|e|\,\mathfrak{H}/(2\pi\hbar)^2\, c]\, dp_z$$

(see *QM*, §112).

For brevity, we shall denote the set of numbers n, p_z, p_x by one symbol q, and write the expansion of the function $\psi(\mathbf{r})$ as

$$\psi = \sum_q c_q\psi_q(\mathbf{r}), \tag{49.5}$$

where $c_q = c_q'+ic_q''$ are arbitrary complex coefficients and the eigenfunctions are assumed normalized by the condition $\int |\psi_q|^2 dV = 1$ (with integration over the volume of the metal).

Substitution of the expansion (49.5) in (49.1) allows, first of all, a change from integration over the volume to summation over q: integrating the first term by parts, we can bring (49.1) to the form

$$\Delta F[\psi] = \int\left\{\psi^*\frac{1}{4m}\left(-i\hbar\nabla-\frac{2e}{c}\mathbf{A}\right)^2\psi+\psi^* a\psi\right\}dV.$$

Substituting (49.5) here and noting that each of the functions ψ_q satisfies equation (49.3) with $E = E_q$ and that the eigenfunctions with different q are orthogonal, we find

$$\Delta F[\psi] = \sum_q |c_q|^2(E_q+a). \tag{49.6}$$

The functional integration in (49.2) denotes integration over all $dc_q'dc_q''$. After the substitution of (49.6), the integrations over all these variables separate, giving

$$\exp(-\Delta F/T) = \prod_q \frac{\pi T}{E_q+a}$$

or

$$\Delta F = -T \sum_q \log \frac{\pi T}{E_q + a}. \tag{49.7}$$

In terms of the quantum numbers n and p_z, this expression becomes

$$\Delta F = -V \frac{2|e|T\mathfrak{H}}{(2\pi\hbar)^2 c} \sum_n \int_{-\infty}^{\infty} \log \frac{\pi T}{E(p_z, n+\frac{1}{2})+a} dp_z. \tag{49.8}$$

This sum diverges for large E, but the divergence is in fact spurious and due only to the fact that the original formula (49.1) is applicable only for slowly varying functions $\psi(\mathbf{r})$: the change in ψ over distances $\sim\xi_0$ must be small. In terms of the eigenvalues E_q, this means that only $E_q \ll \hbar^2/m\xi_0^2$ are allowable. Cutting off the sum over n at some large N which satisfies the condition stated, we use Poisson's formula

$$\sum_{n=0}^{N} f(n+\tfrac{1}{2}) \approx \int_0^N f(x)\,dx - \tfrac{1}{24}[f'(x)]_0^N;$$

see Part 1, (59.10). When applied to (49.8), the integral term here is easily seen to give a contribution to the free energy that is independent of $\mathfrak{H}$; this term is not needed in calculating the magnetic susceptibility, and we shall omit it. In the second term, we can now let $N \to \infty$, so that the cut-off parameter does not appear in the result:†

$$\Delta F = V \frac{e^2 T_c \mathfrak{H}^2}{48\pi^2\hbar m c^2} \int_{-\infty}^{\infty} \frac{dp_z}{a + p_z^2/4m}.$$

Finally, calculation of the integral gives

$$\Delta F = V \frac{e^2 T_c \mathfrak{H}^2}{24\pi\hbar c^2 \sqrt{(ma)}}. \tag{49.9}$$

Hence the magnetic susceptibility is

$$\chi = -\frac{1}{V}\frac{\partial^2 \Delta F}{\partial \mathfrak{H}^2} = -\frac{e^2 T_c}{12\hbar c^2 (m\alpha)^{1/2}(T-T_c)^{1/2}} \tag{49.10}$$

(H. Schmidt 1968, A. Schmid 1969). We see that the susceptibility increases as $(T-T_c)^{-1/2}$ near the transition point. In this range, (49.10) is the principal contribution to the magnetic susceptibility of a normal metal.

† In the coefficient we have put $T \approx T_c$. For T close to T_c, the important values in this integral are $p_z \sim \sqrt{(ma)} \sim \hbar/\xi(T) \ll \hbar/\xi_0$, i.e. satisfy the requirement stated.

PROBLEMS

PROBLEM 1. Determine the magnetic moment of a thin film (thickness $d \ll \xi(T)$) in a weak magnetic field perpendicular to its plane at temperatures $T > T_c$ with $T - T_c \ll T_c$.

SOLUTION. The finite thickness of the film makes the quantum number p_z in (49.4) discrete; for a thin film, we must take in (49.7) only the value $p_z = 0$; even the first non-zero value $p_z \sim \hbar/d$, so that $E \sim \hbar^2/md^2 \gg \hbar^2/m\xi^2 \sim a$. The number of eigenfunctions with given n and p_z (and any possible p_x) is $2|e|\mathfrak{H}S/2\pi\hbar c$, where S is the area of the film; hence the summation over q in (49.7) is to be taken as $(\mathfrak{H}S/\pi\hbar c)\sum_n$. Applying Poisson's formula to the sum, we get

$$\Delta F = S\frac{e^2 T_c \mathfrak{H}^2}{24\pi mc^2 a}.$$

The magnetic moment of the film is

$$M = -\partial\, \Delta F/\partial \mathfrak{H} = -S\frac{e^2 T_c \mathfrak{H}}{12\pi mc^2 \alpha(T-T_c)}.$$

It should be noted that this increases faster, as $T \to T_c$, than for an infinite metal.

PROBLEM 2. The same as Problem 1, but for a sphere of radius $R \ll \xi(T)$ (V. V. Shmidt 1966).

SOLUTION. In this case, of all the eigenvalues of equation (49.3), only the lowest is important, corresponding to the eigenfunction ψ = constant, and equal to $E_0 = e^2R^2\mathfrak{H}^2/10mc^2$; see the discussion in §47, Problem. The sum (49.7) reduces to a single term, and the magnetic moment is

$$M \approx -\frac{T_c}{a}\frac{\partial E_0}{\partial \mathfrak{H}} = \frac{e^2 T_c R^2 \mathfrak{H}}{5mc^2\alpha(T-T_c)}.$$

§ 50. The Josephson effect

Let us consider two superconductors separated by a thin layer of an insulator. For electrons, this layer is a potential barrier, and if it is sufficiently thin there is a finite probability that they will penetrate it by quantum tunnelling. Even if the transmission coefficient of the barrier is small, its difference from zero is of fundamental importance: the two superconductors become one system described by a single condensate wave function. This leads to effects first predicted by B. D. Josephson (1962).

Because there is a single condensate wave function of the system, a superconductivity current can flow through the contact between the two superconductors even when no external potential difference is applied. Just as the current density within the superconductors is determined by the gradient of the phase Φ of the condensate wave function, so the density j of the superconductivity current through the contact depends on the difference of the phase values Φ_1 and Φ_2 on the two sides of the contact.† Since values of the difference $\Phi_2 - \Phi_1$

† In order that the superconductivity current through the contact should have an appreciable value, the thickness of the insulating layer must in fact be very small, $\sim 10^{-7}$ cm. Such distances are small even in comparison with the smallest characteristic length parameter of the superconductor, the coherence length ξ_0. In this sense the layer is to be regarded as of infinitesimal thickness, and the behaviour of the phase within it does not appear in the theory at all.

which differ by an integral multiple of 2π are physically identical, it is clear that the function

$$j = j(\Phi_{21}), \quad \Phi_{21} = \Phi_2 - \Phi_1, \tag{50.1}$$

must be periodic, with period 2π. The operation of time reversal changes the sign of the current j and also that of the phase Φ_{21}, since the wave functions are replaced by their complex conjugates. This means that the function (50.1) must be odd, and is zero when $\Phi_{21} = 0$. Being bounded, of course, $j(\Phi_{21})$ has maximum and minimum values, between which it varies with the phase difference; since the function is odd, these values are equal in magnitude, and will be denoted by $\pm j_m$.

The form (50.1) presupposes that the current is not affected by the magnetic field of the currents within the contact. In the contrary case, the difference Φ_{21} would have to be replaced by the gauge-invariant expression

$$\Phi_2 - \Phi_1 - \frac{2e}{\hbar c}\int_1^2 A_x\, dx.$$

Because the thickness of the insulating layer is very small, the condition for the integral of the continuous function $A_x(x)$ to be negligible is easily satisfied, and the values of A_x itself on either side of the contact may be regarded as equal.

The form of the function $j(\Phi_{21})$ at all temperatures can be established only from the microscopic theory. We shall give here only a phenomenological treatment within the range of applicability of the Ginzburg–Landau theory.

If the contact were entirely impermeable to electrons, the wave functions ψ in each superconductor would satisfy at the boundary of the contact the conditions (45.15):

$$\frac{\partial \psi_1}{\partial x} - \frac{2ie}{\hbar c} A_x \psi_1 = 0, \quad \frac{\partial \psi_2}{\partial x} - \frac{2ie}{\hbar c} A_x \psi_2 = 0.$$

The finite permeability of the barrier and the finite value of ψ at the boundaries of the contact lead to non-zero expressions on the right-hand sides of these conditions, depending on the values of ψ on the other side of the contact. Since ψ is small (near the transition point T_c), we need consider only the terms linear in ψ in these functions, putting

$$\frac{\partial \psi_1}{\partial x} - \frac{2ie}{\hbar c} A_x \psi_1 = -\frac{\psi_2}{\lambda}, \quad \frac{\partial \psi_2}{\partial x} - \frac{2ie}{\hbar c} A_x \psi_2 = \frac{\psi_1}{\lambda}, \tag{50.2}$$

the coefficient $1/\lambda$ being proportional to the permeability of the barrier. The equations (50.2) must satisfy the requirements of symmetry under time reversal, remaining valid under the transformation $\psi \to \psi^*$, $\mathbf{A} \to -\mathbf{A}$, whence it follows that the constant λ is real; then, under the transformation mentioned, equations (50.2) simply become their complex conjugates.

The relation between the superconductivity current through the contact and the difference between the phases of ψ can be determined by applying formula (45.14) to either side of the contact, say side 1:

$$j = -\frac{ie\hbar}{2m}\left(\psi_1^* \frac{\partial \psi_1}{\partial x} - \psi_1 \frac{\partial \psi_1^*}{\partial x}\right) - \frac{2e^2}{mc} A_x \psi_1^* \psi_1.$$

Substituting $\partial\psi_1/\partial x$ from the boundary condition (50.2), we find

$$j = \frac{ie\hbar}{2m\lambda}(\psi_1^*\psi_2 - \psi_1\psi_2^*).$$

For contacts of the same metal, ψ_1 and ψ_2 differ only in phase, and we then have for the current density

$$j = j_m \sin \Phi_{21}, \quad j_m = (e\hbar/m\lambda)\,|\psi|^2. \tag{50.3}$$

As the transition point is approached, $|\psi|^2$ tends to zero as $T_c - T$, and therefore so does the maximum current density through the contact.†

Now let a potential difference be applied from an external source to the tunnel contact, so that there is an electric field **E** in the contact. We shall describe this field by a scalar potential, denoted by V: $\mathbf{E} = -\nabla V$. The influence of this field on the superconductivity current through the contact can be ascertained from the requirements of gauge invariance.

In the absence of the field ($V = 0$), the phase of the wave function is independent of time: $\partial\Phi/\partial t = 0$.‡ To generalize this equation to the case where the electric field is present, we note that the general relation must be invariant under the gauge transformation

$$V \to V - \frac{1}{c}\frac{\partial \chi(t)}{\partial t} \tag{50.4}$$

of the scalar potential, which does not affect the vector potential (assumed independent of time). Just as in the derivation of the transformation (44.3), (44.6), we find that together with V, the phase of the wave function must be transformed, by

$$\Phi \to \Phi + (2e/\hbar c)\,\chi(t). \tag{50.5}$$

† The microscopic theory based on the BCS model shows that a relation of the type (50.3) between j and Φ_{21} is valid at all temperatures. The same theory gives a relation between j_m and the electrical resistance of the contact between two metals in the normal state. This theory is described by I. O. Kulik and I. K. Yanson, *The Josephson Effect in Superconductive Tunneling Structures*, Israel Program for Scientific Translations, Jerusalem 1972.

‡ The time factor $\exp(-2i\mu t/\hbar)$ is eliminated from the wave function because the Hamiltonian $\hat{H}$ of the system is replaced by $\hat{H}' = \hat{H} - \mu\hat{N}$; cf. the sixth footnote to §31.

Hence it is clear that the relation

$$\frac{\partial \Phi}{\partial t}+\frac{2e}{\hbar}V = 0 \tag{50.6}$$

is gauge-invariant; it becomes $\partial\Phi/\partial t = 0$ when $V = 0$.

When the electric field is independent of time, integration of (50.6) gives

$$\Phi = \Phi^{(0)}-(2e/\hbar)\,Vt,$$

where $\Phi^{(0)}$ is independent of time. Hence, if a constant electric potential difference V_{21} is applied to the contact, the phase difference there is

$$\Phi_{21} = \Phi_{21}^{(0)}-(2e/\hbar)\,V_{21}t.$$

Substituting this expression in (50.3), we find the superconductivity current through the contact:

$$j = j_m \sin\left(\Phi_{21}^{(0)}-(2e/\hbar)\,V_{21}t\right). \tag{50.7}$$

We thus arrive at a noteworthy result: the application of a constant potential difference to the tunnel contact causes the appearance of a superconductivity alternating current with frequency

$$\omega_j = 2\,|eV_{21}|/\hbar. \tag{50.8}$$

The power consumed at the contact is jV_{21}; its (time) average value is zero, i.e. there is no systematic expenditure of energy by the external source; this is as it should be for a superconductivity current, which does not involve any dissipation of energy. We must emphasize, however, that when there is an external e.m.f. there will also be a normal current (weak when V_{21} is small) through the contact, and this is accompanied by dissipation.

The conclusion that the superconductivity current through the contact varies periodically with the frequency (50.8) follows, in fact, from the periodic dependence of j on Φ_{21} and the linear time dependence of Φ_{21}; it does not rest on any assumptions about the magnitude of the potential difference. The specific formula (50.7) is valid only if the frequency ω_j is small in comparison with the frequency $\Delta/\hbar$ that characterizes the superconductivity:

$$\hbar\omega_j = 2\,|eV| \ll \Delta(T). \tag{50.9}$$

PROBLEM

Write down the equation for the current in a circuit consisting of a resistor R connected in series with a superconductor having a tunnel contact, with an e.m.f. V_0 acting in the circuit.

SOLUTION. The total voltage drop in the circuit is $V_0 = RJ+V_{21}$, where J is the current in the circuit and V_{21} the potential difference across the contact.† Substituting $J = J_m \sin\Phi_{21}$,

† We neglect the normal current in the superconductor, which is small if V_0 is small.

and V_{21} from (50.6), we find

$$\frac{\hbar}{2|e|}\frac{\partial \Phi_{21}}{\partial t} = V_0 - RJ_m \sin \Phi_{21}.$$

The variable current described by this equation is not sinusoidal.

§ 51. Relation between current and magnetic field in a superconductor

Formulae have been derived in §44 which give the relation between the current and the magnetic field in a superconductor in the limiting (London) case where all quantities vary slowly through the body; the field was assumed to be much less than the critical value. Let us now consider this problem in the general case where the static field, though still assumed weak, varies in any manner in space. The words "varies in any manner" here mean that the field may vary considerably over distances $\sim \xi_0$ (but, of course, will still vary only slightly over distances of the order of the lattice constant; the inhomogeneity of the metal over atomic distances is therefore unimportant).

In the general case, the relation between the current and the magnetic field in a spatially infinite medium is given by an integral formula of the type

$$j_i(\mathbf{r}) = -\int Q_{ik}(\mathbf{r}-\mathbf{r}')\,A_k(\mathbf{r}')\,d^3x', \tag{51.1}$$

where the kernel Q_{ik} depends only on the properties of the medium itself.† The linearity of (51.1) corresponds to the assumption that the field is weak.

The current density may be regarded as the variational derivative of the energy of the system with respect to the vector potential: the change in the Hamiltonian of the system when **A** is varied is

$$\delta H = -(1/c)\int \mathbf{j}.\delta\mathbf{A}\; d^3x;$$

see *QM*, (115.1). Hence the kernel Q_{ik} in (51.1) is the second variational derivative, and the symmetry as regards the order of the twofold differentiation (with respect to $A_i(\mathbf{r})$ and $A_k(\mathbf{r}')$) has the result that

$$Q_{ik}(\mathbf{r}-\mathbf{r}') = Q_{ki}(\mathbf{r}'-\mathbf{r}). \tag{51.2}$$

Expanding $\mathbf{A}(\mathbf{r})$ and $\mathbf{j}(\mathbf{r})$ in Fourier integrals, we can write the relation (51.1) for the Fourier components as

$$j_i(\mathbf{k}) = -Q_{ik}(\mathbf{k})\,A_k(\mathbf{k}), \tag{51.3}$$

where, from (51.2), $Q_{ik}(\mathbf{k}) = Q_{ki}(-\mathbf{k})$.

† The problem of an infinite medium here has only formal significance. Its actual importance lies in the subsequent application of the results to the problem of a finite medium (§52).

Some important properties of the function $Q_{ik}(\mathbf{k})$ follow from the requirements of gauge invariance. The current $\mathbf{j}$ must be unaltered by the gauge transformation $\mathbf{A}(\mathbf{r}) \to \mathbf{A}(\mathbf{r})+\nabla\chi(\mathbf{r})$ or, in Fourier components,

$$\mathbf{A}(\mathbf{k}) \to \mathbf{A}(\mathbf{k})+i\mathbf{k}\chi(\mathbf{k}).$$

This means that the tensor $Q_{ik}(\mathbf{k})$ must be orthogonal to the wave vector:

$$Q_{ik}(\mathbf{k})\,k_k = 0. \tag{51.4}$$

In particular, in a crystal with cubic symmetry, the tensor dependence of Q_{ik} reduces to terms of the forms δ_{ik} and $k_i k_k$; it then follows from (51.4) that

$$Q_{ik} = \left(\delta_{ik} - \frac{k_i k_k}{k^2}\right) Q(\mathbf{k}), \tag{51.5}$$

where $Q(\mathbf{k})$ is a scalar function.

We now choose a potential gauge such that $\operatorname{div} \mathbf{A}(\mathbf{r}) = 0$. This implies that for the Fourier components $\mathbf{k}.\mathbf{A}(\mathbf{k}) = 0$. Hence the relation (51.3) between the current and the potential reduces to

$$\mathbf{j}(\mathbf{k}) = -Q(\mathbf{k})\,\mathbf{A}(\mathbf{k}), \tag{51.6}$$

i.e. is determined only by the scalar function $Q(\mathbf{k})$.

The London case corresponds to the limit of $Q(\mathbf{k})$ as $\mathbf{k} \to 0$. This is easily found by taking the curl of both sides of equation (44.8),

$$\operatorname{curl} \mathbf{j} = -(e^2 n_s/mc) \operatorname{curl} \mathbf{A},$$

and using the fact that $\operatorname{div} \mathbf{A} = 0$. Since the equation of continuity gives $\operatorname{div} \mathbf{j} = 0$, we find

$$\triangle \mathbf{j} = -(e^2 n_s/mc)\,\triangle \mathbf{A}.$$

In infinite space with the functions $\mathbf{j}(\mathbf{r})$ and $\mathbf{A}(\mathbf{r})$ everywhere finite, it then follows that

$$\mathbf{j}(\mathbf{r}) = -(e^2 n_s/mc)\,\mathbf{A}(\mathbf{r}), \tag{51.7}$$

i.e. the value of the current at every point is determined only by the value of the potential at that point. A similar equation is valid between the Fourier components $\mathbf{j}(\mathbf{k})$ and $\mathbf{A}(\mathbf{k})$, and comparison with (51.6) shows that $Q(\mathbf{k})$ is independent of $\mathbf{k}$:†

$$Q(\mathbf{k}) = e^2 n_s/mc\,(= c/4\pi\delta_L^2 \text{ as } \mathbf{k} \to 0). \tag{51.8}$$

The rest of this section will deal with the calculation of $Q(\mathbf{k})$ for the BCS model, which supposes, as already mentioned, an isotropic degenerate Fermi

† In this and the following sections, the London penetration depth is denoted by δ_L.

gas with weak attraction between the particles (electrons). It is also assumed that these particles interact with the magnetic field through their charge e.

In §42 we have given the equations (42.5) for the temperature Green's functions of a Fermi gas in the absence of an external field. The introduction of the magnetic field is achieved by replacing the operator ∇ by $\nabla - ie\mathbf{A}/c$ in the Hamiltonian $\hat{H}^{(0)}$ (7.7).† A similar change therefore occurs in equation (7.8) for $\hat{\Psi}$ and correspondingly the change $\nabla \to \nabla + ie\mathbf{A}/c$ in the similar equation for $\hat{\Psi}^+$; it is evident that the same applies to the equations for $\hat{\Psi}^M$ and $\hat{\bar{\Psi}}^M$. The spin term ($\sim \boldsymbol{\sigma}.\mathbf{H}$), corresponding to the direct interaction of the magnetic moment of the electron with the field, is small and may be neglected in the Hamiltonian and in the equations. When the operator ∇ acts on the functions $\mathcal{G}(\tau, \mathbf{r}; \tau', \mathbf{r}')$ and $\overline{\mathcal{F}}(\tau, \mathbf{r}; \tau', \mathbf{r}')$, the operators $\hat{\Psi}^M(\tau, \mathbf{r})$ and $\hat{\bar{\Psi}}^M(\tau, \mathbf{r})$ respectively are differentiated. Hence, in equations (42.5), the magnetic field is introduced by the same substitutions $\nabla \to \nabla \mp ie\mathbf{A}/c$.

The presence of the external field makes the system no longer homogeneous in space, and the dependence of the Green's functions on the arguments $\mathbf{r}$ and $\mathbf{r}'$ is no longer simply a dependence on $\mathbf{r} - \mathbf{r}'$; but the functions still depend on τ and τ' only through the difference $\tau - \tau'$. We shall write down immediately the equations for the Fourier components with respect to $\tau - \tau'$:

$$\left.\begin{aligned} \left\{i\zeta_s + \frac{1}{2m}\left[\nabla - \frac{ie}{c}\mathbf{A}(\mathbf{r})\right]^2 + \mu\right\} \mathcal{G}(\zeta_s; \mathbf{r}, \mathbf{r}') + g\Xi \overline{\mathcal{F}}(\zeta_s; \mathbf{r}, \mathbf{r}') = \delta(\mathbf{r}), \\ \left\{-i\zeta_s + \frac{1}{2m}\left[\nabla + \frac{ie}{c}\mathbf{A}(\mathbf{r})\right]^2 + \mu\right\} \overline{\mathcal{F}}(\zeta_s; \mathbf{r}, \mathbf{r}') - g\Xi^* \mathcal{G}(\zeta_s; \mathbf{r}, \mathbf{r}') = 0. \end{aligned}\right\} \tag{51.9}$$

For a weak field, the only case we shall consider here, these equations can be linearized; we put

$$\mathcal{G} = \mathcal{G}^{(0)} + \mathcal{G}^{(1)}, \quad \overline{\mathcal{F}} = \overline{\mathcal{F}}^{(0)} + \overline{\mathcal{F}}^{(1)}, \tag{51.10}$$

where the first terms are the values of the functions in the absence of the field, and the second terms are small corrections linear in the field, and we retain in the equations only the terms of the first order of smallness in $\mathbf{A}$.

Here it must be borne in mind that the presence of the field also changes the condensate wave function Ξ, which in this case does not reduce to a constant. This complication, however, does not occur with our choice of the vector potential gauge, in which

$$\operatorname{div} \mathbf{A} = 0. \tag{51.11}$$

This is because the first-order correction (to the constant value $\Xi^{(0)}$) in the scalar function $\Xi(\mathbf{r})$ could only be proportional to div $\mathbf{A}$, and is zero with the condition (51.11). Hence, with the necessary accuracy, we can put $g\Xi =$

† We put $\hbar = 1$ in the rest of this section (in equations (51.9)–(51.19)).

$= g\Xi^{(0)} \equiv \Delta$ in the linearized equations, where Δ, the gap in the energy spectrum of the gas in the absence of the field, is a real quantity.

The linearized equations (51.9) then become

$$\left. \begin{aligned} &\left(i\zeta_s + \frac{\triangle}{2m} + \mu\right) \mathcal{G}^{(1)}(\zeta_s; \mathbf{r}, \mathbf{r}') + \Delta \overline{\mathcal{F}}^{(1)}(\zeta_s; \mathbf{r}, \mathbf{r}') \\ &\qquad = (ie/mc)\, \mathbf{A}(\mathbf{r}).\nabla \mathcal{G}^{(0)}(\zeta_s; \mathbf{r}-\mathbf{r}'), \\ &\left(-i\zeta_s + \frac{\triangle}{2m} + \mu\right) \overline{\mathcal{F}}^{(1)}(\zeta_s; \mathbf{r}, \mathbf{r}') - \Delta \mathcal{G}^{(1)}(\zeta_s; \mathbf{r}, \mathbf{r}') \\ &\qquad = -(ie/mc)\, \mathbf{A}(\mathbf{r}).\nabla \overline{\mathcal{F}}^{(0)}(\zeta_s; \mathbf{r}-\mathbf{r}'). \end{aligned} \right\} \tag{51.12}$$

Since these equations are linear in **A**, it is sufficient to solve them for one Fourier component of the field, i.e.

$$\mathbf{A}(\mathbf{r}) = \mathbf{A}(\mathbf{k})\, e^{i\mathbf{k}.\mathbf{r}}, \quad \mathbf{k}.\mathbf{A}(\mathbf{k}) = 0. \tag{51.13}$$

With this $\mathbf{A}(\mathbf{r})$, the dependence of $\mathcal{G}^{(1)}$ and $\overline{\mathcal{F}}^{(1)}$ on $\mathbf{r}+\mathbf{r}'$ can be separated immediately by putting

$$\left. \begin{aligned} \mathcal{G}^{(1)}(\zeta_s; \mathbf{r}, \mathbf{r}') &= g(\zeta_s; \mathbf{r}-\mathbf{r}')\, e^{i\mathbf{k}.(\mathbf{r}+\mathbf{r}')/2}, \\ \overline{\mathcal{F}}^{(1)}(\zeta_s; \mathbf{r}, \mathbf{r}') &= f(\zeta_s; \mathbf{r}-\mathbf{r}')\, e^{i\mathbf{k}.(\mathbf{r}+\mathbf{r}')/2}. \end{aligned} \right\} \tag{51.14}$$

For example, the first equation (51.12) then becomes

$$\begin{aligned} &\left[i\zeta_s + \frac{1}{2m}\left(\nabla + \frac{1}{2} i\mathbf{k}\right)^2 + \mu\right] g(\zeta_s; \mathbf{r}-\mathbf{r}') + \Delta f(\zeta_s; \mathbf{r}-\mathbf{r}') \\ &\qquad = (ie/mc)\, \mathbf{A}(\mathbf{k}).e^{\,i\mathbf{k}.(\mathbf{r}-\mathbf{r}')/2} \nabla \mathcal{G}^{(0)}(\zeta_s; \mathbf{r}-\mathbf{r}'), \end{aligned}$$

and similarly for the second equation. We now make a Fourier transformation of the functions g and f with respect to $\mathbf{r}-\mathbf{r}'$. We finally arrive at the following pair of algebraic equations:

$$\left. \begin{aligned} &\left[i\zeta_s - \frac{1}{2m}\left(\mathbf{p} + \frac{1}{2}\mathbf{k}\right)^2 + \mu\right] g(\zeta_s, \mathbf{p}) + \Delta f(\zeta_s, \mathbf{p}) \\ &\qquad = -(e/mc)\mathbf{p}.\mathbf{A}(\mathbf{k})\, \mathcal{G}^{(0)}\left(\zeta_s, \mathbf{p} - \frac{1}{2}\mathbf{k}\right), \\ &\left[-i\zeta_s - \frac{1}{2m}\left(\mathbf{p} + \frac{1}{2}\mathbf{k}\right)^2 + \mu\right] f(\zeta_s, \mathbf{p}) - \Delta g(\zeta_s, \mathbf{p}) \\ &\qquad = (e/mc)\, \mathbf{p}.\mathbf{A}(\mathbf{k})\, \overline{\mathcal{F}}^{(0)}\left(\zeta_s, \mathbf{p} - \frac{1}{2}\mathbf{k}\right). \end{aligned} \right\} \tag{51.15}$$

After some simple calculations using (42.7) and (42.8) for $\mathcal{G}^{(0)}$ and $\overline{\mathcal{F}}^{(0)}$, the solution of these equations is found to be

$$g(\zeta_s, \mathbf{p}) = -\frac{e}{mc}\, \mathbf{p}.\mathbf{A}(\mathbf{k}) \frac{(i\zeta_s + \eta_+)(i\zeta_s + \eta_-) + \Delta^2}{(\zeta_s^2 + \varepsilon_+^2)(\zeta_s^2 + \varepsilon_-^2)}, \tag{51.16}$$

where $\varepsilon_\pm = \varepsilon(\mathbf{p}\pm\frac{1}{2}\mathbf{k})$, $\eta_\pm = \eta(\mathbf{p}\pm\frac{1}{2}\mathbf{k})$; the function $f(\zeta_s, \mathbf{p})$ will not be needed below.

Let us now calculate the current. To do so, we start from the expression for the current density operator in the second-quantization representation:†

$$\hat{\mathbf{j}} = \frac{ie}{2m}[(\nabla\hat{\Psi}^+_\alpha)\hat{\Psi}_\alpha - \hat{\Psi}^+_\alpha(\nabla\hat{\Psi}_\alpha)] - \frac{e^2}{mc}\mathbf{A}\hat{\Psi}^+_\alpha\hat{\Psi}_\alpha.$$

To change to the Matsubara representation of this operator, the Heisenberg operators $\hat{\Psi}$, $\hat{\Psi}^+$ are to be replaced by the Matsubara operators $\hat{\Psi}^M$, $\hat{\bar{\Psi}}^M$. Using the definition of the Green's function (37.2), we find that the current density (the diagonal matrix element of the operator $\hat{\mathbf{j}}$, averaged over the Gibbs distribution) may be written

$$\mathbf{j}(\mathbf{r}) = 2\frac{ie}{2m}[(\nabla' - \nabla)\mathcal{G}(\tau, \mathbf{r}; \tau', \mathbf{r}')]_{\mathbf{r}'=\mathbf{r},\, \tau'=\tau+0} - (e^2/mc)\mathbf{A}(\mathbf{r})n, \quad (51.17)$$

where n is the particle number density; the factor 2 comes from $\mathcal{G}_{\alpha\alpha} = 2\mathcal{G}$.

When we substitute $\mathcal{G} = \mathcal{G}^{(0)} + \mathcal{G}^{(1)}$ in (51.17), the term in $\mathcal{G}^{(0)}$ disappears: for a homogeneous isotropic system, $\mathcal{G}^{(0)}(\mathbf{r}-\mathbf{r}')$ is even, and its derivative is zero for $\mathbf{r}-\mathbf{r}' = 0$. Taking a Fourier expansion with respect to $\tau-\tau'$, we obtain

$$\mathbf{j}(\mathbf{r}) = \frac{ie}{m}T\sum_{s=-\infty}^{\infty}[(\nabla' - \nabla)\mathcal{G}^{(1)}(\zeta_s; \mathbf{r}, \mathbf{r}')]_{\mathbf{r}'=\mathbf{r}} - \frac{e^2 n}{mc}\mathbf{A}(\mathbf{r}),$$

and on substituting $\mathbf{A}(\mathbf{r})$ and $\mathcal{G}^{(1)}$ from (51.13) and (51.14),

$$\mathbf{j}(\mathbf{k}) = \frac{2eT}{m}\sum_{s=-\infty}^{\infty}\int \mathbf{p}g(\zeta_s, \mathbf{p})\frac{d^3p}{(2\pi)^3} - \frac{ne^2}{mc}\mathbf{A}(\mathbf{k}).$$

When $g(\zeta_s, \mathbf{p})$ is substituted here from (51.16), it is convenient to use at the same time the fact that the vectors $\mathbf{j}(\mathbf{k})$ and $\mathbf{A}(\mathbf{k})$ are transverse and to average over directions of $\mathbf{p}_\perp$ in a plane perpendicular to $\mathbf{k}$, using the formula

$$\overline{p_{\perp i}p_{\perp k}} = \tfrac{1}{2}p^2\sin^2\theta(\delta_{ik} - k_i k_k/k^2),$$

where θ is the angle between $\mathbf{k}$ and $\mathbf{p}$. We thus obtain the following expression for the function $Q(\mathbf{k})$ which determines the relation between $\mathbf{j}(\mathbf{k})$ and $\mathbf{A}(\mathbf{k})$:

$$\left.\begin{aligned} Q(\mathbf{k}) &= \frac{e^2 T}{m^2 c}\sum_{s=-\infty}^{\infty}\int p^2\sin^2\theta\,\frac{(i\zeta_s+\eta_+)(i\zeta_s+\eta_-)+\Delta^2}{(\zeta_s^2+\varepsilon_+^2)(\zeta_s^2+\varepsilon_-^2)}\frac{d^3p}{(2\pi)^3} + \frac{ne^2}{mc},\\ \varepsilon_\pm^2 &= \eta_\pm^2 + \Delta^2, \quad \eta_\pm = \frac{1}{2m}\left(\mathbf{p}\pm\frac{1}{2}\mathbf{k}\right)^2 - \mu. \end{aligned}\right\} \quad (51.18)$$

† See *QM*, §115. The term giving the contribution to the current from the particle spin is here omitted. For a non-ferromagnetic system, in which the Green's function $\mathcal{G}_{\alpha\beta} = \mathcal{G}\delta_{\alpha\beta}$, this term gives zero on averaging.

The integrals and the sum as written here are formally divergent. Although these divergences are actually spurious, the calculation must be made with caution: until the divergence is removed, the result may depend on the order of integration and summation.

This difficulty can be avoided by making use of the obvious fact that $Q = 0$ when $\Delta = 0$: in a normal metal there is no superconductivity current. We therefore do not alter the result by subtracting from (51.18) the same expression with $\Delta = 0$:

$$Q(\mathbf{k}) = \frac{e^2T}{m^2c} \sum_{s=-\infty}^{\infty} \int p^2 \sin^2\theta \left\{ \frac{(i\zeta_s+\eta_+)(i\zeta_s+\eta_-)+\Delta^2}{(\zeta_s^2+\varepsilon_+^2)(\zeta_s^2+\varepsilon_-^2)} - \right.$$
$$\left. - \frac{1}{(i\zeta_s-\eta_+)(i\zeta_s-\eta_-)} \right\} \frac{d^3p}{(2\pi)^3}. \tag{51.19}$$

This expression is satisfactorily convergent, and the integration and summation can be performed in any order.

First of all, let us note that the relevant values of $\mathbf{k}$ are small in the sense that $k \ll p_F$; this inequality simply expresses the fact that the characteristic distances over which the field and current vary in a superconductor are large in comparison with the distances between the particles, i.e. with $\sim 1/p_F$.

In (51.19), we first integrate with respect to p. The integral comes mainly from a narrow range of momenta near the Fermi surface, $|p-p_F| \sim k$. In this range,

$$\eta_\pm \approx \eta \pm \tfrac{1}{2} v_F k \cos\theta \approx v_F(p-p_F) \pm \tfrac{1}{2} v_F k \cos\theta;$$

the factor p^2 in the integrand may be replaced by p_F^2, and integration over d^3p by one over $2\pi m p_F \, d\eta \, d\cos\theta$. The integral over $d\eta$ of the second term in the braces in (51.19) is then zero: the contour of integration can be closed by an infinite semicircle in the complex η-plane, and the integral vanishes because both poles of the integrand are in the same half-plane (upper or lower, depending on the sign of ζ_s). The integration over $d\eta$ in the first term in (51.19) is elementary, and this leaves only the integral with respect to $x = \cos\theta$. With the density n from the equation $p_F^3 = 3\pi^2 n$, we have as the final result (in ordinary units)

$$\left. \begin{aligned} Q(\mathbf{k}) &= \frac{3\pi T n e^2}{4mc} \sum_{s=-\infty}^{\infty} \int_{-1}^{1} \frac{\Delta^2(1-x^2)\,dx}{[\zeta_s^2+\Delta^2+(\frac{1}{2}\hbar v_F k x)^2](\zeta_s^2+\Delta^2)^{1/2}}, \\ \zeta_s &= (2s+1)\pi T \end{aligned} \right\} \tag{51.20}$$

(J. Bardeen, L. N. Cooper and J. R. Schrieffer 1957).†

In the limit of small k ($k\xi_0 \ll 1$, where $\xi_0 \sim \hbar v_F/\Delta_0 \sim \hbar v_F/T_c$ is the coherence length), it can be shown that (51.20) reduces to the London expression (51.8), which is independent of k, but we shall not pause to prove this.

† The derivation given here, using the temperature Green's functions, is due to A. A. Abrikosov and L. P. Gor'kov (1958).

In the opposite limit, when $k\xi_0 \gg 1$, the important range in the integral (51.20) is $x \lesssim T_c/\hbar k v_F \ll 1$. We can therefore neglect x^2 in comparison with unity in the numerator of the integrand, and then (because of the rapid convergence) extend the integration from $-\infty$ to ∞. The result is

$$Q(\mathbf{k}) = \frac{3\pi^2 n e^2 T}{2mc\hbar v_F k} \sum_{s=-\infty}^{\infty} \frac{\Delta^2}{\zeta_s^2+\Delta^2}.$$

Carrying out the summation by means of (42.10), we can write this as†

$$\left.\begin{aligned} Q(\mathbf{k}) &= c\beta/4\pi k, \\ \beta &= \frac{4\pi n e^2}{mc^2}\,\frac{3\pi^2}{4\hbar v_F}\,\Delta \tanh\frac{\Delta}{2T}, \quad k\xi_0 \gg 1. \end{aligned}\right\} \tag{51.21}$$

When $T \ll T_c$ we have $n_s \approx n$, $\Delta \approx \Delta_0$, and $\beta \sim 1/\delta_L^2\xi_0$. When $T_c - T \ll T_c$, the gap Δ is small, so that $\tanh(\Delta/2T) \approx \Delta/2T_c$; using formulae (40.4) and (40.5), we again find $\beta \sim 1/\delta_L^2\xi_0$. Thus, at all temperatures from 0 to T_c,

$$\beta \sim 1/\delta_L^2\xi_0. \tag{51.22}$$

The function $Q(\mathbf{k})$ therefore remains approximately constant in the range $k \lesssim 1/\xi_0$ (and has a regular expansion in powers of k^2 near $\mathbf{k} = 0$); outside this range, $Q(\mathbf{k})$ decreases, as $1/k$ when $k \gg 1/\xi_0$. This behaviour of $Q(\mathbf{k})$ corresponds to a coordinate function $Q(\mathbf{r})$ that decreases slowly (as $1/r^2$) in the range $r \lesssim \xi_0$ and rapidly (exponentially) outside that range. Thus the correlation between the field and the current always extends to distances $\sim \xi_0$. It should be emphasized that this statement is valid at all temperatures from zero to T_c. We have thus justified the assertion in §44 that ξ_0 is universal as a characteristic length parameter for superconductivity.

§ 52. Depth of penetration of a magnetic field into a superconductor

Let us apply the results of §51 to the problem of the penetration of an external magnetic field into a superconductor, which has been analysed in the London approximation in §44.

Let the superconductor have a plane boundary surface and be in the half-space $x > 0$; let the external field $\mathfrak{H}$ (and therefore the induction $\mathbf{B}$ within the superconductor) be along the z-axis parallel to the surface. Then all quantities depend only on the coordinate x, and the current $\mathbf{j}$ and the vector potential $\mathbf{A}$ (in the gauge with div $\mathbf{A} = 0$) are along the y-axis. Maxwell's equation curl $\mathbf{B} = -\triangle\mathbf{A} = 4\pi\mathbf{j}/c$ reduces to

$$A''(x) = -4\pi j(x)/c, \quad x > 0, \tag{52.1}$$

where the prime denotes differentiation with respect to x.

† A formula of this kind was suggested by A. B. Pippard (1953) from qualitative arguments, before the microscopic theory of superconductivity existed.

The boundary conditions on this equation depend, however, on the physical properties of the metal surface as regards electrons incident on it. The simplest case is that of mirror reflection of electrons from the surface. It is evident that, with this law of reflection, the problem of a half-space is equivalent to that of an infinite medium in which the field $A(x)$ is distributed symmetrically on either side of the plane $x = 0$: $A(x) = A(-x)$. The derivative $A'(x)$, an odd function of x, is continuous at $x = 0$, changing sign as x passes through zero. Thus the condition $B = A' = \mathfrak{H}$ on the surface of the half-space corresponds in the infinite-medium problem to the condition

$$A'(+0)-A'(-0) = 2\mathfrak{H}. \tag{52.2}$$

We multiply (52.1) by e^{-ikx} and integrate with respect to x from $-\infty$ to ∞. On the left-hand side, we write

$$\int_{-\infty}^{\infty} A'' e^{-ikx}\,dx = \int_{-\infty}^{0} (A' e^{-ikx})'\,dx+\int_{0}^{\infty} (A'e^{-ikx})'\,dx+ik \int_{-\infty}^{\infty} A'e^{-ikx}\,dx.$$

The first two integrals together give $-2\mathfrak{H}$, and in the third integral we can simply integrate by parts, since $A(x)$ itself is continuous at $x = 0$. The result is

$$2\mathfrak{H}+k^2A(k) = 4\pi j(k)/c,$$

where $A(k)$ and $j(k)$ are the Fourier components of the functions $A(x)$ and $j(x)$ defined through all space. They are related by $j(k) = -Q(k)\,A(k)$, where $Q(k)$ is given by the formulae derived in §51. Thus we have for the Fourier components of the field

$$A(k) = -\frac{2\mathfrak{H}}{k^2+4\pi Q(k)/c}. \tag{52.3}$$

The penetration depth δ is defined as†

$$\delta = \frac{1}{\mathfrak{H}}\int_{-\infty}^{\infty} B(x)\,dx = -\frac{A(x=0)}{\mathfrak{H}}. \tag{52.4}$$

Expressing $A(x = 0)$ in terms of the Fourier components $A(k)$ and substituting the latter from (52.3), we have

$$\delta = -\frac{1}{\mathfrak{H}}\int_{-\infty}^{\infty} A(k)\frac{dk}{2\pi} = \frac{1}{\pi}\int_{-\infty}^{\infty}\frac{dk}{k^2+4\pi Q(|k|)/c}. \tag{52.5}$$

The important range of values of k in this integral is that for which $k^2 \sim 4\pi Q/c$. In the London case (when $\delta_L \gg \xi_0$), these values are small in the sense that $k\xi_0 \ll 1$. Here $Q(k)$ is given by the expression (51.8), which is independent of k, and the integration in (52.5) leads, of course, to the value $\delta = \delta_L$.

† When the field decays exponentially, this definition agrees with (44.13).

In the opposite (Pippard) case, for which $\delta_L \ll \xi_0$, the important values in the integral are $k \gg 1/\xi_0$. Here $Q(k)$ is given by (51.21), and the integral (52.5) gives

$$\delta = \delta_P = 4/3^{3/2}\beta^{1/3}. \tag{52.6}$$

Using (51.22), we thus find for the Pippard penetration depth

$$\delta_P \sim (\delta_L^2 \xi_0)^{1/3}. \tag{52.7}$$

These calculations have related to the case of mirror reflection of electrons from the surface of the metal. In the London case, however, the penetration depth is independent of the law of reflection, as is clear from the derivation of the value of δ_L in §44; when $\delta \gg \xi_0$, the details of the surface structure are not significant.

In the Pippard case also, however, the dependence of the penetration depth on the law of reflection is in fact very slight. For example, in the opposite case to mirror reflection, namely diffuse reflection, when the velocities of the reflected electrons have directions distributed isotropically for any direction of incidence, the value of δ_P is only a factor of 9/8 larger than for mirror reflection.

§ 53. Superconducting alloys

The presence of impurities has a much more profound effect on the properties of superconductors than on those of normal metals. The corrections to the thermodynamic quantities for a normal metal remain small so long as the impurity atom concentration x is small, and become considerable only when $x \sim 1$, i.e. when the mean distance between impurity atoms becomes comparable with the lattice constant a. We should emphasize that we are speaking here, of course, only of the electronic contributions to the thermodynamic quantities, and only of those that are determined by the mean quantum state distribution density of conduction electrons in momentum space (for example, the specific heat and the magnetic susceptibility in weak fields).

The situation in superconducting metals is different, because of the existence of a characteristic length parameter, which is large compared with a, namely the coherence length ξ_0. Since the scattering of electrons by impurity atoms destroys the correlation between electrons, the properties of the superconductor may alter considerably when the electron mean free path l is comparable with ξ_0, the concentration x being still small. We shall here describe qualitatively the basic results needed for a general understanding of the properties of such low-concentration alloys.†

† The full theory of superconducting alloys by A. A. Abrikosov and L. P. Gor'kov is quite complicated and lies outside the scope of this book. See the original papers in *Soviet Physics JETP* **8**, 1090, 1959; **9**, 220, 1959.

Let the impurity atoms have no angular momentum, and therefore no magnetic moment (non-paramagnetic impurities). In such a case, they only slightly affect the thermodynamic properties of the superconductor in the absence of a magnetic field. The reason is that such impurities do not destroy the symmetry under time reversal: the interaction between impurity atoms (distributed in some manner) and electrons can be described by specifying a potential field $U(\mathbf{r})$. According to Kramers' theorem, the electron energy levels in such a field remain doubly degenerate, and the states corresponding to these levels are changed into each other by time reversal, so that electrons in them can form Cooper pairs. This will happen, as before, near the sharply defined Fermi surface, with the difference that this surface itself now forms the boundary of the occupied states not in momentum space but in the space of quantum numbers in the field $U(\mathbf{r})$; at low impurity concentrations, the density of quantum states near the Fermi surface changes only slightly.

It is therefore clear that, after averaging over the positions of the impurity atoms, we must obtain formulae that differ from those of the theory of pure superconductors only by corrections that are of the same order of smallness as x. When these unimportant corrections are neglected there is, in particular, no change in the temperature T_c of the transition point and the discontinuity of the specific heat at that point (P. W. Anderson 1959). There is therefore also no change in the ratio α^2/b of the coefficients in the Ginzburg–Landau equation (see (45.8)); the form of this equation is independent of the presence or absence of impurities, and it is equally valid for pure superconductors and for superconducting alloys.

On the other hand, the magnetic properties of a superconductor, and in particular the magnetic field penetration depth, change considerably even when $l \sim \xi_0$. Let us estimate the penetration depth, assuming that the mean free path $l \ll \xi_0$ although the concentration $x \ll 1$ (A. B. Pippard 1953).

The collisions of electrons with impurity atoms annihilate the correlation in the motion of electrons at distances $r \gtrsim l$. This means that the kernel $Q(r)$ in the integral relation between the current and the field in the superconductor will decay exponentially at distances $r \sim l \ll \xi_0$. Accordingly, in the momentum representation, the function $Q(k)$ will remain constant in the range $k \lesssim 1/l$. The value of this constant may be found by "joining" at $kl \sim 1$ to formula (51.21), which remains valid for $k \gg 1/l \gg 1/\xi_0$. Thus we find that

$$Q(k) \sim \frac{ne^2}{mc}\frac{l\Delta}{\hbar v_F}\tanh\frac{\Delta}{2T} \quad \text{for} \quad kl \lesssim 1. \tag{53.1}$$

The penetration depth δ is determined by the relation $k^2 \sim Q(k)/c$ when $k \sim 1/\delta$ (see §52). Using (53.1), we find

$$\begin{aligned} \delta &\sim \delta_L^{\text{pure}}(T=0)\left[\frac{\xi_0}{l\tanh(\Delta/2T)}\right]^{1/2} \\ &\sim \delta_L^{\text{pure}}(T)\,(\xi_0/l)^{1/2}; \end{aligned} \tag{53.2}$$

for this formula to be valid, we must have $\delta \gg l$, which justifies the use of (53.1). The superscript "pure" denotes the value in the absence of impurities, and ξ_0 is also assumed to have its value for a pure superconductor. The expression (53.2) can also be represented by a London formula, taking the number density of superconducting electrons in it as

$$n_s \sim n_s^{\text{pure}}\,\xi_0/l. \tag{53.3}$$

In terms of the coefficients α and b in the Ginzburg–Landau equation, the relation (53.2) implies (see (45.16)) that

$$b/\alpha \sim (b/\alpha)_{\text{pure}}\,\xi_0/l.$$

Since also, as mentioned above, α^2/b is independent of the presence of impurities, we find that

$$\alpha \sim \alpha_{\text{pure}}\,\xi_0/l, \quad b \sim b_{\text{pure}}\,(\xi_0/l)^2. \tag{53.4}$$

According to (45.17), we hence have for the correlation radius

$$\xi(T) \sim \xi(T)_{\text{pure}}\,(l/\xi_0)^{1/2} \tag{53.5}$$

and for the parameter $\varkappa$ (45.18)

$$\varkappa \sim \varkappa_{\text{pure}}\,\xi_0/l \gg \varkappa_{\text{pure}}. \tag{53.6}$$

When the mean free path is sufficiently short, $\varkappa$ exceeds $1/\sqrt{2}$, and so sufficiently "dirty" superconductors are of the second kind.

The range of applicability of the Ginzburg–Landau equations to "dirty" superconductors is limited at low temperatures, in practice, only by the condition $T_c - T \ll T_c$. The necessary inequality $\delta(T) \gg l$ is equivalent in this case to the weaker condition

$$\frac{T_c - T}{T_c} \ll \varkappa_{\text{pure}}^2(\xi_0/l)^3 \sim \varkappa^2\xi_0/l.$$

Finally, some comments on the properties of superconductors containing paramagnetic impurities. Such impurities destroy the symmetry of the system under time reversal, and therefore the pairing of electrons (when magnetic moments are present, time reversal implies a change in their sign, i.e. essentially replaces one physical system by another). A quantitative measure of the influence of these impurities on the properties of the superconductor is the mean free path l_s for scattering with change of spin direction (due to the exchange interaction with impurity atoms). The superconductivity disappears when the concentration x reaches the critical value at which $l_s \sim \xi_0$.

Actually, however, there are two critical concentrations, of the same order of magnitude. At the lower (x_1), the gap Δ in the energy spectrum becomes zero; the condensate wave function Ξ, however, becomes zero only at some concentration $x_2 > x_1$. At concentrations between x_1 and x_2, there is *zero-gap* superconductivity. Since, in deriving the London equation in §44, we used only the existence of the condensate function and considerations of gauge invariance, it is clear that the basic properties of the superconductor (the superconductivity current, and the Meissner effect) will continue to exist in this range. The absence of a gap in the spectrum is shown (in the equilibrium properties of the superconductor) by a non-exponential temperature dependence of the specific heat. There is no conflict with the Landau condition for superfluidity (§23), because in disordered systems (such as the alloys in question) this condition does not apply, the elementary excitations having no definite momentum.†

§ 54. The Cooper effect for non-zero orbital angular momenta of the pair

It has been mentioned several times that the occurrence of superfluidity in a Fermi system is due ultimately to the Cooper effect, the formation of bound states (pairing) by mutually attracting particles at the Fermi surface. For a Fermi gas, the condition of attraction is formulated as that of a negative scattering length $a = \int U\, d^3x$, i.e. a positive scattering amplitude for two particles in a state with zero orbital angular momentum of their relative motion, $l = 0$ (this state makes the main contribution to the scattering at low energies).

A much stronger statement is valid, however: pairing (and therefore superfluidity) occurs if the interaction is attractive even for just one value of the angular momentum l (L. D. Landau 1959). We must emphasize that the system in question is an isotropic one (liquid or gas), in which the states can be classified according to the value of l.

The statement will be proved for a Fermi gas by a method that in principle allows us to determine the temperature T_c at which the transition to the superfluid state occurs, using only the properties of the system (normal Fermi gas) at temperatures $T > T_c$.

It has been noted in §18 that, in the mathematical formalism of Green's functions for a normal Fermi system, the energy of the bound state of the pair of particles appears as the pole of the vertex function Γ; the same is true (when $T \neq 0$) of the temperature vertex function, which we denote by τ. When such a pole appears, the whole formalism in fact becomes inapplicable, but it is still applicable at the instant when, with decreasing temperature at $T = T_c$, a pole appears; the binding energy of the pair must then be zero, and the states of the superfluid and normal phases coincide.

In the skeleton diagram

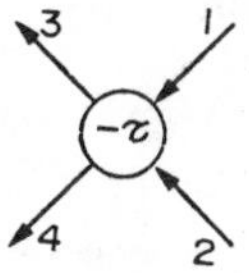

† See also the fourth footnote to §39. The theory of zero-gap superconductivity is described in the original paper by A. A. Abrikosov and L. P. Gor'kov, *Soviet Physics JETP* **12**, 1243, 1961.

the circle denotes $-\mathcal{T}$. The transition point T_c is determined, from the above, as the temperature at which $\mathcal{T}$ has a pole with

$$\zeta_{s1} = \zeta_{s2} = 0, \quad \mathbf{p}_1 + \mathbf{p}_2 = 0. \tag{54.1}$$

The first equation states that the particles paired are on the Fermi surface, and that the binding energy of the pair is zero; the second equation states that the particles paired have opposite momenta.

The pairing of particles occurs if they have any attraction, however slight. It is clear that, if a pole is to occur, the perturbation-theory series for the vertex function must include terms containing integrals that diverge with the condition (54.1) and as $T_c \to 0$ (T_c is small when the attraction is weak); otherwise, all corrections to the (finite) first-approximation term would certainly be small in comparison with that term at all temperatures, and no pole could occur.

This requirement is satisfied by the series of ladder diagrams

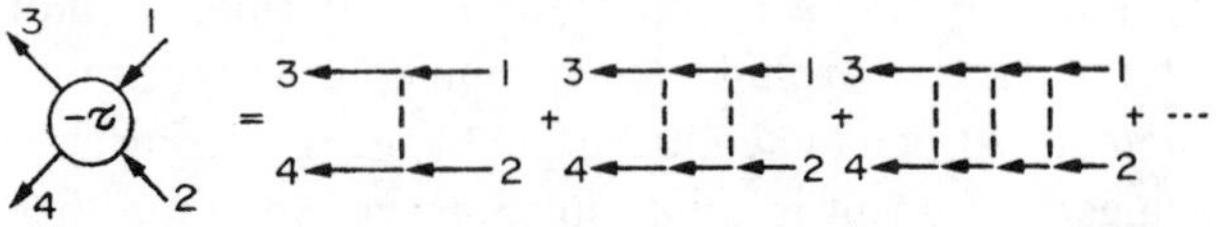

(54.2)

As will be seen below, in all these diagrams (from the second onwards) the smallness with regard to the interaction (from the adding of the broken lines) is compensated, in the sense mentioned, by the divergence of the integrals.†

Applying to this series the procedure used to obtain (17.4) from (17.3), we find that the equation (54.2) is equivalent to the diagram equation

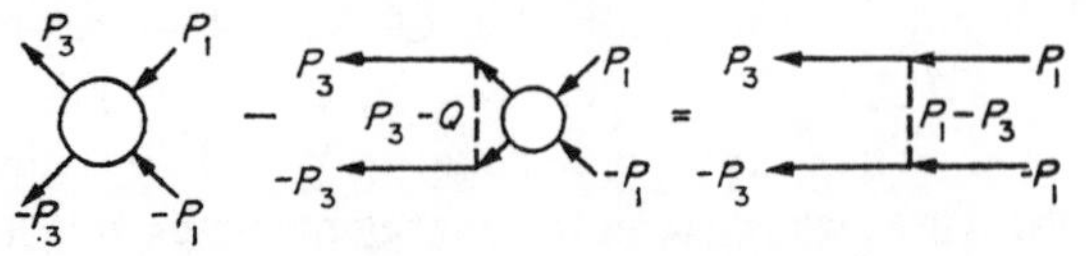

(54.3)

The external and internal lines in the diagrams correspond to the arguments which are shown in (54.3) with (54.1) taken into account:

$$P_1 = (0, \mathbf{p}_1), \quad P_3 = (0, \mathbf{p}_3), \quad Q = (i\zeta_s, \mathbf{q}).$$

The spin dependence of the Green's functions of an ideal gas is separated in the form $\mathcal{G}^{(0)}_{\alpha\beta} = \delta_{\alpha\beta}\mathcal{G}^{(0)}$ and the spin dependence of the vertex function (without antisymmetrization) in the form

$$\mathcal{T}_{\gamma\delta,\alpha\beta}(P_3, P_4; P_1, P_2) = \delta_{\alpha\gamma}\delta_{\beta\delta}\mathcal{T}(P_3, P_4; P_1, P_2).$$

† The diagrams (54.2) should be supplemented by a further series of diagrams with the external lines 3 and 4 interchanged, making the vertex function antisymmetrical with respect to its spin and orbital arguments. However, for the present purpose of determining T_c this is not necessary, since the pole appears simultaneously in both these parts of the vertex function.

Expanding the diagrams (54.3) by the rules given in §38 and cancelling the spin factors, we obtain for $\mathcal{T}$ the integral equation

$$\mathcal{T}(\mathbf{p}_3, -\mathbf{p}_3; \mathbf{p}_1, -\mathbf{p}_1) + T \sum_{s=-\infty}^{\infty} \int U(\mathbf{p}_3-\mathbf{q})\, \mathcal{G}^{(0)}(\zeta_s, \mathbf{q})\, \mathcal{G}^{(0)}(-\zeta_s, -\mathbf{q}) \times \mathcal{T}(\mathbf{q}, -\mathbf{q}; \mathbf{p}_1, -\mathbf{p}_1) \frac{d^3q}{(2\pi)^3} = U(\mathbf{p}_1-\mathbf{p}_3). \tag{54.4}$$

In the sum and integrals here, small values of the discrete variable ζ_s are important, and values of $\mathbf{q}$ near the Fermi surface (see below). Hence we can put $\zeta_s = 0$ and $q = p_F$ in the factors U and $\mathcal{T}$ in the integrand. The vectors $\mathbf{p}_1$ and $\mathbf{p}_3$ are also on the Fermi surface. Thus the functions $\mathcal{T}$ and U in (54.4) will each depend on only one independent variable, the angle between any two of the three vectors $\mathbf{p}_1$, $\mathbf{p}_3$ and $\mathbf{q}$ on the Fermi surface.

Equation (54.4) can now be solved by expanding U and $\mathcal{T}$ in series of Legendre polynomials:

$$\left.\begin{aligned} U(\vartheta) &= \sum_{l=0}^{\infty} (2l+1)\, a_l P_l(\cos\vartheta), \\ \mathcal{T}(\vartheta) &= \sum_{l=0}^{\infty} (2l+1)\, \mathcal{T}_l P_l(\cos\vartheta), \end{aligned}\right\} \tag{54.5}$$

where ϑ is any of the angles mentioned. Substituting these expansions in (54.4) and integrating by the addition theorem for spherical harmonics, we find

$$\mathcal{T}_l(1+a_l\Pi) = a_l, \tag{54.6}$$

where

$$\Pi = T \sum_{s=-\infty}^{\infty} |\mathcal{G}^{(0)}(\zeta_s, \mathbf{q})|^2 \frac{d^3q}{(2\pi)^3} = \frac{T}{(2\pi)^3} \sum_{s=-\infty}^{\infty} \int \frac{d^3q}{\zeta_s^2+\eta_q^2}; \tag{54.7}$$

the function $\mathcal{G}^{(0)}$ is taken from (37.13), and $\eta_q = q^2/2m - \mu \approx v_F(q-p_F)$. According to the summation formula (42.10),

$$\Pi = \frac{1}{2(2\pi)^3} \int \tanh\frac{\eta_q}{2T} \frac{d^3q}{\eta_q}. \tag{54.8}$$

The divergence of the integral over $dq = d\eta/v_F$ at the upper limit is spurious (cf. the last footnote to §39), and the integral must be cut off at some $\eta \approx \tilde{\varepsilon}_l$.† As $T \to 0$, however, the integral diverges logarithmically at the lower limit also, i.e. behaves as $\log(1/T)$.

† Because of the rapid convergence of the sum over s in (54.7), only small values of ζ_s are actually important, and the logarithmic form of the integral over dq justifies the assumption that q is close to p_F.

It is seen from (54.6) that $\mathcal{T}_l$ becomes infinite (i.e. $\mathcal{T}$ has a pole) under the condition

$$1+a_l\Pi = 0. \tag{54.9}$$

This equation has the same form as that which determines the transition point for pairing with $l = 0$, differing from it only in that the "coupling constant" g is replaced by $-a_l$ (cf. (42.11)); taking this as an equation to determine T_c, we must put $\Delta = 0$, and $\varepsilon(p)$ then coincides with η_p. We see, therefore, that the vertex function has a pole if any of the quantities a_l is negative; the transition temperature is then

$$T_c^{(l)} = (\gamma\tilde{\varepsilon}_l/\pi) \exp(-2/|a_l| \nu_F); \tag{54.10}$$

cf. (40.4) and (39.19). If $a_l < 0$ for several different values of l, the transition occurs at a temperature $T_c^{(l)}$ corresponding to the largest $|a_l|$.†

It can be shown that for any Fermi gas (or liquid) consisting of electrically neutral atoms, the quantities a_l must always become negative for sufficiently large l (L. P. Pitaevskiĭ 1959). The reason is that, in the interaction of neutral atoms, there is always a range of (large) distances at which it is an attraction (the van der Waals attraction).

In an actual liquid of this kind, the liquid isotope He^3, the occurrence of superfluidity seems to be due to pairing with $l = 1$.‡ We shall not consider here the structure of the superfluid phase, and shall discuss only briefly the choice of the order parameter which distinguishes this from the normal phase. The anomalous Green's function $F_{\alpha\beta}(t, \mathbf{r}_1; t, \mathbf{r}_2) \equiv F_{\alpha\beta}(\mathbf{r}_1-\mathbf{r}_2)$ is a quantity that is zero above the transition point but not below it; as already mentioned in §41, it acts as the wave function for bound pairs of particles. Its Fourier component $F_{\alpha\beta}(\mathbf{p})$ on the Fermi surface (i.e. for $\mathbf{p} = 2p_F\mathbf{n}$) is a function of the direction $\mathbf{n}$, and not a constant as in pairing with $l = 0$. Since the ψ operators anticommute, the function $F_{\alpha\beta}(\mathbf{n})$ is antisymmetrical, as it should be, with respect to the interchange of particles: $F_{\alpha\beta}(\mathbf{n}) = -F_{\beta\alpha}(-\mathbf{n})$.

In pairing with $l = 1$, as with any odd angular momentum, $F_{\alpha\beta}$ is an odd function of $\mathbf{n}$, and so it is a symmetrical spinor. This means that the spin of the pair is 1, as it should be for a state of two equal fermions with odd l. A symmetrical spinor of rank two is equivalent to a vector, which we denote by $\mathbf{d}$. In the case $l = 1$, the dependence of $\mathbf{d}$ on $\mathbf{n}$ must correspond to the Legendre polynomial $P_1(\cos\theta)$, i.e. must be linear, $d_i = \psi_{ik}n_k$. The complex tensor ψ_{ik} of rank two (not necessarily symmetrical) describes the superfluid phase. There are in fact two different superfluid phases of liquid He^3, with different forms of the tensor ψ_{ik}.

† If all $a_l > 0$ there is no transition, and formula (54.6) for $\mathcal{T}_l$ is valid at all temperatures down to zero. Then all $\mathcal{T}_l$ tend to zero as $1/|\log T|$ when $T \to 0$. This is an instance of the fact, already mentioned in the last footnote to §6, that when $T = 0$ the function $\mathcal{T}$ (and therefore the quasi-particle interaction function f) is zero for particles with opposite momenta.

‡ The transition occurs at a temperature $\sim 10^{-3}$°K. The smallness of T_c ensures the existence of a range in which the theory of the normal Fermi liquid is applicable to liquid He^3.

CHAPTER VI

ELECTRONS IN THE CRYSTAL LATTICE

§ 55. An electron in a periodic field

THE electron clouds of the atoms in a crystal interact strongly with one another, and in consequence we cannot speak of the energy levels of individual atoms, but only of levels for the electron clouds of all the atoms in the body as a whole. The nature of the electron energy spectrum is different for various types of solids. As the first step in the study of these spectra, however, we must consider the more formal problem of the behaviour of a single electron in a spatially periodic external electric field, which serves as a model of a crystal lattice. This subject is dealt with in §§55–60.

The periodicity of the field signifies that it is unaltered by a translation of the form $\mathbf{a} = n_1\mathbf{a}_1+n_2\mathbf{a}_2+n_3\mathbf{a}_3$ (where $\mathbf{a}_1$, $\mathbf{a}_2$, $\mathbf{a}_3$ are the basic lattice vectors and n_1, n_2, n_3 are any integers):

$$U(\mathbf{r}+\mathbf{a}) = U(\mathbf{r}). \tag{55.1}$$

Hence the Schrödinger's equation describing the motion of the electron in such a field is invariant under any transformation $\mathbf{r} \to \mathbf{r}+\mathbf{a}$. Thus, if $\psi(\mathbf{r})$ is the wave function of a stationary state, then $\psi(\mathbf{r}+\mathbf{a})$ is also a solution of Schrödinger's equation describing the same state of the electron. This means that the two functions must be the same apart from a constant factor: $\psi(\mathbf{r}+\mathbf{a})$ = constant $\times\, \psi(\mathbf{r})$. It is evident that the constant must have unit modulus; otherwise, the wave function would tend to infinity when the displacement through $\mathbf{a}$ (or $-\mathbf{a}$) was repeated indefinitely. The general form of a function having this property is

$$\psi_{s\mathbf{k}}(\mathbf{r}) = e^{i\mathbf{k}.\mathbf{r}}u_{s\mathbf{k}}(\mathbf{r}), \tag{55.2}$$

where $\mathbf{k}$ is an arbitrary (real) constant vector and $u_{s\mathbf{k}}$ is a periodic function:

$$u_{s\mathbf{k}}(\mathbf{r}+\mathbf{a}) = u_{s\mathbf{k}}(\mathbf{r}). \tag{55.3}$$

This result was first derived by F. Bloch (1929); wave functions of the form (55.2) are called *Bloch functions*, and in this connection an electron in a periodic field is often called a *Bloch electron*.

For a given value of $\mathbf{k}$, Schrödinger's equation has in general an infinity of different solutions corresponding to an infinity of different discrete values of the electron energy $\varepsilon(\mathbf{k})$; these solutions are labelled by the suffix s in $\psi_{s\mathbf{k}}$.

A similar suffix, the number of the *energy band*, must be assigned to the various branches of the function $\varepsilon = \varepsilon_s(\mathbf{k})$, the electron dispersion relation in the periodic field. In each band, the energy takes values in a certain finite range.

For different bands, these ranges are separated by "energy gaps" or partly overlap; in the latter case, in the overlap region, each value of the energy corresponds to values of **k** that are different for the various bands. Geometrically, this means that the constant-energy surfaces corresponding to two overlapping bands s and s' are in different regions of **k**-space. Formally, the overlapping of the bands signifies degeneracy: different states have the same energy, but this does not cause any peculiarity in the spectrum, because they correspond to different values of **k**. The general case of overlapping is to be distinguished from the intersection of bands, when the values of $\varepsilon_s(\mathbf{k})$ and $\varepsilon_{s'}(\mathbf{k})$ coincide at particular points **k** (the constant-energy surfaces intersect). Degeneracy is usually understood to refer only to such a case; intersection leads to the occurrence of certain peculiarities in the spectrum.

All the functions $\psi_{s\mathbf{k}}$ with different s or **k** are, of course, orthogonal. In particular, the orthogonality of $\psi_{s\mathbf{k}}$ with different s and the same **k** implies that the functions $u_{s\mathbf{k}}$ are orthogonal. Because they are periodic, it is sufficient to integrate over the volume v of one lattice cell; with the appropriate normalization,

$$\int u^*_{s'\mathbf{k}} u_{s\mathbf{k}}\, dv = \delta_{ss'}. \tag{55.4}$$

The significance of the vector **k** is that it determines the behaviour of the wave function under translation: the transformation $\mathbf{r} \to \mathbf{r}+\mathbf{a}$ multiplies it by $e^{i\mathbf{k}.\mathbf{a}}$,

$$\psi_{s\mathbf{k}}(\mathbf{r}+\mathbf{a}) = e^{i\mathbf{k}.\mathbf{a}}\psi_{s\mathbf{k}}(\mathbf{r}). \tag{55.5}$$

Hence it follows immediately that **k** is by definition not unique: values differing by any vector **b** of the reciprocal lattice give the same value of the wave function; the factor $\exp\{i(\mathbf{k}+\mathbf{b}).\mathbf{a}\} = \exp(i\mathbf{k}.\mathbf{a})$. That is, such values of **k** are physically equivalent; they correspond to the same state of the electron, i.e. to the same wave function. We can say that the functions $\psi_{s\mathbf{k}}$ are periodic (in the reciprocal lattice) with respect to the suffix **k**:

$$\psi_{s,\mathbf{k}+\mathbf{b}}(\mathbf{r}) = \psi_{s\mathbf{k}}(\mathbf{r}). \tag{55.6}$$

The energy also is periodic:

$$\varepsilon_s(\mathbf{k}+\mathbf{b}) = \varepsilon_s(\mathbf{k}). \tag{55.7}$$

The functions (55.2) have a certain similarity to the wave functions of a free electron, the plane waves $\psi = \text{constant}\times\exp(i\mathbf{p}.\mathbf{r}/\hbar)$, the conserved momentum being replaced by the constant vector $\hbar\mathbf{k}$. We again arrive, as with the phonon (see Part 1, §71), at the concept of the *quasi-momentum* of an electron in a periodic field. It must be emphasized that in this case there is no actual conserved momentum, since in the external field there is no law of conservation of momentum. It is noteworthy, however, that in a periodic field an electron is still characterized by a certain constant vector.

In a stationary state with a given quasi-momentum $\hbar\mathbf{k}$, the actual momentum can have, with various probabilities, an infinity of values, of the form $\hbar(\mathbf{k}+\mathbf{b})$. This follows because the Fourier-series expansion of a function periodic in space contains terms $e^{i\mathbf{b}.\mathbf{r}}$:

$$u_{s\mathbf{k}}(\mathbf{r}) = \sum_{\mathbf{b}} a_{s,\,\mathbf{k}+\mathbf{b}}\, e^{i\mathbf{b}.\mathbf{r}},$$

and the expansion of the wave function (55.2) in plane waves is therefore

$$\psi_{s\mathbf{k}}(\mathbf{r}) = \sum_{\mathbf{b}} a_{s,\,\mathbf{k}+\mathbf{b}}\, e^{i(\mathbf{k}+\mathbf{b}).\mathbf{r}}. \tag{55.8}$$

The fact that the expansion coefficients depend only on the sums $\mathbf{k}+\mathbf{b}$ expresses the property of periodicity in the reciprocal lattice (55.6). We must emphasize that this fact, like the property (55.6), is not an extra condition imposed on the wave function, but follows necessarily from the periodicity of the field $U(\mathbf{r})$.

All physically different values of the vector $\mathbf{k}$ lie in one unit cell of the reciprocal lattice. The "volume" of this cell is $(2\pi)^3/v$, where v is the volume of the unit cell of the crystal lattice itself. On the other hand, the volume in $(\mathbf{k}/2\pi)$-space determines the number of corresponding states (per unit volume of the body). Thus the number of such states in each energy band is $1/v$, i. e. the number of unit cells per unit volume of the crystal.

In addition to being periodic in $\mathbf{k}$-space, the functions $\varepsilon_s(\mathbf{k})$ are symmetrical under the rotations and reflections that correspond to the symmetry of directions (the crystal class) of the lattice. Independently of the presence or absence of a centre of symmetry in the crystal class concerned, we always have

$$\varepsilon_s(-\mathbf{k}) = \varepsilon_s(\mathbf{k}). \tag{55.9}$$

This property is a consequence of the symmetry under time reversal: because of the symmetry, if $\psi_{s\mathbf{k}}$ is the wave function of a stationary state of the electron, the complex conjugate function $\psi^*_{s\mathbf{k}}$ also describes a state with the same energy. But $\psi^*_{s\mathbf{k}}$ is multiplied by $e^{-i\mathbf{k}.\mathbf{a}}$ on translation, i.e. corresponds to a quasi-momentum $-\mathbf{k}$.†

Next, let us consider two electrons in a periodic field. Regarding them together as one system with the wave function $\psi(\mathbf{r}_1, \mathbf{r}_2)$, we find that this function must be multiplied by a factor $e^{i\mathbf{k}.\mathbf{a}}$ on translation, where $\mathbf{k}$ may be called the quasi-momentum of the system. On the other hand, for large distances between

† In the presence of overlapping bands, it follows from these considerations, strictly speaking, only that $\varepsilon_s(-\mathbf{k}) = \varepsilon_{s'}(\mathbf{k})$, where s and s' are the numbers of certain bands. Equation (55.9) can, however, always be satisfied by appropriately numbering the various branches of the function $\varepsilon(\mathbf{k})$.

the electrons $\psi(\mathbf{r}_1, \mathbf{r}_2)$ reduces to the product of the wave functions of the individual electrons, and is multiplied by $e^{i\mathbf{k}_1.\mathbf{a}}e^{i\mathbf{k}_2.\mathbf{a}}$ on translation. Since the two forms of the factor must be same, we find that

$$\mathbf{k} = \mathbf{k}_1+\mathbf{k}_2+\mathbf{b}. \tag{55.10}$$

In particular, it follows from this that, in a collision of two electrons moving in a periodic field, the sum of their quasi-momenta is conserved to within a reciprocal lattice vector:

$$\mathbf{k}_1+\mathbf{k}_2 = \mathbf{k}_1'+\mathbf{k}_2'+\mathbf{b}. \tag{55.11}$$

A further analogy between the quasi-momentum and the actual momentum is shown by determining the mean velocity of the electron. The calculation requires a knowledge of the velocity operator $\hat{\mathbf{v}} = \hat{\dot{\mathbf{r}}}$ in the $\mathbf{k}$ representation. The operators in this representation act on the coefficients $c_{s\mathbf{k}}$ in the expansion of an arbitrary wave function in terms of the eigenfunctions $\psi_{s\mathbf{k}}$:

$$\psi = \sum_s \int c_{s\mathbf{k}}\psi_{s\mathbf{k}}\, d^3k. \tag{55.12}$$

Let us first find the operator $\hat{\mathbf{r}}$. We have identically

$$\begin{aligned}\mathbf{r}\psi &= \sum_s \int c_{s\mathbf{k}}\mathbf{r}\psi_{s\mathbf{k}}\, d^3k \\ &= \sum_s \int c_{s\mathbf{k}}\left(-i\frac{\partial\psi_{s\mathbf{k}}}{\partial\mathbf{k}}+ie^{i\mathbf{k}.\mathbf{r}}\frac{\partial u_{s\mathbf{k}}}{\partial\mathbf{k}}\right) d^3k.\end{aligned}$$

In the first term we integrate by parts, and in the second we expand the function $\partial u_{s\mathbf{k}}/\partial\mathbf{k}$, which is periodic (like $u_{s\mathbf{k}}$), in terms of the mutually orthogonal functions $u_{s\mathbf{k}}$ with the same $\mathbf{k}$:

$$\partial u_{s\mathbf{k}}/\partial\mathbf{k} = -i\sum_{s'}\langle s\mathbf{k}\,|\,\mathbf{\Omega}\,|\,s'\mathbf{k}\rangle\, u_{s'\mathbf{k}}, \tag{55.13}$$

where $\langle s\mathbf{k}\,|\,\mathbf{\Omega}\,|\,s'\mathbf{k}\rangle$ are constant coefficients. We then find

$$\begin{aligned}\mathbf{r}\psi &= \sum_s \int i\psi_{s\mathbf{k}}\frac{\partial c_{s\mathbf{k}}}{\partial\mathbf{k}}\, d^3k + \sum_{s,\,s'}\int c_{s\mathbf{k}}\langle s\mathbf{k}\,|\,\mathbf{\Omega}\,|\,s'\mathbf{k}\rangle\,\psi_{s'\mathbf{k}}\, d^3k \\ &= \sum_s \int\left\{i\frac{\partial c_{s\mathbf{k}}}{\partial\mathbf{k}}+\sum_{s'}\langle s'\mathbf{k}\,|\,\mathbf{\Omega}\,|\,s\mathbf{k}\rangle\, c_{s'\mathbf{k}}\right\}\psi_{s\mathbf{k}}\, d^3k.\end{aligned}$$

From the definition of the operator $\hat{\mathbf{r}}$, we must have

$$\mathbf{r}\psi = \sum_s \int (\hat{\mathbf{r}}c_{s\mathbf{k}})\,\psi_{s\mathbf{k}}\, d^3k.$$

Comparison with the above expression gives

$$\hat{\mathbf{r}} = i\frac{\partial}{\partial\mathbf{k}}+\hat{\mathbf{\Omega}}, \tag{55.14}$$

where the Hermitian operator $\hat{\mathbf{\Omega}}$ is specified by its matrix $\langle s'\mathbf{k} | \mathbf{\Omega} | s\mathbf{k}\rangle$. It is important to note that this matrix is diagonal in $\mathbf{k}$, and the operator $\hat{\mathbf{\Omega}}$ therefore commutes with the operator $\hat{\mathbf{k}} \equiv \mathbf{k}$.

The velocity operator is found, according to the general rules, by commuting the operator $\hat{\mathbf{r}}$ with the Hamiltonian of the electron. In the $\mathbf{k}$ representation, the Hamiltonian $\hat{H}$ is a diagonal matrix (with respect to $\mathbf{k}$ and with respect to the band numbers s) with elements $\varepsilon_s(\mathbf{k})$.† The operator $\partial/\partial\mathbf{k}$, which acts only on the variable $\mathbf{k}$, is diagonal with respect to s. Hence, in the expression

$$\hat{\mathbf{v}} = \frac{i}{\hbar}(\hat{H}\hat{\mathbf{r}} - \hat{\mathbf{r}}\hat{H})$$

$$= -\frac{1}{\hbar}\left(\hat{H}\frac{\partial}{\partial\mathbf{k}} - \frac{\partial}{\partial\mathbf{k}}\hat{H}\right) + \hat{\dot{\mathbf{\Omega}}},$$

the first term is a diagonal matrix with elements

$$-\frac{1}{\hbar}\left(\varepsilon_s(\mathbf{k})\frac{\partial}{\partial\mathbf{k}} - \frac{\partial}{\partial\mathbf{k}}\varepsilon_s(\mathbf{k})\right) = \frac{1}{\hbar}\frac{\partial\varepsilon_s(\mathbf{k})}{\partial\mathbf{k}}.$$

The matrix elements of $\dot{\mathbf{\Omega}}$ are related to those of $\mathbf{\Omega}$ by

$$\langle s\mathbf{k} | \dot{\mathbf{\Omega}} | s'\mathbf{k}\rangle = (i/\hbar)[\varepsilon_s(\mathbf{k}) - \varepsilon_{s'}(\mathbf{k})]\langle s\mathbf{k} | \mathbf{\Omega} | s'\mathbf{k}\rangle;$$

this is zero when $s = s'$, i. e. $\dot{\mathbf{\Omega}}$ has no elements diagonal with respect to the band number. Thus we have finally for the matrix elements of the electron velocity

$$\langle s\mathbf{k} | \mathbf{v} | s\mathbf{k}\rangle = \partial\varepsilon_s(\mathbf{k})/\hbar\partial\mathbf{k}, \quad \langle s\mathbf{k} | \mathbf{v} | s'\mathbf{k}\rangle = \langle s\mathbf{k} | \dot{\mathbf{\Omega}} | s'\mathbf{k}\rangle \quad (s \neq s'). \tag{55.15}$$

The diagonal elements of this matrix are the mean values of the velocity in the corresponding states. These values are therefore given as functions of the quasi-momentum by

$$\bar{\mathbf{v}}_s = \partial\varepsilon_s(\mathbf{k})/\hbar\,\partial\mathbf{k}, \tag{55.16}$$

which is entirely analogous to the usual classical relation.

So far the discussion has ignored the existence of the electron spin. When relativistic effects (the spin–orbit interaction) are neglected, the inclusion of the spin simply causes a twofold degeneracy of each energy level with a given value of the quasi-momentum $\mathbf{k}$, there being two values of the spin projection on any fixed direction in space. When the spin–orbit interaction is taken into account, the situation differs according as the crystal lattice does or does not have a centre of inversion.

† More precisely, it is the $\mathbf{k}s$ representation. We may recall that the wave functions $c_{s\mathbf{k}}$ in this representation are not completely arbitrary, but must be periodic in $\mathbf{k}$.

The spin–orbit interaction for an electron in a periodic field is described by the operator

$$\hat{H}_{sl} = \frac{i\hbar^2}{4m^2c^2}\boldsymbol{\sigma}\times\nabla U.\nabla, \tag{55.17}$$

where $\boldsymbol{\sigma}$ is the Pauli matrix vector (see *RQT*, §33). The wave functions on which this operator acts are spinors $\psi_{s\mathbf{k}\alpha}$ of rank one, where α is the spinor index. According to Kramers' theorem (see *QM*, §60), which relates to any electric field (periodic or not), the complex conjugate spinors $\psi_{s\mathbf{k}\alpha}$ and $\psi^*_{s\mathbf{k}\alpha}$ always describe two different states with the same energy. Since also the function $\psi^*_{s\mathbf{k}\alpha}$ corresponds to the quasi-momentum $-\mathbf{k}$, we again obtain (and now with inclusion of the spin-orbit interaction) a relation of the type (55.9):

$$\varepsilon_{s\sigma}(-\mathbf{k}) = \varepsilon_{s\sigma'}(\mathbf{k}), \tag{55.18}$$

where σ and σ' denote two different (time-reversed) spin states.†

Equation (55.18) does not, of course, imply degeneracy in the sense considered above, since the energies on the two sides of the equation refer to different values of $\mathbf{k}$. If the lattice has a centre of inversion, however, the states $\mathbf{k}$ and $-\mathbf{k}$ have the same energy. We then get $\varepsilon_{s\sigma}(\mathbf{k}) = \varepsilon_{s\sigma'}(\mathbf{k})$, which again signifies double degeneracy of each level with a given quasi-momentum.

As well as the degeneracy due to the symmetry under time reversal, for an electron in a periodic field there can also be degeneracy due to the spatial symmetry of the lattice. This is dealt with in §68.

PROBLEMS

PROBLEM 1. Find the dispersion relation for one-dimensional motion of an electron in the periodic field shown in Fig. 10 (R. de L. Kronig and W. G. Penney 1930).

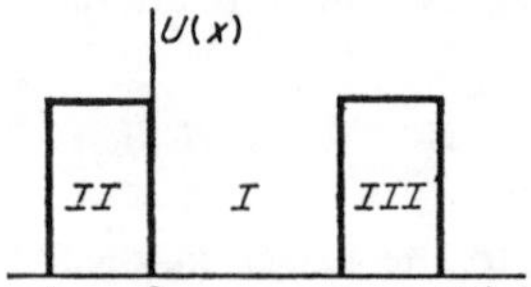

FIG. 10.

SOLUTION. The wave function in the well region I ($0 < x < a$) is of the form

$$\psi = c_1 e^{i\varkappa_1 x} + c_2 e^{-i\varkappa_1 x}, \quad \varkappa_1 = \sqrt{(2m\varepsilon)}/\hbar, \tag{1}$$

and in the barrier region II ($-b < x < 0$)

$$\psi = c_3 e^{i\varkappa_2 x} + c_4 e^{-i\varkappa_2 x}, \quad \varkappa_2 = \sqrt{[2m(\varepsilon - U_0)/\hbar]}\cdot \tag{2}$$

† When the spin–orbit interaction is taken into account, the spin projection operator does not commute with the Hamiltonian; the spin projection is therefore not conserved, and the spin states cannot, strictly speaking, be characterized by this number.

In the next barrier region III, the wave function can differ from (2) only by the phase factor $e^{ik(a+b)}$, $a+b$ being the period of the field:

$$\psi = e^{ik(a+b)}(c_3 e^{i\varkappa_2(x-a-b)} + c_4 e^{-i\varkappa_2(x-a-b)}). \tag{3}$$

The conditions of continuity of ψ and ψ' at $x=0$ and $x=a$ give four equations for $c_1, \ldots, c_4$; the condition for these equations to be compatible leads to the dispersion relation

$$\cos k(a+b) = \cos \varkappa_1 a . \cos \varkappa_2 b - \frac{1}{2}\left(\frac{\varkappa_2}{\varkappa_1} + \frac{\varkappa_1}{\varkappa_2}\right) \sin \varkappa_1 a . \sin \varkappa_2 b, \tag{4}$$

which implicitly determines the required relation $\varepsilon(k)$. For $\varepsilon < U_0$, $\varkappa_2$ is imaginary, and the equation must then be written

$$\cos k(a+b) = \cos \varkappa_1 a \cosh |\varkappa_2 b| + \frac{1}{2}\left(\frac{|\varkappa_2|}{\varkappa_1} - \frac{\varkappa_1}{|\varkappa_2|}\right) \sin \varkappa_1 a \sinh |\varkappa_2 b|. \tag{5}$$

If we take in (5) the limit $U_0 \to \infty$, $b \to 0$ with $U_0 b$ = constant $\equiv Pa$, we obtain the disper sion relation

$$\cos ka = \cos \varkappa_1 a + \frac{Pma^2}{\hbar^2} \frac{\sin \varkappa_1 a}{\varkappa_1 a}. \tag{6}$$

This solves the problem of energy levels in a periodic field composed of delta-function peaks:

$$U(x) = aP \sum_n \delta(x-an).$$

Figure 11 shows a graphical plot of the distribution of roots of equation (6). The right-hand side is represented as a function of $\varkappa_1 a$; when it takes values between ± 1, the roots of the equation take values in the ranges shown by the thick segments of the horizontal axis.

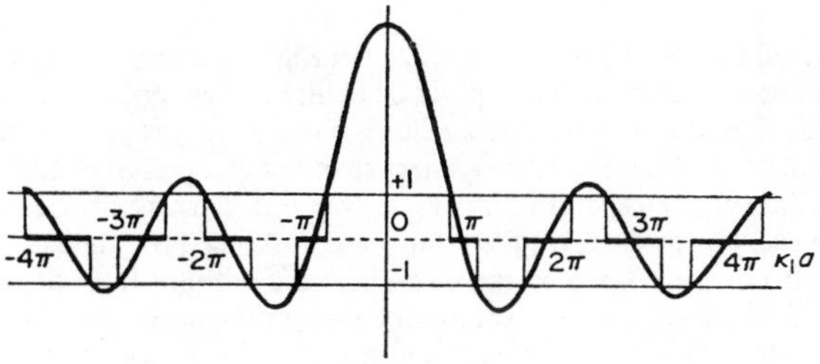

FIG. 11.

PROBLEM 2. Find the dispersion relation for one-dimensional motion of a particle in a weak periodic field $U(x)$.

SOLUTION. Regarding the field as a small perturbation, we start from the zero-order approximation, in which the particle is in free motion described by a plane wave

$$\psi^{(0)}(x) = (Na)^{-1/2} e^{ikx}$$

(with normalization to one particle in the length Na, where a is the period of the field); the particle energy $\varepsilon^{(0)} = \hbar^2 k^2/2m$. We write the periodic function $U(x)$ as a Fourier series:

$$U(x) = \sum_{n=-\infty}^{\infty} U_n e^{2\pi inx/a}.$$

The matrix elements of this field with respect to the plane waves are zero except for transitions between states with wave numbers k and $k' = k + 2\pi n/a$; for these, they are $U_{k'k} = U_n$.

In first-order perturbation theory, the correction to the energy is given by the diagonal matrix element $\varepsilon^{(1)} = U_{kk} = U_0$, a constant independent of k, which merely shifts the origin of energy. An exception occurs, however, for the energy levels near $k = \pi n/a$ ($n = \pm 1, \pm 2, \ldots$). At these points, k differs only in sign from $k' = k - 2\pi n/a$, so that the energies $\varepsilon^{(0)}(k)$ and $\varepsilon^{(0)}(k')$ are equal. Near these values, therefore, the matrix elements are non-zero for transitions between states with similar energies, and to find the correction we must use perturbation theory as applied to the case of close eigenvalues (see *QM*, §79). The answer is given by *QM* (79.4), according to which we have in the present case

$$\varepsilon_n(k) = \tfrac{1}{2}[\varepsilon^{(0)}(k) + \varepsilon^{(0)}(k - K_n) \pm$$
$$\pm \{\tfrac{1}{4}[\varepsilon^{(0)}(k) - \varepsilon^{(0)}(k - K_n)]^2 + |U_n|^2\}^{1/2},$$

where $K_n = 2\pi n/a$, and the additive constant U_0 is omitted; the choice of the sign of the square root is determined by the requirement that, far from the value $k = \pm\frac{1}{2}K_n$, the function $\varepsilon(k)$ becomes $\varepsilon^{(0)}(k)$: the signs + and − refer respectively to the ranges $|k| > |\frac{1}{2}K_n|$ and $|k| < |\frac{1}{2}K_n|$. At the points $k = \pm\frac{1}{2}K_n$ themselves, the function $\varepsilon(k)$ has a discontinuity of $2|U_n|$. In Fig. 12a the energy $\varepsilon(k)$ is shown as a function of the variable k, which runs from $-\infty$ to ∞. If k, the quasi-momentum, is restricted to the range between $\pm\pi/a$, we have Fig. 12b, which shows the first two energy bands.

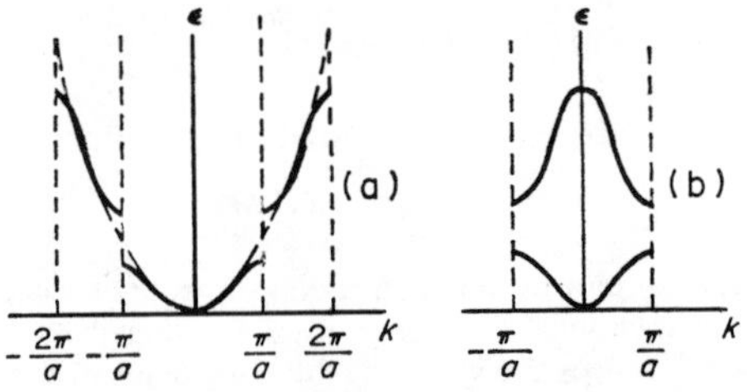

FIG. 12.

It should be noted that the bands in Fig. 12, as in Fig. 11, do not overlap. This is a general property of one-dimensional motion in a periodic field. Each energy level is doubly degenerate (with respect to the sign of k), and no higher degree of degeneracy is possible in one-dimensional motion. Moreover, in the one-dimensional case the limits of each band, the minimum and maximum values of $\varepsilon(k)$, correspond to $k = 0$ and $k = \pi/a$. This is because the wave functions corresponding to energies in the forbidden range are multiplied by some real factor upon displacement through the period a, and therefore increase without limit at infinity. The wave functions in the allowed energy ranges are multiplied by e^{ika} upon such a displacement. At the boundary between the allowed and forbidden ranges, this factor must consequently be a quantity both real and of unit modulus, whence it follows that $ka = 0$ or π.

PROBLEM 3. Find the dispersion relation for a particle in a one-dimensional periodic field consisting of a sequence of symmetrical potential wells satisfying the condition of quasi-classicality (so that the probability that the particle penetrates the barrier between the wells is small).

SOLUTION. This is similar to the solution of the problem of level-splitting in a double well (*QM* §50, Problem 3). Let $\psi_0(x)$ be the normalized wave function describing the motion (with a certain energy ε_0, Fig. 13) in one of the wells, i.e. decaying exponentially beyond the

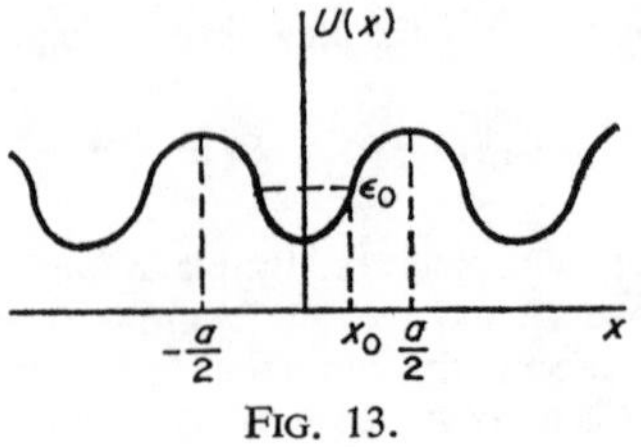

FIG. 13.

limits of this well on either side; it may be either an odd or an even function of x. The correct wave function in the zero-order approximation for the motion of a particle in a periodic field is the sum

$$\psi_k(x) = C \sum_{n=-\infty}^{\infty} e^{ikan}\,\psi_0(x-an), \tag{1}$$

where C is a normalization constant; the displacement $x \to x+a$ multiplies this function by e^{ika}, as it should.

Schrödinger's equations are

$$\psi_k'' + (2m/\hbar^2)[\varepsilon(k) - U(x)]\,\psi_k = 0,$$
$$\psi_0'' + (2m/\hbar^2)[\varepsilon_0 - U(x)]\,\psi_0 = 0;$$

we multiply the former by ψ_0, the latter by ψ_k, subtract corresponding terms, and integrate with respect to x from $-\frac{1}{2}a$ to $\frac{1}{2}a$ (Fig. 13). Since the products $\psi_0(x)\psi_0(x-an)$ with $n \neq 0$ are everywhere negligible, we have

$$\int_{-a/2}^{a/2} \psi_k(x)\,\psi_0(x)\,dx \approx C.$$

The result is

$$\varepsilon(k) - \varepsilon(0) = (\hbar^2/2mC)\,[\psi_0'\psi_k - \psi_0\psi_k']_{-a/2}^{a/2}.$$

At $x = \frac{1}{2}a$, only the terms with $n = 0$ and 1 need be retained in the sum (1); $\psi_0(-\frac{1}{2}a) = \pm\psi_0(\frac{1}{2}a)$, according as the function $\psi_0(x)$ is even or odd:

$$\psi_k(\tfrac{1}{2}a) = C\psi_0(\tfrac{1}{2}a)\,(1 \pm e^{ika}),$$
$$\psi_k'(\tfrac{1}{2}a) = C\psi_0'(\tfrac{1}{2}a)\,(1 \mp e^{ika});$$

similarly, when $x = -\frac{1}{2}a$, only the terms with $n = 0$ and -1 need be retained. The result is

$$\varepsilon(k) - \varepsilon_0 = \pm(2\hbar^2/m)\,\psi_0(\tfrac{1}{2}a)\,\psi_0'(\tfrac{1}{2}a)\cos ka.$$

Here we must substitute

$$\psi_0(\tfrac{1}{2}a) = \left[\frac{m\omega}{2\pi p(\frac{1}{2}a)}\right]^{1/2} \exp\left[-\frac{1}{\hbar}\int_{x_0}^{a/2} |p(x)|\,dx\right],$$

$$\psi_0'(\tfrac{1}{2}a) = \frac{p(\frac{1}{2}a)}{\hbar}\,\psi_0(\tfrac{1}{2}a),$$

where ω is the classical frequency of oscillations of the particle in the well, and x_0 is the turning point, corresponding to the energy ε_0. The final result is

$$\varepsilon(k) - \varepsilon_0 = \pm(\hbar\omega/\pi)\sqrt{D}\cos ka,$$

$$D = \exp\left[-\frac{4}{\hbar}\int_{x_0}^{a/2} |p(x)|\,dx\right].$$

Thus each energy level ε_0 for the motion of a particle in an isolated well is broadened into a band with small width $2\hbar\omega\sqrt{D}/\pi$, dependent on the transmission coefficient of the potential barrier between the two wells.

§ 56. Effect of an external field on electron motion in a lattice

Let us consider the motion of an electron when a constant magnetic field **H** is applied to the lattice. If we start from the Hamiltonian of an electron in a periodic field $U(\mathbf{r})$ in the coordinate representation,

$$\hat{H} = \hat{\mathbf{p}}^2/2m + U(\mathbf{r}), \tag{56.1}$$

where $\hat{\mathbf{p}} = -i\hbar\nabla$ is the operator of the actual momentum, the external magnetic field is introduced in the usual way:

$$\hat{H} = \frac{1}{2m}\left(\hat{\mathbf{p}} - \frac{e}{c}\mathbf{A}\right)^2 + U(\mathbf{r}), \tag{56.2}$$

where $\mathbf{A}(\mathbf{r})$ is the vector potential of the field. The problem is, however, greatly simplified (in the case of a sufficiently weak field) by changing to the quasi-momentum representation.

Because of the great variety of the possible forms of the band structure in the energy spectrum of an electron in a lattice, the condition for the external field to be small can be generally formulated only in a fairly crude way. Let the electron be in some sth band before the field is applied. Let ε_0 denote the smallest of the energy characteristics for this band, such as its characteristic width, or the distance to the neighbouring bands, i. e. the difference $\varepsilon_s(\mathbf{k}) - \varepsilon_{s'}(\mathbf{k})$ for given $\mathbf{k}$. If the magnetic field may be supposed weak, the following condition must certainly be satisfied:

$$\hbar\omega_H \ll \varepsilon_0, \tag{56.3}$$

where the "Larmor frequency" $\omega_H \sim |e|H/m^*c$, and $m^* \sim \hbar k/v$ is the effective mass of the electron.†

In the absence of an external field, the Hamiltonian of the electron in the lattice in the **k** representation is, as already mentioned, a diagonal matrix with elements $\varepsilon_s(\mathbf{k})$. In the presence of the field, the Hamiltonian will also contain the potential $\mathbf{A}(\mathbf{r})$ and its derivatives with respect to the coordinates, the field **H** (and in a non-uniform field also the derivatives of the field); in the **k** representation, the function $\mathbf{A}(\mathbf{r})$ is replaced by the operator $\hat{\mathbf{A}} = \mathbf{A}(\hat{\mathbf{r}})$, where $\hat{\mathbf{r}}$ is the operator (55.14).

The potential $\mathbf{A}(\mathbf{r})$ is an increasing (for a uniform field, linearly increasing) function of the coordinates. Because of this increase, the potential is not, even for a weak field, a small perturbation in the Hamiltonian of an infinite system (an electron in a lattice). For this reason, even a weak magnetic field considerably alters the properties of a large system, making the spectrum discrete

† A more precise definition of the frequency is given in (57.7) below. For conduction electrons in a metal (see §61), the characteristic values of $k \sim 1/a$ (where a is the lattice constant); putting also $\varepsilon_0 \sim \hbar^2/m^*a^2$, we find that the condition (56.3) is equivalent to $r_H \gg a$, where the "orbit radius" $r_H \sim v/\omega_H$.

instead of continuous; it quantizes the levels (see §58). The weak field intensity, in contrast to the potential, gives only small corrections.

We shall show that, when these corrections are neglected, the dependence of the Hamiltonian on the field potential may be ascertained in a general form by using only the requirements of gauge invariance. Since we are considering constant fields, it is sufficient to use the invariance of the equations under time-independent transformations of the potential and the wave functions, of the form

$$\mathbf{A} \rightarrow \mathbf{A} + \nabla f, \quad \psi \rightarrow \psi \exp(ief/\hbar c), \tag{56.4}$$

where $f(\mathbf{r})$ is an arbitrary function of the coordinates; see *QM*, (111.8), (111.9).

In a weak field, the potential $\mathbf{A}(\mathbf{r})$ is a slowly varying function of the coordinates. In order to elucidate the significance of this slowness, let us first consider the limiting case of a constant potential, $\mathbf{A}(\mathbf{r}) = \text{constant} \equiv \mathbf{A}_0$ (the constant potential is of course fictitious, there being then no actual field, so that a formal transformation is in question). The change from $\mathbf{A} = 0$ to $\mathbf{A} = \mathbf{A}_0$ is equivalent to the transformation (56.4) with $f = \mathbf{A}_0 . \mathbf{r}$; hence, instead of the original eigenfunctions (for $\mathbf{A} = 0$)

$$\psi_{s\mathbf{k}} = u_{s\mathbf{k}} e^{i\mathbf{k}.\mathbf{r}}, \tag{56.5}$$

the eigenfunctions of the new Hamiltonian will be

$$u_{s\mathbf{k}}(\mathbf{r}) \exp\left\{i\left(\mathbf{k} + \frac{e}{\hbar c}\mathbf{A}_0\right).\mathbf{r}\right\}.$$

From this we see that, in order to give the quasi-momentum its previous significance as a quantity determining the change in the phase of the wave function in translations, we must put $\mathbf{k} + e\mathbf{A}_0/\hbar c = \mathbf{K}$; the quantity $\mathbf{K}$ thus defined may be called the *generalized quasi-momentum*. Then the new eigenfunctions are

$$\psi_{s\mathbf{K}} = u_{s,\,\mathbf{K} - e\mathbf{A}_0/\hbar c}(\mathbf{r})\, e^{i\mathbf{K}.\mathbf{r}},$$

and the corresponding electron energy values are $\varepsilon_s(\mathbf{k}) = \varepsilon_s(\mathbf{K} - e\mathbf{A}_0/\hbar c)$.

We can now assert that, for a potential $\mathbf{A}(\mathbf{r})$ that is not constant but varies slowly in space, the wave functions of the "zero-order" approximation (with respect to the field) are

$$\psi_{s\mathbf{K}} = u_{s,\,\mathbf{K} - e\mathbf{A}(\mathbf{r})/\hbar c}\, e^{i\mathbf{K}.\mathbf{r}}; \tag{56.6}$$

the functions u are no longer strictly periodic, because of the variability of $\mathbf{A}$.† The energies $\varepsilon_s(\mathbf{K} - e\mathbf{A}/\hbar c)$ must now be regarded as operators forming the

† If the functions (56.6) are expanded in terms of the $\psi_{s\mathbf{k}}$, the expansion will in general include functions with various s. It must be emphasized, however, that this does not signify an actual transition to a different band, but simply expresses the change in the wave function under the action of the constant field. In this connection it may be recalled that a constant field cannot cause an actual transition with change of energy. To understand the position, it should be remarked that, although the field is weak, the resulting change in the classification of states (including the relation between quasi-momentum and energy) is considerable.

Hamiltonian in the **K** representation. In this representation, moreover, $\hat{\mathbf{r}}$ is to be understood as the operator $\hat{\mathbf{r}} = i\partial/\partial\mathbf{K}$, the second term ($\hat{\mathbf{\Omega}}$) in the definition (55.14) being omitted. For, when the operator $i\partial/\partial\mathbf{K}$ acts on the wave function, the effect, in order of magnitude, is to multiply it by the "orbit dimension" r_H, which increases as the field decreases; the result of the action of the operator $\hat{\mathbf{\Omega}}$ on the wave function does not contain such an increasing factor. In this sense, $\hat{\mathbf{\Omega}}$ in a weak field is small compared with $i\partial/\partial\mathbf{K}$. Since, on the other hand, the operator $\partial/\partial\mathbf{K}$ is diagonal with respect to the band numbers, the Hamiltonian also is diagonal.

We thus conclude that the motion of an electron in a weak magnetic field is described by the Hamiltonian (in the **K** representation)

$$\hat{H}_s = \varepsilon_s\left(\mathbf{K} - \frac{e}{\hbar c}\mathbf{A}(\hat{\mathbf{r}})\right), \qquad \hat{\mathbf{r}} = i\frac{\partial}{\partial\mathbf{K}} \tag{56.7}$$

(R. E. Peierls 1933). In this approximation, therefore, there is a complete analogy with the introduction of the magnetic field in the Hamiltonian of a free particle in the momentum representation.

The expression (56.7) is still not fully determinate, as the order of action of the non-commuting operators (the components of the vector $\hat{\mathbf{k}} = \mathbf{K} - e\hat{\mathbf{A}}/\hbar c$) has not been specified. This must be determined in such a way as to make the Hamiltonian Hermitian, which can in principle always be achieved by expressing the function $\varepsilon_s(\mathbf{k})$, which is periodic (in the reciprocal lattice) as a Fourier series

$$\varepsilon_s(\mathbf{k}) = \sum_{\mathbf{a}} A_{s\mathbf{a}}\, e^{i\mathbf{k}\cdot\mathbf{a}}; \tag{56.8}$$

the summation is over all vectors **a** in the original lattice. After the change $\mathbf{k} \to \hat{\mathbf{k}}$, the exponent of each term in this series will contain only one operator (the projection of the vector $\hat{\mathbf{A}}$ on **a**), so that the question of the order of action does not arise, everything reducing to powers of this one operator. This method of "Hermitianization" is, of course, not the only one, but it is important that the difference between the different methods lies beyond the range of the approximation considered, since the commutators of the operators $\hat{k}_x, \hat{k}_y, \hat{k}_z$ are small quantities in this approximation. For instance, in a uniform field the operator

$$\hat{\mathbf{A}} = \tfrac{1}{2}\mathbf{H}\times\hat{\mathbf{r}} = \tfrac{1}{2}i\mathbf{H}\times\partial/\partial\mathbf{K}; \tag{56.9}$$

a direct calculation easily shows that the commutators

$$\hat{k}_x\hat{k}_y - \hat{k}_y\hat{k}_x = i(e/\hbar c)\,H_z, \ \ldots, \tag{56.10}$$

are proportional to the low field strength **H**.

The operators $\hat{\mathbf{r}} = i\partial/\partial\mathbf{K}$ and $\hat{\mathbf{K}} \equiv \mathbf{K}$ obey the same commutation rules as the coordinates and generalized momenta of a "free" particle (without the lattice). It is therefore natural that the calculation of the commutators of these operators with the Hamiltonian gives the operator equations

$$\hbar\hat{\dot{\mathbf{K}}} = -\partial\hat{H}/\partial\mathbf{r}, \quad \hat{\dot{\mathbf{r}}} = \partial\hat{H}/\hbar\,\partial\mathbf{K}, \tag{56.11}$$

which have the form of the usual Hamilton's equations; for the calculation see *QM* (16.4) and (16.5).

We repeat that the Hamiltonian (56.7) is approximate in the sense that it omits all terms depending on the field **H** and not containing large factors of the order of the orbit size r_H. In subsequent approximations, the result may again be expressed as an effective Hamiltonian $\hat{H}_s(\mathbf{K}-e\hat{\mathbf{A}}/\hbar c, \mathbf{H})$, diagonal with respect to band numbers but not expressible in terms of the functions $\varepsilon_s(\mathbf{k})$ only.†

If the spin–orbit interaction is neglected, the inclusion of the electron spin causes the Hamiltonian to contain the usual term describing the interaction of the magnetic moment with the field: $-\beta\boldsymbol{\sigma}.\mathbf{H}$, where $\boldsymbol{\sigma}$ are the Pauli matrices and $\beta = |e|\hbar/2mc$ is the Bohr magneton. If the crystal has a centre of inversion, the spin–orbit interaction simply changes the magnetic moment of the electron, so that the interaction of the spin with the magnetic field becomes

$$-\beta\sigma_i H_k \xi_{ik}(\mathbf{k}). \tag{56.12}$$

For in this case the Hamiltonian must be invariant under the simultaneous operations of time reversal and inversion. The transformation must be accompanied by the changes $\mathbf{H}\to-\mathbf{H}$ and $\boldsymbol{\sigma}\to-\boldsymbol{\sigma}$ with **k** unchanged; (56.12) is the most general expression satisfying the given condition. The tensor $\xi_{ik}(\mathbf{k})$ cannot, of course, be calculated in a general form.

Lastly, let us consider the behaviour of the electron when a weak electric field **E** is applied to the lattice. The weakness condition is that the energy acquired by the electron in the field over a distance $\sim a$ is small compared with the characteristic energy ε_0: $|e|Ea \ll \varepsilon_0$.

As with a magnetic field, the most important terms are those which contain an increasing function of the coordinates, the scalar potential $\phi(\mathbf{r})$ of the electric field. The dependence of the Hamiltonian on ϕ can again be found in a general form by arguments similar to those used above. The application of a fictitious constant potential $\phi = \phi_0$ is equivalent in Schrödinger's equation to the inclusion in the energy of a constant term $e\phi_0$; such a term is also added to each of the eigenvalues $\varepsilon_s(\mathbf{k})$. When the potential $\phi(\mathbf{r})$ is not constant but varies slowly in space, a corresponding operator term is added to the effective Hamiltonian in the **k** representation:

$$\hat{H}_s = \varepsilon_s(\mathbf{k}) + e\phi(\hat{\mathbf{r}}). \tag{56.13}$$

† A simple example of the calculation of the correction term will be given in §59. An account of the regular method of deriving the Hamiltonian as a series in powers of **H**, and general expressions for the first few terms in this series, are given by E. I. Blount, *Physical Review* **126,** 1636, 1962; *Solid State Physics* **13,** 305, 1962. If the crystal has a centre of inversion, the series begins with terms of order $\mathbf{H}^2$ (see §59).

§ 57. Quasi-classical trajectories

Let us apply the results of §56 to the important case where the electron motion in the magnetic field is quasi-classical. The condition for this is, as we know, that the de Broglie wavelength of the particle varies only slightly over distances of the order of itself. In the present case, this condition is equivalent to the inequality

$$r_H \gg \lambda: \tag{57.1}$$

the radius of curvature of the orbit is large in comparison with the wavelength $\lambda \sim 1/k$.†

In the quasi-classical case, the particle trajectory is a meaningful concept. This is determined by the equations of motion, which are obtained from (56.11) by replacing the operators by the corresponding classical quantities:

$$\hbar\dot{\mathbf{K}} = -\partial H/\partial \mathbf{r}, \quad \mathbf{v} = \partial H/\hbar\,\partial \mathbf{K}, \quad H = \varepsilon\left(\mathbf{K} - \frac{e}{\hbar c}\mathbf{A}(\mathbf{r})\right);$$

the suffix s is omitted, for brevity. We expand these equations by changing from the generalized quasi-momentum $\mathbf{K}$ to the "kinetic quasi-momentum"

$$\mathbf{k} = \mathbf{K} - e\mathbf{A}(\mathbf{r})/\hbar c.$$

We have

$$\hbar\frac{d\mathbf{k}}{dt} + \frac{e}{c}\frac{d\mathbf{A}(\mathbf{r})}{dt} = -\frac{\partial H}{\partial \mathbf{r}} = \frac{e}{c}v_i\frac{\partial A_i}{\partial \mathbf{r}}.$$

Writing here $d\mathbf{A}/dt = (\mathbf{v}\cdot\nabla)\mathbf{A}$ and noting that

$$(v_i\nabla)A_i - (\mathbf{v}\cdot\nabla)\mathbf{A} = \mathbf{v}\times\operatorname{curl}\mathbf{A} = \mathbf{v}\times\mathbf{H},$$

we obtain the equation of motion

$$\hbar\frac{d\mathbf{k}}{dt} = \frac{e}{c}\mathbf{v}\times\mathbf{H}, \quad \mathbf{v} = \frac{\partial\varepsilon(\mathbf{k})}{\hbar\,\partial\mathbf{k}}. \tag{57.2}$$

This equation differs from the ordinary classical Lorentz equation only as regards the function $\varepsilon(\mathbf{k})$, which instead of a simple quadratic function is a complicated periodic function, and the function $\mathbf{v}(\mathbf{k})$ is therefore also of this kind. There is consequently, of course, a considerable change in the motion of the electron.

Let us consider this motion in a uniform magnetic field. Multiplying equation (57.2) by $\mathbf{v}$, we find in the usual way that $\hbar\mathbf{v}\cdot d\mathbf{k}/dt = d\varepsilon/dt = 0$. Multiplying

† This condition is in general stronger than (56.3). But, if $k \sim 1/a$, as occurs for conduction electrons in a metal, the two conditions coincide and are in practice always satisfied: for $r_H \sim c\hbar k/|e|H \sim c\hbar/a|e|H$, the condition $r_H \gg a$ leads to $H \ll c\hbar/|e|a^2 \sim 10^8-10^9$ Oe.

(57.2) by **H** gives $d(\mathbf{H}.\mathbf{k})/dt = 0$. Thus for the motion of the electron in the lattice, as in the motion of a free electron in a magnetic field,

$$\varepsilon = \text{constant}, \quad k_z = \text{constant}, \tag{57.3}$$

the z-axis being in the direction of the field **H**. The equations (57.3) define the path of the electron in **k**-space. Geometrically, it is the outline of a cross-section of the constant-energy surface $\varepsilon(\mathbf{k}) = $ constant by a plane perpendicular to the magnetic field.

The constant-energy surfaces may have many different forms. They may include (in each cell of the reciprocal lattice) several disconnected sheets, which may be singly or multiply connected, open or closed. To illustrate the latter difference, it is useful to consider a constant-energy surface which continues periodically throughout the reciprocal lattice. In each cell, there will be equal closed cavities; the open surfaces pass continuously through the whole lattice to infinity.†

The cross-sections of the constant-energy surface are made up of an infinity of contours. These include both cross-section contours of different sheets of the constant-energy surface within one reciprocal lattice cell and those of sheets repeated in different cells. If a sheet of the constant-energy surface is closed, all its cross-sections are closed curves. If it is open, its cross-sections may be either closed, or open (i.e. continuous throughout the reciprocal lattice).

The quasi-classicality of the motion also presupposes the smallness of the probability of *magnetic breakdown*, i.e. a discontinuous change in the electron quasi-momentum as it passes from one contour to another; we shall give the condition for this smallness at the end of the section. If this probability is neglected, therefore, the electron moves only along one contour of the constant-energy surface.

Let us consider in more detail the motion along closed trajectories in quasi-momentum space. It is evident that such a motion is periodic in time, and we can determine its period as follows. Taking the projection of equation (57.2) on the $k_x k_y$-plane perpendicular to the field, we obtain

$$\frac{dl_k}{dt} = \frac{|e|H}{c\hbar} v_\perp, \quad v_\perp = \sqrt{(v_x^2 + v_y^2)},$$

where $dl_k = \sqrt{(dk_x^2 + dk_y^2)}$ is an element of length of the **k** orbit. Hence

$$t = \frac{c\hbar}{|e|H} \int \frac{dl_k}{v_\perp}.$$

† To avoid misunderstanding, we should mention that it may be impossible to choose the reciprocal lattice cell in such a way that all the essentially different (i.e. not periodic repetitions) closed cavities are within one cell and not cut by its faces.

If the trajectory is closed, the period of the motion is given by the integral

$$T = \frac{c\hbar}{|e|H} \oint \frac{dl_k}{v_\perp}, \tag{57.4}$$

taken along the whole contour. This expression may now be converted into a more obviously significant form.

We use the area $S(\varepsilon, k_z)$ of the cross-section of the constant-energy surface $\varepsilon =$ constant by the plane $k_z =$ constant. The width of the annulus in this plane between the contours $\varepsilon =$ constant and $\varepsilon + d\varepsilon =$ constant is at each point

$$\frac{d\varepsilon}{|\partial\varepsilon/\partial\mathbf{k}_\perp|} = \frac{d\varepsilon}{\hbar v_\perp},$$

so that the area of the annulus is

$$dS = d\varepsilon \oint dl_k/\hbar v_\perp.$$

Hence we see that the integral in (57.4) is just the partial derivative $\partial S/\partial\varepsilon$. Thus the period of the motion is

$$T = \frac{c\hbar^2}{|e|H} \frac{\partial S(\varepsilon, k_z)}{\partial\varepsilon} \tag{57.5}$$

(W. Shockley 1950). Here it is natural to introduce the quantity

$$m^* = (\hbar^2/2\pi)\,\partial S/\partial\varepsilon, \tag{57.6}$$

called the *cyclotron mass* of the electron in the lattice. The orbital revolution frequency of the electron is expressed in terms of this quantity by

$$\omega_H = |e|H/m^*c, \tag{57.7}$$

which differs from the usual formula for the Larmor frequency of free electrons in that their mass is replaced by m^*.†

We must emphasize, however, that for electrons in a lattice the cyclotron mass is not a constant, but a function of ε and k_z, and is therefore different for different electrons. Moreover, it may be either positive or negative, the electron moving in its orbit as a negatively charged particle or positively charged hole respectively. Accordingly we speak of *electron* and *hole trajectories*.

Hitherto we have discussed the trajectory of an electron in **k**-space. It is easy to see, however, that there is a close relationship between the trajectories in quasi-momentum space and ordinary space. The equation of motion (57.2), written in the form

$$\hbar d\mathbf{k} = -|e|\,d\mathbf{r}\times\mathbf{H}/c,$$

† For a free electron, the constant-energy surface is the sphere $\varepsilon = \hbar^2k^2/2m$. Its cross-sections are circles with area $S = \pi(2m\varepsilon/\hbar^2 - k_z^2)$, and so the derivative $\partial S/\partial\varepsilon = 2\pi m/\hbar^2$, and $m^* = m$.

gives on integration (with the appropriate choice of the origin of the coordinates **r** and the quasi-momenta **k**)

$$\hbar\mathbf{k} = -|e|\,\mathbf{r}\times\mathbf{H}/c. \tag{57.8}$$

From this we see that the xy-projection of the orbit in ordinary space essentially repeats the **k** trajectory, differing from it only in orientation and scale: it is obtained from the **k** trajectory by the changes

$$k_x \to -|e|\,Hy/\hbar c, \quad k_y \to |e|\,Hx/\hbar c.$$

Moreover, in ordinary space there is a motion along the z-axis with velocity $v_z = \partial\varepsilon/\hbar\partial k_z$. If the trajectory in **k**-space is closed, in ordinary space it is a helix with the axis along the field. If the trajectory is open, so is its projection on the xy-plane in ordinary space, and the motion in this plane is therefore infinite.

Some comments may be added about the quasi-classical motion of an electron when a constant and uniform electric field **E** is applied to the lattice. The quasi-classical equation $\hbar\dot{\mathbf{k}} = e\mathbf{E}$ gives

$$\mathbf{k} = \mathbf{k}_0 + e\mathbf{E}t/\hbar. \tag{57.9}$$

The law of conservation of energy gives

$$\varepsilon(\mathbf{k}) - e\mathbf{E}\cdot\mathbf{r} = \text{constant}. \tag{57.10}$$

But the energy $\varepsilon(\mathbf{k})$ takes values in a finite range $\Delta\varepsilon$ (the band width); it therefore follows from (57.10) that the motion of an electron in a uniform electric field is finite in the direction of the field: the electron oscillates in that direction with amplitude $\Delta\varepsilon/|e|E$. If the field is parallel to any basic vector **b** of the reciprocal lattice, the motion is periodic with frequency $\omega = 2\pi|e|E/\hbar b$; when $b \sim 1/a$, we have $\hbar\omega_E \sim |e|Ea$. In the general case of an arbitrary direction of the field, the motion is quasi-periodic.

Lastly, let us consider the condition for the magnetic breakdown described above to be negligible. The probability of transition from one trajectory (in **k**-space) to another is, of course, large if these trajectories anywhere pass unusually close to each other. Such a situation arises when the trajectory is close to a self-intersecting one, or if it passes near the intersection of two sheets of the constant-energy surface (i.e. near a point of degeneracy). Figure 14 shows a

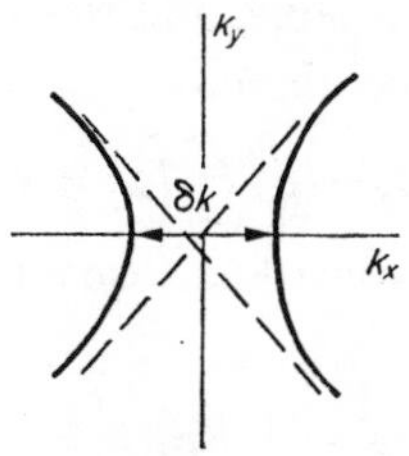

FIG. 14.

typical pattern of trajectories in such cases; the gap δk between the trajectories is small compared with the characteristic dimensions of the orbits as a whole, and the radius of curvature R_k of the trajectories near their points of closest approach is in general of the order of δk. The transition from one orbit to the other takes place by quantum tunnelling. The probability of this process is (exponentially) small if δk is large compared with the distance Δk_x over which the wave function decays in the classically inaccessible region between the trajectories.

An estimate of Δk_x can be obtained by using the analogy between the motion of the electron in the magnetic field and a one-dimensional motion in a potential field $U(x)$. This analogy is based on the fact that, according to (56.10), the operators $\hat{q} \equiv \hat{k}_x \hbar c/|e|H$ and $\hat{p} \equiv \hbar \hat{k}_y$ obey a commutation rule that is the same as for the coordinate and momentum. Near the points of closest approach, the trajectories are parabolic, like the (x, p) phase trajectory of one-dimensional motion in a uniform field ($U = -Fx$), whose equation is $p^2/2m = Fx$ if the coordinate x is measured from the turning point. In the latter case, the wave function decays over a distance $\Delta x \sim (\hbar^2/mF)^{1/3}$ beyond the turning point (see *QM*, §24); with the radius of curvature of the phase trajectory $R \sim (d^2x/dp^2)^{-1} \sim mF$, we find $\Delta x \sim (\hbar^2/R)^{1/3}$. By the analogy mentioned, the Δk_x sought can be found by making the changes $\Delta x \rightarrow \hbar c \Delta k_x/|e|H$, $R \rightarrow R_k \hbar |e| H/c$. Thus we have $\Delta k_x \sim (|e|H/\hbar c)^{2/3}(\delta k)^{-1/3}$, and the condition $\Delta k_x \ll \delta k$ becomes

$$|e|H/\hbar c \ll (\delta k)^2. \tag{57.11}$$

§ 58. Quasi-classical energy levels

We have seen that classical motion of an electron in a lattice in a magnetic field, along a closed trajectory in **k**-space, corresponds in ordinary space to a motion that is finite in a plane perpendicular to the direction of the field **H**. When we change to quantum mechanics, there are discrete energy levels for any fixed value of the longitudinal quasi-momentum k_z. These are determined by the general rules of quasi-classical quantization.

We choose the vector potential of the uniform magnetic field (along the z-axis) in the form $A_x = -Hy$, $A_y = A_z = 0$. Then the components of the generalized quasi-momentum are

$$K_x = k_x + |e|Hy/c\hbar, \quad K_y = k_y, \quad K_z = k_z. \tag{58.1}$$

The coordinate x is a cyclic variable, and the x-component of the generalized quasi-momentum is therefore conserved:

$$K_x = k_x + |e|Hy/c\hbar = \text{constant}. \tag{58.2}$$

According to Bohr and Sommerfeld's quantization rule (see *QM*, §48), we can write the condition

$$\frac{1}{2\pi}\left|\oint K_y\, dy\right| = n, \tag{58.3}$$

where the integration is taken over the period of the motion, and n is a positive integer, assumed large.† Substituting here from (58.1) and (58.2) $K_y = k_y$ and $dy = -(c\hbar/e|H|)\,dk_x$, we obtain

$$\frac{c\hbar}{2\pi|e|H}\left|\oint k_y\,dk_x\right| = n, \tag{58.4}$$

where the integral is now taken along a closed trajectory in **k**-space. This integral is just the area enclosed by the trajectory, i.e. the area $S(\varepsilon, k_z)$ of the cross-section of the constant-energy surface by the plane k_z = constant, as defined in §57.

Thus we have finally

$$S(\varepsilon, k_z) = 2\pi(|e|H/c\hbar)\,n \tag{58.5}$$

(I. M. Lifshitz 1951, L. Onsager 1952). This condition implicitly determines the energy levels $\varepsilon_n(k_z)$. Thus the energy band (whose number s is omitted, for brevity) consists of a discrete set of *Landau sub-bands*, each of which is a range of energy levels distinguished by the value of the continuous variable k_z.

The quasi-classical quantization condition can be refined by including a correction which amounts to adding a number of the order of unity to the large quantum number n. The determination of this correction involves a consideration of the motion near the "turning points" which form the limits of the range of integration in (58.3).

The dependence of $K_y = k_y$ on y on the electron trajectory is determined by the equation

$$\varepsilon(\mathbf{k}) = \left(K_x - \frac{|e|H}{c\hbar}y,\ k_y,\ k_z\right) = \text{constant} \tag{58.6}$$

for given k_z and with K_x = constant; the turning point $y = y_0$ is determined by the condition for the velocity $v_y = \partial\varepsilon/\hbar\partial k_y$ to be zero. Near this point, an expansion of (58.6) in powers of $y - y_0$ gives

$$-\frac{|e|H}{c\hbar}\left(\frac{\partial\varepsilon}{\partial k_x}\right)_0 (y-y_0) + \frac{1}{2}\left(\frac{\partial^2\varepsilon}{\partial k_y^2}\right)_0 (k_y - k_{y0})^2 = 0,$$

where $k_{y0} = k_y(y_0)$. Hence we see that the turning point is approached according to a square-root law:

$$k_y - k_{y0} = \pm A\sqrt{(y-y_0)};$$

† For motion in a uniform magnetic field, the integral $\oint \mathbf{K}_t . d\mathbf{r}/2\pi$ is an adiabatic invariant independent of the choice of the vector potential, $\mathbf{K}_t$ being the projection of the generalized quasi-momentum on the plane perpendicular to the field; cf. *Fields*, §21. With our choice of **A**, the integral $\oint K_z dx = K_z \oint dx = 0$, so that the adiabatic invariant is the same as the integral in (58.3).

we take the particular case where the classically inaccessible region is in $y < y_0$. This law is the same as the one covered by the usual derivation of the correction to the quasi-classical quantization (see *QM* §§47 and 48). The corrected rule (58.5) is consequently

$$S(\varepsilon, k_z) = 2\pi(|e|H/c\hbar)(n+\tfrac{1}{2}). \tag{58.7}$$

As is clear from the derivation, which is based on the expansion of the function (58.6), it is necessary for the validity of the corrected quantization rule that the trajectory should pass sufficiently far from the singular points of the function $\varepsilon(\mathbf{k})$ (including the complex branch-points). It is also necessary that the quasi-classicality condition should be satisfied at all points near the trajectory, and in particular that the xy-projection of the velocity $\partial\varepsilon/\partial\mathbf{k}$ should not be zero.† Lastly, it must be remembered that the Hamiltonian (56.7), on which the whole of the derivation is based, is itself approximate. If the lattice has a centre of inversion, the corrections to the Hamiltonian are quadratic in the field, and do not affect the condition (58.7), but if there is no centre of inversion they are linear in $\mathbf{H}$; in the latter case the correction term $\frac{1}{2}$ in (58.7) is meaningless, since the approximate nature of the Hamiltonian implies an error of the same order.‡

The interval $\Delta\varepsilon$ between two successive levels corresponds to a change of unity in the large number n. It is therefore determined by

$$\Delta S = (\partial S/\partial\varepsilon)\Delta\varepsilon = 2\pi|e|H/\hbar c. \tag{58.8}$$

With the classical frequency ω_H of periodic motion from (57.7), we have

$$\Delta\varepsilon = \hbar\omega_H. \tag{58.9}$$

It must be emphasized that the frequency ω_H is itself a function of ε (and of k_z). Hence the successive energy levels ε_n (for given k_z) are not strictly equidistant as in the case of free electrons (where ω_H is a constant).

The fact that the energy levels are independent of the conserved quantity K_x means that they are degenerate (as are those of free electrons in a magnetic field; see *QM* §112). If we imagine a lattice with a large but finite volume V, the degree of degeneracy is finite. The number of states in the range dk_z with a given value of n is $V\Delta S\,.\,dk_z/(2\pi)^3$, where ΔS is the area in the plane $k_z =$ = constant between the trajectories with quantum numbers n and $n+1$. This

† Near points of anomalous approach of two trajectories, these conditions coincide with the requirement that the probability of magnetic breakdown is small.

‡ For free electrons (see the last footnote to §57), the condition (58.7) gives

$$\varepsilon = \hbar\omega_H(n+\tfrac{1}{2})+\hbar^2k_z^2/2m, \quad \omega_H = |e|H/mc$$

in agreement with the familiar Landau formula for a free electron in a magnetic field (*QM*, §112).

area is given by (58.8), and we thus find for the required number of states

$$\frac{V\,dk_z}{(2\pi)^2}\frac{|e|H}{c\hbar}, \tag{58.10}$$

the same as in the case of free electrons.

The intuitive reason for the degeneracy of levels in a magnetic field is that the energy is independent of the spatial position of the centre of the Larmor orbit of the electron. For a free electron, the degeneracy is exact, but for an electron in a lattice it may be only approximate: because of the presence of the non-uniform (periodic) electric field, the different positions of the centre of the orbit in the unit cell of the lattice are not equivalent. This must cause some splitting of the Landau levels.

Taking account of the electron spin causes each level to be split into two components; if the spin–orbit coupling is neglected, these components are separated (like those for a free electron) by a constant interval $2\beta H$, where β is the Bohr magneton:

$$\varepsilon_{n\sigma}(k_z) = \varepsilon_n(k_z)+\sigma\beta H, \qquad \sigma = \pm 1. \tag{58.11}$$

The same situation exists when the spin–orbit interaction is not neglected, if the crystal has a centre of inversion. In this case, the states of the electron in the absence of the field are degenerate with respect to the spin, and the magnetic field removes this degeneracy. The result is the same formula (58.11) with β replaced by $\beta\xi_n(k_z)$, where $\xi_n(k_z)$ represents the change in the electron magnetic moment.

§ 59. The electron effective mass tensor in the lattice

Let us consider a point $\mathbf{k} = \mathbf{k}_0$ in $\mathbf{k}$-space where the electron energy $\varepsilon_s(\mathbf{k})$ has an extremum; for example, the points corresponding to the top and bottom of a band. If there is no degeneracy at this point (apart from the possible Kramers degeneracy with regard to the spin; see the end of §55), the function $\varepsilon_s(\mathbf{k})$ near it has a regular expansion in powers of the difference $\mathbf{q} = \mathbf{k}-\mathbf{k}_0$. The first terms in this expansion are quadratic:

$$\varepsilon_s(\mathbf{k}) = \varepsilon_s(\mathbf{k}_0)+\tfrac{1}{2}\hbar^2 m^{-1}{}_{ik}q_i q_k. \tag{59.1}$$

The tensor m_{ik}, the inverse of the tensor of the coefficients $m^{-1}{}_{ik}$ in (59.1), is called the electron *effective mass tensor* in the lattice. We shall show how this tensor may be expressed in terms of the matrix elements with respect to the Bloch functions $\psi_{s\mathbf{k}_0}$ at the point $\mathbf{k}_0$.

If the spin–orbit interaction is neglected, the electron Hamiltonian has the form (56.1). We substitute in Schrödinger's equation with this Hamiltonian the wave function in the form

$$\psi_{s\mathbf{k}} = e^{i(\mathbf{k}_0+\mathbf{q})\cdot\mathbf{r}}\,u_{s\mathbf{k}} \equiv e^{i\mathbf{q}\cdot\mathbf{r}}\,\phi_{s\mathbf{k}}. \tag{59.2}$$

The equation then becomes

$$\left\{-\frac{\hbar^2}{2m}\triangle+U(\mathbf{r})+\left[\frac{\hbar}{m}\mathbf{q}\cdot\hat{\mathbf{p}}+\frac{\hbar^2q^2}{2m}\right]\right\}\phi_{s\mathbf{k}}=\varepsilon_s(\mathbf{k})\,\phi_{s\mathbf{k}}, \tag{59.3}$$

where $\hat{\mathbf{p}} = -i\hbar\nabla$ is the operator of the actual momentum.

In the neighbourhood of the point $\mathbf{k} = \mathbf{k}_0$, the vector $\mathbf{q}$ is a small quantity, and the expression in the square brackets in (59.3) may be regarded as the operator of a perturbation. In the zero-order approximation, when $\mathbf{q} = 0$, the functions $\phi_{s\mathbf{k}}$ are the same as the $\psi_{s\mathbf{k}_0}$. The usual perturbation theory therefore enables us to express the correction to the energy in terms of the matrix elements with respect to these functions.

Since $\mathbf{k}_0$ is an extremum point, there is no correction linear in $\mathbf{q}$. This means that the diagonal matrix elements are

$$\langle s\mathbf{k}_0\,|\,\mathbf{p}\,|\,s\mathbf{k}_0\rangle = 0. \tag{59.4}$$

To determine the correction quadratic in $\mathbf{q}$, we must take into account the q^2 term in the perturbation operator in first-order perturbation theory, and the $\mathbf{q}$ term in the second order. This leads to formula (59.1) for $\varepsilon_s(\mathbf{k})$, where

$$m^{-1}{}_{ik}=\frac{\delta_{ik}}{m}+\frac{1}{m^2}\sum_{s'}{}'\frac{(p_i)_{ss'}(p_k)_{s's}+(p_k)_{ss'}(p_i)_{s's}}{\varepsilon_{s'}(\mathbf{k}_0)-\varepsilon_s(\mathbf{k}_0)}; \tag{59.5}$$

the summation is over all $s' \neq s$.† To simplify the notation, the matrix elements are written here and henceforward without the diagonal suffix $\mathbf{k}_0$: $p_{ss'} \equiv \langle s\mathbf{k}_0\,|\,\mathbf{p}\,|\,s'\mathbf{k}_0\rangle$. When there are close bands (i.e. small differences $\varepsilon_{s'}-\varepsilon_s$), the second term in (59.5) may be large in comparison with the first, and the effective masses are then small compared with m.

Now let a uniform magnetic field $\mathbf{H}$ be applied to the crystal. According to (56.7), the Hamiltonian acting on functions of the generalized quasi-momentum $\mathbf{Q}$ is obtained from (59.1) on replacing $\mathbf{q}$ by the operator

$$\hat{\mathbf{q}} = \mathbf{Q}-e\hat{\mathbf{A}}/\hbar c, \quad \hat{\mathbf{A}} = \tfrac{1}{2}\mathbf{H}\times i\,\partial/\partial\mathbf{Q}. \tag{59.6}$$

The resulting Hamiltonian

$$\hat{H}_s^{(0)} = \varepsilon_s(\mathbf{k}_0)+\tfrac{1}{2}\hbar^2 m^{-1}{}_{ik}\,\hat{q}_i\hat{q}_k \tag{59.7}$$

is, of course, valid only in the same energy range as the original formula (59.1). This means that, as well as the condition (56.3) for the field to be weak, it is assumed that the Landau levels considered are not too high. In this sense, the quantities $\mathbf{q}$ and $\mathbf{Q}$ are to be regarded as small; the increase of the potential $\mathbf{A}$ has the effect that, even in a weak field, we cannot suppose that $\mathbf{A}$ is small compared with $\mathbf{Q}$.

† The summation over $\mathbf{k}'$ does not appear, since, according to (55.15), the momentum $\mathbf{p} = m\mathbf{v}$ has only matrix elements diagonal in $\mathbf{k}$, so that all the intermediate states pertain to the same quasi-momentum $\mathbf{k}_0$.

The next terms in the Hamiltonian after (59.7) contain the field **H** in the "pure" form (i.e. without acompanying operators $\partial/\partial\mathbf{Q}$). Such terms cannot be found simply from considerations of gauge invariance. Let us determine the first of these terms, which is linear in **H**. Because of the relative smallness of this correction, we can calculate it with $\mathbf{Q} = 0$.

Let us first investigate this problem without the spin–orbit interaction. The term of interest, linear in **H**, can only arise from the term linear in **A** in the original exact Hamiltonian (56.2) of the electron, i.e. by averaging with respect to the wave function $\psi_{s\mathbf{k}_0}$ the expression

$$-(e/2mc)(\hat{\mathbf{p}}\cdot\mathbf{A}+\mathbf{A}\cdot\hat{\mathbf{p}}) = -(e/mc)\,\mathbf{A}\cdot\hat{\mathbf{p}}; \tag{59.8}$$

the equality depends on the gauge already chosen, with $\operatorname{div}\mathbf{A} = 0$. This adds to the Hamiltonian (59.7) the term

$$H_s^{(1)} = -\mathbf{M}\cdot\mathbf{H}, \tag{59.9}$$

where

$$\mathbf{M} = (e/2mc)\langle s\mathbf{k}_0\,|\,\mathbf{r}\times\mathbf{p}\,|\,s\mathbf{k}_0\rangle \tag{59.10}$$

is just the mean value of the magnetic moment of the electron in the state $s\mathbf{k}_0$. We must emphasize that the correction (59.9) may be added to the Hamiltonian (59.7) without any fear that this effect has already been partly taken into account by the substitution (59.6): the terms linear in **H** in (59.7) do not occur when $\mathbf{Q} = 0$.

We expand the expression (59.10) by the matrix multiplication rule, using the fact that, from (59.4), **p** has no diagonal matrix elements:

$$M_x = \frac{e}{2mc}\sum_{s'}{}' \left[(\Omega_y)_{ss'}(p_z)_{s's} - (\Omega_z)_{ss'}(p_y)_{s's}\right]$$

(and similarly for M_y and M_z); the correction to the Hamiltonian (59.7) is, as it should be, expressed in terms of matrix elements of the operator $\mathbf{\Omega}$. Using the relation

$$\mathbf{\Omega}_{s's} = \mathbf{p}_{s's}/i(\varepsilon_{s'}-\varepsilon_s),$$

we can put **M** in the form

$$M_x = \frac{ie}{2mc}\sum_{s'}{}' \frac{(p_z)_{ss'}(p_y)_{s's} - (p_y)_{ss'}(p_z)_{s's}}{\varepsilon_{s'}(\mathbf{k}_0)-\varepsilon_s(\mathbf{k}_0)}, \ \ldots. \tag{59.11}$$

If the crystal has a centre of inversion, **M** is zero, and therefore so is the whole correction (59.9): under simultaneous time reversal and inversion, the state of the electron is unchanged (if its spin is ignored), and the right-hand side of (59.11) is therefore unchanged, whereas the magnetic moment must change sign under this transformation.

We now take into account the spin–orbit interaction in the crystal by adding to the Hamiltonian (56.1) the spin–orbit term $\hat{H}_{sl}$ from (55.17). This changes the term linear in $\mathbf{q}$ in equation (59.3): the operator $\hat{\mathbf{p}}$ in this term is replaced by

$$\hat{\boldsymbol{\pi}} = \hat{\mathbf{p}} + (\hbar^2/4m^2c^2)\,\boldsymbol{\sigma}\times\nabla U. \tag{59.12}$$

The operator $\hat{\boldsymbol{\pi}}$ has a simple physical significance: on directly commuting the Hamiltonian (including $\hat{H}_{sl}$) with $\mathbf{r}$, we find, in the absence of a magnetic field,

$$\hat{\dot{\mathbf{r}}} = \hat{\boldsymbol{\pi}}/m. \tag{59.13}$$

Similarly, in the presence of a magnetic field, by making the usual substitution $\hat{\mathbf{p}} \rightarrow \hat{\mathbf{p}} - e\mathbf{A}/c$ in the original Hamiltonian (including $\hat{H}_{sl}$), we find that the term linear in $\mathbf{A}$ has the form $-e\hat{\boldsymbol{\pi}}.\mathbf{A}/mc$, which differs from (59.8) only in that $\hat{\mathbf{p}}$ is replaced by $\hat{\boldsymbol{\pi}}$. The magnetic moment (59.11) must be supplemented by the spin magnetic moment of the free electron, giving

$$M_x = \beta\langle s\mathbf{k}_0 \,|\, \sigma_x \,|\, s\mathbf{k}_0\rangle + \frac{ie}{2mc}\sum_{s'}{}' \frac{(\pi_z)_{ss'}(\pi_y)_{s's} - (\pi_y)_{ss'}(\pi_z)_{s's}}{\varepsilon_{s'} - \varepsilon_s}. \tag{59.14}$$

With the spin–orbit interaction included, the second term in this expression is not zero even if the crystal has a centre of inversion: simultaneous time reversal and inversion yield a state with the opposite spin direction, so that the whole expression (59.14), if it changes sign under this transformation, must reduce to the mean value of the operator $\beta\sigma_i\xi_{ik}(\mathbf{k})$; cf. (56.12).

Let us calculate the tensor ξ_{ik} for the case where the spin–orbit interaction may be regarded as a perturbation.† We can rewrite (55.17) as

$$\hat{H}_{sl} = \boldsymbol{\sigma}.\hat{\boldsymbol{\chi}}, \quad \hat{\boldsymbol{\chi}} = (i\hbar^2/4m^2c^2)\,\nabla U\times\nabla. \tag{59.15}$$

Regarding (59.9) and (59.15) as a perturbation, we find the correction to the energy in second-order perturbation theory, retaining only the cross-terms from (59.9) and (59.15). This correction (which still is an operator, a matrix with respect to the spin variables) has the form (56.12), with the tensor ξ_{ik} given by

$$\xi_{ik} = \delta_{ik} + \frac{1}{2}\sum_{s'}{}' \frac{(\chi_i)_{ss'}(L_k)_{s's} + (L_k)_{ss'}(\chi_i)_{s's}}{\varepsilon_{s'} - \varepsilon_s} \tag{59.16}$$

where $\hbar\hat{\mathbf{L}} = \mathbf{r}\times\hat{\mathbf{p}}$.

The above discussion has related to states that are not degenerate (except with respect to the spin). If there is degeneracy at $\mathbf{k} = \mathbf{k}_0$, the energy has to be determined by setting up the secular equation which takes account of the perturbation (the square brackets in (59.3)) as far as the second-order terms, i. e. according to *QM*, (39.4). The properties of the resulting secular equation depend on the symmetry at the point $\mathbf{k}_0$. We shall return to this topic in §68.

† The expression (55.17) for $\hat{H}_{sl}$ is the first term in an expansion in powers of the relativistic ratio $(v/c)^2$, and is therefore, in a certain sense, always small. This smallness, however, is irrelevant to the applicability of perturbation theory in any particular band. Hence $\hat{H}_{sl}$, in the problem under consideration, cannot always be regarded as a small perturbation.

PROBLEM

Find the quasi-classical energy levels for a particle with the quadratic dispersion relation (59.1) in a magnetic field acting in any direction.

SOLUTION. We reduce the tensor m_{ik} to diagonal form, and measure energy and momentum from an extremum point, say the minimum. Then

$$\varepsilon(\mathbf{k}) = \frac{1}{2}\hbar^2\left(\frac{k_1^2}{m_1}+\frac{k_2^2}{m_2}+\frac{k_3^2}{m_3}\right), \tag{1}$$

where m_1, m_2, m_3 are the principal values of the tensor m_{ik} (and are positive quantities). Let $\mathbf{n}$ be a unit vector in the direction of the field $\mathbf{H}$; then

$$k_z = \mathbf{n}\,.\,\mathbf{k} = n_1k_1+n_2k_2+n_3k_3, \tag{2}$$

where n_1, n_2, n_3 are the direction cosines of the field relative to the principal axes of the tensor m_{ik}. We have to find the area S of the part of the plane (2) that lies within the ellipsoid (1); it may be written as the integral

$$S = \int \delta(\mathbf{n}\,.\,\mathbf{k}-k_z)\,d^3k, \tag{3}$$

taken over the volume of the ellipsoid (1).† By the change of variables $\hbar k_i = (2\varepsilon m_i)^{1/2}q_i$, the integral is brought to the form

$$S = (2\varepsilon)^{3/2}\hbar^{-3}(m_1m_2m_3)^{1/2}\int \delta(\boldsymbol{\nu}\,.\,\mathbf{q}-k_z)\,d^3q,$$

where the vector $\boldsymbol{\nu}$ in $\mathbf{q}$-space has the components $\nu_i = (2\varepsilon m_i)^{1/2}n_i/\hbar$, and the integration is taken over the volume of the sphere $\mathbf{q}^2 = 1$. The integration is easily carried out in cylindrical polar coordinates with the axis along $\boldsymbol{\nu}$, and gives

$$S(\varepsilon, k_z) = (2\pi/\hbar^2)\,m_\perp(\varepsilon-\hbar^2k_z^2/2m_{\|}),$$

where

$$m_{\|} = m_1n_1^2+m_2n_2^2+m_3n_3^2,$$
$$m_\perp = (m_1m_2m_3/m_{\|})^{1/2}. \tag{4}$$

Substituting in (58.7), we find the energy levels

$$\varepsilon_n(k_z) = \frac{|e|\hbar H}{m_\perp c}\left(n+\frac{1}{2}\right)+\frac{\hbar^2k_z^2}{2m_{\|}}. \tag{5}$$

§ 60. Symmetry of electron states in a lattice in a magnetic field

In this section we shall consider the exact general properties of translational symmetry of the wave functions of a Bloch electron in a magnetic field, which do not depend on any approximation (such as the condition for the field to be weak or the quasi-classicality condition).

The application of a uniform magnetic field does not affect the physical translational symmetry of the system, which remains spatially periodic. A

† Let $f(x, y, z)$ = constant be a family of surfaces occupying a certain volume. The distance dl between two of these surfaces, infinitely close to each other, is $dl = df/|\nabla f|$, and the volume between them is $dV = S(f)dl$, where $S(f)$ is the area of the surface having a given value of f. Multiplying the equation $S(f)df = |\nabla f|\,dV$ by the delta function $\delta(f)$ and integrating over the volume and df, we find the area of the surface $f(x, y, z) = 0$ as $S(0) = \int |\nabla f|\delta(f)\,d^3x$. In our case $|\nabla f| = 1$, and this gives equation (3).

distinctive feature is, however, that the electron Hamiltonian (56.2) does lose its symmetry, because it involves not the constant field strength **H** but the vector potential **A**(**r**), which depends on the coordinates and is not periodic.

The non-invariance of the Hamiltonian naturally complicates the transformation law for wave functions under translation. We take the vector potential of the uniform field in the gauge

$$\mathbf{A} = \tfrac{1}{2}\mathbf{H}\times\mathbf{r}, \tag{60.1}$$

and let $\psi(\mathbf{r})$ be an eigenfunction of the Hamiltonian $\hat{H}(\mathbf{r})$. Under a translation $\mathbf{r}\to\mathbf{r}+\mathbf{a}$ (where $\mathbf{a}$ is any of the basic lattice vectors), this function becomes $\psi(\mathbf{r}+\mathbf{a})$, and is then an eigenfunction of the Hamiltonian $\hat{H}(\mathbf{r}+\mathbf{a})$, which is not the same as $\hat{H}(\mathbf{r})$, because the vector potential has changed:

$$\mathbf{A}(\mathbf{r}) \to \mathbf{A}(\mathbf{r}+\mathbf{a}) = \mathbf{A}(\mathbf{r})+\tfrac{1}{2}\mathbf{H}\times\mathbf{a}.$$

To find the required transformation law, we must go back to the original Hamiltonian by the gauge transformation

$$\mathbf{A}\to\mathbf{A}+\nabla f, \quad f = -\tfrac{1}{2}\mathbf{H}\times\mathbf{a}\cdot\mathbf{r}.$$

The wave function is then transformed according to (56.4): $\psi\to\psi\exp(ief/\hbar c)$. Denoting the result of all these operations by $\hat{T}_a\psi(\mathbf{r})$, we thus find

$$\hat{T}_a\psi(\mathbf{r}) = \psi(\mathbf{r}+\mathbf{a})\exp(\tfrac{1}{2}i\mathbf{r}\cdot\mathbf{h}\times\mathbf{a}), \tag{60.2}$$

where $\mathbf{h} = |e|\,\mathbf{H}/\hbar c$; $\hat{T}_a$ is called the *magnetic translation operator*. If $\psi(\mathbf{r})$ is a solution of Schrödinger's equation $\hat{H}(\mathbf{r})\psi = \varepsilon\psi$, then (60.2) is a solution of that equation for the same energy ε (R. E. Peierls 1933).

From the definition (60.2), we easily find that

$$\hat{T}_{\mathbf{a}}\hat{T}_{\mathbf{a}'} = \hat{T}_{\mathbf{a}+\mathbf{a}'}\omega(\mathbf{a},\mathbf{a}'), \quad \omega(\mathbf{a},\mathbf{a}') = \exp(-\tfrac{1}{2}i\mathbf{h}\cdot\mathbf{a}\times\mathbf{a}'). \tag{60.3}$$

When $\mathbf{a}$ and $\mathbf{a}'$ are interchanged, the exponent in the factor $\omega(\mathbf{a},\mathbf{a}')$ changes sign, and the operators $\hat{T}_{\mathbf{a}}$ and $\hat{T}_{\mathbf{a}'}$ therefore do not in general commute:

$$\hat{T}_{\mathbf{a}}\hat{T}_{\mathbf{a}'} = \hat{T}_{\mathbf{a}'}\hat{T}_{\mathbf{a}}\exp(-i\mathbf{h}\cdot\mathbf{a}\times\mathbf{a}'). \tag{60.4}$$

Thus the product of two operators $\hat{T}_{\mathbf{a}}$ and $\hat{T}_{\mathbf{a}'}$ generally differs by a phase factor from the operator $\hat{T}_{\mathbf{a}+\mathbf{a}'}$. In mathematical terms, this means that the operators $\hat{T}_{\mathbf{a}}$ give not an ordinary but a projective representation of the translation group; the basis of these representations is formed by the wave functions of the stationary states of a Bloch electron in a magnetic field.† The classifica-

† Projective representations of groups have already been met with in Part 1, §134. The projective representations of a group G are those given by operators $\hat{G}$ such that the relations between them are the same as those between the corresponding elements of G only to within phase factors: if $G_1G_2 = G_3$, we have for the operators $\hat{G}_1\hat{G}_2 = \omega_{12}\hat{G}_3$, where ω_{12} need only have unit modulus.

tion of the energy levels must consequently be based on the irreducible projective representations of the translation group, just as, in the absence of the field, they are based on the irreducible ordinary representations of this group.

The translation group is Abelian (all its elements commute), and therefore all its irreducible ordinary representations are one-dimensional. The base function ψ of each such representation is simply multiplied by a phase factor under translation; for two successive translations, this factor must be equal to the product of the factors for each translation separately. This means that

$$\hat{T}_{\mathbf{a}}\psi = e^{i\mathbf{k}\cdot\mathbf{a}}\psi,$$

where **k** is a constant vector, the quasi-momentum of the electron, which is the parameter classifying the irreducible representations.

A complete classification of the irreducible projective representations of the translation group can be made (E. Brown 1964, J. Zak 1964) if the magnetic field satisfies the condition

$$\mathbf{h} = 4\pi p\mathbf{a}_3/qv, \tag{60.5}$$

where p and q are any two relatively prime integers, and $\mathbf{a}_3$ is one of the three arbitrary chosen basic vectors $\mathbf{a}_1$, $\mathbf{a}_2$, $\mathbf{a}_3$ of the lattice; $v = \mathbf{a}_1\times\mathbf{a}_2\cdot\mathbf{a}_3$ is the volume of a unit cell of the lattice. Thus the magnetic field must be along one of the lattice vectors, and $hv/4\pi a_3$ must be a rational number. Multiplying (60.5) by $\mathbf{a}_1\times\mathbf{a}_2$, we can also write this condition as

$$\mathbf{h}\cdot\mathbf{a}_1\times\mathbf{a}_2 = 4\pi p/q. \tag{60.6}$$

To classify the irreducible projective representations of the translation group it is important to note that we can select from this group a sub-group (the *magnetic sub-group*) with respect to which the representation is not projective but ordinary. With the condition (60.6), such a sub-group is the set of translations of the form

$$\mathbf{a}_m = n_1\mathbf{a}_1 + n_2 q\mathbf{a}_2 + n_3\mathbf{a}_3 \tag{60.7}$$

with integral coefficients n_1, n_2, n_3. For, when the vector **h** is along $\mathbf{a}_3$ and satisfies the condition (60.6) for all translations of this form, the exponent in (60.3) is zero or a multiple of 2π, so that all the factors $\omega(\mathbf{a}, \mathbf{a}') = 1$.† The set of translations (60.7) forms a lattice with basic vectors $\mathbf{a}_1$, $q\mathbf{a}_2$, $\mathbf{a}_3$, which we call the *magnetic lattice*. The magnetic reciprocal lattice correspondingly has basic vectors $\mathbf{b}_1$, $\mathbf{b}_2/q$, $\mathbf{b}_3$, where $\mathbf{b}_1$, $\mathbf{b}_2$, $\mathbf{b}_3$ are basic vectors of the original reciprocal lattice.

The irreducible ordinary representations of the magnetic sub-group, like those of the whole translation group, are one-dimensional; they are charac-

† The choice of the magnetic sub-group is in general not unique: instead of (60.7) we can take any set of translations of the form $\mathbf{a}_m = n_1 q_1\mathbf{a}_1 + n_2 q_2\mathbf{a}_2 + n_3\mathbf{a}_3$, where q_1, q_2 are integers such that $q_1 q_2 = q$.

terized by wave vectors (quasi-momenta) **K**, all non-equivalent values of which are in one cell of the magnetic reciprocal lattice.

Let $\psi^{(1)}$ be the base function of one such representation with quasi-momentum $\mathbf{k}^{(1)} \equiv \mathbf{K}$. For this function,

$$\hat{T}_{\mathbf{a}_m}\psi^{(1)}(\mathbf{r}) = e^{i\mathbf{k}^{(1)}\cdot\mathbf{a}_m}\psi^{(1)}(\mathbf{r}). \tag{60.8}$$

In a translation through the basic vector $\mathbf{a}_2$ (which is not in the magnetic sub-group), we obtain from $\psi^{(1)}$ a function $\psi^{(2)}$ with a different quasi-momentum. To determine the latter, we use (60.4) and (60.8) to write

$$\begin{aligned}\hat{T}_{\mathbf{a}_m}\psi^{(2)} &= \hat{T}_{\mathbf{a}_m}\hat{T}_{\mathbf{a}_2}\psi^{(1)}(\mathbf{r})\\ &= \exp(-i\mathbf{h}\cdot\mathbf{a}_m\times\mathbf{a}_2)\hat{T}_{\mathbf{a}_2}\hat{T}_{\mathbf{a}_m}\psi^{(1)}(\mathbf{r})\\ &= \exp\{-i\mathbf{a}_m\cdot\mathbf{a}_2\times\mathbf{h}+i\mathbf{a}_m\cdot\mathbf{k}^{(1)}\}\hat{T}_{\mathbf{a}_2}\psi^{(1)}(\mathbf{r})\end{aligned}$$

or finally

$$\hat{T}_{\mathbf{a}_m}\psi^{(2)}(\mathbf{r}) = e^{i\mathbf{k}^{(2)}\cdot\mathbf{a}_m}\psi^{(2)}(\mathbf{r}),$$

where

$$\mathbf{k}^{(2)} = \mathbf{k}^{(1)} - \mathbf{a}_2\times\mathbf{h} = \mathbf{K} - 2(p/q)\,\mathbf{b}_1;$$

in the last equation we have substituted (60.5) and the reciprocal lattice vector $\mathbf{b}_1 = 2\pi\mathbf{a}_2\times\mathbf{a}_3/v$. We must now take separately the cases of odd and even q.†

Let q be odd. Repeating the translation through $\mathbf{a}_2$ a further $q-2$ times, we obtain a total of q different functions with quasi-momenta

$$\mathbf{k}^{(1)} = \mathbf{K},\quad \mathbf{k}^{(2)} = \mathbf{K} - 2(p/q)\,\mathbf{b}_1,\ \ldots,\ \mathbf{k}^{(q)} = \mathbf{K} - 2\,\frac{p(q-1)}{q}\,\mathbf{b}_1. \tag{60.9}$$

By subtracting an appropriate integral multiple of the vector $\mathbf{b}_1$, these values are converted (in some order) to the values

$$\mathbf{k} = \mathbf{K},\quad \mathbf{K}+\mathbf{b}_1/q,\quad \mathbf{K}+2\mathbf{b}_1/q,\ \ldots,\ \mathbf{K}+(q-1)\,\mathbf{b}_1/q. \tag{60.10}$$

These q functions form a q-dimensional irreducible projective representation of the translation group. We obtain all the non-equivalent representations when K takes values in a cell with sides $\mathbf{b}_1/q$, $\mathbf{b}_2/q$, $\mathbf{b}_3$; the quasi-momenta $\mathbf{k}^{(1)}$, $\mathbf{k}^{(2)}, \ldots$ then take values in a cell with sides $\mathbf{b}_1$, $\mathbf{b}_2/q$, $\mathbf{b}_3$.

Now let q be even. Then, in the sequence (60.9), the $(\frac{1}{2}q+1)$th value, equal to $\mathbf{K}-p\mathbf{b}_1$, differs from $\mathbf{K}$ only by an integral multiple of the reciprocal lattice vector $\mathbf{b}_1$. Thus there are only $\frac{1}{2}q$ nonequivalent values of $\mathbf{k}$, which are given by (60.10) with $\frac{1}{2}q$ instead of q. In this case, therefore, the irreducible represen-

† When $q = 1$, the magnetic sub-group is the same as the complete translation group. Thus, if $\mathbf{h}$ is an integral multiple of $4\pi\mathbf{a}_3/v$, the irreducible projective representations of the translation group are the same as the ordinary irreducible representations, and the electron states are classified in the same way as in the absence of the field.

tations are $\frac{1}{2}q$-dimensional, and $\mathbf{K}$ takes values in a cell with edges $2\mathbf{b}_1/q$, $\mathbf{b}_2/q$, $\mathbf{b}_3$.

These results enable us to formulate the following conclusion about the manner of variation of the electron energy spectrum in the lattice when a magnetic field is applied to it that satisfies the condition (60.5). In the absence of the field, the spectrum consists of discrete energy bands, in each of which the energy $\varepsilon(\mathbf{k})$ is a function of the quasi-momentum, the latter taking values in one reciprocal lattice cell. When the field is applied, the band splits into q sub-bands, in each of which all the energy levels are q-fold or $\frac{1}{2}q$-fold degenerate, according as q is odd or even. The energy in the sub-band may be expressed as a function $\varepsilon(\mathbf{K})$ of the vector $\mathbf{K}$, the latter taking values in $1/q^2$ or $2/q^2$ of the reciprocal lattice cell, according as q is odd or even.

The picture given above is, in a certain sense, extremely sensitive to the magnitude and direction of the magnetic field. For, there are values of $\mathbf{H}$ arbitrarily close to a value satisfying (60.5) with some p and q, which satisfy the same condition with much larger q, so that by an infinitesimal change of the field the number of sub-bands can be increased indefinitely. It must be emphasized, however, that this does not imply a similar instability in observable physical properties. These are determined not by the specific band structure, but by the distribution of the number of states among small but finite energy ranges, and this distribution changes only slightly with the field, because what is greatly changed is not the energy of the states but only their classification as a result of the altered range of definition of the quasi-momentum.

§ 61. Electronic spectra of normal metals

In actual crystals of normal (not superconducting) metals, the electrons form a quantum Fermi liquid of the type described in Chapter I. Certain differences arise, however, because this is not a "free" isotropic liquid but a liquid in the anisotropic periodic field of the lattice.

Just as the energy spectrum of a free Fermi liquid has a similar structure to that of an ideal Fermi gas, so the spectrum of the electron Fermi liquid in a metal has a similar structure to the of an ideal "gas in the lattice". The occurrence of the quasi-momentum as a conserved quantity is due only to the spatial periodicity of the system (just as the conservation of the actual momentum is a consequence of the complete spatial homogeneity). It is therefore natural that the properties enumerated in §55 apply to the classification of levels in the spectrum of an electron liquid in a metal, the role of the particles (electrons) being taken by the quasi-particles.

At absolute zero, the particles of an ideal Fermi gas in a periodic field occupy all the lowest levels, with energies ε up to a limiting value ε_F (which is equal to the chemical potential μ at $T = 0$), determined by the condition that the number of states with $\varepsilon \leqslant \varepsilon_F$ is equal to the total number of electrons. The energy bands such that $\varepsilon_s(\mathbf{k}) < \varepsilon_F$ for all $\mathbf{k}$ are fully occupied; those with $\varepsilon_s(\mathbf{k}) > \varepsilon_F$ are empty, and those for which the equation

$$\varepsilon_s(\mathbf{k}) = \varepsilon_F \tag{61.1}$$

has a solution are partly occupied. Equation (61.1) determines in $\mathbf{k}$-space the limiting Fermi surface which (for each band) separates the filled and empty states.

Similarly, in an actual metal there is a surface in $\mathbf{k}$-space which separates the region of quasi-particle states that are filled (at $T = 0$) from the unoccupied states; on one side of this surface the quasi-particle energies $\varepsilon > \varepsilon_F$, and on the other side $\varepsilon < \varepsilon_F$. However, the concept of quasi-particles in a Fermi liquid has a real physical significance only near the Fermi surface, where the decay of the elementary excitations is relatively slight (see §1). Hence the idea of occupied energy bands which occurs in the description of the spectrum of an ideal Fermi gas has no literal meaning in an actual electron liquid.

The quasi-particles near the Fermi surface are called *conduction electrons*. Their energy is in general a linear function of the quasi-momentum; similarly to (1.12), we have

$$\varepsilon(\mathbf{k}) - \varepsilon_F \approx \hbar(\mathbf{k} - \mathbf{k}_F) \cdot \mathbf{v}_F, \tag{61.2}$$

where $\mathbf{k}_F$ is a point on the Fermi surface and

$$\hbar \mathbf{v}_F \approx (\partial \varepsilon / \partial \mathbf{k})_{\mathbf{k} = \mathbf{k}_F} \tag{61.3}$$

gives the conduction electron velocity at such a point.†

Near the Fermi surface there must also be a "transitional zone" in the conduction electron distribution at non-zero temperatures. Hence we have as the condition for the Fermi-liquid theory to be valid $T \ll \hbar k_F v_F$, where k_F and v_F are the characteristic values for the dimensions of the Fermi surface and the velocity on it. The dimensions k_F are usually of the same order of magnitude as those of the reciprocal lattice cell, so that $k_F \sim 1/a$; an exception occurs for the semi-metals (see below). Putting also as an estimate $v_F \sim \hbar k_F/m$, we get the condition $T \ll 10^4 - 10^5$°K, which is always satisfied in practice.

Almost all metals have crystal lattices with a centre of inversion. According to the discussion at the end of §55, all energy levels of the conduction electrons (with given $\mathbf{k}$) are doubly degenerate with respect to the spin (the metals referred to are neither ferromagnetic nor antiferromagnetic).

The shape and configuration of the Fermi surface are important characteristics of any particular metal. In various metals they are quite different and in general complicated. The Fermi surface may consist of several separate

† Formulae such as (2.11) for the effective mass, derived in §2 for the "free" Fermi liquid from considerations of Galilean invariance, of course do not apply to the electron liquid in a crystal lattice.

sheets, which may be singly or multiply connected, open or closed; cf. the discussion of constant-energy surfaces in §55.

The closed sheets of the Fermi surface can be divided into two classes, according as they form the boundary of regions of filled (at $T = 0$) or empty states of quasi-particles (in the former case $\varepsilon < \varepsilon_F$ within the cavity, in the latter case $\varepsilon > \varepsilon_F$). Both cases can, however, be described in a similar manner if we suppose in the second case that the "empty" cavity is filled with "quasi-holes"; the transition of the system to the excited state is then described as a transition of quasi-holes from inside to outside the Fermi surface. The Fermi surface itself is then said to be *hole-type*, in contrast to *electron-type* in the first case.† The physical difference between the two types of quasi-particles (electrons and holes) is clearly seen when they move in external fields. For example, all cross-sections of a hole-type (or electron-type) Fermi surface, which determine the quasi-classical trajectories for motion in a magnetic field, are of the hole (or electron) type as defined in §57.

In an isotropic "free" Fermi liquid, as considered in §1, the Fermi surface was a sphere whose radius was determined by the density of the liquid in accordance with Landau's theorem (1.1). An analogous relation holds for the electron liquid in a metal, but the specific nature of the properties resulting from the periodicity of the lattice causes some changes in its formulation.

The number of electrons in the metal is conveniently expressed per unit lattice cell. Let n be the total number of electrons in the atoms in one cell, and τ_F the total volume per cell of the reciprocal lattice on the occupied side of the Fermi surface (i.e. the side where $\varepsilon < \varepsilon_F$). The word "total" here signifies that if the occupied regions corresponding to different sheets of the Fermi surface partly overlap, they must still be combined independently. The volume τ_F will be measured in units of the reciprocal lattice cell volume itself, and the remark just made about overlapping regions means that τ_F thus defined may exceed unity.

The requisite proposition (*Luttinger's theorem*), which replaces Landau's theorem (§20) for a metal, is expressed by

$$n_c \equiv 2\tau_F = n - 2l \tag{61.4}$$

where l is an integer ($\geqslant 0$). In the model of an ideal gas in a lattice, this number has a simple significance: complete occupation of each band corresponds to two electrons in the reciprocal lattice cell (because of the two spin states), so that $2l$ is the number of electrons occupying the l lowest bands, and $n-2l$ is the number of electrons in partly filled bands. Formula (61.4) expresses the far from trivial result that a similar situation continues to hold when the

† We must emphasize, however, to avoid misunderstanding, that the term "hole" is not used here in the same sense as in the alternative method (see the end of §1) for describing the spectrum of a Fermi liquid; there, the holes were simply empty places formed in the filled region on excitation of the system.

interaction between electrons is taken into account.† By the definition of a metal, the integer n_c is not zero.

Let there be only closed sheets (electron-type and hole-type) of the Fermi surface in a metal; let $\tau_-^{(s)}$ and $\tau_+^{(s)}$ be the contributions to τ_F from the electron and hole cavities individually:

$$\tau_F = \sum_s \tau_-^{(s)} + \sum_s \tau_+^{(s)};$$

the summation is over all electron-type and hole-type sheets respectively. The quantity $\tau_-^{(s)}$ is the volume of the electron-type cavity, and $1-\tau_+^{(s)}$ is that of the hole-type cavity. The numbers of electron and hole quasi-particles are

$$n_- = 2\sum_s \tau_-^{(s)}, \quad n_+ = 2\sum_s (1-\tau_+^{(s)}).$$

With n even (and therefore n_c even), cases can occur where n_c is equal to twice the number of hole-type cavities. Equation (61.4) is then easily found to reduce to

$$n_- = n_+. \tag{61.5}$$

Such metals with equal numbers of quasi-particles and quasi-holes are said to be *compensated.*

It should be pointed out that, when equation (61.5) is exactly satisfied, n_- and n_+ themselves may be arbitrary, and in particular may be arbitrarily small. In such cases, when the volumes of all the cavities of the Fermi surface are very small (in comparison with the volume of one reciprocal lattice cell), the substance is said to be a *semi-metal.*‡ There is, however, a lower limit to the number of conduction electrons, beyond which an electron spectrum of the metal type becomes unstable and cannot exist (see the end of §66).

The thermodynamic quantities for a metal consist of lattice and electron parts. The temperature dependence of the latter is determined by the quasi-particles in the neighbourhood of the Fermi surface (the dispersion relation (61.2)). The nature of this dependence is of course the same as for an ideal Fermi gas or an isotropic Fermi liquid (cf. §1); the only difference in the formulae is due to the different number of quasi-particle states near the Fermi surface, which is no longer a sphere.

Let $\nu\, d\varepsilon$ be the number of states (per unit volume of the metal) in an energy range $d\varepsilon$. The volume element in **k**-space between infinitely close constant-energy surfaces corresponding to energies ε_F and $\varepsilon_F+d\varepsilon$ is $df\, d\varepsilon/\hbar v_F$, where df is an element of area on the Fermi surface and v_F the magnitude of the vector $\mathbf{v}_F = (1/\hbar)\, \partial\varepsilon/\partial\mathbf{k}$ normal to the surface. Hence

$$\nu_F = \frac{2}{(2\pi)^3}\int \frac{df}{\hbar v_F}, \tag{61.6}$$

† For a rigorous derivation of this result, see J. M. Luttinger, *Physical Review* **119**, 1153, 1960.

‡ For example, in bismuth $n_- = n_+ \sim 10^{-5}$.

where the integration is over all the sheets of the Fermi surface within one reciprocal lattice cell; for an open Fermi surface the faces of the cell itself are, of course, not part of the range of integration.

The quantity (61.6) replaces in the thermodynamic quantities the expression which for a gas of free particles (with a spherical Fermi surface) was

$$\frac{2}{(2\pi\hbar)^3}\frac{4\pi p_F^2}{p_F/m} = \frac{mp_F}{\pi^2\hbar^3}.$$

For example, the electron part of the thermodynamic potential Ω of a metal is (cf. Part 1, §58)

$$\Omega_e = \Omega_{0e} - \tfrac{1}{6}\pi^2\nu_F VT^2, \tag{61.7}$$

where Ω_{0e} is the value of the potential at $T = 0$. Regarding the second term in (61.7) as a small correction to Ω_{0e}, by the theorem of small increments, we can write down a similar formula for the thermodynamic potential Φ:

$$\Phi_e = \Phi_{0e} - \tfrac{1}{6}\pi^2\nu_F VT^2, \tag{61.8}$$

where ν_F and V are now assumed to be expressed in terms of P (in the "zero" approximation, i.e. at $T = 0$).

Determining from (61.8) the entropy and hence the specific heat, we find

$$C_e = \tfrac{1}{3}\pi^2\nu_F VT. \tag{61.9}$$

The lattice part of the specific heat is proportional to T^3 (at temperatures small compared with the Debye temperature Θ); hence, at sufficiently low temperatures, the electron contribution to the specific heat becomes predominant.†

For the same reason, the electron contribution to the thermal expansion of the metal becomes predominant in this temperature range. Determining from (61.8) the volume $V = \partial\Phi/\partial P$ and hence the thermal expansion coefficient α, we find

$$\alpha = \frac{1}{V}\left(\frac{\partial V}{\partial T}\right)_P = -T\frac{\pi^2}{3V}\frac{\partial(V\nu_F)}{\partial P}. \tag{61.10}$$

Here, as also in the range $T \gg \Theta$ (see Part 1, §67), the ratio

$$\alpha V/C = -\partial \log (V\nu_F)/\partial P$$

s independent of the temperature.

§ 62. Green's function of electrons in a metal

The discussion in §§56–58 related to the motion of one electron in a lattice to which an external magnetic field is applied. We shall now show that the results obtained remain essentially valid for quasi-particles (conduction electrons)

† The small parameter of the expansion in (61.9) is the ratio T/ε_F; in the lattice specific heat it is T/Θ; thus the two parts of the specific heat become comparable when $T^2 \sim \Theta^3/\varepsilon_F$.

Electrons in the Crystal Lattice

in the electron liquid in an actual metal; there is only a certain change in the definitions of the quantities appearing in the relations (Yu. A. Bychkov and L. P. Gor'kov 1961, J. M. Luttinger 1961). The Green's function formalism is suitable for a general treatment of the electron liquid.

In Chapter II, this formalism has been developed for a "free" Fermi liquid. We shall now show how it must be modified for a liquid in a lattice.

The Green's function for an electron liquid (at $T = 0$) is defined in terms of the Heisenberg ψ operators of the electrons by the same formula (7.9), where the averaging is with respect to the ground state of the metal. Because of the homogeneity of time, this function depends on the arguments t_1 and t_2 only through the difference $t = t_1 - t_2$. The spatial homogeneity, however, is now destroyed by the presence of the lattice field, which is external to the liquid. Hence the Green's function does not depend only on the difference $r_1 - r_2$. All we can say is that it is invariant under a simultaneous shift of r_1 and r_2 by the same arbitrary basic lattice vector. In the following we shall consider the Green's function in the ω, r representation, i.e. use its Fourier component with respect to t, $G_{\alpha\beta}(\omega; r_1, r_2)$. This function enables us, in principle, to determine the energy spectrum of the electron liquid in the metal. We shall repeat (without going through all the calculations again) the arguments of §8 as they apply to the present case.

It has been shown in §8 that the homogeneity of the system allows a complete determination of the coordinate dependence of the matrix elements of the ψ operators and thus the formulation of a general expression for the Green's function in the space-time representation in the form (8.5), (8.6); from this we can go to the momentum representation in the form of the expansion (8.7).

For the electron liquid in a lattice, the invariance of the matrix elements, expressed by equation (8.3), occurs only for translations through basic vectors of the lattice, i.e. for $r = a$. This naturally leads to less definiteness in the dependence on the coordinates: instead of (8.4), we can only say that

$$\left.\begin{array}{l} \langle 0 | \hat{\Psi}_\alpha(t, r) | mk \rangle = \chi^{(+)}_{\alpha mk}(r) \exp\left[-i\omega_{m0}(k)\, t\right], \\ \langle mk | \hat{\Psi}_\alpha(t, r) | 0 \rangle = \chi^{(-)}_{\alpha m, -k}(r) \exp\left[i\omega_{m0}(k)\, t\right], \end{array}\right\} \tag{62.1}$$

where

$$\left.\begin{array}{l} \chi^{(+)}_{\alpha mk}(r) = e^{ik \cdot r} u_{\alpha mk}(r), \\ \chi^{(-)}_{\alpha mk}(r) = e^{ik \cdot r} v_{\alpha mk}(r), \end{array}\right\} \tag{62.2}$$

k is the quasi-momentum of the state, m is the set of all other quantum numbers describing the state, and u and v are some functions of the coordinates, periodic in the lattice; we have written the matrix elements only for transitions from the ground state, i.e. state 0. The properties of the functions $\chi^{(+)}$ and $\chi^{(-)}$ are similar to those of the Bloch wave functions of an electron in a periodic field. Expressing the Green's function in terms of these matrix elements and then changing to the Fourier components with respect to time (as in §8), we now obtain instead of (8.7) the expansion

$$G_{\alpha\beta}(\omega;\mathbf{r}_1,\mathbf{r}_2) = \sum_{m,\mathbf{k}} \left\{ \frac{\chi^{(+)}_{\alpha m\mathbf{k}}(\mathbf{r}_1)\,\chi^{(+)*}_{\beta m\mathbf{k}}(\mathbf{r}_2)}{\omega+\mu-\varepsilon^{(+)}_{m\mathbf{k}}+i0} + \frac{\chi^{(-)}_{\alpha m\mathbf{k}}(\mathbf{r}_1)\,\chi^{(-)*}_{\beta m\mathbf{k}}(\mathbf{r}_2)}{\omega+\mu-\varepsilon^{(-)}_{m\mathbf{k}}-i0} \right\}, \quad (62.3)$$

with the same notation $\varepsilon^{(+)}$ and $\varepsilon^{(-)}$ as previously, and the change $\mathbf{k} \to -\mathbf{k}$ in the second term.

The presence of non-decaying one-particle elementary excitations near the Fermi surface of a metal has the result that, when ε is near μ, the energy of the state depends only on $\mathbf{k}$. For such states the function $G_{\alpha\beta}(\omega;\mathbf{r}_1,\mathbf{r}_2)$ has a pole at $\omega = \varepsilon(\mathbf{k})-\mu$. Near the pole, it has the form

$$G_{\alpha\beta}(\omega;\mathbf{r}_1,\mathbf{r}_2) = \frac{\chi_{\alpha\mathbf{k}}(\mathbf{r}_1)\,\chi^{*}_{\beta\mathbf{k}}(\mathbf{r}_2)}{\omega+\mu-\varepsilon(\mathbf{k})+i0.\mathrm{sgn}\,\omega}. \quad (62.4)$$

When there is degeneracy with respect to spin, we must also sum over the two spin states.

The determination of the energy spectrum from the Green's function reduces, in principle, to an eigenvalue problem for a certain linear integro-differential operator.

The basic ideas of the diagram technique in coordinate space remain the same in this case as for an ordinary Fermi liquid. In particular, with the self-energy function $\Sigma_{\alpha\beta}(t,\mathbf{r}_1,\mathbf{r}_2)$ as the sum of the set of diagrams determined in §14, we can write the Green's function $G_{\alpha\beta}(t,\mathbf{r}_1,\mathbf{r}_2)$ as the series (14.3), which is summed as the diagram equation (14.4). The thin continuous line in these diagrams represents the Green's function $G^{(0)}_{\alpha\beta}(t,\mathbf{r}_1-\mathbf{r}_2)$ of free electrons that interact neither with one another nor with the lattice. According to (9.6), this function satisfies the equation

$$\left(i\frac{\partial}{\partial t}+\frac{\triangle_1}{2m}+\mu\right) G^{(0)}_{\alpha\beta}(t,\mathbf{r}_1-\mathbf{r}_2) = \delta_{\alpha\beta}\delta(t)\,\delta(\mathbf{r}_1-\mathbf{r}_2).$$

Applying the operator (...) to the left of (14.4) and then changing to Fourier components with respect to time, we obtain the required equation

$$(\omega+\mu+\triangle_1/2m)\,G_{\alpha\beta}(\omega;\mathbf{r}_1,\mathbf{r}_2)-\int \Sigma_{\alpha\gamma}(\omega;\mathbf{r}_1,\mathbf{r}')\,G_{\gamma\beta}(\omega;\mathbf{r}',\mathbf{r}_2)\,d^3x' = \delta_{\alpha\beta}\delta(\mathbf{r}_1-\mathbf{r}_2). \quad (62.5)$$

Near the pole of G (with respect to the variable ω) the right-hand side of the equation may be omitted, leaving a homogeneous integro-differential equation whose eigenvalues determine the energy spectrum of the system. The suffix β and the variable $\mathbf{r}_2$ are not affected by any operations here, i.e. they act as unimportant parameters in the equation. To determine the spectrum, we can therefore use the equation†

$$(\omega+\mu+\triangle_1/2m)\,\chi_\alpha(\mathbf{r})-\int \Sigma_{\alpha\gamma}(\omega;\mathbf{r},\mathbf{r}')\,\chi_\gamma(\mathbf{r}')\,d^3x' \equiv (\omega-\hat{L})\,\chi(\mathbf{r}) = 0. \quad (62.6)$$

† For a microscopically homogeneous Fermi liquid, this equation in the momentum representation reduces to (14.13):

$$\omega+\mu = \varepsilon^{(0)}(\mathbf{p})+\Sigma(\omega,\mathbf{p}).$$

For an electron Fermi liquid in a metal, it replaces the ordinary Schrödinger's equation. Its eigenvalues determine the spectrum, as already mentioned, with $\omega = \varepsilon(\mathbf{k})-\mu$; the corresponding eigenfunctions are $\chi_{\alpha\mathbf{k}}(\mathbf{r})$ from (62.4), as is evident from a direct substitution of (62.4) in (62.5). Since the decay of excitations near the Fermi surface is only slight, the operator $\hat{L}$ is Hermitian for small ω (up to and including terms of order ω).

To go to the case where a weak external magnetic field is present, we must note that, in a gauge transformation of the vector potential, the ψ operators are transformed as wave functions (cf. (44.3), (44.4)), and therefore the Green's function $G_{\alpha\beta}(\omega; \mathbf{r}_1, \mathbf{r}_2)$ is transformed as a product of ψ functions, $\psi(\mathbf{r}_1)\psi^*(\mathbf{r}_2)$. This means that the function $\chi(\mathbf{r})$ in (62.6) must also be transformed as an ordinary ψ function. If the arguments of §56 are followed through, we easily find that they make use only of the periodicity of the crystal lattice, the general properties of the gauge transformation, and the fact that the energy spectrum is determined by the eigenvalues of a certain Hamiltonian; in the present case, the latter is the operator $\hat{L}$ in (62.6).† It is therefore clear that the result, that is the rule for changing from the spectrum in the absence of the field to that in the presence of the field, will be the same; the new spectrum is determined from the eigenvalues of the Hamiltonian

$$\varepsilon\left(\mathbf{K}-\frac{e}{\hbar c}\mathbf{A}(\hat{\mathbf{r}})\right), \quad \hat{\mathbf{r}} = i\frac{\partial}{\partial \mathbf{K}}, \tag{62.7}$$

where $\varepsilon(\mathbf{k})$ is the spectrum in the absence of the field. The significance of $\varepsilon(\mathbf{k})$ itself is now, of course, different from that in (56.7), since it takes account of the collective interaction of all the electrons in the system.

Next, since the treatment of the quasi-classical case in §§57 and 58 was entirely based on the existence of a Hamiltonian of the form (62.7), the results obtained there are also directly applicable to an electron liquid. The question arises, however, of what is to be regarded as the field strength acting on a conduction electron (and thus what is to be taken as the vector potential $\mathbf{A}$). Strictly speaking, it should be the exact microscopic value of the field at the point $\mathbf{r}$ due to all the electrons (and to the external field). In the quasi-classical case, however, the characteristic dimension r_H of the region in which interaction occurs (the Larmor radius of the orbits) is large compared with the order of magnitude of the distances between electrons, i.e. the lattice constant a. This causes an automatic averaging of the microscopic field. The origin of the averaging may be explained as follows.

Let us represent the microscopic field strength as the sum of its mean value (which is, according to customary terminology of macroscopic electrodynamics,

† There may seem to be an important difference here in that the operator $\hat{L}$ in (62.6) itself depends on ω. In fact, this merely causes the Hamiltonian to be written in an implicit form. For small ω (near the Fermi surface) we can change to an explicit form by expanding $\hat{L} \approx \hat{L}_0+\omega\hat{L}_1$ and then multiplying the equation $\hat{L}_0\chi = \omega(1-\hat{L}_1)\chi$ on the left by the operator $(1-\hat{L}_1)^{-1}$.

the magnetic induction **B**) and a rapidly varying part $\tilde{\mathbf{H}}$. The vector potential corresponding to the uniform field **B** increases over the whole extent of the orbit, taking characteristic values $\sim Br_H$. The potential corresponding to the field $\tilde{\mathbf{H}}$ which oscillates over distances $\sim a$ does not increase steadily, and only reaches values $\sim Ba$, which are negligible in comparison with Br_H. But, as shown in §56, it is the potential of the field which determines the quantization of the motion of the electrons. Thus we conclude that it is sufficient to take account of just the potential **A** of the uniform induction **B** = curl **A**, which will play the part of the field acting on the electron (D. Shoenberg 1962). We shall see later (at the end of §63) that this situation may lead to some new phenomena in the magnetization of metals.

Thus the quasi-classical quantization rule (58.7) for the electron liquid in a metal is written

$$S(\varepsilon, k_z) = (2\pi\,|e|/\hbar c)\,B(n+\tfrac{1}{2}), \tag{62.8}$$

where $S(\varepsilon, k_z)$ is now the cross-sectional area of the actual constant-energy surfaces of the conduction electrons in a metal (near its Fermi surface).

As in the problem of one electron in a lattice having a centre of inversion†, the inclusion of the conduction electron spin causes a splitting of the levels in the magnetic field into two components:

$$\varepsilon_{n\sigma}(k_z) = \varepsilon_n(k_z) + \sigma\beta\xi(k_z)\,B, \quad \sigma = \pm 1. \tag{62.9}$$

The quantity $\xi(k_z)$ results from the averaging of a function $\xi(\mathbf{k})$ over the quasi-classical trajectory. We can with sufficient accuracy regard all trajectories as lying on the Fermi surface itself, so that the result of the averaging depends only on k_z. We should emphasize that, for electrons in the Fermi liquid, the difference of $\xi(k_z)$ from the value unity which it has for free electrons is due not only to the spin–orbit interaction but also to the exchange interaction between electrons.

§ 63. The de Haas–van Alphen effect

The magnetic susceptibility of a metal in weak magnetic fields ($\beta B \ll T$, where β is the Bohr magneton and B the magnetic induction) cannot be calculated in general form. The reason is that in the theory of the Fermi liquid we can only deal with the paramagnetic (spin) part of the susceptibility: this part is determined by the conduction electrons near the Fermi surface, since the electron spins within the distribution compensate one another. The diamagnetic (orbital) part of the susceptibility, however, contains contributions from all the electrons, including those within the distribution, where the concept of quasi-particles in the theory of the Fermi liquid has no meaning. The two parts of the conductivity are in general of the same order of magnitude, and only their sum has an actual physical significance.

† As in fact do the crystal lattices of all metals.

Let us now consider "strong" fields, for which

$$T \lesssim \beta B \ll \mu, \tag{63.1}$$

i.e. the intervals between the Landau levels are comparable with the temperature but still small in comparison with the chemical potential. In this case, the paramagnetic and diamagnetic parts of the magnetization cannot be separated at all, but the situation is different in that the magnetization of the metal has an oscillatory dependence on the field (*the de Haas–van Alphen effect*).† The monotonic part of the magnetization depends in this case also on all the electrons in the metal and cannot be calculated within the theory of the Fermi liquid. But the oscillatory part of the magnetization is, as we shall see, determined only by the conduction electrons near the Fermi surface and can be considered in a general form (I. M. Lifshitz and A. M. Kosevich 1965); we shall now discuss this part.

The oscillatory dependence of the magnetization on the field is a consequence of the quantization of the energy levels of the orbital motion of the electrons. But the quantization affects only states corresponding to the motion of electrons in trajectories that are closed (in **k**-space). Hence the contribution to the oscillatory part of the thermodynamic quantities comes only from the conduction electrons on closed cross-sections of the constant-energy surfaces by planes perpendicular to the given direction of the field. We shall assume that the quasi-classicality condition is satisfied on these cross-sections, i.e. that the numbers n determined by equation (62.8) are large:

$$\hbar c S/|e|B \gg 1. \tag{63.2}$$

For typical Fermi surfaces in metals, the linear dimensions of the cross-sections are $\sim 1/a$, so that $S \sim a^{-2}$, and the condition (63.2) is certainly satisfied; cf. the first footnote to §57.

The quasi-classical levels are given (with allowance for spin) by the expression (62.9), where $\varepsilon_n(k_z)$ are the solutions of equation (62.8); for each level there is a number of states given by formula (58.10). Hence the partition function which determines the thermodynamic potential Ω (a function of μ, T and the volume V of the system) is as shown in

$$\Omega = -T\frac{|e|BV}{4\pi^2\hbar c}\sum_n \int \sum_{s,\,\sigma} \log\left\{1+\exp\frac{\mu-\varepsilon_{n\sigma}^{(s)}(k_z)}{T}\right\} dk_z. \tag{63.3}$$

The suffix s numbers the various sheets of the constant-energy surface; this suffix, and the sign of summation with respect to it, will be omitted for brevity. The integration with respect to k_z is taken over an interval such as to include all different cross-sections (i.e. excluding periodic repetitions) of all sheets of the constant-energy surfaces.

† Cf. Part 1, §60, where this effect has been discussed for an ideal electron gas.

Let us first separate from Ω the part $\tilde{\Omega}$ that is an oscillatory function of the field, transforming the sum (63.3) by means of Poisson's formula:†

$$\tfrac{1}{2}F(0)+\sum_{n=1}^{\infty}F(n) = \int_0^{\infty}F(x)\,dx+2\,\mathrm{re}\sum_{l=1}^{\infty}\int_0^{\infty}F(x)\,e^{2\pi ilx}\,dx. \qquad (63.4)$$

The first term in this formula when applied to (63.3) gives the non-oscillatory contribution to Ω; we omit this, and write

$$\tilde{\Omega} = -\frac{|e|\,BVT}{4\pi^2 c\hbar}2\,\mathrm{re}\sum_{l=1}^{\infty}\sum_{\sigma=\pm 1}\tilde{I}_{l\sigma}, \qquad (63.5)$$

where $\tilde{I}_{l\sigma}$ is the oscillatory part of the integral

$$I_{l\sigma} = \int_0^{\infty}dn\int\log\left\{1+\exp\frac{\mu_\sigma-\varepsilon_n(k_z)}{T}\right\}e^{2\pi iln}\,dk_z \qquad (63.6)$$

and we have used the notation $\mu_\sigma = \mu-\sigma\beta\xi B$.

For the further analysis we define the function

$$n(\varepsilon, k_z) = \frac{c\hbar S(\varepsilon, k_z)}{2\pi\,|e|\,B}-\frac{1}{2} \qquad (63.7)$$

(cf. (62.8)), and change from integration over n in (63.6) to integration over ε:

$$I_{l\sigma} = \int_0^{\infty}\int\log\left\{1+\exp\frac{\mu_\sigma-\varepsilon}{T}\right\}e^{2\pi iln}\frac{\partial n}{\partial\varepsilon}\,dk_z\,d\varepsilon; \qquad (63.8)$$

the choice of the lower limit of integration over ε, which has been arbitrarily taken as zero, is immaterial, since in any case only the neighbourhood of $\varepsilon = \mu_\sigma$ is important in the integral.

Since the function $n(\varepsilon, k_z)$ is large, the exponential factor in the integrand in (63.8) is a rapidly oscillating function of k_z. These oscillations reduce to nothing the integral with respect to k_z, and so the main contribution comes from the ranges of the variable k_z in which $n(\varepsilon, k_z)$ varies least rapidly and the oscillations are therefore slowest. That is, the main contribution to the integral comes from the regions near the extrema of n as a function of k_z for any given ε. Let $k_{z,\,\mathrm{ex}}(\varepsilon)$ be one such point; near it, we calculate the integral by the saddle-point method, putting in the exponent of the exponential

$$n(\varepsilon, k_z) \approx n_{\mathrm{ex}}(\varepsilon)+\frac{1}{2}\left(\frac{\partial^2 n}{\partial k_z^2}\right)_{\mathrm{ex}}(k_z-k_{z,\,\mathrm{ex}})^2,$$

$$n_{\mathrm{ex}}(\varepsilon) = n(\varepsilon-k_{z,\,\mathrm{ex}}(\varepsilon)),$$

† See Part 1, §60. It is not important that the term $F(0)$ in the sum in (63.4) has a coefficient $\frac{1}{2}$, since only the terms with large n are significant in the sum (63.3).

and in the non-exponential factors taking the value at $k_z = k_{z,\,\mathrm{ex}}$. The result is that each of the extrema contributes to the integral the term

$$\int\limits_0^\infty \log\left\{1+\exp\frac{\mu_\sigma-\varepsilon}{T}\right\}\frac{dn_{\mathrm{ex}}}{d\varepsilon}\cdot\frac{1}{\sqrt{l}}\left|\frac{\partial^2 n}{\partial k_z^2}\right|_{\mathrm{ex}}^{-1/2}\exp\left\{2\pi i l n_{\mathrm{ex}}\pm\frac{1}{4}i\pi\right\}d\varepsilon.$$

The replacement of $\partial n(\varepsilon, k_z)/\partial\varepsilon$ by $dn_{\mathrm{ex}}/d\varepsilon$ is legitimate, since at the extremum $\partial n/\partial k_z = 0$. The plus and minus signs in the exponent refer respectively to the cases where $k_{z,\,\mathrm{ex}}$ is a minimum and a maximum of $n(\varepsilon, k_z)$.† We transform this expression by integration by parts, with

$$\frac{dn_{\mathrm{ex}}}{d\varepsilon}\exp(2\pi i l n_{\mathrm{ex}})\,d\varepsilon = \frac{1}{2\pi i l}\,d\exp(2\pi i l n_{\mathrm{ex}}(\varepsilon))$$

and using the fact that the slowly varying function $|\partial^2 n/\partial k_z^2|_{\mathrm{ex}}$ need not be differentiated. The integrated term does not give an oscillatory dependence on the field; omitting it, we have

$$\tilde{I}_{l\sigma} = \sum_{\mathrm{ex}}\frac{e^{\pm i\pi/4}}{2\pi i T l^{3/2}}\int\limits_0^\infty\frac{\exp(2\pi i l n_{\mathrm{ex}})\,d\varepsilon}{\left[1+\exp\left(\dfrac{\varepsilon-\mu_\sigma}{T}\right)\right]|\partial^2 n/\partial k_z^2|_{\mathrm{ex}}^{1/2}}, \tag{63.9}$$

where the summation is over all the extrema (whose significance is further discussed below).

The factor $\exp(2\pi i l n_{\mathrm{ex}})$ in the numerator of the integrand is a rapidly oscillating function of ε. These oscillations reduce to nothing the integral with respect to ε everywhere except in the region $\varepsilon-\mu_\sigma \sim T$, where the denominator varies rapidly. The function $n_{\mathrm{ex}}(\varepsilon)$ itself varies smoothly in this region, and may therefore be represented as

$$n_{\mathrm{ex}}(\varepsilon) \approx n_{\mathrm{ex}}(\mu_\sigma) + n'_{\mathrm{ex}}(\mu_\sigma)(\varepsilon-\mu_\sigma);$$

the factor $|\partial^2 n/\partial k_z^2|_{\mathrm{ex}}^{-1/2}$ is simply replaced by its value at $\varepsilon=\mu_\sigma$. Then, changing from integration with respect to ε to one with respect to $x = (\varepsilon-\mu_\sigma)/T$ and replacing the lower limit $-\mu_\sigma/T$ by $-\infty$ (since $\mu/T \gg 1$), we obtain‡

$$\tilde{I}_{l\sigma} = -\sum_{\mathrm{ex}}\frac{\exp\left[2\pi i l n_{\mathrm{ex}}(\mu_\sigma)\pm\frac{1}{4}i\pi\right]}{2l^{3/2}\,|\partial^2 n/\partial k_z^2|_{\mathrm{ex},\,\mu_\sigma}^{1/2}}\sinh^{-1}[2\pi^2 l T n'_{\mathrm{ex}}(\mu_\sigma)].$$

† The saddle-point integral of the form $\int e^{iaz^2}\,dz$ is calculated by putting $z = ue^{i\pi/4}$ or $z = ue^{-i\pi/4}$ for $a > 0$ or $a < 0$, after which the integration with respect to u extends from $-\infty$ to ∞.

‡ We use the value of the integral

$$I \equiv \int\limits_{-\infty}^{\infty}\frac{e^{i\alpha z}}{e^z+1}\,dz = -\frac{i\pi}{\sinh\pi\alpha}.$$

This formula may be derived by considering the integral along a closed contour in the complex z-plane, consisting of the real axis, the straight line im $z = 2\pi$, and two "sides" at infinity; to ensure convergence on the latter, the real parameter α is replaced by $\alpha - i0$. The integral along this contour is determined by the residue at the pole $z = i\pi$, whence we find $I - e^{-2\pi\alpha}I = -2\pi i e^{-\pi\alpha}$.

In summing this expression over $\sigma = \pm 1$, we can everywhere (except in the exponential factor) replace μ_σ by μ, since by the hypothesis (63.1) $\beta B \ll \mu$. In the phase (exponential) factor, however, this replacement is not allowable: since $n_{ex}(\varepsilon)$ is large, even a relatively small change in its argument causes a considerable change of the phase. It is sufficient here, nevertheless, to expand $n_{ex}(\mu \pm \beta B)$ in powers of βB, taking only the linear terms. The result is

$$\sum_\sigma \tilde{I}_{l\sigma} = -\sum_{ex} \frac{\exp [2\pi i l n_{ex}(\mu) \pm \frac{1}{4} i\pi]}{l^{3/2} \left| \partial^2 n/\partial k_z^2 \right|^{1/2}_{ex,\mu}} \times$$

$$\times \sinh^{-1} [2\pi^2 l T n'_{ex}(\mu)] \cos [2\pi l \beta B \xi_{ex} n'_{ex}(\mu)], \qquad (63.10)$$

where $\xi_{ex} = \xi(k_{z,ex})$. It remains to elucidate the significance of the quantities appearing in this expression, and to substitute it in (63.5).

According to the definition (63.7), the function $n_{ex}(\varepsilon)$ is related to the extremum value $S_{ex}(\varepsilon)$ of the cross-sectional area of the constant-energy surface $S(\varepsilon, k_z)$ as a function of k_z, and its value at $\varepsilon = \mu$ is the area of the extremal cross-section of the Fermi surface. As an illustration, Fig. 15 shows the extremal

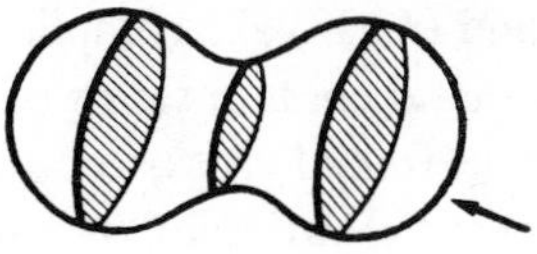

FIG. 15.

(two maxima and one minimum) cross-sections of a dumb-bell-shaped Fermi surface; they are perpendicular to the direction of the field, which is shown by the arrow. The summation over ex in (63.10) is over all extremal closed cross-sections of all sheets of the Fermi surface. To simplify the notation, we shall also use the cyclotron mass of the conduction electron in its motion along an extremal closed trajectory. According to the definition (57.6), this mass is

$$m^* = \frac{\hbar^2}{2\pi} \left[\frac{\partial S(\varepsilon, k_z)}{\partial \varepsilon} \right]_{\mu,\, k_z,\, ex} = \frac{\hbar^2}{2\pi} S'_{ex}(\mu),$$

where $S_{ex}(\varepsilon) = S(\varepsilon, k_{z,ex}(\varepsilon))$; the second equation again follows because at the extremum point $\partial S(\varepsilon, k_z)/\partial k_z = 0$.

We thus have the final formula for the oscillatory part of the thermodynamic potential:

$$\tilde{\Omega} = \sum_{ex} \sum_{l=1}^{\infty} (-1)^l \Omega_l \cos \left(\frac{l\hbar^2 S_{ex}}{2m\beta B} \pm \frac{1}{4}\pi \right),$$

$$\Omega_l = \frac{2V(m\beta B)^{5/2}}{\pi^{7/2}\hbar^3 m^* l^{5/2}} \left| \frac{\partial^2 S(\mu, k_z)}{\partial k_z^2} \right|^{-1/2}_{ex} \frac{\lambda}{\sinh \lambda} \cos \pi l m^* \xi_{ex}/m, \qquad (63.11)$$

$$\lambda = l\pi^2 T m^*/m\beta B,$$

where m is the actual mass of the electron, and the plus and minus signs in the argument of the cosine relate to minimum and maximum cross-sections respectively.†

The magnetization **M** (the magnetic moment per unit volume) is calculated as the derivative‡

$$\mathbf{M} = -\frac{1}{V}\frac{\partial \Omega}{\partial \mathbf{B}}. \tag{63.12}$$

In (63.11), only the most rapidly varying factors, the cosines, are to be differentiated. Because of the anisotropy of the Fermi surface (m^* and S_{ex} depend on the direction of the field) the direction of **M** is not in general the same as that of **B**. For the oscillatory part of the longitudinal (along the field) magnetization we find

$$M_z = \sum_{ex} (-1)^{l+1} M_l \sin\left(\frac{l\hbar^2 S_{ex}}{2m\beta B} \pm \frac{1}{4}\pi\right),$$

$$M_l = \frac{B^{1/2}(m\beta)^{3/2} S_{ex}}{\pi^{7/2} m^* l^{3/2} \hbar} \left|\frac{\partial^2 S(\mu, k_z)}{\partial k_z^2}\right|_{ex}^{-1/2} \frac{\lambda}{\sinh \lambda} \cos \pi l m^* \xi_{ex}/m. \tag{63.13}$$

The expressions (63.11) and (63.13) are complicated oscillatory functions of the magnetic field, and in general contain terms of various periodicities: terms originating from each of the extremal cross-sections of the Fermi surface have different periods with respect to the variable $1/B$, namely

$$\Delta \frac{1}{B} = \frac{4\pi m\beta}{\hbar^2 S_{ex}} = \frac{2\pi |e|}{c\hbar S_{ex}}. \tag{63.14}$$

These periods do not depend on the temperature.

The temperature dependence of the amplitude of the oscillations is given by the factor $\lambda/\sinh \lambda$. When $\lambda \gg 1$, the amplitudes decrease exponentially and the oscillations almost disappear. When $\lambda \lesssim 1$, the factor $\lambda/\sinh \lambda \sim 1$, and the order of magnitude of the amplitudes is determined by the remaining factors in Ω_l and M_l; all the following estimates refer to this case.

† For a gas of free electrons, the Fermi surface is a sphere with radius $k_F = \sqrt{(2m\mu)}/\hbar$, $S_{ex} = \pi k_F^2$, and formula (63.11) becomes that in Part 1, (60.5).

‡ The differentiation with respect to **B** needs explaining. Formula (63.12) may be derived as follows. The change in the Hamiltonian of the system due to an infinitesimal change in the vector potential of the field is

$$\delta \hat{H} = -\int \hat{\mathbf{j}} \cdot \delta \mathbf{A}\, dV/c,$$

where $\hat{\mathbf{j}}$ is the current density operator; see *QM*, (115.1). The change in the thermodynamic potential Ω is found by averaging $\delta \hat{H}$ for given values of μ, T and V. Since the quantization of the system is determined (as shown in §62) not by the exact microscopic field **H** but by its macroscopic mean value **B**, this means that in $\delta \hat{H}$ also **A** is to be taken as the vector potential of the mean field **B**. The variation $\delta \mathbf{A}$ can therefore be taken outside the averaging, after which

$$\delta \Omega = \langle \delta H \rangle = -\int \langle \mathbf{j} \rangle \cdot \delta \mathbf{A}\, dV/c.$$

Now with the magnetic moment defined by $\langle \mathbf{j} \rangle = c \operatorname{curl} \mathbf{M}$, integration by parts gives

$$\delta \Omega = -\delta \mathbf{B} \cdot \int \mathbf{M}\, dV.$$

For a rough estimate we put $m^* \sim m$, $\mu \sim \hbar^2 k_F^2/m$, $S \sim k_F^2$, where $k_F \sim 1/a$ is the linear dimension of the Fermi surface. Then

$$\tilde{\Omega} \sim V(m\beta B)^{5/2}/\hbar^3 \sim Vn\mu(\beta B/\mu)^{5/2}, \quad \tilde{M} \sim n\beta(\beta B/\mu)^{1/2}, \quad (63.15)$$

where $n \sim k_F^3$ is the number density of electrons. The part $\bar{M}$ of the magnetization that varies monotonically with the field may be estimated by putting

$$\bar{M} \sim \bar{\chi}B \sim \beta^2 mk_F B/\hbar^2 \sim n\beta . \beta B/\mu, \quad (63.16)$$

where $\bar{\chi}$ is the "monotonic" part of the magnetic susceptibility, estimated, for example, by means of the formula for the susceptibility of an electron gas in weak fields (see Part 1, §59). Accordingly, the monotonic part of the thermodynamic potential is $\bar{\Omega} \ll V\bar{M}B \sim Vn\mu(\beta B/\mu)^2$. A comparison of the above expressions shows that the oscillatory part of the thermodynamic potential is small in comparison with its monotonic magnetic part:

$$\tilde{\Omega}/\bar{\Omega} \sim (\beta B/\mu)^{1/2} \ll 1,$$

and therefore in comparison with its value $\Omega_0 \sim Vn\mu$ in the absence of the field: $\tilde{\Omega}/\Omega_0 \ll (\beta B/\mu)^{5/2}$. The oscillatory part of the magnetization, on the other hand, is large in comparison with the monotonic part:

$$\tilde{M}/\bar{M} \sim (\mu/\beta B)^{1/2} \gg 1.$$

Regarding the whole of the above theory of the oscillations of the magnetization, it should be noted that this applies to an electron liquid in an ideal crystal, and takes no account of a possible influence from processes of conduction electron scattering by phonons and by lattice defects (e.g. impurity atoms). These processes cause an uncertainty in the electron energy, $\Delta\varepsilon \sim \hbar/\tau \sim \hbar v_F/l$, where τ is the time between collisions, l is the mean free path, and v_F is the electron velocity. The blurring of the sharp energy levels in turn smooths out the oscillations of the magnetization. The condition for it to be permissible to neglect scattering processes is that the uncertainty $\Delta\varepsilon$ is small in comparison with the intervals between the levels:

$$\hbar\omega_B \gg \hbar v_F/l. \quad (63.17)$$

When $T \to 0$ the values of B allowed by the condition (63.1) become arbitrarily small, and only the condition (63.17) sets a limit. The magnetization $\tilde{M}$ may in principle become comparable with the induction B itself (since $\tilde{M}/B \sim \bar{\chi}(\mu/\beta B)^{1/2}$), but the magnetic susceptibility $\chi = \partial M/\partial H$ becomes large (in modulus) sooner:† since again only the oscillating factors need be differentiated, we have

$$|\tilde{\chi}| \sim \bar{\chi}(\mu/\beta B)^{3/2}. \quad (63.18)$$

† To avoid unnecessary complications, the influence of anisotropy will be ignored in the following qualitative treatment.

In such a situation, the oscillations of the magnetization cause the curve of the macroscopic field $H = B-4\pi M(B)$ as a function of the induction B to have a series of bends, as shown diagrammatically in Fig. 16 (A. B. Pippard 1963). The condition of thermodynamic stability requires that†

$$(\partial H/\partial B)_{T,\mu} > 0.$$

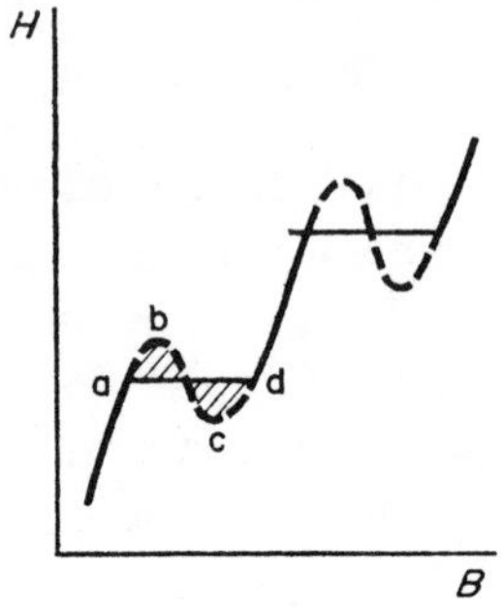

FIG. 16.

Hence the states corresponding to parts of the curve such as *bc* cannot occur. The resulting situation is exactly similar to the one which causes a phase transition in a substance when there is such a bend on the curve of pressure against volume (cf. Part 1, §§84 and 152). The equilibrium $H(B)$ curve will in fact include a straight horizontal section *ad*, drawn so that the two shaded areas in Fig. 16 are equal; the sections *ab* and *cd* correspond to metastable states.

Let a metal sample have the form of a cylinder with its axis along the external field $\mathfrak{H}$. Then the field H within the cylinder is the same as $\mathfrak{H}$, and as the latter increases the body will undergo successive phase transitions with discontinuous changes of the induction: each time a point such as a is reached, the induction changes discontinuously from B_a to B_d.‡ If, however, the sample is a flat plate in a magnetic field perpendicular to it, the body separates into alternating layers (*diamagnetic domains*) with different values of the induction, exactly like the separation of a superconductor in the intermediate state into normal and superconducting layers (J. H. Condon 1966). The external field $\mathfrak{H}$ is in this case equal to the magnetic induction averaged over all the layers. For example, in the range $B_a < \mathfrak{H} < B_d$ the plate separates into layers with induction B_a and B_d and, as $\mathfrak{H}$ increases, the volume of the latter increases at the expense of the volume of the former.

§ 64. Electron–phonon interaction

So far, we have considered the conduction electrons in a crystal without reference to their interaction with the lattice vibrations, i.e. the phonons. This interaction represents the fact that the deformation of the lattice alters the field

† Cf. *ECM*, §18, where a similar condition is derived for the electric case.

‡ We assume that the surface energy of the interface between the phases is positive.

in which the electron moves; this change in the field is called the *deformation potential.*

The electron–phonon interaction plays a decisive role in transport phenomena in semiconductors and metals, but here we shall be interested only in the qualitative influence of this interaction on the energy spectrum of the electrons. To study it, we may usefully ignore the complications due to the anisotropy of the lattice and its microscopic inhomogeneity. Thus we regard the medium as a microscopically homogeneous isotropic liquid, and accordingly only longitudinal acoustic vibrations can occur in it.

In the first approximation with respect to the deformation, the potential corresponding to this simplified model may be written

$$U_{\text{def}}(\mathbf{r}) = \frac{1}{\varrho}\int W(\mathbf{r}-\mathbf{r}')\,\varrho'(\mathbf{r}')\,d^3x', \tag{64.1}$$

where ϱ' is the variable part of the density of the medium (and ϱ is its constant equilibrium value). The function $W(\mathbf{r}-\mathbf{r}')$ decreases over distances of the order of the interatomic distances a. We shall simplify the expression (64.1) further by noting that, for interaction with phonons with wave numbers $k \ll 1/a$, these distances may be taken as zero, i.e. we assume that $W = w\delta(\mathbf{r}-\mathbf{r}')$ with w a constant. Then $U_{\text{def}} = w\varrho'(\mathbf{r})/\varrho$. In the quantum theory in the second-quantization representation, this potential is written as the Hamiltonian of the electron–phonon interaction,

$$\hat{H}_{\text{ep}} = (w/\varrho)\int \hat{\Psi}^+_\alpha(t, \mathbf{r})\,\hat{\varrho}'(t, \mathbf{r})\,\hat{\Psi}_\alpha(t, \mathbf{r})\,d^3x, \tag{64.2}$$

where the operators $\hat{\Psi}$ and $\hat{\Psi}^+$ refer to the electrons, and $\hat{\varrho}'$ is the Heisenberg density operator describing the phonon field; for free phonons (not interacting with electrons), it is given by (24.10).

In the mathematical formalism of Green's functions, as applied to the electron–phonon interaction, we have not only the electron Green's function G but also the phonon Green's function defined by

$$D(X_1, X_2) \equiv D(X_1 - X_2) = -i\langle \mathrm{T}\,\hat{\varrho}'(X_1)\,\hat{\varrho}'(X_2)\rangle, \tag{64.3}$$

the chronological product being expanded by the rule (31.2), which corresponds to the case of bosons. For free phonons, the Green's function in the momentum representation is

$$\begin{aligned} D^{(0)}(\omega, \mathbf{k}) &= \frac{\varrho k}{2u}\left\{\frac{1}{\omega - uk + i0} - \frac{1}{\omega + uk - i0}\right\} \\ &= \frac{\varrho k^2}{\omega^2 - u^2k^2 + i0}\,; \end{aligned} \tag{64.4}$$

see §31, Problem. (In these intermediate formulae, we put $\hbar = 1$.)

If the electron–phonon interaction is regarded as a small perturbation, we can set up a diagram technique based on the operator (64.2), as was done in §13 for a pair interaction of fermions. Without repeating all the arguments, we shall formulate the resulting rules for the construction of the diagrams (in the momentum representation).†

The basic elements of the diagrams are electron (continuous) lines and phonon (broken) lines, each having assigned to it a certain "4-momentum". An electron line with 4-momentum P corresponds to a factor $iG^{(0)}_{\alpha\beta} = i\delta_{\alpha\beta}G^{(0)}(P)$, the Green's function of free electrons. A phonon line with 4-momentum K corresponds to a factor $iD^{(0)}(K)$, the Green's function of free phonons. Two continuous lines and one broken line meet at each vertex of the diagram; such a point has an additional factor $-iw/\varrho$.

For example, the first correction to the electron Green's function is represented by the diagram‡

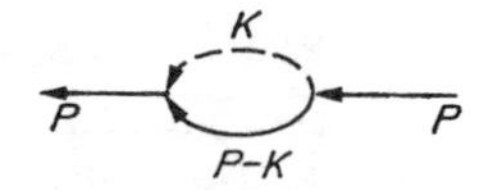

(64.5)

with the corresponding analytical expression

$$i\delta G(P) = -(w^2/\varrho^2)[G^{(0)}(P)]^2 \int G^{(0)}(P-K)\, D^{(0)}(K)\, d^4K/(2\pi)^4. \quad (64.6)$$

The first correction to the phonon Green's function is represented by the diagram

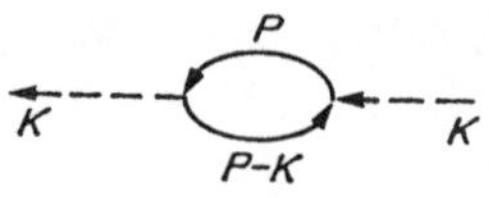

(64.7)

or, in analytical form,

$$i\delta D(K) = 2(w^2/\varrho^2)[D^{(0)}(K)]^2 \int G^{(0)}(P)\, G^{(0)}(P-K)\, d^4P/(2\pi)^4; \quad (64.8)$$

the coefficient 2 comes from the contraction of the spin factors ($\delta_{\alpha\beta}\delta_{\beta\alpha} = 2$), and we have also included the factor -1 due to the presence of one closed fermion loop (cf. §13).

We shall show that the electron–phonon interaction in a metal leads to the occurrence of an "effective attraction" between electrons near the Fermi sur-

† The structure of the expression (64.2) for the electron–phonon interaction operator is similar to that of the electron–photon interaction operator in quantum electrodynamics. The rules of the diagram technique are therefore also analogous in the two cases.

‡ There is no diagram with a closed electron line, similar to (13.13a), because $D^{(0)}(0) = 0$. Here it is assumed that the limit $\mathbf{k} \to 0$ is taken before $\omega \to 0$. This corresponds to the fact that in coordinate space the integration over d^3x (which in the present case signifies the passage to $\mathbf{k} \to 0$) is present in the definition of the Hamiltonian (64.2), and is therefore carried out before the integration with respect to time which arises when perturbation theory is applied to this Hamiltonian.

face. It can be intuitively described as resulting from the emission of a virtual phonon by one electron and its absorption by another (J. Bardeen 1950, H. Fröhlich 1950).

Let us consider the diagram

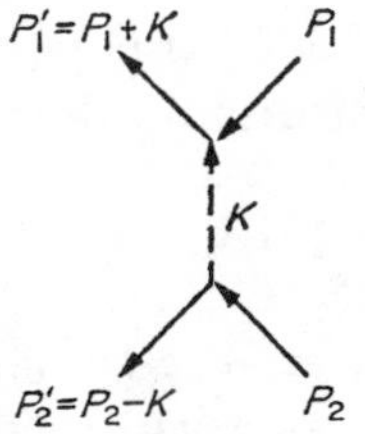

(64.9)

representing the scattering of two electrons by the exchange of virtual phonons; the 4-momenta $P = (\varepsilon-\mu, \mathbf{p})$, $K = (\omega, \mathbf{k})$, and μ is the chemical potential of the electrons at $T = 0$, which is equal to the limiting energy ε_F. This diagram corresponds to the vertex function

$$\Gamma_{\gamma\delta,\alpha\beta} = \Gamma\delta_{\alpha\gamma}\delta_{\beta\delta}, \quad i\Gamma = (-iw/\varrho)^2 iD^{(0)}(K),$$

or

$$\Gamma = -\frac{w^2k^2}{\varrho(\omega^2-u^2k^2+i0)}, \tag{64.10}$$

with $\hbar\omega = \varepsilon_1'-\varepsilon_1$, $\hbar\mathbf{k} = \mathbf{p}_1'-\mathbf{p}_1$.

The momenta of electrons near the Fermi surface are in order of magnitude $p \sim p_F \sim \hbar/a$. Scattering of electrons through an angle ~ 1 corresponds to a phonon momentum $\hbar k \sim \hbar/a$ and phonon energy $\hbar uk \sim \hbar u/a \sim \hbar\omega_D$, where ω_D is the Debye frequency; for metals, $\hbar\omega_D \ll \varepsilon_F$. On the other hand, the electron cannot transfer an amount of energy greater than $\varepsilon-\varepsilon_F$. Hence, if for both electrons $|\varepsilon-\varepsilon_F| \ll \omega_D$, we certainly have

$$\Gamma \approx w^2/\varrho u^2 > 0. \tag{64.11}$$

Taking into account the significance of Γ as a scattering amplitude (§16), we see that its sign corresponds to attraction between particles. It must be emphasized that this result is valid only for electrons in a comparatively narrow shell (with energy thickness $\sim \hbar\omega_D$) in momentum space near the Fermi surface. This fact has already been used in §43 to determine the value of the cut-off parameter in the theory of the superconductivity of metals†.

† For a rough estimate of the constant w in metals, we may note that the change in the electron energy must be of the order of the energy itself ($\sim \varepsilon_F$) when the density change $\varrho' \sim \varrho$; hence $w \sim \varepsilon_F$.

§ 65. Effect of the electron–phonon interaction on the electron spectrum in a metal

Let us consider the influence of the electron–phonon interaction on the energy spectrum of electrons in a metal.†

It has been shown in §14 that for a Fermi-type spectrum the correction to the dispersion relation $\varepsilon(\mathbf{p})$ (in comparison with the spectrum of a system of free fermions) is

$$\delta\varepsilon(\mathbf{p}) = \Sigma(\varepsilon-\mu, \mathbf{p}) - \Sigma(0, \mathbf{p}), \tag{65.1}$$

where $\Sigma = [G^{(0)}]^{-1} - G^{-1}$ is the self-energy function. In the present case, the correction is due to the interaction with phonons, and the "unperturbed" spectrum is that which takes account of the "direct" interaction of the particles (electrons). According to (64.6)‡,

$$\Sigma(P) = -\delta G^{-1} = \delta G/[G^{(0)}]^2 = i(w^2/\varrho^2)\int G^{(0)}(P-K)\,D^{(0)}(K)\,d^4K/(2\pi)^4, \tag{65.2}$$

but $G^{(0)}$ must now be taken as the Green's function for electrons interacting with one another. Near its pole, such a function has the form

$$G^{(0)}(\varepsilon-\mu, \mathbf{p}) = Z[\varepsilon-\mu-v_F^{(0)}(p-p_F)+i0.\mathrm{sgn}\,(\varepsilon-\mu)]^{-1}; \tag{65.3}$$

see (10.2). The superscript (0) to v_F denotes that this quantity does not yet include the effect of the electron–phonon interaction.

Our object is now to obtain an estimate of the quantity (65.1), i.e. of the integral

$$\delta\varepsilon = \frac{iw^2}{\varrho^2}\int \{G^{(0)}(\varepsilon-\mu-\omega, \mathbf{p}-\mathbf{k}) - G^{(0)}(-\omega, \mathbf{p}-\mathbf{k})\}\,D^{(0)}(\omega, \mathbf{k})\,d^4K/(2\pi)^4. \tag{65.4}$$

It will be seen from the subsequent calculations that the main contribution to this integral comes from the range in which the momentum $\mathbf{p}-\mathbf{k}$ and the energy $\varepsilon-\omega$ (like $\mathbf{p}$ and ε) lie within the Fermi surface, i.e. $k \ll p_F$, $\omega \ll \mu$. For this reason we can take (65.3) as the functions $G^{(0)}$.

In spherical polar coordinates in $\mathbf{k}$-space, with the polar axis parallel to $\mathbf{p}$, we have $d^4K = 2\pi k^2\,dk\,d\omega\,d\cos\theta$, where θ is the angle between $\mathbf{k}$ and $\mathbf{p}$. Instead of $\cos\theta$ we use the variable $p_1 = |\mathbf{p}-\mathbf{k}|$; since $p_1^2 = p^2+k^2-2pk\cos\theta$, we have

$$d^4K = 2\pi k^2\,dk\,d\omega\,p_1\,dp_1/pk \approx 2\pi k\,dk\,d\omega\,dp_1,$$

putting $p_1 \approx p \approx p_F$.

In the integrand in (65.4), only the factor in the braces depends on p_1; it is

$$\{\ldots\} = -(\varepsilon-\mu)\,Z[\varepsilon-\mu-\omega-v_F^{(0)}(p_1-p_F)+i0.\mathrm{sgn}\,(\varepsilon-\mu-\omega)]^{-1}\times$$
$$\times[-\omega-v_F^{(0)}(p_1-p_F)-i0.\mathrm{sgn}\,\omega]^{-1}.$$

† The results in this section are due to A. B. Migdal (1958).

‡ In the intermediate formulae, we put $\hbar = 1$.

Because of the rapid convergence of the integral with respect to $p_1 - p_F$, we can extend the integration to $\pm\infty$; with the variable $\eta = v_F^{(0)}(p_1 - p_F)$, we obtain the integral

$$\int \{\ldots\} dp_1$$

$$= -\frac{(\varepsilon-\mu)Z}{v_F^{(0)}} \int_{-\infty}^{\infty} \frac{d\eta}{[\eta-(\varepsilon-\mu-\omega)-i0.\mathrm{sgn}\,(\varepsilon-\mu-\omega)][\eta+\omega+i0.\mathrm{sgn}\,\omega]}.$$

If the two poles of the integrand are on the same side of the real axis, the integral is zero (as may be shown by closing the contour of integration in the other half-plane). The integral is therefore non-zero only if $\varepsilon-\mu > \omega > 0$ or $\varepsilon-\mu < \omega < 0$; it is $-2\pi i Z/v_F^{(0)}$ in the first case and $2\pi i Z/v_F^{(0)}$ in the second. Thus, using also the fact that $D^{(0)}(\omega, \mathbf{k})$ is an even function of ω, we find

$$\delta\varepsilon = \frac{Zw^2}{8\pi^2 \varrho u v_F^{(0)}} \int \int_0^{|\varepsilon-\mu|} \left[\frac{1}{\omega - uk + i0} - \frac{1}{\omega + uk - i0}\right] k^2\, d\omega\, dk. \quad (65.5)$$

The real and imaginary parts of this expression determine respectively the correction to the spectrum of quasi-particles (conduction electrons) and their decay. Let us first consider the latter.

Separating from (65.5) the imaginary part by the rule (8.11), we find

$$-\mathrm{im}\,\delta\varepsilon = \frac{Zw^2}{8\pi \varrho u v_F^{(0)}} \int k^2\, dk; \quad (65.6)$$

the integration with respect to k is taken from 0 to $|\varepsilon-\mu|/u$, for which the pole $\omega = uk$ of the integrand in (65.5) lies in the range from 0 to $|\varepsilon-\mu|$. Thus (in ordinary units)

$$-\mathrm{im}\,\delta\varepsilon = \frac{Zw^2 |\varepsilon-\mu|^3}{24\pi\hbar^3 \varrho u^4 v_F^{(0)}}. \quad (65.7)$$

For a rough estimate of this quantity, we note that the parameters $v_F^{(0)}$ and w are of electronic origin and are expressed, in order of magnitude, in terms of only the interatomic distances a and the electron mass m: $v_F^{(0)} \sim p_F/m \sim \hbar/ma$, $w \sim \varepsilon_F \sim \hbar^2/ma^2$; see the last footnote to §64. The density ϱ and the velocity of sound u depend also on the ion mass M: $\varrho \propto M$, $u \propto M^{-1/2}$, and so $\varrho u^4 \propto 1/M$. Hence an estimate of the decay may be written

$$-\mathrm{im}\,\delta\varepsilon \sim |\varepsilon-\mu|^3/(\hbar\omega_D)^2, \quad (65.8)$$

where the Debye frequency $\omega_D \sim u/a \propto M^{-1/2}$.

Strictly speaking, the estimate (65.8) relates to values $|\varepsilon-\mu| \ll \hbar\omega_D$, for which the integration in (65.6) is taken over the range $k < |\varepsilon-\mu|/u\hbar \ll \omega_D/u$, in which the phonon dispersion relation $\omega = ku$ which we have used is in fact valid. For a rough order-of-magnitude estimate, however, we can apply (65.8) even at the limit of the range where $\varepsilon-\mu \sim \hbar\omega_D$, obtaining

$$-\operatorname{im}\delta\varepsilon \sim \hbar\omega_D \sim |\varepsilon-\mu|. \tag{65.9}$$

Lastly, when $\varepsilon-\mu \gg \hbar\omega_D$, the range of integration in (65.6) is independent of $\varepsilon-\mu$, since the pole $\omega = uk \lesssim \omega_D$ is always between 0 and $\varepsilon-\mu$. In this case, $\int k^2\,dk \sim (\omega_D/u)^3$, and the decay is

$$-\operatorname{im}\delta\varepsilon \sim \hbar\omega_D \ll \varepsilon-\mu. \tag{65.10}$$

The expressions (65.8)–(65.10) give the specific decay due to the emission of phonons by electrons.† We see that, in the immediate neighbourhood of the Fermi surface, when $|\varepsilon-\mu| \ll \hbar\omega_D$, according to (65.8), the decay is slight ($|\operatorname{im}(\varepsilon-\mu)| \ll |\varepsilon-\mu|$), so that the concept of quasi-particles (conduction electrons) has an entirely definite meaning. In the range $|\varepsilon-\mu| \sim \hbar\omega_D$, however, the decay of the quasi-particle becomes comparable with its energy, and the spectrum is blurred and loses much of its meaning. At still larger distances above the Fermi surface, when $\varepsilon-\mu \gg \hbar\omega_D$ (but of course still $\varepsilon-\mu \ll \mu$), according to (65.10), the decay remains the same in absolute magnitude but again becomes small in comparison with the energy $\varepsilon-\mu$, so that the quasi-particles recover a certain significance. Of course, as well as the phonon decay of the conduction electrons, there is always also the decay due to collisions between electrons. The latter decay, characteristic of any normal Fermi liquid (§1), is proportional to $(\varepsilon-\mu)^2$ and in order of magnitude is $\sim(\varepsilon-\mu)^2/\mu$, i.e. is always small in the range where the theory is applicable.

Let us now estimate the correction to the real part of ε, i.e. to the spectrum itself. The real part of the integral over ω in (65.5) is given by its principal value:

$$\operatorname{re}\int_0^{|\varepsilon-\mu|} D^{(0)}(\omega,\mathbf{k})\,d\omega = \frac{\varrho k}{2u}\,\mathrm{P}\int_0^{|\varepsilon-\mu|}\left\{\frac{1}{\omega-uk}-\frac{1}{\omega+uk}\right\}d\omega$$
$$= \frac{\varrho k}{2u}\log\left|\frac{\varepsilon-\mu-uk}{\varepsilon-\mu+uk}\right|.$$

We therefore have for re $\delta\varepsilon$ (in ordinary units)

$$\operatorname{re}\delta\varepsilon = \frac{Z\omega^2}{8\pi^2\varrho u v_F^{(0)}}\int k^2\log\left|\frac{\varepsilon-\mu-\hbar uk}{\varepsilon-\mu+\hbar uk}\right|dk. \tag{65.11}$$

When $\varepsilon-\mu \gg \hbar\omega_D$, the logarithm in the integrand is $\sim \hbar uk/(\varepsilon-\mu)$, and the whole integral is estimated as $\hbar uk^3_{\max}/(\varepsilon-\mu) \sim \hbar u/a^3(\varepsilon-\mu)$. Noting also that the

† The conservation of energy in the creation of a low-frequency phonon by a quasi-particle is expressed by $(\partial\varepsilon/\partial\mathbf{k})\cdot\delta\mathbf{k} \equiv \mathbf{v}\cdot\delta\mathbf{k} = u\delta k$; this can be true only if $v > u$. In a metal this condition is always satisfied, since $v_F \gg u$.

factor ϱ in the denominator of (65.11) makes the whole expression $\propto 1/M$, we arrive at the estimate

$$\text{re}\,\delta\varepsilon \sim (\hbar\omega_D)^2/(\varepsilon-\mu) \ll \varepsilon-\mu.$$

Thus in this case the correction to the spectrum is relatively small, and so the spectrum is given by

$$\varepsilon-\mu \approx v_F^{(0)}(p-p_F) \quad \text{for} \quad \varepsilon-\mu \gg \hbar\omega_D, \tag{65.12}$$

with the "unperturbed" value $v_F^{(0)}$ of the velocity on the Fermi surface.

In the range $\varepsilon-\mu \ll \hbar\omega_D$, the logarithm in (65.11) is $\sim(\varepsilon-\mu)/\hbar uk$, and the integral is estimated as $(\varepsilon-\mu)k_{\max}^2/\hbar u \sim (\varepsilon-\mu)/\hbar ua^2$. The whole expression (65.11) is thus proportional to $\varepsilon-\mu$, with a coefficient independent of the ion mass M (since the product ϱu^2 is independent of M). This means that the spectrum in that range is again of the same type:

$$\varepsilon-\mu \approx v_F(p-p_F) \quad \text{for} \quad \varepsilon-\mu \ll \hbar\omega_D, \tag{65.13}$$

but with the velocity v_F, which differs from $v_F^{(0)}$ by an amount of the order of $v_F^{(0)}$ itself.†

Thus the Fermi-type spectrum for electrons in a metal has two different values of the velocity, v_F and $v_F^{(0)}$, one in the immediate neighbourhood of the Fermi surface ($\varepsilon-\mu \ll \hbar\omega_D$) and the other at $\varepsilon-\mu \gg \hbar\omega_D$. The thermodynamic properties of the metal at low temperatures ($T \ll \hbar\omega_D$) involve the parameter v_F from (65.13). Such phenomena as the optical properties of the metal at frequencies $\omega \gg \omega_D$ are determined by the velocity $v_F^{(0)}$.

PROBLEM

Determine the decay of long-wavelength ($k \ll p_F$) phonons in a metal due to their absorption by electrons.

SOLUTION. The correction to the Green's function of the phonons is given, according to (64.8), by

$$i\delta D^{-1}(K) = -(2w^2/\varrho^2)\int G^{(0)}(P)\,G^{(0)}(P-K)\,d^4P/(2\pi)^4,$$
$$P = (p_0, \mathbf{p}), \quad K = (\omega, \mathbf{k}).$$

In the G functions, however, we must also include corrections due to the interaction of electrons with short-wavelength photons. According to the discussion in the text, these changes simply replace $G^{(0)}$ by a function G that differs from (65.3) only in that the velocity $v_F^{(0)}$ is changed to v_F, and the renormalization constant Z to another one Z'. For small K, formula (17.10) may be used for the product $G^{(0)}(P)G^{(0)}(P-K)$. The integration over dp_0dp reduces to the removal of the delta functions, after which there remains the integration over $d\cos\theta$ (where θ is the angle between $\mathbf{p}$ and $\mathbf{k}$):

$$\delta D^{-1}(\omega, k) = -\frac{Z'^2w^2p_F^2k}{2\pi^2\varrho^2}\int_{-1}^{1}\frac{\cos\theta\, d\cos\theta}{\omega - v_Fk\cos\theta + i0}$$

† In these conditions, of course, the use of the first approximation of perturbation theory is not, strictly speaking, legitimate. The use of higher approximations, however, cannot alter the nature of the result: when the first-order correction becomes of the order of unity, the remaining corrections are also of that order.

(we take $\omega > 0$). The pole at $\cos\theta = \omega/kv_F$ is within the range of integration (since $v_F > u$), and the imaginary part of the integral is

$$\text{im}\,\delta D^{-1} = Z'^2 w^2 p_F^2 \omega / 2\pi \varrho^2 v_F^2 k.$$

The phonon dispersion relation is found from the equation $[D^{(0)}]^{-1} + \delta D^{-1} = 0$, whence (in ordinary units)

$$\omega = uk(1 - i\alpha), \quad \alpha = Z'^2 w^2 p_F^2 / 4\pi\hbar^3 \varrho u v_F^2;$$

the correction to the real part of ω is not of interest here. The product $\varrho u \propto \sqrt{M}$; hence, as a rough estimate, $\alpha \sim \sqrt{(m/M)}$, i.e. the decay is always small.

§ 66. The electron spectrum of solid insulators

A characteristic feature of the electron energy spectrum of a non-magnetic insulating crystal is that even the first excited level is at a finite distance from the ground level, i.e. there is an energy gap (which in ordinary insulators is of the order of a few electron-volts) between the ground level and the spectrum of excited levels.

An elementary excitation in an insulating crystal may be visualized as an excited state of an atom but one that cannot be assigned to any particular atom; the translational symmetry of the lattice leads, as always, to a "collectivization" of the excitation propagated in the crystal as if it were hopping from one atom to another. As in other cases, these excitations may be regarded as quasi-particles (which are here called *excitons*) with definite energies and quasi-momenta. Like all quasi-particles that can occur singly, the excitons have integral angular momentum and obey Bose statistics.†

For a given quasi-momentum **k**, the exciton energy can take a discrete series of values $\varepsilon_s(\mathbf{k})$. When the quasi-momentum takes values in one cell of the reciprocal lattice, each of the functions $\varepsilon_s(\mathbf{k})$ covers a certain *band* of exciton energy values; different bands may partly overlap. The minimum value of each function $\varepsilon_s(\mathbf{k})$ is non-zero.

The insulator may contain, as well as excitons, electron excitations of another kind. These may be regarded as resulting from the ionization of individual atoms. Each such ionization causes the occurrence in the insulator of two independently propagated quasi-particles, a conduction electron and a "hole". The latter is an absence of an electron in an atom, and so behaves as a positively charged particle. Here again, in speaking of the motion of an electron and a hole, we are really referring to certain collective excited states of electrons in the insulator, accompanied (unlike exciton states) by a transfer of a negative or positive unit charge.

The electrons and holes have half-integral spin and obey Fermi statistics. We must emphasize, however, that the electron–hole spectrum of the insulator is not at all like the Fermi-type electron spectrum in metals. The latter is char-

† The concept of excitons was first introduced by Ya. I. Frenkel' (1931).

acterized by the existence of a limiting Fermi surface in **k**-space, near which the electron quasi-momenta lie. In the present case, however, there is no such surface, and the electron and hole that appear simultaneously can have any quasi-momenta.

The distinction between the two types of spectra can be more fully understood by considering the decay of elementary excitations. In a Fermi liquid, any quasi-particle outside the Fermi surface can create pairs of new excitations (a particle and a hole), and therefore has a finite lifetime which decreases rapidly away from the Fermi surface (an electron in a metal can also emit phonons; see §65). The decay of an individual electron (or hole) in an insulator, in an ideal lattice (at $T = 0$) is, however, precisely zero in a finite range of energies above its minimum value:† the formation of an electron–hole pair always requires a finite expenditure of energy, because of the energy gap Δ (see below). The emission of an acoustic phonon by a quasi-particle is possible only if the latter's velocity $\boldsymbol{v}$ is not less than the velocity of sound u; see the third footnote to §65.

The possible values of the conduction electron energy $\varepsilon^{(e)}(\mathbf{k})$ and hole energy $\varepsilon^{(h)}(\mathbf{k})$ also occupy bands. The width of the energy gap in an insulator is usually taken as the sum $\Delta = \varepsilon^{(e)}_{\min} + \varepsilon^{(h)}_{\min}$ of the smallest possible values of the electron and hole energies. Since the electron and the hole appear or disappear together, it is this sum which has a real significance, not $\varepsilon^{(e)}_{\min}$ or $\varepsilon^{(h)}_{\min}$ separately; it is usual to take arbitrarily $\varepsilon^{(h)}_{\min} = 0$. The minimum energy values can be reached for electrons and holes with the same or different values of the quasi-momentum $\mathbf{k} = \mathbf{k}_0$; the gap is said to be respectively *direct* and *indirect*. If the energy levels in the band are not degenerate (or have only double degeneracy with respect to the spin because of the symmetry under time reversal), $\varepsilon(\mathbf{k})$ near its minimum has the form

$$\varepsilon^{(e)}(\mathbf{k}) = \Delta + \tfrac{1}{2} m^{(e)-1}{}_{ik} q_i q_k, \quad \varepsilon^{(h)}(\mathbf{k}) = \tfrac{1}{2} m^{(h)-1}{}_{ik} q_i q_k, \tag{66.1}$$

where $\mathbf{q} = \mathbf{k} - \mathbf{k}_0$, and $m^{(e)}_{ik}$ and $m^{(h)}_{ik}$ are the electron and hole effective mass tensors.

In the literature, the electron band is often called just the *conduction band*, and the hole band is referred to as the *valency band*, which in the ground state of the crystal is completely filled with electrons. The formation of a pair of quasi-particles (electron and hole) is then regarded as the result of the passage of an electron from the valency band to the conduction band, leaving a hole at its original position.

At large distances (compared with those between the atoms), the electron and the hole attract each other in accordance with Coulomb's law. They can therefore form bound states. The electron and the hole together constitute an electrically neutral quasi-particle, i. e. an exciton. For a given quasi-momentum, the bound states correspond to discrete energy levels of the electron–hole system,

† At finite temperatures there is, of course, always a decay due to scattering by other quasi-particles.

with one exciton energy band for each level. The energies of the excitons thus lie below those of electron–hole excitations; the energy gap in the sense mentioned at the beginning of this section is consequently not the same as Δ, but is less by an amount equal to the maximum binding energy of the exciton.†

The exciton energy levels are easily calculated in the limiting case of weakly bound states, when the mean distances between the electron and the hole are large compared with the lattice constant a; this is called a *Wannier–Mott exciton.* In the opposite limiting case, when the electron–hole distance is of the order of atomic distances, we have a *Frenkel' exciton*, which can of course be regarded only formally as a bound state of electron and hole.

Let us consider an insulator crystal having cubic symmetry. For a Wannier–Mott exciton, we can suppose that the electron and the hole attract each other according to Coulomb's law, the remaining atoms in the lattice serving only to create a uniform dielectric background which weakens the interaction by a factor ε, the permittivity of the crystal (taken for the frequencies that correspond in order of magnitude to the exciton binding energy); thus the electron–hole interaction energy is written as $U = -e^2/\varepsilon r$. Let the gap in the spectrum be direct, and let us assume for simplicity that the electron and hole energy minima are at $\mathbf{k} = 0$. In a cubic crystal, the effective mass tensors reduce to scalar constants m_e and m_h, so that

$$\varepsilon^{(e)}(\mathbf{k}) = \Delta + \hbar^2k^2/2m_e, \qquad \varepsilon^{(h)}(\mathbf{k}) = \hbar^2k^2/2m_h. \tag{66.2}$$

It has been mentioned at the end of §56 that the motion of the particle in a crystal lattice subjected to an external electric field varying slowly in space is described by Schrödinger's equation with a Hamiltonian in which $\varepsilon(\mathbf{k})$ plays the part of the kinetic energy. Since in the present case the functions $\varepsilon^{(e)}(\mathbf{k}) - \Delta$ and $\varepsilon^{(h)}(\mathbf{k})$ are the same in form as the kinetic energies of ordinary free particles, Schrödinger's equation for the system in question is the same in form as that for a system of two ordinary particles with Coulomb interaction, i.e. Schrödinger's equation for the hydrogen atom. We can therefore write down immediately the energy levels of the system, i.e. the exciton energy, as

$$\varepsilon_n^{\text{ex}}(\mathbf{k}) - \Delta = \frac{\hbar^2k^2}{2(m_e+m_h)} - \frac{me^4}{2\varepsilon^2\hbar^2n^2} \tag{66.3}$$

(G. H. Wannier 1937). The first term in this expression is the energy of the exciton moving "as a whole" with quasi-momentum $\mathbf{k}$. The second term gives the binding energy of the electron and hole in the exciton; $m = m_e m_h/(m_e+m_h)$ is the reduced mass of the system. For a given $\mathbf{k}$, the discrete energy levels of the system become closer as the energy increases to the limit of the continuous spectrum. The condition for (66.3) to be valid is that the orbit radius should be

† Exciton states, however, have a finite lifetime, since the electron and hole can recombine with the emission of, for example, a phonon or a photon.

sufficiently large, $r_{ex} \sim \hbar^2 \varepsilon n^2/me^2 \gg a$. This is certainly true for large n, but in crystals with large ε it may also be satisfied for $n \sim 1$.†

To conclude this section, let us return to the assertion in §61 that there is a lower limit to the number density of conduction electrons in a semi-metal.

In an insulator, where there are no electrons and holes at $T = 0$, the possibility that they can form bound states means only the occurrence of new branches of the energy spectrum. In a compensated metal, this possibility would mean that a state with free electrons and holes is not the lowest state, i.e. a spectrum of the metallic type would be unstable. The possibility of forming bound states is eliminated by the screening of the Coulomb interaction between the electron and the hole by other quasi-particles lying "between them". That is, the mean distance between the quasi-particles must be of the order of or less than the exciton size r_{ex} (in its ground state). The lower limit of electron and hole number densities in a metal, set by this condition, decreases with their effective mass.

§ 67. Electrons and holes in semiconductors

The energy spectrum of pure (or *intrinsic*) crystalline semiconductors differs from that of insulators only quantitatively: the gap Δ is smaller, and there is therefore a considerable carrier density (in comparison with an insulator) in a semiconductor at ordinary temperatures. The distinction is obviously arbitrary, and also depends on the temperature range concerned.‡ In *impurity* (or *doped*) semiconductors, the impurity atoms are a further source of electrons or holes, the energy gap of these atoms for electron transfer to the lattice (*donor impurity*) or from the lattice (*acceptor impurity*) being less than the energy gap in the original spectrum.

Let us consider further the relation between the gap Δ and the density of conduction electrons and holes in a semiconductor (or insulator).

The formation or disappearance of an electron (e) and hole (h) pair may be thermodynamically regarded as a "chemical reaction" $e+h \rightleftarrows 0$, the ground state of the crystal acting as a "vacuum". From the general rules (see Part 1, §101), the condition of thermodynamic equilibrium in this reaction is

$$\mu_e + \mu_h = 0, \tag{67.1}$$

where μ_e and μ_h are the chemical potentials of the electrons and holes. Because of the comparatively low densities of electrons (n_e) and holes (n_h) in a semicon-

† It is noteworthy that, near the upper edge (maximum) of the band, where the effective masses are negative, bound states of two electrons (or two holes) may be formed. The energy of such states lies in the forbidden region above the maximum total energy of the electrons.

‡ The values of the energy gap Δ for several semiconductors are: silicon 1.17 eV, germanium 0.74 eV, indium antimonide 0.24 eV, gallium arsenide 1.52 eV, lead sulphide 0.29 eV. For diamond, a typical insulator, $\Delta = 5.4$ eV.

ductor (at $T \ll \Delta$), the Fermi distribution for them is almost exactly the Boltzmann distribution, so that the electrons and holes form a classical gas.† Then the law of mass action follows in the usual way (see Part 1, §101) from (67.1); according to this law, the product of the equilibrium densities is

$$n_e n_h = K(T), \tag{67.2}$$

where the right-hand side is a function of temperature depending only on the properties of the host lattice, at whose atoms the electrons and holes are created and annihilated; this function is independent of the presence or absence of impurities. Let us calculate the function $K(T)$ for the particular case where the electron and hole energies are quadratic functions of the quasi-momentum (66.1).

The quasi-momentum distribution of the electrons (per unit volume) is given by the Boltzmann formula

$$\exp\left[\frac{\mu_e - \varepsilon_e(\mathbf{k})}{T}\right] 2\frac{d^3k}{(2\pi)^3},$$

the factor 2 taking account of the two directions of the spin. The energy distribution is given by the substitution

$$2\frac{d^3k}{(2\pi)^3} \to \frac{\sqrt{2}m_e^{3/2}}{\pi^2\hbar^3}\sqrt{(\varepsilon_e - \Delta)}\, d\varepsilon_e,$$

where $m_e = (m_1 m_2 m_3)^{1/3}$ and m_1, m_2, m_3 are the principal values of the effective mass tensor $m_{ik}^{(e)}$. The total number of electrons per unit volume is thus

$$n_e = \frac{\sqrt{2}m_e^{3/2}}{\pi^2\hbar^3} e^{\mu_e/T} \int_\Delta^\infty \sqrt{(\varepsilon_e - \Delta)}\, e^{-\varepsilon_e/T}\, d\varepsilon_e;$$

the integration can be extended to infinity, because of the rapid convergence. Calculation of the integral gives

$$n_e = 2\left(\frac{m_e T}{2\pi\hbar^2}\right)^{3/2} e^{(\mu_e - \Delta)/T}. \tag{67.3}$$

Similarly,

$$n_h = 2\left(\frac{m_h T}{2\pi\hbar^2}\right)^{3/2} e^{\mu_h/T}. \tag{67.4}$$

Finally, multiplying the two expressions and using (67.1), we obtain the required result:

$$n_e n_h = \frac{(m_e m_h)^{3/2}}{2\pi^3\hbar^6} T^3 e^{-\Delta/T}. \tag{67.5}$$

† The electron and hole densities in semiconductors at ordinary temperatures are 10^{13}–10^{17} cm^{-3}, whereas in metals they are 10^{22}–10^{23} cm^{-3}.

In an intrinsic semiconductor, where all the electrons and holes are formed in pairs,

$$n_e = n_h = \frac{(m_e m_h)^{3/4}}{\sqrt{2}\pi^{3/2}\hbar^3} T^{3/2} e^{-\Delta/2T}. \tag{67.6}$$

Equating (67.6) and (67.3), we find the chemical potential of the electrons:†

$$\mu_e = \tfrac{1}{2}\Delta + \tfrac{3}{4} T \log (m_h/m_e). \tag{67.7}$$

The contribution of the electrons and holes to the thermodynamic quantities in a semiconductor is exponentially small when $T \ll \Delta$. Since the creation of one electron–hole pair requires an energy close to Δ, we have as the electron–hole contribution to the internal energy $E_{\text{eh}} \approx V n_e \Delta$ with n_e from (67.6). This may usually be neglected in comparison with the lattice contribution to the crystal energy.

§ 68. The electron spectrum near the degeneracy point

In this section, we shall use simple examples to show how symmetry arguments may give the form of the energy spectrum of electrons or holes in a semiconductor or insulator near certain points in **k**-space (the reciprocal lattice) that are distinguished by their symmetry.‡

Let us consider a lattice of the cubic crystal class $\boldsymbol{O}_h$, and the properties of the energy spectrum near the point $\mathbf{k} = 0$, a vertex of a cubic cell in the reciprocal lattice; this point has the symmetry intrinsic to the complete point group $\boldsymbol{O}_h$.

As a first example, let us take the spectrum without allowance for the electron spin, and let the energy level in the band, belonging to the irreducible representation E_g of the group $\boldsymbol{O}_h$, be doubly degenerate at the point $\mathbf{k} = 0$ itself.§ Away from the point $\mathbf{k} = 0$, the degeneracy is removed, and the problem is to find all branches of the dispersion relation $\varepsilon(\mathbf{k})$ near this point.

It has been shown in §59 how the departure from some point $\mathbf{k} = \mathbf{k}_0$ in **k**-space may be regarded as a perturbation. The specific form of the perturbation operator is, in the present case, unimportant. It is sufficient to know just the structure of the expressions that give the correction to the energy in each order relative to the small quantity $\mathbf{q} = \mathbf{k} - \mathbf{k}_0$; in our case $\mathbf{k}_0 = 0$, and so $\mathbf{q} \equiv \mathbf{k}$. To the first order, the corrections are determined by the secular equation formed from the matrix elements (for transitions between states belonging to the same degenerate level) of an operator having the form $\mathbf{k}.\hat{\boldsymbol{\gamma}}$, where $\hat{\boldsymbol{\gamma}}$ is some vector

† In the literature this is often called the Fermi level, but the chemical potential of the electrons in a semiconductor does *not* have the significance of a limiting energy as in metals.

‡ When the electron spin is neglected, this problem is formally identical with that of the energy spectrum of phonons in a crystal; see Part 1, §136.

§ For the notation for representations of point groups see *QM*, §§95 and 99.

operator. In the present case, because the symmetry group contains a centre of inversion, all the matrix elements of $\hat{\gamma}$ are certainly zero, so that there is no effect of the first order in **k** (cf. Part 1, §136). To the second order in **k**, the corrections to the energy are determined by the secular equation formed from the matrix elements of an operator having the form

$$\hat{V} = \hat{\gamma}_{ik}k_i k_k, \tag{68.1}$$

where $\hat{\gamma}_{ik}$ is an Hermitian tensor operator (symmetrical in the suffixes i and k). These include the corrections from the terms linear in **k** in the Hamiltonian in second-order perturbation theory and the corrections from the terms quadratic in **k** in first-order perturbation theory. The matrix elements of the operator (68.1) certainly include some that are not zero, but relations exist between them because of the requirements of symmetry.

As regards their transformation law under the symmetry operations, the wave functions which form the basis of the representation E_g can be taken in the form

$$\psi_1 \sim x^2+\omega y^2+\omega^2 z^2, \quad \psi_2 \sim x^2+\omega^2 y^2+\omega z^2,$$

where

$$\omega = e^{2\pi i/3}, \quad \omega^2 = \omega^*, \quad 1+\omega+\omega^2 = 0,$$

and $\sim$ denotes "is transformed as". The rotation C_3 about a diagonal of the cube transforms the coordinates by $x, y, z \to z, x, y$; the functions ψ_1 and ψ_2 are transformed as shown by

$$C_3: \quad \psi_1 \to \omega\psi_1, \quad \psi_2 \to \omega^2\psi_2.$$

The rotation C_4^x about an edge of the cube $(x, y, z \to x, -z, y)$ transforms the functions as follows:

$$C_4^x: \quad \psi_1 \to \psi_2, \quad \psi_2 \to \psi_1;$$

and so on. Under inversion, the coordinates x, y, z change sign; ψ_1 and ψ_2 are unchanged.

From this, we easily conclude that all the matrix elements of the non-diagonal components $\hat{\gamma}_{ik}$ are zero, and the matrix elements of the diagonal components reduce to two independent real constants:

$$\langle 1|\gamma_{xx}|1\rangle = \langle 2|\gamma_{xx}|2\rangle = \langle 1|\gamma_{yy}|1\rangle = \ldots \equiv A,$$
$$\langle 1|\gamma_{xx}|2\rangle = \langle 2|\gamma_{xx}|1\rangle \equiv B,$$
$$\langle 1|\gamma_{yy}|2\rangle = \omega B, \quad \langle 1|\gamma_{zz}|2\rangle = \omega^2 B.$$

The matrix elements of the operator (68.1) are now

$$\langle 1|V|1\rangle = \langle 2|V|2\rangle = Ak^2,$$
$$\langle 1|V|2\rangle = \langle 2|V|1\rangle^* = B(k_x^2+\omega k_y^2+\omega^2 k_z^2).$$

Forming and solving the secular equation from these matrix elements, we obtain two branches of the spectrum:

$$\varepsilon_{1,2}(k)-\varepsilon(0) = Ak^2 \pm B[k^4-3(k_x^2k_y^2+k_x^2k_z^2+k_y^2k_z^2)]^{1/2}. \tag{68.2}$$

The degeneracy is removed on departure from the point $\mathbf{k} = 0$ in any direction except the diagonal of the cube ($k_x = k_y = k_z$).†

As a second example, let us consider the spectrum taking account of electron spin; the energy levels then correspond to two-valued (spinor) representations of the symmetry group. Let the level be fourfold degenerate at the point $\mathbf{k} = 0$, corresponding to the irreducible representation D'_u (or D'_g) of the group $\boldsymbol{O}_h$.‡

The base functions of this representation may be chosen so as to transform like the eigenfunctions ψ_m^j ($m = -j, \ldots, j$) of the angular momentum $j = 3/2$.§ This enables us to use the following procedure, which considerably simplifies the solution of the problem (J. M. Luttinger 1956).

For a four-dimensional representation, the matrix of the operator (68.1) is 4×4, with 16 elements. Any such matrix can be represented as a linear combination of 16 given linearly independent 4×4 matrices, which we take as the 15 matrices comprising $\hat{j}_x, \hat{j}_x^2, [\hat{j}_x, \hat{j}_y]_+, \hat{j}_x^3, [\hat{j}_x, \hat{j}_y^2-\hat{j}_z^2]_+$ and those obtained from these five by cyclic permutations of the suffixes x, y, z, together with $[\hat{j}_x, [\hat{j}_y, \hat{j}_z]_+]_+$; $[\ldots]_+$ denotes the anticommutator. Here $\hat{j}_x, \hat{j}_y, \hat{j}_z$ are the matrices of the Cartesian components of the angular momentum $j = 3/2$, taken with respect to the four functions $\psi_m^{3/2}$. On the other hand, with that choice of the base functions we must assume that the operators $\hat{j}_x, \hat{j}_y, \hat{j}_z$ are transformed under rotations and reflections as the components of an axial vector. This enables us to write the operator $\hat{V}$, quadratic in k_x, k_y and k_z, as a combination of expressions invariant under all transformations in the group $\boldsymbol{O}_h$:

$$\begin{aligned}\hat{V} &= \beta_1\mathbf{k}^2+4\beta_2(k_x^2\hat{j}_x^2+k_y^2\hat{j}_y^2+k_z^2\hat{j}_z^2)\\ &\quad+\beta_3(k_xk_y[\hat{j}_x, \hat{j}_y]_+ +k_yk_z[\hat{j}_y, \hat{j}_z]_+ +k_xk_z[\hat{j}_z, \hat{j}_x]_+),\end{aligned} \tag{68.3}$$

where β_1, β_2, β_3 are real constants.

The matrix elements of the operator (68.3) with respect to the functions

$$\psi_1 \sim \psi_{3/2}^{3/2}, \quad \psi_2 \sim \psi_{1/2}^{3/2}, \quad \psi_3 \sim \psi_{-1/2}^{3/2}, \quad \psi_4 \sim \psi_{-3/2}^{3/2}$$

† The same result (68.2) is obtained for the representation E_u (at the point $\mathbf{k} = 0$). The dispersion relation near a given point is indeed always the same for representations differing only by a multiplication by any one-dimensional representation of the group (here, $E_u = E_g\times A_{1u}$). It is evident that in such cases the matrix elements for transitions between different base function are in one-to-one relation.

‡ Such a situation occurs for the bottom of the hole band in diamond, silicon and germanium, which all have the same type of lattice.

§ In *QM*, §99, Problem, it is shown that the irreducible representation $D^{(3/2)}$ of the complete rotation group remains irreducible with respect to the group $\boldsymbol{O}$, and is the same as its representation D'.

are now easily calculated from the well-known matrix elements of the angular momentum, given in *QM*, (29.7)–(29.10). The calculation gives

$$\left.\begin{aligned} V_{11} &= V_{44} = (\beta_1+3\beta_2)(k_x^2+k_y^2)+(\beta_1+9\beta_2)\,k_z^2,\\ V_{22} &= V_{33} = (\beta_1+7\beta_2)(k_x^2+k_y^2)+(\beta_1+\beta_2)\,k_z^2,\\ V_{12} &= -V_{34} = \tfrac{1}{2}\sqrt{3}\beta_3 k_z(k_y+ik_x),\\ V_{13} &= V_{24} = 2\sqrt{3}\beta_2(k_y^2-k_x^2)+\tfrac{1}{2}\sqrt{3}\beta_3 ik_xk_y,\\ V_{14} &= V_{23} = 0. \end{aligned}\right\} \qquad (68.4)$$

The construction of the secular equation can be simplified by noting that the splitting of the level certainly cannot be complete; there must remain a twofold (Kramers) degeneracy. This means that each root $\lambda \equiv \varepsilon(\mathbf{k})-\varepsilon(0)$ of the secular equation (each eigenvalue of the matrix $\hat{V}$) is double. Thus to each eigenvalue λ there will correspond two linearly independent sets of quantities ϕ_n ($n = 1, 2, 3, 4$), solutions of the equations

$$\sum_m V_{nm}\phi_m = \lambda\phi_n. \qquad (68.5)$$

By combining these two sets, we can therefore impose one further condition on the ϕ_n, in particular make one of them vanish, say $\phi_4 = 0$. Then equation (68.5) with $n = 4$ gives

$$V_{41}\phi_1+V_{42}\phi_2+V_{43}\phi_3 = 0.$$

Substituting ϕ_3 from this in the equations with $n = 1$ and 2, we obtain two homogeneous equations in two unknowns ϕ_1 and ϕ_2:

$$\begin{pmatrix} V_{11}-V_{41}V_{13}/V_{43} & V_{12}-V_{42}V_{13}/V_{43} \\ V_{21}-V_{41}V_{23}/V_{43} & V_{22}-V_{42}V_{23}/V_{43} \end{pmatrix}\begin{pmatrix}\phi_1\\ \phi_2\end{pmatrix} = \lambda\begin{pmatrix}\phi_1\\ \phi_2\end{pmatrix};$$

the equation with $n = 3$ gives nothing new. Thus the problem of the eigenvalues of the 4×4 matrix is reduced to the problem for a 2×2 matrix. Forming and solving its secular equation, with the values of V_{nm} from (68.4), we find

$$\lambda = \tfrac{1}{2}(V_{11}+V_{22})\pm[\tfrac{1}{4}(V_{11}-V_{22})^2+|V_{12}|^2+|V_{13}|^2]^{1/2},$$

or finally

$$\varepsilon_{1,2}(\mathbf{k})-\varepsilon(0) = Ak^2\pm[Bk^4+C(k_x^2k_y^2+k_x^2k_z^2+k_y^2k_z^2)]^{1/2}, \qquad (68.6)$$

where

$$A = \beta_1+5\beta_2, \quad B = 16\beta_2^2, \quad C = 3(\tfrac{1}{4}\beta_3^2-16\beta_2^2)$$

(G. Dresselhaus, A. F. Kip and C. Kittel 1955). The level is split on departing from the point $\mathbf{k} = 0$ in any direction.†

† The application of perturbation theory to the states of only one degenerate level presupposes that the intervals $\varepsilon(\mathbf{k})-\varepsilon(0)$ in the resulting splitting are small in comparison with the distances of the adjacent bands, including those which are split off by the spin–orbit interaction.

Let us briefly consider the form of the equations describing the behaviour of particles near the degenerate bottom of a band in a magnetic field. We shall take a particular case, the second of those considered in this section, i.e. the spectrum (68.6).

The direct application of the Hamiltonian formed from (68.6) by the general rule (56.7) would encounter difficulties because the spectrum is not analytic near $\mathbf{k} = 0$. These difficulties can be avoided by making the change $\mathbf{k} \to \hat{\mathbf{k}} = \mathbf{K} - e\hat{\mathbf{A}}/\hbar c$ not in (68.6) but in the matrix Hamiltonian (68.3) (with symmetrization with respect to the components of $\mathbf{k}$, in order to preserve the Hermitian property). Each matrix element of the Hamiltonian is then converted into a linear differential operator which acts not only on the spin indices but also on the arguments of the functions $\phi_n(\mathbf{K})$ in the equations (68.5); these are thereby transformed into a set of four linear differential equations.

To take account of spin effects in the presence of a magnetic field, we must add to the Hamiltonian (68.3) terms which depend directly on $\mathbf{H}$ and are not determined by considerations of gauge invariance. Since the field is assumed weak, the terms added must be linear in $\mathbf{H}$; also, because $\mathbf{k}$ is assumed small, they must be independent of $\mathbf{k}$ (cf. §59). In the present case, the general form of such terms, invariant under all symmetry transformations of the crystal, is

$$\beta_4 \mathbf{H} \cdot \hat{\mathbf{j}} + \beta_5 (H_x \hat{j}_x^3 + H_y \hat{j}_y^3 + H_z \hat{j}_z^3). \tag{68.7}$$

To conclude this section, we may mention the interesting situation that arises if one of the bands meeting at the degeneracy point $\mathbf{k}_0$ is the conduction band and the other the valency band. The energy gap in such a spectrum is zero; an infinitesimal amount of energy is enough to create an electron and a hole with momenta close to $\mathbf{k}_0$. Such crystals are in a certain sense intermediate between insulators and metals. There is no energy gap, but the electron and hole states are separated except at one point in $\mathbf{k}$-space. We may say that it is a metal in which the Fermi surface is shrunk to one point $\mathbf{k}_0$. At $T = 0$, such a *gapless semiconductor*† has no carriers, but at low temperatures their number increases by a power law (not exponentially). The form of the spectrum near the point $\mathbf{k}_0$ cannot be established from symmetry arguments alone; the Coulomb interaction of the electrons and holes causes a singularity in the matrix elements of the perturbation at that point.

† One example is a modification of tin called *grey tin*.

CHAPTER VII

MAGNETISM

§ 69. Equation of motion of the magnetic moment in a ferromagnet

THE magnetic structure of crystals generates specific branches of the energy spectrum. In proceeding to investigate such spectra, let us first recall some features of interactions in magnetic bodies.

The chief form of interaction in ferromagnetic substances is the exchange interaction of the atoms, which establishes the spontaneous magnetization. A characteristic property of this interaction is that it is independent of the orientation of the magnetization relative to the lattice: the exchange interaction results from the electrostatic interaction of the electrons and the symmetry of the wave function of the system, and does not depend on the direction of the total spin.†

The simplest ferromagnetic system is an insulator whose crystal lattice contains atoms having a magnetic moment, the sign of the exchange interaction being "ferromagnetic", i.e. such that the parallel position of the moments is energetically favourable. The ground state of the system is then the one in which all the spins are parallel. More precisely, in this state the projection of the total spin of the system on some direction has its greatest possible value Σs_a (summed over all atoms), where s_a is the spin of one atom. For the exchange interaction Hamiltonian $\hat{H}_{\text{exch}}$ commutes with the operator $\hat{\mathbf{S}}$ of the total spin of the system, and therefore with its component $\hat{S}_z$, as follows from the facts that $\hat{H}_{\text{exch}}$ is independent of the direction of the spins, and $\hat{\mathbf{S}}$ is the rotation operator in spin space. Hence the ground state must have a definite value of S_z, and the maximum S_z corresponds to the minimum energy. Then s_a and the component s_z of the spin of each atom have their maximum values, so that the magnetic moment in the ground state is equal to its "nominal" value $\Sigma\mu_a$, where μ_a is the magnetic moment of one atom. This property however, is destroyed by weaker (relativistic) interactions.

In more complex cases, the magnetization of the body is not equal to its nominal value. In particular when the interaction is not ferromagnetic between all atoms, structures can be formed of two oppositely magnetized sub-lattices whose magnetizations are different and which therefore are not completely compensated; substances with such a structure are called *ferrites*, while the case of complete compensation corresponds to the antiferromagnets.

† Experimental results on the gyromagnetic ratios g, which are very close to 2 for ferromagnets, indicate that ferromagnetism is of spin origin.

Lastly, in a ferromagnetic metal we cannot treat the spins of the atoms independently of the conduction electrons, which are never (because of Fermi degeneracy effects) completely magnetized even at $T = 0$.

As in any macroscopic system, weakly excited states of a ferromagnet may be regarded as an assembly of elementary excitations, a quasi-particle gas. The elementary excitations in an ordered distribution of atomic magnetic moments are called *magnons*. Since we are dealing with quasi-particles in a crystal lattice with translational symmetry, the magnons have definite quasi-momenta (taking values in one cell of the reciprocal lattice), not actual momenta. In the classical picture, the magnons become *spin waves*, i.e. oscillations of the magnetic moments, propagated through the lattice. The magnons obey Bose statistics, and large occupation numbers of the magnon states correspond to the classical limiting case of spin waves.

If the spin wavelength is large compared with the lattice constant a, i.e. the wave number $k \ll 1/a$, the spin wave may be treated macroscopically; the dispersion relation $\omega(\mathbf{k})$ of the waves is then expressed in terms of phenomenological parameters (material constants) that occur in the macroscopic equations of motion of the magnetic moments. Thus the magnon spectrum $\varepsilon = \hbar\omega(\mathbf{k})$ is also expressed in terms of these parameters. Such a method of defining the magnon spectrum is exactly analogous to the definition of the long-wave phonon spectrum in terms of macroscopic parameters (elastic moduli) which appear in the macroscopic equations of vibrations in sound waves. To carry out this programme, we must first derive the relevant equations of motion.†

Let us first consider just exchange interactions. Since we are concerned with weakly excited states of a ferromagnet (and only their properties can be elucidated in a general form), we must take just the "slow" (low-frequency) motions of the magnetic moment. These are movements in which the direction of the magnetic moment varies slowly in space, its magnitude remaining constant: the equilibrium magnetization is fixed by the exchange interaction, and so its variation must involve a finite expenditure of energy at any wavelength (we assume that the body is sufficiently far from its Curie point, at which the spontaneous magnetization becomes zero). On the other hand, the energy is unchanged when the magnetic moment of the body as a whole is rotated; hence a rotation of the magnetization that is not uniform throughout the body requires an amount of energy that decreases with increasing wavelength (i.e. long-wave vibrations have a low frequency). The equation for the magnetic moment density (magnetization) $\mathbf{M}$ must therefore be such as to conserve $|\mathbf{M}|$:

$$\partial\mathbf{M}/\partial t = \mathbf{\Omega}\times\mathbf{M}, \tag{69.1}$$

† The subsequent results in this section are due to L. D. Landau and E. M. Lifshitz (1935). They are valid for "exchange" ferromagnets. We shall not deal here with what is called weak ferromagnetism, where the ferromagnetic moment is due only to relativistic interactions.

where $\mathbf{\Omega}$ is the angular velocity of precession of the moment, which we shall now determine. We write the equation of motion as a differential equation of the first order in the time, since at low frequencies the higher derivatives may be neglected.

To determine $\mathbf{\Omega}$, we must use the fact that, at long wavelengths and low temperatures, the dissipation of energy when the magnetization varies is small and may be neglected; this assumption will be justified at the end of §70. To determine the form of the condition for dissipation to be absent, we shall regard the magnetic moment of the magnet as an independent parameter whose equilibrium distribution is found by minimizing the free energy. We shall carry out the minimization for constant temperature, constant volume of the body, and constant field $\mathbf{H}$ at every point of the body; the thermodynamic potential with respect to these variables is the free energy $\tilde{F}$.† The variation $\delta\tilde{F}$ for an infinitesimal change of $\mathbf{M}$ may be written

$$\delta\tilde{F} = -\int \mathbf{H}_{\text{eff}} \cdot \delta\mathbf{M}\, dV, \tag{69.2}$$

with the notation $\mathbf{H}_{\text{eff}}$ for the "effective field" by analogy with the expression for the energy of the magnetic moment in an external magnetic field. In equilibrium, $\mathbf{H}_{\text{eff}} = 0$.

The dissipation of energy when the magnetization varies with time is calculated as the derivative

$$Q = T\,\partial S/\partial t = -\partial R_{\text{min}}/\partial t = -\partial\tilde{F}/\partial t,$$

where S is the entropy of the body and R_{min} the minimum work needed to bring the body into a given non-equilibrium state. From (69.1) we thus have

$$Q = \int \mathbf{H}_{\text{eff}} \cdot (\partial\mathbf{M}/\partial t)\, dV = \int \mathbf{H}_{\text{eff}} \cdot \mathbf{\Omega}\times\mathbf{M}\, dV. \tag{69.3}$$

Hence it is seen that in the absence of dissipation the vector $\mathbf{\Omega}$ must be parallel to the vector $\mathbf{H}_{\text{eff}}$, so that we can put $\mathbf{\Omega} = \text{constant}\times\mathbf{H}_{\text{eff}}$. Then equation (69.1) becomes

$$\dot{\mathbf{M}} = \text{constant}\times\mathbf{H}_{\text{eff}}\times\mathbf{M}; \tag{69.4}$$

he significance and value of the constant will be ascertained below.

According to the definition (69.2), the explicit form of the effective field is found by varying the total free energy of the body. This energy is

$$\tilde{F} = \int \{f_0(M) + U_{\text{non-u}} - \mathbf{M}\cdot\mathbf{H} - H^2/8\pi\}\, dV; \tag{69.5}$$

see *ECM*, §36. Here $f_0(M)$ is the free energy density of a uniformly magnetized body at $H = 0$, which takes account only of the exchange interactions and is

† See *ECM*, §36, where the thermodynamic quantities relating to the body as a whole are denoted by script letters. For an inhomogeneous distribution, it is more correct to speak of the free energy of the body (for a given volume), not of the thermodynamic potential $\tilde{\Phi}$. We shall not be concerned here with striction effects, i.e. stresses and strains that occur in the crystal when the magnetization varies.

therefore independent of the direction of **M**; $U_{\text{non-u}}$ is the additional exchange energy density due to the slow change of direction of **M** through the non-uniformly magnetized body.

The first terms in the expansion of this energy in powers of the derivatives of the moment **M** with respect to the coordinates are

$$U_{\text{non-u}} = \frac{1}{2}\alpha_{ik}\frac{\partial \mathbf{M}}{\partial x_i}\cdot\frac{\partial \mathbf{M}}{\partial x_k}; \tag{69.6}$$

this form (quadratic in the derivatives) is positive-definite. The expression (69.6) is constructed so that it is independent of the absolute direction of **M** (in accordance with the properties of the exchange interaction). In uniaxial crystals, the symmetrical tensor α_{ik} of rank two has components $\alpha_{xx} = \alpha_{yy} \equiv \alpha_1$, $\alpha_{zz} \equiv \alpha_2$ (where the z-axis is the axis of symmetry of the crystal); in cubic crystals, $\alpha_{ik} = \alpha\delta_{ik}$.

The order of magnitude of the coefficients α_{ik} may be estimated by noting that the non-uniformity energy relative to the volume of one unit cell in the crystal lattice would have to equal the characteristic atomic values of the exchange interaction energy if the direction of the moment varied considerably over distances of the order of the lattice constant a. The characteristic exchange energy is equal, in order of magnitude, to the Curie temperature T_c (the point at which the ferromagnetism disappears). From the condition $T_c/a^3 \sim \alpha M^2/a^2$ we find

$$\alpha \sim T_c/aM^2. \tag{69.7}$$

Varying the integral (69.5) for given values of **H** at every point of the body, and integrating by parts in the second term, we obtain

$$\delta\tilde{F} = \int\left\{f_0'(M)\,\mathbf{M}/M - \alpha_{ik}\frac{\partial^2\mathbf{M}}{\partial x_i\,\partial x_k} - \mathbf{H}\right\}.\,\delta\mathbf{M}\,dV.$$

According to the definition (69.2), the expression in the braces is $-\mathbf{H}_{\text{eff}}$. The first term is parallel to **M**, but disappears on substitution in the equation of motion (69.4), and may therefore be omitted.† Thus we find

$$\mathbf{H}_{\text{eff}} = \alpha_{ik}\frac{\partial^2\mathbf{M}}{\partial x_i\,\partial x_k} + \mathbf{H}. \tag{69.8}$$

When only the exchange interactions, independent of the direction of the magnetic moment, are taken into account, for a uniformly magnetized body equation (69.4) must reduce to the equation of motion of a freely precessing moment:

$$\frac{\partial \mathbf{M}}{\partial t} = \frac{g\,|e|}{2mc}\mathbf{H}\times\mathbf{M},$$

† But then $\mathbf{H}_{\text{eff}}$ need not be zero at equilibrium.

where $e = -|e|$ and m are the electron charge and mass, and g the gyromagnetic ratio of the ferromagnet (cf. *Fields*, §45). On the other hand, for uniform magnetization $\mathbf{H}_{\text{eff}} = \mathbf{H}$; hence it follows that the constant coefficient in (69.4) is $g|e|/2mc$, and the equation of motion is therefore

$$\frac{\partial \mathbf{M}}{\partial t} = \frac{g|e|}{2mc}\mathbf{H}_{\text{eff}} \times \mathbf{M}, \tag{69.9}$$

with $\mathbf{H}_{\text{eff}}$ from (69.8).

To form a complete set of equations, we must add Maxwell's equation relating the field $\mathbf{H}$ to the distribution of the magnetization $\mathbf{M}$. The spin waves, to be discussed in §70, are low-frequency waves in the sense that $\omega \ll ck$. Under these conditions, the field is quasi-steady; the time derivatives may be neglected in Maxwell's equations, which become

$$\operatorname{curl}\mathbf{H} = 0, \quad \operatorname{div}\mathbf{B} = \operatorname{div}(\mathbf{H}+4\pi\mathbf{M}) = 0. \tag{69.10}$$

The question may arise here of the legitimacy of varying the integral (69.5) with respect to $\mathbf{M}$ with $\mathbf{H}$ constant despite the fact that they are related by the second equation (69.10). The answer is that, if we put $\mathbf{H} = -\nabla\phi$ (from the first equation) and calculate the variation of the integral with respect to ϕ, it is zero by the second equation, so that the variation with respect to $\mathbf{H}$ makes no contribution to $\delta\tilde{F}$.

If the body is not in an external magnetic field, the field within it is due entirely to the distribution of the magnetization, and is in general of the same order as $\mathbf{M}$. In this sense, the term $\mathbf{H}$ in the effective field (69.8) is a relativistic effect (the atomic magnetic moments, and therefore the spontaneous magnetization $\mathbf{M}$, are determined by the Bohr magneton $\beta = |e|\hbar/2mc$, which has c in the denominator). In the purely exchange approximation so far considered, therefore, the second term in (69.8) is to be omitted, and the equation of motion is

$$\frac{\partial \mathbf{M}}{\partial t} = \frac{g|e|}{2mc}\alpha_{ik}\frac{\partial^2 \mathbf{M}}{\partial x_i \partial x_k} \times \mathbf{M}. \tag{69.11}$$

This equation is not linear.

Equation (69.11) may be rewritten as the equation of continuity for the magnetic moment:

$$\frac{\partial M_i}{\partial t} + \frac{\partial \Pi_{il}}{\partial x_l} = 0,$$

where the moment flux tensor is

$$\Pi_{il} = \frac{g|e|}{2mc}\alpha_{ik}\left(\mathbf{M}\times\frac{\partial \mathbf{M}}{\partial x_k}\right)_l.$$

This was to be expected, since in the exchange approximation the total magnetic moment of the body is conserved.

Let us now take into account the fact that in a ferromagnet there are not only exchange interactions but the considerably weaker relativistic interactions (spin–spin and spin–orbit) of the electron angular momenta. In the macroscopic theory, they are described by the magnetic anisotropy energy, whose density U_{an} depends on the direction of the magnetization vector relative to the crystal lattice; these interactions determine the equilibrium direction of the spontaneous magnetization in the ferromagnet. The relativistic interactions include, as already mentioned, that of **M** with the magnetic field **H**.

In a uniaxial crystal, the anisotropy energy has the form

$$U_{an} = -\tfrac{1}{2}KM_z^2. \tag{69.12}$$

If $K > 0$, the equilibrium magnetization is along the axis of symmetry (the z-axis; a ferromagnet of the "easy axis" type); if $K < 0$, the direction of the spontaneous magnetization lies in the xy-plane (a ferromagnet of the "easy plane" type). In a cubic crystal, the anisotropy energy may be written

$$U_{an} = (K'/M^2)(M_x^2M_y^2+M_x^2M_z^2+M_y^2M_z^2), \tag{69.13}$$

where the axes x, y, z are along the three fourth-order axes of symmetry (the edges of the cubic cells). If $K' > 0$, the equilibrium vector **M** is along one edge of the cubic cells; if $K' < 0$, it is along one of the spatial diagonals of the cells.†

We shall consider the particular case of a uniaxial ferromagnet. Adding to the integrand in (69.5) the term U_{an} (69.12), we obtain on variation an additional term $-KM_z\boldsymbol{\nu}.\delta\mathbf{M}$, where $\boldsymbol{\nu}$ is a unit vector along the axis of symmetry of the crystal. Thus we find for the effective field

$$\mathbf{H}_{eff} = \alpha_{ik}\frac{\partial^2\mathbf{M}}{\partial x_i\,\partial x_k}+KM_z\boldsymbol{\nu}+\mathbf{H}. \tag{69.14}$$

It is easily seen that beyond this change in the effective field there is no alteration of the equation of motion (69.9) due to the inclusion of relativistic effects: the neglect of dissipation signifies as before that the right-hand side of the equation of motion must be perpendicular to $\mathbf{H}_{eff}$, i.e. must have the form $\mathbf{M}'\times\mathbf{H}_{eff}$, where $\mathbf{M}'$ can differ from **M** only by the inclusion of relativistic effects, which are always small in comparison with the large quantity **M** and are therefore unimportant. The relativistic terms in $\mathbf{H}_{eff}$ are added to a quantity that is small because **M** varies only slowly through the body; these terms may become significant at sufficiently long wavelengths.

§ 70. Magnons in a ferromagnet. The spectrum

Let us apply the equations derived in §69 to the propagation of waves in which the magnetic moment density executes small oscillations, precessing about its equilibrium value $\mathbf{M}_0$. We shall consider a one-domain sample with $\mathbf{M}_0$

† The dimensionless quantities K and K' for various ferromagnetic substances have values in a wide range from several tenths to several tens. The order of magnitude of the ratio of the relativistic and exchange interactions is given by a^3U_{an}/T_c, which is usually 10^{-4}–10^{-5}.

constant throughout its volume, and take only the case of wavelengths much less than the size of the sample. The medium may then be regarded as infinite.

Let us first take account of the exchange interactions only, i.e. use equation (69.11). We put $\mathbf{M} = \mathbf{M}_0 + \mathbf{m}$, where $\mathbf{m}$ is small, and linearize the equation by omitting the terms of the second order in $\mathbf{m}$; since $M = M_0$, in this approximation $\mathbf{m} \perp \mathbf{M}_0$. The result is

$$\dot{\mathbf{m}} = \frac{|e|}{mc}\alpha_{ik}\frac{\partial^2 \mathbf{m}}{\partial x_i \, \partial x_k}\times\mathbf{M}_0; \tag{70.1}$$

here and henceforward we take $g = 2$. For a coordinate and time dependence of $\mathbf{m}$ in the form $\exp[i(\mathbf{k}.\mathbf{r}-\omega t)]$, we find

$$i\omega\mathbf{m} = (|e|/mc)\,\alpha k^2 \mathbf{m}\times\mathbf{M}_0, \tag{70.2}$$

where $\alpha = \alpha(\mathbf{n}) = \alpha_{ik}n_i n_k$ and $\mathbf{n}$ is the unit vector in the direction of the wave vector $\mathbf{k}$. Expanding this equation in components gives

$$i\omega m_x = (|e|/mc)\alpha M k^2 m_y,$$
$$i\omega m_y = -(|e|/mc)\,\alpha M k^2 m_x,$$

the z-axis being in the direction of $\mathbf{M}_0$. Hence the dispersion relation for spin waves is†

$$\omega = (|e|\, M/mc)\,\alpha(\mathbf{n})\, k^2. \tag{70.3}$$

We see that, in accordance with the discussion at the beginning of §69, the frequency in the exchange approximation tends to zero as $k \to 0$. The vector $\mathbf{m}$ in the spin wave rotates in the xy-plane with constant angular velocity ω, remaining constant in magnitude.

In the quantum picture, the formula (70.3) determines the energy spectrum of magnons ($\varepsilon = \hbar\omega$):‡

$$\varepsilon(\mathbf{k}) = 2\beta M\alpha(\mathbf{n})\, k^2. \tag{70.4}$$

In the second-quantization formalism, the macroscopic quantities describing the ferromagnet are replaced by operators expressed in terms of the magnon annihilation and creation operators. We shall show how this is to be done for the magnons (70.4).

We assign to the classical quantity $\mathbf{M}$ a vector operator $\hat{\mathbf{M}}$ whose components satisfy certain commutation rules. Let $\hat{\mathbf{S}}(\mathbf{r})\,\delta V$ be the operator of the total spin of the atoms in a physically infinitesimal volume element δV at the point $\mathbf{r}$. The operators $\hat{\mathbf{S}}(\mathbf{r}_1)\,\delta V_1$ and $\hat{\mathbf{S}}(\mathbf{r}_2)\,\delta V_2$ relating to different elements

† A quadratic dispersion relation for spin waves was first obtained by F. Bloch (1930), using a microscopic theory. The expression of this spectrum in terms of macroscopic parameters was given by L. D. Landau and E. M. Lifshitz (1945).

‡ In this chapter, β everywhere denotes the Bohr magneton, $\beta = |e|\,\hbar/2mc$.

δV_1 and δV_2 commute, but the components of one and the same operator $\hat{\mathbf{S}}(\mathbf{r})\delta V$ obey the usual commutation rules for angular momentum:

$$\hat{S}_x\,\delta V.\hat{S}_y\,\delta V - \hat{S}_y\,\delta V.\hat{S}_x\,\delta V = i\hat{S}_z\,\delta V,$$

or $\hat{S}_x\hat{S}_y - \hat{S}_y\hat{S}_x = i\hat{S}_z/\delta V$ (and similarly for the other commutators). In the limit as $\delta V \to 0$ these rules can be written for any $\mathbf{r}_1$ and $\mathbf{r}_2$

$$\hat{S}_x(\mathbf{r}_1)\,\hat{S}_y(\mathbf{r}_2) - \hat{S}_y(\mathbf{r}_2)\,\hat{S}_x(\mathbf{r}_1) = i\hat{S}_z(\mathbf{r}_1)\,\delta(\mathbf{r}_1 - \mathbf{r}_2).$$

Now multiplying this equation by $4\beta^2$ and noting that the magnetization operator $\hat{\mathbf{M}} = -2\beta\hat{\mathbf{S}}$, we obtain

$$\hat{M}_x(\mathbf{r}_1)\,\hat{M}_y(\mathbf{r}_2) - \hat{M}_y(\mathbf{r}_2)\,\hat{M}_x(\mathbf{r}_1) = -2i\beta\hat{M}_z(\mathbf{r}_1)\,\delta(\mathbf{r}_1 - \mathbf{r}_2). \tag{70.5}$$

For spin waves in which $\mathbf{M}$ executes small oscillations round the z-axis, in the first approximation with respect to the small quantities m_x and m_y, we can replace the operator $\hat{M}_z$ by the number $M_z \approx M$; then

$$\hat{m}_x(\mathbf{r}_1)\,\hat{m}_y(\mathbf{r}_2) - \hat{m}_y(\mathbf{r}_2)\,\hat{m}_x(\mathbf{r}_1) = -2i\beta M\delta(\mathbf{r}_1 - \mathbf{r}_2). \tag{70.6}$$

From this it is clear that m_y and m_x are in this case (apart from constant factors) the canonically conjugate "generalized coordinates and momenta", like ϕ and ϱ' in the quantization of sound waves in a liquid (§24). There is, however, an essential difference between the two cases. The commutation rule (24.7) for the phonon operators is exact, and does not depend on the smallness of the oscillations (i.e. the smallness of the phonon state occupation numbers). The rule (70.6) is approximate and is valid only in the first approximation with respect to the small quantity $\mathbf{m}$.

From the commutation rule (70.6) and the relation between the operators $\hat{m}_x$ and $\hat{m}_y$ that corresponds to the linear equations (70.1), we can find expressions for these operators in terms of the magnon annihilation and creation operators, as was done for phonons in §24 (see §71, Problem 4).

Let us return to the study of the magnon spectrum and the influence on it of relativistic effects. It is now necessary to take account also of the magnetic field $\mathbf{H}$ that results from the oscillations of $\mathbf{M}$. This will be of the same order of smallness as $\mathbf{m}$; it will be denoted here by $\mathbf{h}$.

Maxwell's equations (69.10) give

$$\mathbf{k}\times\mathbf{h} = 0, \quad \mathbf{k}\,.\,\mathbf{h} = -4\pi\mathbf{k}\,.\,\mathbf{m}.$$

From this we see that $\mathbf{h}$ is parallel to the wave vector, and is

$$\mathbf{h} = -4\pi(\mathbf{n}\,.\,\mathbf{m})\,\mathbf{n}. \tag{70.7}$$

Substitution of (70.7) in the last two terms of the integrand in (69.5) leads to

$$-\mathbf{m}\,.\,\mathbf{h} - h^2/8\pi = 2\pi(\mathbf{n}\,.\,\mathbf{m})^2; \tag{70.8}$$

the term $\mathbf{M}_0 . \mathbf{h}$ is here omitted, since on integration over the whole volume it becomes a surface integral (owing to the fact that $\mathbf{h}$ is a potential field) and vanishes. The part (70.8) of the anisotropy energy in a spin wave is sometimes called the *magnetostatic part*.

Let the ferromagnet be uniaxial and of the "easy axis" type, so that $\mathbf{M}_0$ is parallel to the axis of symmetry of the crystal (the z-axis): $\mathbf{M}_0 = \boldsymbol{\nu} M$. With a view to later applications, we shall allow the presence also of an external field $\mathfrak{H}$ parallel to the same direction $\boldsymbol{\nu}$; the sample is to be regarded as a cylinder with its axis along $\boldsymbol{\nu}$. Then the field within the body is $\mathbf{H} = \mathfrak{H} + \mathbf{h}$. The linearized equation of motion, which is here written after multiplication by $\hbar$, is

$$-i\varepsilon\mathbf{m} = 2\beta\{(\alpha k^2 + K + \mathfrak{H}/M)\,\boldsymbol{\nu}\times\mathbf{m} - \boldsymbol{\nu}\times\mathbf{h}\}. \tag{70.9}$$

For a uniaxial crystal $\alpha = \alpha_1 \sin^2\theta + \alpha_2 \cos^2\theta$, where θ is the angle between $\mathbf{k}$ and $\boldsymbol{\nu}$.

Substituting $\mathbf{h}$ from (70.7), we write out the equation in components; it is convenient to take the x-axis in the plane through the directions of $\boldsymbol{\nu}$ and $\mathbf{n}$. From the condition for the resulting two equations for m_x and m_y to be compatible, we find the dispersion relation

$$\varepsilon(\mathbf{k}) = 2\beta M[(\alpha k^2 + K + \mathfrak{H}/M)(\alpha k^2 + K + \mathfrak{H}/M + 4\pi\sin^2\theta)]^{1/2}. \tag{70.10}$$

Because of the term in $\sin^2\theta = k_x^2/k^2$, the expansion of $\varepsilon(\mathbf{k})$ in powers of the components of $\mathbf{k}$ is not a simple power series. This is due to the long range of the magnetic interactions.

An expression of the form (70.10), which has been derived here for a uniaxial ferromagnet (of the "easy axis" type), is valid for cubic crystals. This follows because the change in the anisotropy energy for small deviations of the vector $\mathbf{M}$ from its equilibrium direction has the same form in either case. For example, in a cubic crystal with $K' > 0$ the change δU_{an} when $\mathbf{M}$ deviates from the direction of $\mathbf{M}_0$ along an edge of the cube depends only on the angle ϑ between $\mathbf{M}$ and $\mathbf{M}_0$, and is $\delta U_{\text{an}} = K'M^2\vartheta^2$. Comparing this with the corresponding expression $\delta U_{\text{an}} = \frac{1}{2}KM^2\vartheta^2$ for a uniaxial crystal, we see that, to change to the case of a cubic crystal with $K' > 0$, we need only replace K by $2K'$ in (70.10). Similarly, it is easy to see that, to change to the case of a cubic crystal with $K' < 0$ ($\mathbf{M}_0$ along a spatial diagonal of the cube), we must replace K by $4|K'|/3$. Moreover, in a cubic crystal $\alpha(\mathbf{n})$ reduces to a constant. For a uniaxial ferromagnet of the "easy plane" type ($K < 0$), the situation is different: the change δU_{an} when $\mathbf{M}$ deviates from $\mathbf{M}_0$ depends both on the polar angle and on the azimuth of the direction of $\mathbf{M}$ relative to $\mathbf{M}_0$; this case therefore needs special consideration (see the Problem).

The result (70.10) relates only to the initial part of the spectrum, in which the quasi-momenta $k \ll 1/a$ and a macroscopic treatment is permissible. At large k that still satisfy this condition ($\alpha k^2 \gg 4\pi$, K) the expression (70.10) becomes

$$\varepsilon(\mathbf{k}) = 2\beta M\alpha(\mathbf{n})\,k^2 + 2\beta\mathfrak{H}. \tag{70.11}$$

The first term here is the same as the purely exchange expression (70.4). The external field simply adds the term $2\beta\mathfrak{H}$ to the magnon energy. In this approximation, therefore, the magnon has a moment projection on $\mathbf{M}_0$ equal to -2β. The excitation of each magnon reduces the total magnetic moment of the body by 2β.

In the opposite case, as $\mathbf{k} \to 0$, the expression (70.10) tends to a non-zero quantity, which for $\mathfrak{H} = 0$ is

$$\varepsilon(0) = 2\beta MK\left(1+\frac{4\pi}{K}\sin^2\theta\right)^{1/2}. \tag{70.12}$$

Thus the inclusion of the magnetic anisotropy causes an energy gap in the magnon spectrum. (The corresponding frequency $\omega(0) = \varepsilon(0)/\hbar$ is called the *ferromagnetic resonance frequency*.) This is natural, since in the presence of anisotropy even a rotation of the magnetic moment as a whole (i.e. for $\mathbf{k} = 0$) involves a finite energy. We see that for small **k** the relativistic effects, though small, cause relatively large corrections to the spectrum.

The concept of magnons as elementary excitations applies to weakly excited states of the body, and therefore to low temperatures. Hence, in the formulae relating to magnons, the values of all the constants of the material (including the magnetization M) must be taken for $T = 0$.

Let us now return to the assumption made in §69 that the dissipation is weak. In the quantum picture, the dissipation means that the lifetime of the magnons is finite because they interact with one another and with other quasi-particles.

If we take first of all the interaction between magnons, we must begin by noting that in the exchange approximation the number of magnons is constant: each magnon makes the same contribution -2β to M_z, and the exchange interaction conserves M_z. In this approximation, therefore, only scattering processes can occur. Their probability, however, decreases with the temperature, simply because of the decrease in the number of scatterers, so that the exchange decay always tends to zero at $T = 0$. We shall see later (§72) that a state with one magnon in the exchange approximation is in fact a strictly stationary state of the system.†

At $T = 0$ the decay of magnons is due only to their disintegration. Such processes can occur only because of relativistic interactions, and their probability is therefore small. Moreover, for small **k** the disintegration probability is always decreased by the smallness of the statistical weights (phase volumes) of the final states of the process.

The decay of magnons is also caused by their interaction with phonons; here the part of the exchange interaction Hamiltonian that depends on the

† It may also be noted that the cross-section for mutual scattering of two magnons in the exchange approximation tends to zero as their energy decreases (see §73). This further reduces the exchange decay of magnons at low temperatures. At sufficiently low temperatures, relativistic effects are important also for scattering processes.

deformation of the crystal acts as the perturbation operator. At $T = 0$ the creation of a phonon by a magnon is possible, but the magnon quasi-momentum must be sufficiently large: the magnon velocity $\partial\varepsilon/\hbar\, \partial\mathbf{k}$ must exceed the velocity of sound (cf. the third footnote to §65). The probability of the process is small, because the statistical weight of the final state is small.

Finally, in a ferromagnetic metal there can always occur excitation by a magnon of an electron from below the Fermi surface, because of the exchange interaction with conduction electrons. Here again, the probability of the process at small **k** is small, because the statistical weight of the final states is small.

PROBLEM

Find the magnon spectrum in a uniaxial ferromagnet of the "easy plane" type ($K < 0$).

SOLUTION. The equilibrium magnetization $\mathbf{M}_0$ lies in the plane perpendicular to the axis of symmetry of the crystal (the z-axis); let its direction be taken as the x-axis. The linearized equation of motion of the magnetic moment in this case has the form

$$-i\varepsilon\mathbf{m} = 2\beta\{\alpha k^2\, \mathbf{n}_x \times \mathbf{m} - |K|\, m_z \mathbf{n}_y - \mathbf{n}_x \times \mathbf{h}\},$$

where $\mathbf{n}_x$ and $\mathbf{n}_y$ are unit vectors along the coordinate axes, and the vector **m** lies in the yz-plane, which is perpendicular to $\mathbf{M}_0$. Substituting **h** from (70.7), writing out the equation in components, and equating to zero the determinant of the resulting system, we obtain the magnon spectrum

$$\varepsilon(\mathbf{k}) = 2\beta M[\alpha k^2(\alpha k^2 + |K|) + 4\pi \sin^2\theta\, (\alpha k^2 + |K| \sin^2\phi)]^{1/2},$$

where θ and ϕ are the polar angle and the azimuth of the direction of **k** relative to that of $\mathbf{M}_0$, the azimuth being measured from the xz-plane. When $\alpha k^2 \gg 1$, we return to the same quadratic spectrum (70.4), and when $k \to 0$ the magnon energy tends to

$$\varepsilon(0) = 4(\pi |K|)^{1/2} \beta M\, |\sin\theta \sin\phi|,$$

which is zero when the vector **k** is in the xz-plane which contains the axis of symmetry and the spontaneous magnetization of the crystal. This vanishing is, however, only approximate: the inclusion in the anisotropy energy of terms of higher order causes anisotropy in the xy-plane and therefore a finite energy gap for all directions of **k**.†

§ 71. Magnons in a ferromagnet. Thermodynamic quantities

Magnons excited in a ferromagnet make a certain contribution to the thermodynamic quantities in it. The results obtained in §70 enable us to calculate this contribution at temperatures that are low in the sense that $T \ll T_c$. For, in thermal equilibrium at a temperature T, most of the magnons have energy $\varepsilon \sim T$. For a quadratic spectrum

$$\varepsilon(\mathbf{k}) = 2\beta M \alpha(\mathbf{n})\, k^2, \tag{71.1}$$

† It may be recalled (see *ECM*, §37), that the expansion of the anisotropy energy in powers of **M** is really an expansion in terms of the relativistic ratio v/c (and is not due to the smallness of **M**, i.e. to the closeness of the Curie point).

this means that at temperatures $T \ll T_c$ magnons are excited with quasi-momenta $k \ll (T_c/\beta M\alpha)^{1/2}$. Using the estimate (69.7) for α and estimating the magnetization as $M \sim \beta/a^3$ (the magnetic moment per unit cell is of the order of a few times β), we find $ak \ll 1$, i.e. the condition for the results of §70 to be valid.

The "magnon" parts of the thermodynamic quantities in a ferromagnet are calculated as the thermodynamic quantities in an ideal Bose gas with zero chemical potential. For example, the magnon part of the thermodynamic potential Ω is

$$\Omega_{\text{mag}} = T \int \log\left(1-e^{-\varepsilon/T}\right) V\, d^3k/(2\pi)^3; \tag{71.2}$$

see Part 1, (54.4). The magnon contribution to the internal energy is then†

$$\begin{aligned} E_{\text{mag}} &= \Omega_{\text{mag}} - T\,\partial\Omega_{\text{mag}}/\partial T \\ &= \int \frac{\varepsilon}{e^{\varepsilon/T}-1}\,\frac{V\,d^3k}{(2\pi)^3}. \end{aligned} \tag{71.3}$$

The magnon contribution to the spontaneous magnetization determines its temperature dependence. It is calculated as the derivative

$$M_{\text{mag}} \equiv M(T) - M(0) = -\frac{1}{V}\left[\frac{\partial \Omega_{\text{mag}}}{\partial \mathfrak{H}}\right]_{\mathfrak{H}\to 0}$$

with respect to the external magnetic field; cf. *ECM*, (31.4). Differentiating the expression (71.2), we obtain

$$M_{\text{mag}} = -\int \left[\frac{\partial \varepsilon}{\partial \mathfrak{H}}\right]_{\mathfrak{H}\to 0} \frac{1}{e^{\varepsilon/T}-1}\,\frac{d^3k}{(2\pi)^3}. \tag{71.4}$$

The derivative $-\partial\varepsilon/\partial\mathfrak{H}$ is the intrinsic magnetic moment of the magnon.

Let us calculate the integrals (71.3) and (71.4) for temperatures $T \gg 2\hbar\beta M$;‡ then the limiting expression (71.1) may be used for the magnon spectrum. In view of the rapid convergence of the integrals, the integration may be extended over all **k**-space instead of over one reciprocal lattice cell. Taking α as constant (for cubic crystals) and replacing d^3k by $4\pi k^2\, dk$, we get after an obvious substitution

$$\begin{aligned} E_{\text{mag}} &= \frac{VT^{5/2}}{4\pi^2 A^{3/2}} \int_0^\infty \frac{x^{3/2}\,dx}{e^x - 1} \\ &= VT^{5/2}\Gamma(5/2)\,\zeta(5/2)/4\pi^2 A^{3/2}, \end{aligned}$$

† With the chemical potential $\mu = 0$ (and therefore also $\Phi = N\mu = 0$) we have $E = \Phi + TS - PV = TS + \Omega$; the entropy $S = -\partial\Omega/\partial T$. The expression (71.3) could also, of course, be written down directly without using (71.2).

‡ For the typical value $M = 2\times 10^3$ G, this condition gives $T \gg 1$°K.

where for brevity A denotes $2\beta M\alpha$ (so that $\varepsilon = Ak^2$).† For the specific heat $C_{mag} = \partial E_{mag}/\partial T$, we hence find

$$C_{mag} = V\frac{5\Gamma(5/2)\zeta(5/2)}{8\pi^2 A^{3/2}}T^{3/2}$$

$$= 0.113(T/A)^{3/2}V. \qquad (71.5)$$

This expression gives only the magnon part of the specific heat; the specific heat of the crystal also contains the usual phonon part.

Returning to the integral (71.4), we substitute the value -2β for the magnetic moment of the magnon according to (70.11). The result is, for $T \gg 2\pi\beta M$,

$$M_{mag} = -\frac{\beta T^{3/2}}{2\pi^2 A^{3/2}}\int_0^\infty \frac{x^{1/2}\,dx}{e^x-1}, \qquad (71.6)$$

whence

$$M(T) = M(0) - \beta T^{3/2}\Gamma(3/2)\zeta(3/2)/2\pi^2 A^{3/2}$$
$$= M(0) - 0.117\beta(T/A)^{3/2}; \qquad (71.7)$$

the magnon contribution is, of course, the whole of the change in the magnetization, since the phonons carry no magnetic moment. Thus the change in the spontaneous magnetization in the temperature range $2\pi\beta M \ll T \ll T_c$ follows a $T^{3/2}$ law (F. Bloch 1930).

The presence of the gap (70.10) in the magnon spectrum leads to an exponential dependence of C_{mag} and M_{mag} on T at still lower temperatures: when $T \ll \beta KM$,

$$C_{mag}, M_{mag} \propto \exp(-2\beta KM/t). \qquad (71.8)$$

The quantity in the numerator of the exponent is the least value of the energy gap, reached at $\theta = 0$ and $\theta = \pi$; see also Problem 1.

If the spontaneous magnetization of a ferromagnet in the ground state is equal to the greatest possible (*nominal*) value, corresponding to the case where all the atomic moments in the body are parallel, this value is unchanged by the application of an external magnetic field in that direction, i.e. the susceptibility χ in that direction is zero.

The inclusion of relativistic interactions reduces the spontaneous magnetization (at $T = 0$) in comparison with its "exchange" value, and causes the appearance of a non-zero susceptibility (T. Holstein and H. Primakoff 1940). Although this effect is very small, its calculation is of fundamental interest.

In the foregoing calculation of the magnetic part of the thermodynamic quantities, we have omitted the zero-point energy of the "magnetic oscillators", which makes no contribution to the temperature dependence of these quantities. The zero-point energy corresponds to magnon state occupation numbers of $\frac{1}{2}$:

$$[E(0)]_{mag} = \int \tfrac{1}{2}\varepsilon(\mathbf{k})\,V\,d^3k/(2\pi)^3.$$

† See Part 1, §58, regarding the calculation of such integrals.

Accordingly, the "zero-point" magnetization is

$$M(0) = -\int \tfrac{1}{2}(\partial\varepsilon/\partial\mathfrak{H})\, d^3k/(2\pi)^3. \tag{71.9}$$

This integral diverges for large k, i.e. is determined mainly by the short-wavelength magnons ($ka \sim 1$), which cannot be treated macroscopically. However, the change in the magnetization under the influence of relativistic effects is determined, as we shall see, by the long-wave region of the magnon spectrum, and can be calculated by means of the formulae derived in §70.

For simplicity, we shall consider a cubic crystal, and neglect the anisotropy constant, which is small in that case, i.e. write the magnon spectrum (70.10) in the form

$$\varepsilon(\mathbf{k}) = 2\beta[(bk^2+\mathfrak{H})(bk^2+\mathfrak{H}+4\pi M\sin^2\theta)]^{1/2}, \tag{71.10}$$

where $b = \alpha M$; in this expression, the term $4\pi M \sin^2\theta$ coming from the magnetostatic energy corresponds to the relativistic effects. The required change δM in the magnetization under the influence of relativistic effects is found by subtracting from (71.9) a similar integral with $\varepsilon_{\text{exch}}(\mathbf{k}) = 2\beta bk^2+2\beta\mathfrak{H}$ in place of $\varepsilon(\mathbf{k})$:

$$\delta M = -\frac{1}{2}\int \frac{\partial}{\partial\mathfrak{H}}[\varepsilon(\mathbf{k})-\varepsilon_{\text{exch}}(\mathbf{k})]\frac{d^3k}{(2\pi)^3}. \tag{71.11}$$

This integral is convergent for large $\mathbf{k}$.†

In the calculation, it is convenient first to differentiate with respect to M at constant b, and the notation b has been used in (71.10) for this purpose. A simple rearrangement gives

$$\frac{\partial\delta M}{\partial M} = -\frac{4\pi^2\beta M}{(2\pi)^3}\int_0^{\pi}\int_0^{\infty}\frac{\sin^4\theta\,.\,2\pi k^2\,dk\,.\sin\theta\,d\theta}{(bk^2+\mathfrak{H})^{1/2}(bk^2+\mathfrak{H}+4\pi M\sin^2\theta)^{3/2}}.$$

Since the integration with respect to k converges, it can be extended to ∞.

The integral with $\mathfrak{H} = 0$ is easily evaluated; then, integrating with respect to M, we obtain

$$\delta M = -\sqrt{\pi}\,\beta/8\alpha^{3/2}. \tag{71.12}$$

This is a very small quantity: $\delta M/M \sim 10^{-6}$.

If the external field is large ($\mathfrak{H} \gg 4\pi M$), we can neglect the term $4\pi M\sin^2\theta$ in the denominator of the integrand. The result is then

$$\delta M = -2\pi\beta M^{1/2}/15\alpha^{3/2}\mathfrak{H}^{1/2}. \tag{71.13}$$

When $\mathfrak{H} \to \infty$, δM tends to zero, as it should.

† To avoid misunderstanding, we may note that the correction to the ground-state energy cannot be determined in this way: without the differentiation with respect to $\mathfrak{H}$, the integral of $\varepsilon-\varepsilon_{\text{exch}}$ is divergent when the long-wavelength expressions for the magnon spectrum are used.

To conclude, we may note that, if we had attempted to investigate the temperature dependence of the magnetization in a two-dimensional ferromagnet by the same method as has been used in this section for the three-dimensional case, then (in the purely exchange approximation) we should have obtained instead of (71.6) a logarithmically divergent integral. This means that the spontaneous magnetization in a two-dimensional system with exchange interaction is actually zero for all $T \neq 0$. This is analogous to the result found in §27 for a two-dimensional Bose liquid, and in Part 1, §137, for a two-dimensional crystal. Since the energy of the system is independent of the direction of the magnetic moment, the expression for it contains only the derivatives of the vector **M**; this in turn leads ultimately to the divergence of the fluctuations (in the two-dimensional case) that destroy the magnetization. The inclusion of the relativistic interactions, which depend on the direction of **M**, stabilizes the fluctuations and makes possible the existence of a two-dimensional ferromagnet.

PROBLEMS

PROBLEM 1. Calculate the magnon parts of the thermodynamic quantities at temperatures $T \ll \varepsilon(0)$.

SOLUTION. The important magnons are those with small quasi-momenta **k** propagated in a direction where the gap is least, i.e. near $\theta = 0$ or $\theta = \pi$; these two directions give equal contributions. For example, when θ is small we have, with the necessary accuracy,

$$\varepsilon(\mathbf{k}) = 2\beta KM + Ak^2 + 4\pi\beta M\theta^2,$$

where $A = 2\beta M\alpha$ for cubic crystals and $A = 2\beta M\alpha_2$ for uniaxial crystals of the "easy axis" type. The magnon distribution at these temperatures may be taken as a Boltzmann distribution (i.e. we may neglect unity in the denominators of the integrands) and $\varepsilon(\mathbf{k})$ may be replaced by $\varepsilon(0)$ everywhere in the coefficients of the exponentials. Integrations with respect to k and θ are extended to infinity; the result is

$$E_{\text{mag}} = V\frac{KT^{5/2}}{32\pi^{5/2}A^{3/2}}\exp\left(-\frac{2\beta KM}{T}\right), \quad M_{\text{mag}} = -\frac{E_{\text{mag}}}{VKM}.$$

In calculating the specific heat, only the exponential factor need be differentiated:

$$C_{\text{mag}} = (2\beta KM/T^2)\,E_{\text{mag}}.$$

PROBLEM 2. Determine the dependence of the magnetization on the external field with the conditions $\mathfrak{H} \gg 4\pi M$, $T \gg \beta\mathfrak{H}$.

SOLUTION. Under these conditions, the relativistic terms may be neglected, and $\varepsilon(\mathbf{k})$ written in the form (70.11). Differentiating the expression (71.4), we find

$$\frac{\partial M}{\partial \mathfrak{H}} = \frac{4\beta^2}{T}\int \frac{e^{\varepsilon/T}}{(e^{\varepsilon/T}-1)^2}\frac{d^3k}{(2\pi)^3}.$$

The important values of **k** in the integral are small. Hence

$$\frac{\partial M}{\partial \mathfrak{H}} \approx 4\beta^2 T\int \frac{1}{\varepsilon^2}\frac{d^3k}{(2\pi)^3} = \frac{T}{2\pi^2}\int_0^\infty \frac{k^2\,dk}{(\alpha k^2 M_0+\mathfrak{H})^2};$$

we take α = constant, and M_0 is the value of M for $\mathfrak{H} = 0$. The final result is

$$\frac{\partial M}{\partial \mathfrak{H}} = \frac{T}{8\pi(\alpha M_0)^{3/2}\mathfrak{H}^{1/2}}.$$

Thus, under these conditions, $M - M_0 \propto \mathfrak{H}^{1/2}$.

PROBLEM 3. Determine the dependence of the magnetization at $T = 0$ on a weak external field.

SOLUTION. Differentiating the integral (71.11), with $\varepsilon(\mathbf{k})$ from (71.10), with respect to $\mathfrak{H}$, we obtain

$$\frac{\partial M}{\partial \mathfrak{H}} = \int \frac{4\pi^2 M_0^2 \beta \sin^4\theta}{[(\alpha M_0 k^2 + 4\pi M_0 \sin^2\theta + \mathfrak{H})(\alpha M_0 k^2 + \mathfrak{H})]^{3/2}} \frac{d^3k}{(2\pi)^3}.$$

As $\mathfrak{H} \to 0$, the integral over dk diverges logarithmically for small k. Hence, with logarithmic accuracy, we can put $k = 0$, $\mathfrak{H} = 0$ in the first factor in the denominator, and $\mathfrak{H} = 0$ in the second factor, but cut off the integral below at $k^2 \sim \mathfrak{H}/\alpha M_0$ and above at $k^2 \sim 4\pi/\alpha$. The result is

$$\frac{\partial M}{\partial \mathfrak{H}} = \frac{\beta}{32\sqrt{\pi M_0}\,\alpha^{3/2}} \log \frac{4\pi M_0}{\mathfrak{H}}.$$

In (71.10), K is neglected. When $\mathfrak{H} \ll KM$, $\mathfrak{H}$ is replaced by KM_0 in the logarithm.

PROBLEM 4. In the exchange approximation, determine the spatial correlation function of the fluctuations in the magnetization at distances $r \gg a$.

SOLUTION. The operators $\hat{m}_x$ and $\hat{m}_y$ satisfying the commutation rule (70.6) and expressed in terms of the magnon annihilation and creation operators have the form (in the Schrödinger representation)

$$\hat{m}_x(\mathbf{r}) = (\beta M/V)^{1/2} \sum_{\mathbf{k}} (\hat{a}_{\mathbf{k}} e^{i\mathbf{k}\cdot\mathbf{r}} + \hat{a}_{\mathbf{k}}^{+} e^{-i\mathbf{k}\cdot\mathbf{r}}),$$

$$\hat{m}_y(\mathbf{r}) = i(\beta M/V)^{1/2} \sum_{\mathbf{k}} (\hat{a}_{\mathbf{k}} e^{i\mathbf{k}\cdot\mathbf{r}} - \hat{a}_{\mathbf{k}}^{+} e^{-i\mathbf{k}\cdot\mathbf{r}}).$$

With these operators, we calculate the correlation function

$$\phi_{ik}(\mathbf{r}) = \tfrac{1}{2}\langle \hat{m}_i(\mathbf{r}_1)\,\hat{m}_k(\mathbf{r}_2) + \hat{m}_k(\mathbf{r}_2)\,\hat{m}_i(\mathbf{r}_1)\rangle, \qquad \mathbf{r} = \mathbf{r}_1 - \mathbf{r}_2,$$

the suffixes i and k taking the values x and y. Since only the products $\langle \hat{a}_{\mathbf{k}}^{+}\hat{a}_{\mathbf{k}}\rangle = n_{\mathbf{k}}$ and $\langle \hat{a}_{\mathbf{k}}\hat{a}_{\mathbf{k}}^{+}\rangle = n_{\mathbf{k}} + 1$ (where $n_{\mathbf{k}}$ are the magnon state occupation numbers) have non-zero diagonal matrix elements, we find

$$\phi_{ik}(\mathbf{r}) = \delta_{ik} \int 2\beta M (n_{\mathbf{k}} + \tfrac{1}{2})\, e^{i\mathbf{k}\cdot\mathbf{r}}\, d^3k/(2\pi)^3.$$

The integrand gives directly the Fourier component of the correlation function. The constant term may be omitted: it corresponds to a delta-function term in $\phi_{ik}(\mathbf{r})$, whereas the whole analysis relates only to distances $r \gg a$. Thus

$$\phi_{ik}(\mathbf{k}) = 2\beta M n_{\mathbf{k}} \delta_{ik} = 2\beta M [e^{\varepsilon(\mathbf{k})/T} - 1]^{-1} \delta_{ik}.$$

In the classical limit, when $\varepsilon \ll T$, we find

$$\phi_{ik}(\mathbf{k}) = \delta_{ik} T/\alpha k^2.$$

In a cubic ferromagnet, α = constant, and

$$\phi_{ik}(\mathbf{r}) = \delta_{ik} T/4\pi\alpha r, \quad r \gg (\beta M\alpha/T)^{1/2}.$$

§ 72. The spin Hamiltonian

To derive the magnon dispersion relation for all values of the quasi-momentum, and not only in the long-wave limit, we must naturally make use of more detailed information about the microscopic structure of a ferromagnet.

Let us consider an insulator consisting of atoms with zero orbital angular momentum but non-zero spin S. If we are not concerned with highly excited states resulting from excitation of the electron shells of the atoms, we can average the Hamiltonian of the system over the orbital variables of the electrons of atoms in the ground state (with the atomic nuclei fixed at the lattice sites). This gives the spin Hamiltonian of the system, containing only the operators of the total spins of the atoms.†

If we take into account just the exchange interaction, which depends only on the relative orientations of the spins, the operators of the atomic spin vectors must appear in the Hamiltonian as scalar combinations. It is of considerable methodological interest to investigate a system described by the simple Hamiltonian

$$\hat{H}_{\text{exch}} = -\tfrac{1}{2}\sum_{\mathbf{m}\neq\mathbf{n}} J_{\mathbf{nm}}\hat{\mathbf{S}}_{\mathbf{n}}\hat{\mathbf{S}}_{\mathbf{m}}, \qquad J_{\mathbf{nm}} = J(\mathbf{r}_{\mathbf{n}}-\mathbf{r}_{\mathbf{m}}), \tag{72.1}$$

where the summation is over all atoms; the "vector" suffixes $\mathbf{m}$ and $\mathbf{n}$ (with integral components) label the lattice sites, and $\mathbf{r}_{\mathbf{n}}$ are their position vectors. The numbers $J_{\mathbf{nm}}$ are called *exchange integrals*; cf. *QM*, §62, Problems.‡ With independent summation over $\mathbf{m}$ and $\mathbf{n}$, each pair of atoms appears twice in the sum (72.1), and of course $J_{\mathbf{nm}} = J_{\mathbf{mn}}$.

In (72.1), all the magnetic atoms in the lattice are assumed to be the same (one in each unit cell). The basic assumption underlying this Hamiltonian is, however, that the atoms in the lattice are sufficiently far apart. The exchange integral is determined by the "overlap" of the wave functions of two atoms, and decreases very rapidly (exponentially) as the distance between them increases. For a system of atoms far apart, we can therefore assume a pair interaction, and (72.1) then contains no terms with products of spin operators of more than two atoms. To the same accuracy, we may suppose that the exchange interaction between two atoms comes about through a single pair of electrons, one in each atom. Then the interaction operator is formed bilinearly from the electron spin operators, and after averaging over the states of the atoms, bilinearly with respect to the atomic spins (C. Herring 1966).§

† The procedure is similar to the construction of the Hamiltonians of individual atoms describing the fine structure of the atomic levels; cf. *QM*, §72.

‡ The description of the exchange interaction by means of the spin Hamiltonian is due to P. A. M. Dirac (1929). The Hamiltonian (72.1) was introduced by J. H. van Vleck (1931), and is usually called the *Heisenberg Hamiltonian*, because it corresponds to a ferromagnet model first discussed by Heisenberg.

§ In such conditions, the summation in (72.1) must of course be taken only over pairs of adjacent atoms. This, however, does not simplify the formulae, and it will therefore not be shown explicitly.

A system described by the Hamiltonian (72.1) is ferromagnetic if the exchange integrals $J_{mn} > 0$. Let us determine the ground-state energy of such a system, and suppose that an external magnetic field $\mathfrak{H}$ is also present, by adding to (72.1) the operator

$$\hat{V} = -2\beta\mathfrak{H} \sum_{\mathbf{m}} \hat{S}_{\mathbf{m}z}, \tag{72.2}$$

the z-axis being parallel to the field. The operator $\Sigma\hat{S}_{\mathbf{m}z}$ of the projection of the total spin of the system commutes both with $\hat{H}_{\mathrm{exch}}$ and with $\hat{V}$; the states of the system may therefore be classified by the eigenvalues of this quantity.

In the ferromagnetic case, the ground state corresponds to the greatest possible value NS of the total-spin projection, where N is the number of atoms in the system; this result is, of course, independent of the presence of an external field, which simply distinguishes the direction of the axis.

Let χ_0 be the normalized spin wave function of the ground state. The maximum value NS of the total-spin projection can be reached only if the spin projection of every atom also has its maximum value S. Hence χ_0 is an eigenfunction of each of the operators $\hat{S}_{\mathbf{n}z}$:

$$\hat{S}_{\mathbf{n}z}\chi_0 = S\chi_0. \tag{72.3}$$

In the following, we shall need the operators $\hat{S}_{\pm} = \hat{S}_x \pm i\hat{S}_y$, which obey the commutation rules

$$\hat{S}_+\hat{S}_- - \hat{S}_-\hat{S}_+ = 2\hat{S}_z, \quad \hat{S}_z\hat{S}_\pm - \hat{S}_\pm\hat{S}_z = \pm\hat{S}_\pm; \tag{72.4}$$

see *QM*, (26.12). Their matrix elements are

$$\langle S_z | S_+ | S_z - 1\rangle = \langle S_z - 1 | S_- | S_z\rangle = \sqrt{[(S+S_z)(S-S_z+1)]}; \tag{72.5}$$

see *QM*, (27.12). The operator $\hat{S}_+$ increases by one the value of the component S_z, and $\hat{S}_-$ decreases it by one. We also write

$$\hat{S}_{\mathbf{m}}\hat{S}_{\mathbf{n}} = \hat{S}_{\mathbf{m}z}\hat{S}_{\mathbf{n}z} + \tfrac{1}{2}(\hat{S}_{\mathbf{m}+}\hat{S}_{\mathbf{n}-} + \hat{S}_{\mathbf{m}-}\hat{S}_{\mathbf{n}+})$$

and

$$\hat{H} = -\tfrac{1}{2}\sum_{\mathbf{m}\neq\mathbf{n}} J_{\mathbf{mn}}(\hat{S}_{\mathbf{m}z}\hat{S}_{\mathbf{n}z} + \hat{S}_{\mathbf{m}-}\hat{S}_{\mathbf{n}+}) - 2\beta\mathfrak{H}\sum_{\mathbf{m}} \hat{S}_{\mathbf{m}z}, \tag{72.6}$$

using the symmetry $J_{\mathbf{mn}} = J_{\mathbf{nm}}$ and the commutativity of operators pertaining to different atoms.

Since the operators $\hat{S}_{\mathbf{n}+}$ have matrix elements only for transitions with increase of the numbers S_z, for the state with the maximum values of these numbers we have

$$\hat{S}_{\mathbf{n}+}\chi_0 = 0, \tag{72.7}$$

as is seen also from the explicit expressions for the matrix elements (72.5). Hence, when the Hamiltonian (72.6) acts on the wave function χ_0, the result is

$$\hat{H}\chi_0 = \Big\{-\tfrac{1}{2}\sum_{\mathbf{m}\neq\mathbf{n}} J_{\mathbf{mn}}S^2 - 2\beta\mathfrak{H}NS\Big\}\chi_0.$$

The expression in the braces is the energy E_0 of the ground state. Replacing the summation over **m** and **n** by summation over **m** and $\mathbf{q} = \mathbf{n}-\mathbf{m}$, we can finally write

$$E_0 = -\tfrac{1}{2}NS^2 \sum_{\mathbf{q}\neq 0} J_{\mathbf{q}} - 2\beta SN\mathfrak{H}. \tag{72.8}$$

The total magnetic moment of the system in this state is $2\beta SN$.

The next state of the system, in order of decreasing total-spin projection, corresponds to the value $NS-1$ of the latter, and to the excitation of one magnon with magnetic moment -2β. This value of the total-spin projection occurs for a state with wave function

$$(2S)^{-1/2}\hat{S}_{\mathbf{n}-}\chi_0, \tag{72.9}$$

in which the effect of the operator $\hat{S}_{\mathbf{n}-}$ reduces by one the spin projection of one atom.† This function is, however, not an eigenfunction of the Hamiltonian of the system; it does not yet take account of the translational symmetry of the lattice. The eigenfunction of the Hamiltonian must be constructed as a linear combination of the functions (72.9) with all **n**. Similar arguments to those which in §55 gave the Bloch functions for an electron in a periodic field show that, to take correct account of the translational symmetry, this linear combination must have the form

$$\chi_{\mathbf{k}} = (2NS)^{-1/2}\sum_{\mathbf{n}} e^{i\mathbf{k}\cdot\mathbf{r}_{\mathbf{n}}}\hat{S}_{\mathbf{n}-}\chi_0; \tag{72.10}$$

the factor $N^{-1/2}$ is for normalization. The constant vector **k** is just the quasi-momentum of the magnon.

The energy $\varepsilon(\mathbf{k})$ of the magnon is the difference $E_{\mathbf{k}}-E_0$ between the energies of the excited and ground states of the system. Hence

$$(\hat{H}-E_0)\chi_{\mathbf{k}} = \varepsilon(\mathbf{k})\chi_{\mathbf{k}}.$$

Substituting on the left-hand side the expression (72.10) and replacing $E_0\chi_0$ by $\hat{H}\chi_0$, we obtain

$$\varepsilon(\mathbf{k})\chi_{\mathbf{k}} = (2NS)^{-1/2}\sum_{\mathbf{n}} e^{i\mathbf{k}\cdot\mathbf{r}_{\mathbf{n}}}(\hat{H}\hat{S}_{\mathbf{n}-}-\hat{S}_{\mathbf{n}-}\hat{H})\chi_0. \tag{72.11}$$

The commutator here is easily calculated by writing $\hat{H}$ in the form (72.6) and using the commutation rules (72.4). Again taking into account the symmetry of the coefficients $J_{\mathbf{mn}}$, we find

$$\hat{H}\hat{S}_{\mathbf{n}-}-\hat{S}_{\mathbf{n}-}\hat{H} = \sum_{\mathbf{m}}{}' J_{\mathbf{mn}}(\hat{S}_{\mathbf{m}z}\hat{S}_{\mathbf{n}-}-\hat{S}_{\mathbf{n}z}\hat{S}_{\mathbf{m}-})+2\beta\mathfrak{H}\hat{S}_{\mathbf{n}-}. \tag{72.12}$$

† The normalization of the function (72.9) is easily verified by noting that

$$\begin{aligned}(\hat{S}_{\mathbf{n}-}\chi_0)^*(\hat{S}_{\mathbf{n}-}\chi_0) &= \chi_0^*\hat{S}_{\mathbf{n}-}^+\hat{S}_{\mathbf{n}-}\chi_0\\ &\equiv \langle S\,|\,S_{\mathbf{n}+}S_{\mathbf{n}-}\,|\,S\rangle\\ &= \langle S\,|\,S_{\mathbf{n}+}\,|\,S-1\rangle\langle S-1\,|\,S_{\mathbf{n}-}\,|\,S\rangle\\ &= 2S.\end{aligned}$$

Finally, substituting this expression in (72.11), using (72.3), and changing to summation over $\mathbf{q} = \mathbf{n}-\mathbf{m}$, we have

$$\varepsilon(\mathbf{k})\,\chi_{\mathbf{k}} = \Big\{S \sum_{\mathbf{q}\neq 0} J_{\mathbf{q}}(1-e^{i\mathbf{k}\cdot\mathbf{r}_{\mathbf{q}}})+2\beta\mathfrak{H}\Big\}\,\chi_{\mathbf{k}}.$$

The expression in the braces is the required magnon energy. The imaginary part of the expression under the summation sign, being an odd function of $\mathbf{r}_{\mathbf{q}}$, gives zero on summation, and the result is

$$\varepsilon(\mathbf{k}) = S \sum_{\mathbf{q}\neq 0} J_{\mathbf{q}}(1-\cos \mathbf{k}\,.\,\mathbf{r}_{\mathbf{q}})+2\beta\mathfrak{H} \tag{72.13}$$

(F. Bloch 1930).

This formula gives the exact dispersion relation for magnons in a system described by the Hamiltonian (72.1). In the limit of small $\mathbf{k}$, it becomes, of course, the quadratic relation

$$\varepsilon(\mathbf{k}) = \tfrac{1}{2}Sk_ik_k \sum_{\mathbf{q}\neq 0} J_{\mathbf{q}}x_{\mathbf{q}i}x_{\mathbf{q}k}+2\beta\mathfrak{H}. \tag{72.14}$$

The Curie point of the system in question is at a temperature $T_c \sim J$, and so at temperatures $T \gg J$ the system is certainly paramagnetic. At such temperatures we can, as a first approximation, altogether neglect the interaction between the atoms. In this approximation, the magnetic susceptibility of the system will be the same as that of an ideal gas of atoms with spin S, namely

$$\chi = \frac{N}{V}\frac{4\beta^2S(S+1)}{3T} \tag{72.15}$$

(see Part 1, §52); the susceptibility is per unit volume. This expression is the first term in the expansion of the function $\chi(T)$ in powers of $1/T$. The subsequent terms do depend on the interaction of the atoms; we shall determine the first such term.

The susceptibility (in a zero field) is defined as the derivative $\chi = \partial M/\partial\mathfrak{H}$ as $\mathfrak{H} \to 0$, and the magnetization M is calculated as the derivative of the free energy: $VM = -\partial F/\partial\mathfrak{H}$. To solve the problem, we need to calculate F as far as the terms in $1/T^2$.

We start from the formula $F = -T\log Z$, where Z is the partition function

$$Z = \sum_n e^{-E_n/T}$$
$$\approx \sum_n \left(1-\frac{E_n}{T}+\frac{E_n^2}{2T^2}-\frac{E_n^3}{6T^3}\right);$$

the summation is over all the energy levels of the system.† The total number of levels in the spectrum of the system is finite, and equal to the number of possible

† The following calculation of the free energy corresponds to that in Part 1, §73, continued to the next term in the expansion.

combinations of orientations of the atomic spins relative to the lattice. Each spin has $2S+1$ different orientations; this number is therefore $(2S+1)^N$. Denoting by a bar the arithmetic mean, we can write

$$Z = (2S+1)^N\left[1-\frac{1}{T}\overline{E}+\frac{1}{2T^2}\overline{E^2}-\frac{1}{6T^3}\overline{E^3}\right].$$

The mean value $\overline{E^m} = \operatorname{tr} \hat{H}^m/(2S+1)^N$. According to a well-known property, the trace of the operator can be calculated with any complete set of wave functions; we take as these the functions corresponding to all possible combinations of orientations of the atomic spins. Then the averaging reduces to an independent averaging of each spin over its directions, and $\overline{E} = 0$. Now taking the logarithm of Z and again expanding in powers of $1/T$, we have to the same accuracy

$$F = -TN\log(2S+1)-\frac{1}{2T}\overline{E^2}+\frac{1}{6T^2}\overline{E^3}. \tag{72.16}$$

In this expression, only the terms containing $\mathfrak{H}^2$ contribute to the susceptibility. Omitting all other terms, and noting that the odd powers of the spin components give zero on averaging, we find

$$F = -\frac{(2\beta\mathfrak{H})^2}{2T}\sum_{\mathbf{n}}\overline{S^2_{\mathbf{n}z}}-\frac{(2\beta\mathfrak{H})^2}{2T^2}\cdot\frac{1}{2}\sum_{\mathbf{n}\neq\mathbf{m}}2J_{\mathbf{mn}}(\overline{S_{\mathbf{n}}S_{\mathbf{n}z}})(\overline{S_{\mathbf{m}}S_{\mathbf{m}z}}).$$

The mean values are

$$\overline{S_{\mathbf{n}z}S_{\mathbf{n}x}} = \overline{S_{\mathbf{n}z}S_{\mathbf{n}y}} = 0, \quad \overline{S^2_{\mathbf{n}z}} = \tfrac{1}{3}S(S+1).$$

Thus

$$F = -\frac{2}{3T}\beta^2\mathfrak{H}^2NS(S+1)-\frac{2}{9T^2}\beta^2\mathfrak{H}^2NS^2(S+1)^2\sum_{\mathbf{q}\neq 0}J_{\mathbf{q}},$$

and hence we have finally for the susceptibility

$$\chi = \frac{4\beta^2S(S+1)N}{3TV}\left[1+\frac{S(S+1)}{3T}\sum_{\mathbf{q}\neq 0}J_{\mathbf{q}}\right]. \tag{72.17}$$

It should be noted that the sign of the correction term in the square brackets depends on the sign of the exchange integral.

PROBLEMS

PROBLEM 1. Find the magnetic part of the specific heat of a system described by the Hamiltonian (72.1) at temperatures $T \gg J$.

SOLUTION. The first term in the expansion of the specific heat in powers of $1/T$ comes from the term $-\overline{E^2}/2T$ in the free energy (72.16). Averaging by the same method the square of the

Hamiltonian (72.1), we get

$$\overline{E^2} = \tfrac{1}{4}.2 \sum_{\mathbf{m} \neq \mathbf{n}} J_{\mathbf{mn}}^2 \overline{S_{\mathbf{m}i}S_{\mathbf{m}k}}\,\overline{S_{\mathbf{n}i}S_{\mathbf{n}k}}$$

$$= 3.\frac{S^2(S+1)^2}{9}.\frac{1}{2}N \sum_{\mathbf{q} \neq 0} J_{\mathbf{q}}^2,$$

since $\overline{S_i S_k} = \frac{1}{3}S(S+1)\delta_{ik}$. The specific heat is then

$$C_{\text{mag}} = \frac{NS^2(S+1)^2}{6T^2} \sum_{\mathbf{q} \neq 0} J_{\mathbf{q}}^2,$$

in agreement with Part 1, (73.4).

Problem 2. Neglecting the interaction between spins, calculate the magnetization of a paramagnet with any relation between $\beta\mathfrak{H}$ and T.

Solution. The partition function (for one spin in the field) is

$$Z = \sum_{S_z=-S}^{S} \exp(-2\beta\mathfrak{H}S_z/T)$$

$$= \frac{\sinh 2\beta\mathfrak{H}(S+\frac{1}{2})/T}{\sinh \beta\mathfrak{H}/T}.$$

A calculation of the free energy, followed by differentiation with respect to $\mathfrak{H}$, gives the magnetization

$$M = \frac{N}{V}T\frac{\partial}{\partial \mathfrak{H}} \log Z$$

$$= \frac{2\beta N}{V}\left\{\left(S+\frac{1}{2}\right)\coth\frac{2\beta\mathfrak{H}(S+\frac{1}{2})}{T} - \frac{1}{2}\coth\frac{\beta\mathfrak{H}}{T}\right\}$$

(L. Brillouin 1927). When $\beta\mathfrak{H} \ll T$, this expression becomes (72.15). In the opposite limit, when $\beta\mathfrak{H} \gg T$, the magnetization tends to its nominal value according to

$$M = \frac{2\beta NS}{V}\left\{1 - \exp\left(-\frac{2\beta\mathfrak{H}}{T}\right)\right\}.$$

§ 73. Interaction of magnons

There is considerable methodological interest in the problem of the contribution from magnon interaction to the magnetic part of the thermodynamic quantities in a ferromagnet. The calculations in §71 were based on an ideal gas of non-interacting magnons. Let us consider this problem for a system described by the exchange spin Hamiltonian (72.1).

Having in view the determination of only the contribution of the lowest order in the small ratio T/T_c, we need only consider the pair interaction of magnons. This means that we have to consider two-magnon states of the system in which the total-spin projection is $NS-2$.

This corresponds to the wave functions

$$\left.\begin{aligned} \chi_{\mathbf{nn}} &= [4S(2S-1)]^{-1/2}\hat{S}_{\mathbf{n}-}\hat{S}_{\mathbf{n}-}\chi_0, \\ \chi_{\mathbf{mn}} &= (2S)^{-1}\hat{S}_{\mathbf{m}-}\hat{S}_{\mathbf{n}-}\chi_0, \qquad \mathbf{m} \neq \mathbf{n}; \end{aligned}\right\} \tag{73.1}$$

since the spin operators of different atoms commute, $\chi_{mn} = \chi_{nm}$.† The functions (73.1) are normalized by the condition $\chi^*_{mn}\chi_{mn} = 1$, as is easily seen by expanding the product in the same way as was done in verifying the normalization in (72.9). The same procedure can be used to show that different functions χ_{mn} are mutually orthogonal.

The functions (73.1) are not themselves eigenfunctions of the Hamiltonian. The wave functions of the two-magnon stationary states of the system must be certain linear combinations of the functions χ_{mn}, which we write as

$$\chi = \sum_{m \neq n} \frac{1}{\sqrt{2}} \psi_{mn}\chi_{mn} + \sum_{n} \psi_{nn}\chi_{nn} \tag{73.2}$$

(since χ_{mn} and χ_{nm} are the same, we must also take $\psi_{mn} \equiv \psi_{nm}$). The set of coefficients ψ_{mn} forms the wave function in the representation in which the independent variables are the numbers labelling the atoms in the lattice. The factor $1/\sqrt{2}$ in the first sum in (73.2) is included so that the squared modulus $|\chi|^2$ shall be equal to the sum $\Sigma|\psi_{mn}|^2$, in which each of the different ψ_{mn} appears once only.

The same method as was used to derive equation (72.11) for the wave functions of one-magnon stationary states shows that the functions (73.2) must satisfy a similar equation:

$$\begin{aligned}\mathscr{E}\chi = &\sum_{m \neq n} \frac{\psi_{mn}}{2^{3/2}S} [\hat{H}, \hat{S}_{m-}\hat{S}_{n-}]\,\chi_0 \\ &+\sum_{n} \frac{\psi_{nn}}{2[S(2S-1)]^{1/2}} [\hat{H}, \hat{S}_{n-}\hat{S}_{n-}]\,\chi_0,\end{aligned} \tag{73.3}$$

where now $\mathscr{E} = E - E_0$ is the energy of the two interacting magnons, and the brackets [ ] denote the commutator.

To expand the commutators on the right of (73.3), we note that

$$[\hat{H}, \hat{S}_{m-}\hat{S}_{n-}] \equiv [\hat{H}, \hat{S}_{m-}]\,\hat{S}_{n-} + \hat{S}_{m-}[\hat{H}, \hat{S}_{n-}],$$

and use the expressions (72.12) for the commutators $[\hat{H}, \hat{S}_{n-}]$. Then, taking account of the commutation rules (72.4), we transpose the operators $\hat{S}_z$ to the extreme right-hand position, where, acting on the function χ_0, they multiply it by S. The result is

$$\begin{aligned}[\hat{H}, \hat{S}_{m-}\hat{S}_{n-}]\,\chi_0 = S &\sum_{l} \{J_{ml}(\hat{S}_{m-} - \hat{S}_{l-})\,\hat{S}_{n-} + J_{nl}(\hat{S}_{n-} - \hat{S}_{l-})\,\hat{S}_{m-}\}\,\chi_0 \\ &+ \delta_{mn} \sum_{l} J_{nl}\hat{S}_{n-}\hat{S}_{l-}\chi_0 - J_{mn}\hat{S}_{m-}\hat{S}_{n-}\chi_0 + 4\beta\mathfrak{H}\hat{S}_{m-}\hat{S}_{n-}\chi_0;\end{aligned} \tag{73.4}$$

to simplify the formulae, the limitations imposed on the summation variables are not shown. The summations are over all values of **l**, it being implied that all the "diagonal" $J_{ll} = 0$.

† If the spin $S = \frac{1}{2}$, a twice repeated application of the same operator $\hat{S}_{n-}$ to the ground-state function χ_0 gives zero. In this case, therefore, all the "diagonal" functions $\chi_{nn} \equiv 0$.

The remaining steps are to substitute (73.4) in (73.3) and equate the coefficients of like functions $\chi_{\mathbf{mn}}$ on both sides of the equation. The calculations are laborious but elementary. They lead to the following set of equations for the $\psi_{\mathbf{mn}}$:

$$(2JS-\mathscr{E})\,\psi_{\mathbf{mn}} = S\sum_{\mathbf{l}}(J_{\mathbf{lm}}\psi_{\mathbf{ln}}+J_{\mathbf{ln}}\psi_{\mathbf{lm}})+J_{\mathbf{mn}}\psi_{\mathbf{mn}}$$
$$-A_S\Big[J_{\mathbf{mn}}(\psi_{\mathbf{mm}}+\psi_{\mathbf{nn}})+2\delta_{\mathbf{mn}}\sum_{\mathbf{l}}J_{\mathbf{lm}}\psi_{\mathbf{lm}}\Big], \qquad (73.5)$$

where

$$A_S = S\left[1-\left(\frac{2S-1}{2S}\right)^{1/2}\right]$$

and J denotes the sum $\sum_{\mathbf{l}} J_{\mathbf{nl}}$, which is obviously independent of the suffix $\mathbf{n}$.†

In this equation, let us change from the coordinate representation (in which the independent variables are the coordinates $\mathbf{r}_{\mathbf{n}}, \mathbf{r}_{\mathbf{m}}$ of the atoms) to the momentum representation, putting

$$\psi_{\mathbf{mn}} = \frac{1}{N}e^{i\mathbf{K}\cdot(\mathbf{r}_{\mathbf{m}}+\mathbf{r}_{\mathbf{n}})/2}\sum_{\mathbf{k}}\psi(\mathbf{K},\mathbf{k})\,e^{i\mathbf{k}\cdot(\mathbf{r}_{\mathbf{m}}-\mathbf{r}_{\mathbf{n}})}. \qquad (73.6)$$

The vector $\mathbf{K}$ plays the role of the total quasi-momentum of the two magnons, and $\mathbf{k}$ that of the quasi-momentum of their relative motion; the summation is over N discrete values of $\mathbf{k}$ admissible for a lattice with volume Nv (where N is the number of atoms in the lattice and v the volume of the unit cell). As well as the $\psi_{\mathbf{mn}}$, the exchange integrals also have to be represented in the form of Fourier series:

$$\left.\begin{aligned} J_{\mathbf{mn}} &= \frac{1}{N}\sum_{\mathbf{k}}e^{i\mathbf{k}\cdot(\mathbf{r}_{\mathbf{m}}-\mathbf{r}_{\mathbf{n}})}\,J(\mathbf{k}),\\ J(\mathbf{k}) &= \sum_{\mathbf{n}}J_{0\mathbf{n}}\,e^{i\mathbf{k}\cdot(\mathbf{r}_0-\mathbf{r}_{\mathbf{n}})}; \end{aligned}\right\} \qquad (73.7)$$

since $J_{\mathbf{mn}} = J_{\mathbf{nm}}$, we have $J(\mathbf{k}) = J(-\mathbf{k})$.

Omitting the straightforward intermediate steps, we shall pass immediately to the final result derived from (73.5):

$$[\varepsilon(\tfrac{1}{2}\mathbf{K}+\mathbf{k})+\varepsilon(\tfrac{1}{2}\mathbf{K}-\mathbf{k})-\mathscr{E}]\,\psi(\mathbf{K},\mathbf{k})$$
$$+\int U(\mathbf{K},\mathbf{k},\mathbf{k}')\,\psi(\mathbf{K},\mathbf{k}')\,V\,d^3k'/(2\pi)^3 = 0, \qquad (73.8)$$

where

$$NU(\mathbf{K},\mathbf{k},\mathbf{k}') = A_S[J(\tfrac{1}{2}\mathbf{K}+\mathbf{k})+J(\tfrac{1}{2}\mathbf{K}-\mathbf{k})+J(\tfrac{1}{2}\mathbf{K}+\mathbf{k}')+J(\tfrac{1}{2}\mathbf{K}-\mathbf{k}')]$$
$$-\tfrac{1}{2}[J(\mathbf{k}-\mathbf{k}')+J(\mathbf{k}+\mathbf{k}')], \qquad (73.9)$$

† These equations are valid also for spin $S = \frac{1}{2}$, when all the $\psi_{\mathbf{nn}}$ are arbitrary. It should be noted that for $S = \frac{1}{2}$ all the "diagonal" quantities $\psi_{\mathbf{nn}}$ disappear from the equations with $\mathbf{m} \neq \mathbf{n}$. The equations with $\mathbf{m} = \mathbf{n}$ in this case are simply to be regarded as non-existent.

and $\varepsilon(\mathbf{k})$ is the energy of one magnon, determined by (72.13); the summation over $\mathbf{k}'$ is replaced by integration over one reciprocal lattice cell.

Thus the exact problem (for the Hamiltonian (72.1)) of the two-magnon states of the system reduces to the solution of an equation exactly similar to Schrödinger's equation for a system of two particles in the momentum representation; cf. *QM*, (130.4). The functions $\varepsilon(\mathbf{k})$ play the part of the kinetic energies of the particles, and the kernel $U(\mathbf{K}, \mathbf{k}, \mathbf{k}')$ of the integral equation that of the matrix element of the energy U of their interaction for a transition (scattering) from states with momenta $\mathbf{k}_1$, $\mathbf{k}_2$ to states with momenta $\mathbf{k}_1'$, $\mathbf{k}_2'$, where

$$\mathbf{k}_1 = \tfrac{1}{2}\mathbf{K}+\mathbf{k}, \quad \mathbf{k}_2 = \tfrac{1}{2}\mathbf{K}-\mathbf{k}, \quad \mathbf{k}_1' = \tfrac{1}{2}\mathbf{K}+\mathbf{k}', \quad \mathbf{k}_2' = \tfrac{1}{2}\mathbf{K}-\mathbf{k}'.$$

Then $U(\mathbf{K}, \mathbf{k}, \mathbf{k}')$ is suitably written in the form

$$\begin{aligned} NU(\mathbf{k}_1', \mathbf{k}_2'; \mathbf{k}_1, \mathbf{k}_2) = {} & A_S[J(\mathbf{k}_1)+J(\mathbf{k}_2)+J(\mathbf{k}_1')+J(\mathbf{k}_2')] \\ & -\tfrac{1}{2}[J(\mathbf{k}_1-\mathbf{k}_1')+J(\mathbf{k}_1-\mathbf{k}_2')]. \end{aligned} \tag{73.10}$$

In the general case, equation (73.8) with (73.9) is very complicated. We shall calculate only the correction to the thermodynamic quantities when $S \gg 1$. The simplicity of this case is due to the fact that the magnon energy $\varepsilon(\mathbf{k})$ is proportional to S, and the interaction U of the magnons is independent of S; for $S \gg 1$, the coefficient in (73.9) is $A_S \approx \frac{1}{4}$. Hence U may be regarded as a small perturbation, and the correction Ω_{int} (from the magnon interaction) to the thermodynamic potential Ω is given by just the mean value of U. By taking the "diagonal matrix element"

$$U(\mathbf{k}_1, \mathbf{k}_2; \mathbf{k}_1, \mathbf{k}_2) = \frac{1}{2N}[J(\mathbf{k}_1)+J(\mathbf{k}_2)-J(\mathbf{k}_1-\mathbf{k}_2)-J(0)], \tag{73.11}$$

we average over a state with given magnon quasi-momenta. The statistical averaging over the equilibrium distribution of magnons is then carried out by integration:

$$\Omega_{\text{int}} = \int n(\mathbf{k}_1)\, n(\mathbf{k}_2)\, U(\mathbf{k}_1, \mathbf{k}_2; \mathbf{k}_1, \mathbf{k}_2) \frac{V^2\, d^3k_1\, d^3k_2}{(2\pi)^6}, \tag{73.12}$$

where $n(\mathbf{k}) = [\exp(\varepsilon(\mathbf{k})/T)-1]^{-1}$ is the Bose distribution function.

At low temperatures, the integral is governed by the range of small $\mathbf{k}_1$ and $\mathbf{k}_2$, and accordingly all the $\varepsilon(\mathbf{k})$ and $J(\mathbf{k})$ are to be expanded in powers of $\mathbf{k}$. Then $\varepsilon(\mathbf{k})$ is given by the quadratic expression (72.14). Since $J(\mathbf{k})$ is an even function of $\mathbf{k}$, the first terms in its expansion are also quadratic:

$$J(\mathbf{k}) \approx J(0)+a_{ik}k_ik_k.$$

Then

$$U(\mathbf{k}_1, \mathbf{k}_2; \mathbf{k}_1, \mathbf{k}_2) = \frac{1}{N} a_{ik}k_{1i}k_{2k}.$$

When this expression, an odd function of $\mathbf{k}_1$ and $\mathbf{k}_2$, is substituted in (73.12), the integral is zero as a result of the averaging over the directions of $\mathbf{k}_1$ and $\mathbf{k}_2$.

In the expansion of $J(\mathbf{k})$, therefore, we have to take into account the fourth-order terms, and then in the integral (73.12) the function $U(\mathbf{k}_1, \mathbf{k}_2; \mathbf{k}_1, \mathbf{k}_2)$ is a quartic form, in which the terms quadratic in $\mathbf{k}_1$ and in $\mathbf{k}_2$ make a non-zero contribution to the integral. Because of the rapid convergence, the integration may be extended over all $\mathbf{k}$-space. We then see, by making the change of variables $\mathbf{k} = \bar{\mathbf{k}}\sqrt{T}$, that the dependence of Ω_{int} on T and $\mathfrak{H}$ has the form

$$\Omega_{\text{int}} = VT^5 f(\mathfrak{H}/T), \tag{73.13}$$

$f(0)$ and $f'(0)$ being finite. Hence it follows that the correction term in the magnetization is

$$M_{\text{int}} = -\frac{1}{V}\left[\frac{\partial\Omega_{\text{int}}}{\partial\mathfrak{H}}\right]_{\mathfrak{H}=0} = \text{constant}\times T^4. \tag{73.14}$$

The correction term in the specific heat has the same dependence.†

We see that the interaction of magnons leads to corrections to the thermodynamic quantities only in a higher approximation with respect to T/T_c. The leading terms in the magnetization and in the magnetic part of the specific heat are proportional to $T^{3/2}$. Between these and the corrections arising from Ω_{int} there are others proportional to $T^{5/2}$ and $T^{7/2}$, which come from the next terms in the expansion of the magnon energy $\varepsilon(\mathbf{k})$ in powers of $\mathbf{k}^2$.

By means of the equations derived above, we can also consider the question of bound states of two magnons. These states appear as discrete (for a given $\mathbf{K}$) eigenvalues of equation (73.8). As functions of the variable $\mathbf{K}$, these eigenvalues $\mathscr{E}(\mathbf{K})$ form new branches of elementary excitations in the system. Analysis shows, however, that such states exist only for fairly large values of $\mathbf{K}$, and they can therefore never affect the thermodynamic quantities in a ferromagnet at low temperatures.‡

PROBLEM

Assuming that $S \gg 1$, find the correction terms due to the magnon interaction in the magnetization and the specific heat for a cubic lattice in which the exchange integrals are zero except for pairs of atoms that are adjacent (along cubic axes).

SOLUTION. Each atom has six nearest neighbours. From the definition (73.7) we find

$$J(\mathbf{k}) = 2J_0(\cos k_x a + \cos k_y a + \cos k_z a),$$

where J_0 is the exchange integral for a pair of adjacent atoms, and a the edge length of a cubic cell. For small $\mathbf{k}$,

$$J(\mathbf{k}) \approx J_0[2 - a^2k^2 + \tfrac{1}{12}a^4(k_x^4 + k_y^4 + k_z^4)].$$

† These results (in the general case of arbitrary spin) were first obtained by F. J. Dyson (1956). In the derivation of (73.5) given here, we have mainly followed R. J. Boyd and J. Callaway (1965).

‡ See M. Wortis, *Physical Review* **132**, 85, 1963. The lattice concerned is three-dimensional. In one and two dimensions, bound states of magnons exist for all values of $\mathbf{k}$.

Hence

$$U(\mathbf{k}_1, \mathbf{k}_2; \mathbf{k}_1, \mathbf{k}_2) = -\frac{a^7 J_0}{4V}(k_{1x}^2 k_{2x}^2 + k_{1y}^2 k_{2y}^2 + k_{1z}^2 k_{2z}^2);$$

the terms that are odd functions of $\mathbf{k}_1$ and $\mathbf{k}_2$ are omitted. The magnon energy is, according to (72.14),

$$\varepsilon(\mathbf{k}) = SJ_0 a^2 k^2 + 2\beta\mathfrak{H}.$$

The calculation of the integral (73.12) gives the results

$$\frac{M_{\text{int}}}{M} = -\frac{3\pi\zeta(3/2)\,\zeta(5/2)}{2S^2}\left(\frac{T}{4\pi SJ_0}\right)^4,$$

$$C_{\text{int}} = \frac{15\pi\zeta^2(5/2)\,N}{S}\left(\frac{T}{4\pi SJ_0}\right)^4,$$

where ζ denotes the zeta function.

§ 74. Magnons in an antiferromagnet

Antiferromagnets have the characteristic that the magnetic moments of all the electrons in each unit cell in the crystal lattice compensate one another (in the equilibrium state in the absence of a magnetic field). The magnetic moment density is, strictly speaking, distributed throughout the volume of the cell. However, in antiferromagnetic insulator crystals, we can suppose with good accuracy that this density is practically localized at the individual atoms, to each of which a certain magnetic moment may be assigned. These moments, periodically repeated in all cells, form the *magnetic sub-lattices* of the antiferromagnet.

Different antiferromagnets vary greatly in structure. We shall discuss the question of their magnetic energy spectrum for the typical example of a crystal with two magnetic atoms at equivalent points in each unit cell (i.e. at points that are changed into each other by some symmetry transformations of the crystal). The mean magnetic moment densities of the sub-lattices formed by these atoms are denoted by $\mathbf{M}_1$ and $\mathbf{M}_2$, and we shall use the two vectors

$$\mathbf{M} = \mathbf{M}_1 + \mathbf{M}_2, \quad \mathbf{L} = \mathbf{M}_1 - \mathbf{M}_2. \tag{74.1}$$

In the ground state of the antiferromagnet $\mathbf{M} = 0$ and $\mathbf{L} \neq 0$, whereas for a ferromagnet $\mathbf{M} \neq 0$ and $\mathbf{L} = 0$. We must emphasize a fundamental difference between the ground states in the two cases. In the exchange approximation, in the ground state of the ferromagnet, the spin components of all the magnetic atoms have definite (the largest possible) values $S_z = S$, corresponding to the nominal value of the magnetization $\mathbf{M}$. In the ground state of the antiferromagnet, the magnetizations of the sub-lattices certainly cannot have their nominal values, since the total projections of the spins of each sub-lattice separately are not conserved (even in the exchange approximation), and therefore do not have definite values (in a stationary state). The spin components of the individual atoms therefore also do not have definite values.

The form of the macroscopic "equations of motion" of the vectors **L** and **M** is established similarly to that for a ferromagnet (§69). The condition for absence of dissipation leads to the requirement that, because of the equations of motion,

$$-\frac{d\tilde{F}}{dt} = \int \left\{ \mathbf{H}_L \cdot \frac{\partial \mathbf{L}}{\partial t} + \mathbf{H}_M \cdot \frac{\partial \mathbf{M}}{\partial t} \right\} dV = 0, \tag{74.2}$$

where the "effective fields" $\mathbf{H}_L$ and $\mathbf{H}_M$ are determined by the expression

$$\delta\tilde{F} = -\int (\mathbf{H}_L \cdot \delta\mathbf{L} + \mathbf{H}_M \cdot \delta\mathbf{M})\, dV \tag{74.3}$$

for the change in the free energy when **L** and **M** are varied; in equilibrium $\mathbf{H}_L = \mathbf{H}_M = 0$.

In the exchange approximation, the required equations must be invariant under simultaneous rotation of all the magnetic moments relative to the crystal lattice. This implies not only the crystallographic equivalence of the positions of the two magnetic atoms in the cell but also the invariance under exchange of $\mathbf{M}_1$ and $\mathbf{M}_2$, i.e. under the transformation $\mathbf{L} \to -\mathbf{L}$, $\mathbf{M} \to \mathbf{M}$. Because the free energy is unaffected by this transformation, we have also $\mathbf{H}_L \to -\mathbf{H}_L$, $\mathbf{H}_M \to \mathbf{H}_M$.

Considering small oscillations of the magnetic moments, we put $\mathbf{L} = \mathbf{L}_0 + \mathbf{l}$ $\mathbf{M} \equiv \mathbf{m}$, where **l** and **m** are small quantities. In the linear approximation, the equations of motion satisfying the conditions stated are

$$\frac{\partial \mathbf{l}}{\partial t} = \gamma \mathbf{H}_M \times \boldsymbol{\nu}, \qquad \frac{\partial \mathbf{m}}{\partial t} = \gamma \mathbf{H}_L \times \boldsymbol{\nu}, \tag{74.4}$$

where $\boldsymbol{\nu}$ is a unit vector in the equilibrium direction of the vector $\mathbf{L}_0$; the transformation $\mathbf{L} \to -\mathbf{L}$ implies that $\boldsymbol{\nu} \to -\boldsymbol{\nu}$. Here we have used the fact that $\mathbf{H}_L$ and $\mathbf{H}_M$, which are zero in equilibrium, are themselves linear in **l** and **m**, and that $\boldsymbol{\nu}$ is the only disposable constant vector. By analogy with §69 the coefficient γ could be written as $\gamma = (g|e|/2mc)L_0$, but unlike the ferromagnetic case we now have $g \neq 2$ even if relativistic effects are neglected. For monochromatic oscillations $\partial\mathbf{l}/\partial t = -i\omega\mathbf{l}$, etc., and the vectors **l** and **m** defined by (74.4) are then perpendicular to $\boldsymbol{\nu}$. In the approximation considered, this means that the vector **L** precesses about the direction of $\boldsymbol{\nu}$ with a constant magnitude $L \approx L_0$.

To determine the effective fields $\mathbf{H}_L$ and $\mathbf{H}_M$, we must establish the form of the free energy of the crystal. The terms needed are those of the second order in the small quantities **l** and **m**, or in the case of terms containing derivatives of these quantities with respect to the coordinates, those not above the second order in the wave number of the oscillations, whose wavelength is assumed large compared with the lattice constant (as in §70). In the exchange approximation, the free energy must be invariant under simultaneous rotation of all the magnetic moments, and also under a change in the sign of $\mathbf{L}$. An expression satisfying all the conditions stated is

$$F_{\text{exch}} = \int \left\{ \frac{1}{2} a\mathbf{m}^2 + \frac{1}{2} b \left(\mathbf{m} \cdot \frac{\partial \mathbf{l}}{\partial z} - \mathbf{l} \cdot \frac{\partial \mathbf{m}}{\partial z} \right) + \frac{1}{2} \alpha_{ik} \frac{\partial \mathbf{l}}{\partial x_i} \cdot \frac{\partial \mathbf{l}}{\partial x_k} \right\} dV, \quad (74.5)$$

where the z-axis can be taken parallel to $\boldsymbol{\nu}$, so that a change in the sign of $\boldsymbol{\nu}$ implies a change in the sign of z. The coefficient $a > 0$, in accordance with the fact that $\mathbf{m}$ must be zero in equilibrium. The term in $\mathbf{l}^2$ is absent here, since it would imply a dependence of the energy on the direction of the vector $\mathbf{L} = \mathbf{L}_0 + \mathbf{l}$ in the crystal which does not occur in the exchange approximation. The term containing the sum $\mathbf{m} . \partial \mathbf{l}/\partial z + \mathbf{l} . \partial \mathbf{m}/\partial z$ reduces to a total derivative and would vanish on integration over the volume. Lastly, the terms quadratic in the derivatives $\partial \mathbf{m}/\partial x_i$ need not be considered, since they are certainly small in comparison with the term in $\mathbf{m}^2$.

Varying the integral (74.5) (and integrating by parts), we obtain

$$\mathbf{H}_L = b \frac{\partial \mathbf{m}}{\partial z} + \alpha_{ik} \frac{\partial^2 \mathbf{l}}{\partial x_i \partial x_k}, \quad \mathbf{H}_M = -a\mathbf{m} - b \frac{\partial \mathbf{l}}{\partial z}. \quad (74.6)$$

For a plane monochromatic spin wave, the equations of motion (74.4) now give

$$\left. \begin{aligned} -i\omega \mathbf{l} &= -\gamma a \mathbf{m} \times \boldsymbol{\nu} - ik_z \gamma b \mathbf{l} \times \boldsymbol{\nu}, \\ -i\omega \mathbf{m} &= ik_z \gamma b \mathbf{m} \times \boldsymbol{\nu} - \gamma \alpha(\mathbf{n}) k^2 \mathbf{l} \times \boldsymbol{\nu}, \end{aligned} \right\} \quad (74.7)$$

where again (as in §70) $\alpha(\mathbf{n}) = \alpha_{ik} n_i n_k$, and $\mathbf{n}$ is a unit vector in the direction of $\mathbf{k}$. Multiplying the first equation vectorially by $\boldsymbol{\nu}$ gives

$$\gamma a \mathbf{m} = -i\omega \mathbf{l} \times \boldsymbol{\nu} - ik_z \gamma b \mathbf{l}, \quad (74.8)$$

and substitution of this in the second equation leads directly to the following dispersion relation for spin waves:

$$\omega = \gamma k [a\alpha(\mathbf{n}) - b^2 (\boldsymbol{\nu} . \mathbf{n})^2]^{1/2}. \quad (74.9)$$

Thus the frequency of the spin waves, and thus also the magnon energy $\varepsilon = \hbar\omega$, in an antiferromagnet in the exchange approximation are proportional to k, not to k^2 as in a ferromagnet.†

Equations (74.7) establish a one-to-one relation between $\mathbf{l}$ and $\mathbf{m}$, but the two components of $\mathbf{l}$ in the plane perpendicular to $\boldsymbol{\nu}$ remain arbitrary. This means that the spin waves in the antiferromagnet considered have two independent directions of polarization.

† This dispersion relation for antiferromagnets was first derived by L. Hulthén (1936). The derivation using the macroscopic treatment of the magnetization of the sub-lattices was given by M. I. Kaganov and V. M. Tsukernik (1958).

To take account of magnetic anisotropy, we must make more specific assumptions as to the symmetry of the crystal. Let this be uniaxial, and let the equilibrium direction of **L** be parallel to the axis of symmetry.†

It is seen from (74.8) that the vector **m** in the spin wave is small compared with **l**, containing an extra power of the small wave vector **k**. In the same way, the effective field $\mathbf{H}_M \gg \mathbf{H}_L$. For this reason, it is sufficient to consider the anisotropy due to the vector **l**. With the assumptions made, the density of this energy is

$$U_{\text{an}} = \tfrac{1}{2} K l^2, \tag{74.10}$$

where $K > 0$. When it is taken into account, there is a further term $-K\mathbf{l}$ in the effective field $\mathbf{H}_L$, which for a plane wave is then

$$\mathbf{H}_L = ik_z b\mathbf{m} - [\alpha(\mathbf{n})k^2 + K]\mathbf{l}. \tag{74.11}$$

Hence we find that, with allowance for the anisotropy, the spin wave dispersion relation is obtained from (74.9) by adding K to αk^2. The result is that, when $k \to 0$, the magnon energy tends not to zero, but to the finite value

$$\varepsilon(0) = \hbar\gamma\sqrt{(aK)} \tag{74.12}$$

(C. Kittel 1951). The frequency $\omega(0) = \varepsilon(0)/\hbar$ is called the *antiferromagnetic resonance frequency*. It should be noted that the gap in the spectrum is proportional to the square root of the anisotropy constant, not to this constant itself as in (70.12). Since the smallness of the relativistic effects is expressed by the relative smallness of the anisotropy constant, we see that these effects are in general more important in an antiferromagnet than in a ferromagnet.

The magnon contribution to the internal energy of an antiferromagnet is calculated from (71.3). In the temperature range $\varepsilon(0) \ll T \ll T_N$ (T_N being the Néel point, the temperature at which the antiferromagnetism disappears), we can use the spectrum (74.9). In a uniaxial crystal,

$$\omega = \gamma a^{1/2}[\alpha_1(k_x^2+k_y^2)+\alpha_2' k_z^2]^{1/2}, \quad \alpha_2' = \alpha_2 - b^2/a.$$

The calculation of the integral (71.3) gives the following result for the magnon contribution to the specific heat:

$$C_{\text{mag}} = V\frac{4\pi^2 T^3}{15\gamma^3 a^{3/2}(\alpha_2'\alpha_1^2)^{1/2}\hbar^3}. \tag{74.13}$$

At temperatures $T \ll \varepsilon(0)$, on the other hand, the magnon contribution to the thermodynamic quantities is exponentially small.

† This type includes the antiferromagnet $FeCO_3$, with a rhombohedral lattice (crystal class D_{3d}) and two iron ions in the unit cell. The magnetic moments of these ions are in opposite directions along the third-order axis of symmetry (the z-azis).

CHAPTER VIII

ELECTROMAGNETIC FLUCTUATIONS

§ 75. **Green's function of a photon in a medium**

TURNING now to study the statistical properties of an electromagnetic field in material media, let us first recall the significance of the averagings of electromagnetic quantities in macroscopic electrodynamics.

If we start, for ease of visualization, from the classical standpoint, we can distinguish the averaging over a physically infinitesimal volume with a given arrangement of all the particles in it, followed by averaging the result with respect to the motion of the particles. Maxwell's equations of macroscopic electrodynamics involve fully averaged quantities. In considering field fluctuations, however, we are concerned with the oscillations in time of quantities averaged only over physically infinitesimal volumes.

From the quantum-mechanical standpoint, we can of course speak of averaging over a volume only for the operator of a physical quantity, not for the quantity itself; the second step is to find the mean value of this operator by the use of quantum-mechanical probabilities. The operators that occur in this chapter will be understood to be averaged only in the first sense.

The statistical properties of electromagnetic radiation in a material medium are described by the Green's function of a photon in the medium. For photons, the ψ operators are replaced by the operators of the electromagnetic field potentials. The photon Green's functions are defined in terms of these operators in the same way as they are for particles in terms of the ψ operators.

The field potentials form a 4-vector $A^{\mu} = (A^0, \mathbf{A})$, where $A^0 \equiv \phi$ is the scalar potential and $\mathbf{A}$ the vector potential. The choice of these potentials is not unique in classical electrodynamics: they allow a gauge transformation, which has no effect on any observable quantities (see *Fields*, §18). Correspondingly, in quantum electrodynamics there is a similar non-uniqueness in the choice of the field operators, and therefore in the definition of the Green's functions of the photon. We shall use the gauge in which the scalar potential is zero:

$$A^0 \equiv \phi = 0, \tag{75.1}$$

so that the field is determined by the vector potential only. This gauge is usually convenient for problems involving the interaction of the electromagnetic field with non-relativistic particles, as in the case of a field in ordinary material media.

In this gauge, the Green's function is a three-dimensional tensor of rank two,

$$D_{ik}(X_1, X_2) = i\langle \mathrm{T}\hat{A}_i(X_1)\hat{A}_k(X_2)\rangle, \tag{75.2}$$

where $i, k = x, y, z$ are three-dimensional vector suffixes and the angle brackets denote (as in (36.1)) averaging over the Gibbs distribution for a system consisting of the medium and the radiation present in it in equilibrium; since the photons are bosons, there is no change in the sign of the product when the operators $\hat{A}_i$, $\hat{A}_k$ are interchanged by the chronological operator. Moreover, the operators $\hat{A}_i$ are self-conjugate (since the photon is a strictly neutral particle); no distinction is therefore made between $\hat{A}_i$ and $\hat{A}_i^+$ in (75.2).†

The basic concept in the construction of all photon Green's functions must, however, be not (75.2) but the retarded Green's function, defined by

$$iD^R_{ik}(X_1, X_2) = \begin{cases} \langle \hat{A}_i(X_1)\hat{A}_k(X_2) - \hat{A}_k(X_2)\hat{A}_i(X_1)\rangle, & t_1 > t_2, \\ 0 & t_1 < t_2; \end{cases} \tag{75.3}$$

the minus sign between the two terms in the angle brackets corresponds to the definition (36.19) for Bose statistics.

For a closed system, the Green's function depends on the times t_1 and t_2 only through their difference $t = t_1 - t_2$. The coordinates $\mathbf{r}_1$ and $\mathbf{r}_2$, in the general case of an inhomogeneous medium, appear in the Green's function independently: $D^R_{ik}(t;\, \mathbf{r}_1, \mathbf{r}_2)$. Accordingly, the Fourier expansion will be made only with respect to time, the component being

$$D^R_{ik}(\omega; \mathbf{r}_1, \mathbf{r}_2) = \int_0^\infty e^{i\omega t} D^R_{ik}(t; \mathbf{r}_1, \mathbf{r}_2)\, dt. \tag{75.4}$$

By considering quantities averaged over physically infinitesimal volumes, we restrict ourselves to considering only the long-wavelength part of the radiation, in which the photon wave numbers satisfy the condition

$$ka \ll 1, \tag{75.5}$$

where a denotes the interatomic distances in the medium. In this frequency range, the photon Green's function can be expressed in terms of other macroscopic characteristics of the medium, the permittivity $\varepsilon(\omega)$ and the permeability $\mu(\omega)$.

To do so, we write the operator of the interaction of the electromagnetic field with the medium,

$$\hat{V} = -\frac{1}{c}\int \hat{\mathbf{j}}\cdot\hat{\mathbf{A}}\, d^3x, \tag{75.6}$$

† In the general case of an arbitrary gauge of the potentials, the photon Green's function is a 4-tensor $D_{\mu\nu}$; in the gauge (75.1), $D_{00} = 0$, $D_{0i} = 0$. The general tensor and gauge properties of the photon Green's function in statistical physics are the same as in quantum electrodynamics of a field in the vacuum. The definition (75.2) differs in sign from that used in *RQT*. It is chosen here so as to correspond to the definition of the Green's functions of other bosons, including phonons.

where $\hat{\mathbf{j}}$ is the electric current density operator due to the particles of the medium.† If a classical "external" current $\mathbf{j}(t, \mathbf{r})$ is put into the medium, there is a corresponding interaction operator

$$\hat{V} = -\frac{1}{c}\int \mathbf{j}(t, \mathbf{r}).\hat{\mathbf{A}}\, d^3x. \tag{75.7}$$

This expression enables us to establish a connection with the general theory of the response of a macroscopic system to an external interaction.

In this theory (see Part 1, §125) there appears a discrete set of quantities x_a ($a = 1, 2, \ldots$) which describe the behaviour of the system under certain external interactions. These interactions are described by "perturbing forces" $f_a(t)$ such that the interaction energy operator has the form

$$\hat{V} = -\sum_a f_a \hat{x}_a,$$

where $\hat{x}_a$ are the operators of the quantities x_a. The mean values $\overline{x_a}(t)$ brought about by the perturbation are linear functionals of the forces $f_a(t)$. For the Fourier components of the quantities, this relation is written

$$\overline{x_{a\omega}} = \sum_b \alpha_{ab}(\omega) f_{b\omega}$$

(we assume that $\overline{x_a} = 0$ in the absence of the perturbation). The coefficients α_{ab} in these relations are called the *generalized susceptibilities* of the system. If the two quantities x_a and x_b behave in the same way under time reversal, and the body is not *magnetoactive* (has no magnetic structure and is not in a magnetic field), the α_{ab} are symmetrical with respect to their suffixes.

Here we are concerned with quantities f_a and x_a that are functions of the coordinates $\mathbf{r}$ of a point in the body and thus have distributions. In this case, the expression for $\hat{V}$ is to be written

$$\hat{V} = -\sum_a \int f_a(t, \mathbf{r})\, \hat{x}_a(t, \mathbf{r})\, d^3x, \tag{75.8}$$

and the relation between the mean values $\overline{x_a}$ and the forces f_a is

$$\overline{x_{a\omega}}(\mathbf{r}) = \sum_b \int \alpha_{ab}(\omega; \mathbf{r}, \mathbf{r}') f_{b\omega}(\mathbf{r}')\, d^3x'. \tag{75.9}$$

The generalized susceptibilities now become functions of the coordinates of two points in the body, and their symmetry is expressed by

$$\alpha_{ab}(\omega; \mathbf{r}, \mathbf{r}') = \alpha_{ba}(\omega; \mathbf{r}', \mathbf{r}). \tag{75.10}$$

† See *RQT*, §53 (where the current is denoted by $e\mathbf{j}$, i.e. the unit charge e is excluded from the definition of $\mathbf{j}$). The operator (75.6) assumes the use of the relativistic expression for the current operator. In non-relativistic problems, we can neglect in the ψ operators (from which the current operator $\hat{\mathbf{j}}$ is constructed) the parts due to negative frequencies, i.e. to antiparticles. This means, in particular, neglecting the radiative corrections that alter the photon Green's function in the vacuum because of the virtual electron–positron pairs. These corrections are negligible at wavelengths $\lambda \gg \hbar/mc$, a condition that is certainly satisfied in the range (75.5).

According to Kubo's formula (see Part 1, (126.9)), the susceptibilities can be expressed in terms of the mean values of the commutators of the Heisenberg operators $\hat{x}_a(t, \mathbf{r})$:

$$\alpha_{ab}(\omega; \mathbf{r}, \mathbf{r}') = \frac{i}{\hbar} \int_0^\infty e^{i\omega t} \langle \hat{x}_a(t, \mathbf{r}) \hat{x}_b(0, \mathbf{r}') - \hat{x}_b(0, \mathbf{r}') \hat{x}_a(t, \mathbf{r}) \rangle \, dt. \tag{75.11}$$

We shall now regard the components of the current vector $\mathbf{j}$ as the "forces" f_a. Then it is seen by comparing (75.7) and (75.8) that the corresponding quantities x_a are the components of the vector potential $\mathbf{A}/c$ of the field. A comparison of (75.11) with the definitions (75.3), (75.4) shows that the generalized susceptibilities $\alpha_{ab}(\omega; \mathbf{r}, \mathbf{r}')$ are the same as the components of the tensor $-D^R_{ik}(\omega; \mathbf{r}, \mathbf{r}')/\hbar c^2$. By (75.10), this immediately gives (for non-magnetoactive media)

$$D^R_{ik}(\omega; \mathbf{r}, \mathbf{r}') = D^R_{ki}(\omega; \mathbf{r}', \mathbf{r}). \tag{75.12}$$

The relations (75.9) become

$$\bar{A}_{i\omega}(\mathbf{r}) = -\frac{1}{\hbar c} \int D^R_{ik}(\omega; \mathbf{r}, \mathbf{r}') j_{k\omega}(\mathbf{r}') \, d^3x'. \tag{75.13}$$

The mean value $\bar{\mathbf{A}}$ is just the vector potential of the macroscopic (fully averaged; see the beginning of this section) electromagnetic field in the medium; henceforward the bar will be omitted from $\bar{\mathbf{A}}$ and the other macroscopic quantities. We now use the fact that the macroscopic field due to the classical current $\mathbf{j}$ satisfies Maxwell's equation

$$\text{curl}\, \mathbf{H}_\omega = \frac{4\pi}{c} \mathbf{j}_\omega - \frac{i\omega}{c} \mathbf{D}_\omega,$$

where $\mathbf{D}$ is the electric induction; in the general case of an anisotropic medium, $\mathbf{D}_\omega$ is related to the field $\mathbf{E}_\omega$ by $D_{i\omega} = \varepsilon_{ik}(\omega) E_{k\omega}$; if the medium is inhomogeneous, the permittivity tensor is also a function of the coordinates, $\varepsilon_{ik}(\omega, \mathbf{r})$.

In the gauge chosen for the potentials (75.1) we have

$$\mathbf{B}_\omega = \text{curl}\, \mathbf{A}_\omega, \quad \mathbf{E}_\omega = i\omega \mathbf{A}_\omega / c, \tag{75.14}$$

where $\mathbf{B}$ is the magnetic induction, related to the field $\mathbf{H}$ by $B_{i\omega} = \mu_{ik} H_{k\omega}$.† The potential therefore satisfies the equation‡

$$[\text{curl}_{im} (\mu^{-1}{}_{mn} \, \text{curl}_{nk}) - (\omega^2/c^2) \varepsilon_{ik}] A_{k\omega} = 4\pi j_{i\omega}/c.$$

† In macroscopic electrodynamics the mean value of the microscopic electric field is usually denoted by $\mathbf{E}$, and the mean value of the magnetic field by $\mathbf{B}$, called the magnetic induction.

‡ Here and henceforward we use the notation $\text{curl}_{il} = e_{ikl} \partial/\partial x_k$, where e_{ikl} is the unit antisymmetrical pseudo-tensor; $(\text{curl}\, \mathbf{A})_i = \text{curl}_{il} A_l$.

Substituting $\mathbf{A}_\omega$ in the form (75.13), we find that the function D^R_{ik} must satisfy the equation

$$[\text{curl}_{im}(\mu^{-1}{}_{mn}\,\text{curl}_{nl})-(\omega^2/c^2)\,\varepsilon_{il}]\,D^R_{lk}(\omega;\mathbf{r},\mathbf{r}')=-4\pi\hbar\delta_{ik}\delta(\mathbf{r}-\mathbf{r}'). \quad (75.15)$$

This equation is considerably simpler for media that are isotropic (in each volume element), where the tensors ε_{ik} and μ_{ik} reduce to scalars. The permeability is usually close to unity, and we shall take $\mu = 1$ in the rest of this section. Putting $\varepsilon_{ik} = \varepsilon\delta_{ik}$ and $\mu_{ik} = \delta_{ik}$, we get the equation

$$\left[\frac{\partial^2}{\partial x_i \partial x_l}-\delta_{il}\triangle-\delta_{il}\frac{\omega^2}{c^2}\,\varepsilon(\omega;\mathbf{r})\right]D^R_{lk}(\omega;\mathbf{r},\mathbf{r}')=-4\pi\hbar\delta_{ik}\delta(\mathbf{r}-\mathbf{r}'). \quad (75.16)$$

Thus the calculation of the retarded Green's function for an inhomogenous medium amounts to the solution of a certain differential equation (I. E. Dzyaloshinskiĭ and L. P. Pitaevskiĭ 1959).†

At the interfaces between different media, the components of the tensor D^R_{lk} must satisfy certain conditions. In equation (75.16), the second variable $\mathbf{r}'$ and the second suffix k are not involved in the differential and algebraic operations on the tensor D^R_{lk}, i.e. they act only as parameters. Hence the boundary conditions are to be imposed only with respect to the coordinates $\mathbf{r}$ for the function $D^R_{lk}(\omega;\mathbf{r},\mathbf{r}')$, regarded as a vector with the suffix l. These conditions correspond to the requirement that the tangential components of $\mathbf{E}$ and $\mathbf{H}$ be continuous, as given by macroscopic electrodynamics.‡ Since $\mathbf{E} = -\dot{\mathbf{A}}/c$, the role of the vector $\mathbf{E}$ is here played by the derivative $-(1/c)\,\partial D^R_{lk}(t;\mathbf{r},\mathbf{r}')/\partial t$ or, in Fourier components,

$$i(\omega/c)D^R_{lk}(\omega;\mathbf{r},\mathbf{r}'). \quad (75.17)$$

Similarly, the role of the vector $\mathbf{H}$ (which is the same as $\mathbf{B}$ when $\mu=1$) is played by

$$\text{curl}_{li}\,D_{ik}(\omega;\mathbf{r},\mathbf{r}'). \quad (75.18)$$

For a spatially homogenous infinite medium, the function D^R_{ik} depends only on the difference $\mathbf{r}-\mathbf{r}'$. For the components of the Fourier expansion with respect to this difference, the differential equation (75.16) becomes a set of algebraic equations

$$\frac{1}{4\pi\hbar}\left[k_i k_l-\delta_{il}k^2+\delta_{il}\frac{\omega^2}{c^2}\,\varepsilon(\omega)\right]D^R_{lk}(\omega,\mathbf{k}) = \delta_{ik}. \quad (75.19)$$

† The function D^R_{ik} is the Green's function of Maxwell's equations in the sense familiar in mathematical physics, namely the solution of the field equations with a point source, satisfying the retardation condition. The advanced function D^A_{ik} satisfies the same equation with ε^* instead of ε.

‡ The boundary conditions for the normal components of $\mathbf{B}$ and $\mathbf{D}$ give nothing new in this case, because in a field that varies with time as $e^{-i\omega t}$ the equations div $\mathbf{D} = 0$ and div $\mathbf{B} = 0$ are implied by the equations curl $\mathbf{E} = i\omega\mathbf{B}/c$ and curl $\mathbf{H} = -i\omega\mathbf{D}/c$.

Their solution is

$$D^R_{ik}(\omega, \mathbf{k}) = \frac{4\pi\hbar}{\omega^2\varepsilon(\omega)/c^2 - k^2}\left[\delta_{ik} - \frac{c^2k_ik_k}{\omega^2\varepsilon(\omega)}\right], \tag{75.20}$$

According to (36.21), the Green's function D_{ik} for a homogeneous medium is expressed in terms of the retarded function D^R_{ik} by

$$D_{ik}(\omega, \mathbf{k}) = \text{re } D^R_{ik}(\omega, \mathbf{k}) + i \coth(\hbar\omega/2T) \text{ im } D^R_{ik}(\omega, \mathbf{k}). \tag{75.21}$$

As $T \to 0$, this becomes

$$D_{ik}(\omega, \mathbf{k}) = \text{re } D^R_{ik}(\omega, \mathbf{k}) + i \operatorname{sgn} \omega \cdot \text{im } D^R_{ik}(\omega, \mathbf{k}). \tag{75.22}$$

The function D^R_{ik} is given by (75.20); since re $\varepsilon(\omega)$ is an even function of ω, and im $\varepsilon(\omega)$ an odd function, we find at $T = 0$

$$D_{ik}(\omega, \mathbf{k}) = D^R_{ik}(|\omega|, \mathbf{k}). \tag{75.23}$$

In a vacuum, $\varepsilon(\omega) = 1$. Since in any material medium im $\varepsilon(\omega) > 0$ when $\omega > 0$, the vacuum corresponds to the limit $\varepsilon \to 1 + i0$. The resulting expression is

$$D^{(0)}_{ik}(\omega, \mathbf{k}) = \frac{4\pi\hbar}{\omega^2/c^2 - k^2 + i0}\left(\delta_{ik} - \frac{c^2k_ik_k}{\omega^2}\right),$$

in agreement with the known result in quantum electrodynamics (see *RQT*, §77).

§ 76. Electromagnetic field fluctuations

As already mentioned at the beginning of §75, in the treatment of electromagnetic field fluctuations we are concerned with the oscillations in time of quantities averaged only over physically infinitesimal volume elements (not over the motion of particles within them). The quantum-mechanical operators of these quantities are to be taken in the same sense.

The basic formulae of the theory of electromagnetic fluctuations can be written down directly from the general formulae of the fluctuation–dissipation theorem (Part 1, §125). For a discrete set of fluctuating quantities x_a, the spectral distribution of the fluctuations is expressed in terms of the generalized susceptibilities $\alpha_{ab}(\omega)$ by

$$(x_ax_b)_\omega = \tfrac{1}{2}i\hbar(\alpha^*_{ba} - \alpha_{ab}) \coth(\hbar\omega/2T),$$

where $(x_ax_b)_\omega$ is a component of the Fourier expansion with respect to time of the correlation function

$$\phi_{ab}(t) = \tfrac{1}{2}\langle\hat{x}_a(t)\,\hat{x}_b(0) + \hat{x}_b(0)\,\hat{x}_a(t)\rangle,$$

and $\hat{x}_a(t)$ are the Heisenberg operators of the quantities x_a. For distributed quantities $x_a(\mathbf{r})$ (functions of the coordinates of the point in the body), this formula becomes

$$(x_a^{(1)}x_b^{(2)})_\omega = \tfrac{1}{2}i\hbar \coth(\hbar\omega/2T)[\alpha_{ba}^*(\omega;\mathbf{r}_2,\mathbf{r}_1)-\alpha_{ab}(\omega;\mathbf{r}_1,\mathbf{r}_2)], \qquad (76.1)$$

where the superscripts (1) and (2) denote that the values are taken at the points $\mathbf{r}_1$ and $\mathbf{r}_2$.

In §75 it has been shown that, if the quantities x_a are the components of the vector potential $\mathbf{A}(\mathbf{r})/c$, the corresponding generalized susceptibilities are the components of the tensor $-D_{ik}^R(\omega;\mathbf{r}_1,\mathbf{r}_2)/\hbar c^2$. We therefore find immediately

$$(A_i^{(1)}A_k^{(2)})_\omega = \tfrac{1}{2}i \coth(\hbar\omega/2T)\{D_{ik}^R(\omega;\mathbf{r}_1,\mathbf{r}_2)-[D_{ki}^R(\omega;\mathbf{r}_2,\mathbf{r}_1)]^*\}. \qquad (76.2)$$

The spectral functions of the field fluctuations are easily found from (76.2). Let $\phi_{ik}^A(t_1,\mathbf{r}_1;t_2,\mathbf{r}_2)$ be the correlation function of the fluctuations of the vector potential; the expression (76.2) is the component of the Fourier expansion of this function with respect to $t = t_1-t_2$. Since the electric field $\mathbf{E} = -\dot{\mathbf{A}}/c$, the corresponding function for the components of $\mathbf{E}$ is

$$\phi_{ik}^E = \frac{1}{c^2}\frac{\partial^2}{\partial t_1\,\partial t_2}\phi_{ik}^A = -\frac{1}{c^2}\frac{\partial^2}{\partial t^2}\phi_{ik}^A,$$

or, in Fourier components,

$$(E_i^{(1)}E_k^{(2)})_\omega = (\omega^2/c^2)(A_i^{(1)}A_k^{(2)})_\omega. \qquad (76.3)$$

Similarly, since $\mathbf{B} = \operatorname{curl}\mathbf{A}$, we have

$$(B_i^{(1)}B_k^{(2)})_\omega = \operatorname{curl}_{il}^{(1)}\operatorname{curl}_{km}^{(2)}(A_l^{(1)}A_m^{(2)})_\omega, \qquad (76.4)$$

$$(E_i^{(1)}B_k^{(2)})_\omega = (i\omega/c)\operatorname{curl}_{km}^{(2)}(A_i^{(1)}A_m^{(2)})_\omega. \qquad (76.5)$$

By expressing the correlation functions of the electromagnetic fluctuations in terms of the retarded Green's function, formulae (76.2)–(76.5) reduce their calculation to the solution of the differential equation (75.15) or (75.16) wth the appropriate boundary conditions at the specified interfaces.†

We shall suppose henceforth that the medium is not magnetoactive. Then the function D_{ik}^R has the symmetry property (75.12), and (76.2) becomes

$$(A_i^{(1)}A_k^{(2)})_\omega = -\coth(\hbar\omega/2T)\,\operatorname{im} D_{ik}^R(\omega;\mathbf{r}_1,\mathbf{r}_2). \qquad (76.6)$$

It should be noted that the expression (76.6) is real, and therefore so are (76.3) and (76.4), while (76.5) is imaginary. This means that the time correlation functions between components of $\mathbf{E}$, and between those of $\mathbf{B}$, are even functions of the time $t = t_1-t_2$ (as they should be for a correlation between quantities that are both even or both odd under time reversal). But the time correlation function of the components of $\mathbf{E}$ with those of $\mathbf{B}$ is an odd function of the

† The theory of electromagnetic fluctuations was developed in another form by S. M. Rytov (1953), and in a form equivalent to (76.2)–(76.5) by M. L. Levin and S. M. Rytov (1967).

time (as it should be for two quantities of which one is even and the other odd under time reversal). Hence it follows that the values of **E** and **B** at any one instant are not correlated (an odd function of t is zero when $t = 0$). Together with the correlation function, the mean values of any expressions bilinear in **E** and **B** (taken at the same instant), for example the Poynting vector, are zero. This is in fact obvious, since, in a body in thermal equilibrium and invariant under time reversal, there cannot be any internal macroscopic energy fluxes.

§ 77. Electromagnetic fluctuations in an infinite medium

In a homogeneous infinite medium, the functions $D^R_{ik}(\omega; \mathbf{r}_1, \mathbf{r}_2)$ depend only on the difference $\mathbf{r} = \mathbf{r}_1 - \mathbf{r}_2$ and are even functions of $\mathbf{r}$; equation (75.15) contains only second derivatives with respect to the coordinates, so that $D^R_{ik}(\omega; \mathbf{r})$ and $D^R_{ik}(\omega; -\mathbf{r})$ satisfy the same equation. Taking the Fourier components with respect to $\mathbf{r}$ of both sides of (76.2), we find

$$(A_i^{(1)}A_k^{(2)})_{\omega\mathbf{k}} = \tfrac{1}{2} i \coth(\hbar\omega/2T)\{D^R_{ik}(\omega, \mathbf{k}) - [D^R_{ki}(\omega, \mathbf{k})]^*\}. \tag{77.1}$$

For non-magnetoactive media, with (75.12), this becomes

$$(A_i^{(1)}A_k^{(2)})_{\omega\mathbf{k}} = -\coth(\hbar\omega/2T)\, \mathrm{im}\, D^R_{ik}(\omega, \mathbf{k}). \tag{77.2}$$

In an isotropic non-magnetic ($\mu = 1$) medium, the function $D^R_{ik}(\omega, \mathbf{k})$ is given by (75.20). The problem of finding the spatial correlation function of the fluctuations reduces to the calculation of the integral

$$D^R_{ik}(\omega; \mathbf{r}) = \int D^R_{ik}(\omega, \mathbf{k})\, e^{i\mathbf{k}\cdot\mathbf{r}}\, d^3k/(2\pi)^3. \tag{77.3}$$

The integration is effected by means of the formulae

$$\int \frac{e^{i\mathbf{k}\cdot\mathbf{r}}}{k^2+\varkappa^2}\frac{d^3k}{(2\pi)^3} = \frac{e^{-\varkappa r}}{4\pi r},$$

$$\int \frac{k_i k_k\, e^{i\mathbf{k}\cdot\mathbf{r}}}{k^2+\varkappa^2}\frac{d^3k}{(2\pi)^3} = -\frac{\partial^2}{\partial x_i \partial x_k}\frac{e^{-\varkappa r}}{4\pi r}, \tag{77.4}$$

the first of which is obtained by taking Fourier components of the known equation

$$(\triangle - \varkappa^2)\frac{e^{-\varkappa r}}{r} = -4\pi\delta(\mathbf{r}), \tag{77.5}$$

and the second by differentiating the first. The result is

$$D^R_{ik}(\omega; \mathbf{r}) = -\hbar\left\{\delta_{ik} - \frac{c^2}{\omega^2\varepsilon}\frac{\partial^2}{\partial x_i \partial x_k}\right\}\frac{1}{r}\exp\left(-\frac{\omega r}{c}\sqrt{-\varepsilon}\right), \tag{77.6}$$

where $r = |\mathbf{r}_2 - \mathbf{r}_1|$ and $\sqrt{-\varepsilon}$ is to be taken with the sign that makes $\mathrm{re}\sqrt{-\varepsilon} > 0$; for a vacuum we must put $\varepsilon = 1$, $\sqrt{-\varepsilon} = -i$ (see below).

Hence, using (76.6) and (76.3), we have immediately

$$(E_i^{(1)}E_k^{(2)})_\omega = \hbar \coth\frac{\hbar\omega}{2T}\,\mathrm{im}\left\{\frac{1}{\varepsilon}\left[\frac{\varepsilon\omega^2}{c^2}\delta_{ik}+\frac{\partial^2}{\partial x_i \partial x_k}\right]\frac{1}{r}\exp\left(-\frac{\omega r}{c}\sqrt{-\varepsilon}\right)\right\} \tag{77.7}$$

(S. M. Rytov 1953). Contracting this expression with respect to the suffixes i and k, and using formula (77.5), we obtain

$$(\mathbf{E}^{(1)}\cdot\mathbf{E}^{(2)})_\omega = 2\hbar \coth\frac{\hbar\omega}{2T}\,\mathrm{im}\left\{\frac{1}{\varepsilon}\left[\frac{\varepsilon\omega^2}{c^2 r}\exp\left(-\frac{\omega r}{c}\sqrt{-\varepsilon}\right)+2\pi\delta(\mathbf{r})\right]\right\}. \tag{77.8}$$

Similarly, a calculation from (76.4) gives expressions for the correlation functions of the magnetic field which differ from (77.7) and (77.8) by the absence of the factor $1/\varepsilon$ in front of the square bracket; the delta-function term in (77.8) then has no imaginary part and does not appear in the result.

The occurrence of the imaginary part of ε in the expressions (77.7) and (77.8) shows clearly the relation between the electromagnetic fluctuations and the absorption in the medium. But if we take the limit im $\varepsilon \to 0$ in these expressions, we obtain non-zero results. This is connected with the order in which two limits are taken, those of an infinite medium and zero im ε. Since in an infinite medium an arbitrarily small im ε eventually gives rise to absorption, with our order of taking the limits the result pertains to a physically transparent medium in which, as in any actual medium, there is still some absorption.

For example, let us take these limits in (77.8). To do so, we note that for a small positive im ε (with $\omega > 0$)

$$\sqrt{-\varepsilon} \approx -i\sqrt{\mathrm{re}\,\varepsilon}\,.\left(1+i\frac{\mathrm{im}\,\varepsilon}{2\,\mathrm{re}\,\varepsilon}\right)$$

(using the condition that re $\sqrt{-\varepsilon} > 0$). Hence, in the limit as im $\varepsilon \to 0$, we obtain

$$(\mathbf{E}^{(1)}\cdot\mathbf{E}^{(2)})_\omega = \frac{1}{n^2}(\mathbf{H}^{(1)}\cdot\mathbf{H}^{(2)})_\omega = \frac{2\omega^2\hbar}{c^2 r}\sin\frac{\omega n r}{c}\coth\frac{\hbar\omega}{2T}, \tag{77.9}$$

where $n = \sqrt{\varepsilon}$ is the real refractive index. Because the delta-function term is not present, this expression remains finite even when the points $\mathbf{r}_1$ and $\mathbf{r}_2$ coincide:

$$(\mathbf{E}^2)_\omega = \frac{1}{n^2}(\mathbf{H}^2)_\omega = \frac{2\omega^3\hbar n}{c^3}\coth\frac{\hbar\omega}{2T}. \tag{77.10}$$

The passage to the limit of a transparent medium could also be made at an earlier point in the calculation, namely in the Green's function. Since the sign of im $\varepsilon(\omega)$ is the same as that of ω, we find that in this limit the function (75.20) becomes

$$D_{ik}^R(\omega, \mathbf{k}) = \frac{4\pi\hbar}{\omega^2 n^2/c^2 - k^2 + i0\,.\,\mathrm{sgn}\,\omega}\left[\delta_{ik}-\frac{c^2 k_i k_k}{\omega^2 n^2}\right]. \tag{77.11}$$

The imaginary part of this function depends only on the way in which the poles $\omega = \pm ck/n$ are avoided; separating it by means of (8.11) and substituting in (77.2), we obtain

$$(E_i^{(1)}E_k^{(2)})_{\omega\mathbf{k}} = \frac{2\pi^2\hbar}{k}\left(\frac{\omega^2}{c^2}\delta_{ik}-\frac{k_ik_k}{n^2}\right)\left\{\delta\left(\frac{n\omega}{c}-k\right)-\delta\left(\frac{n\omega}{c}+k\right)\right\}\coth\frac{\hbar\omega}{2T}. \tag{77.12}$$

The arguments of the delta functions in this expression have a simple physical significance: they show that the field fluctuations with a given value of $\mathbf{k}$ are propagated in space with velocity c/n, equal to that of propagation of electromagnetic waves in the same medium. The inverse Fourier transformation of (77.12) leads us back to (77.7), of course.

The energy of the fluctuation electromagnetic field in a transparent medium (with $\mu = 1$) in the spectrum range $d\omega$ is, per unit volume of space,

$$\frac{1}{8\pi}\left[2(\mathbf{E}^2)_\omega\frac{d(\omega\varepsilon)}{d\omega}+2(\mathbf{H}^2)_\omega\right]\frac{d\omega}{2\pi};$$

see *ECM*, §61.† Substitution of (77.10) readily gives

$$\left[\frac{1}{2}\hbar\omega+\frac{\hbar\omega}{e^{\hbar\omega/T}-1}\right]\frac{\omega^2n^2}{\pi^2c^3}\frac{d(n\omega)}{d\omega}d\omega. \tag{77.13}$$

The first term in the square brackets is due to the zero-point oscillations of the field; the second term gives the energy of electromagnetic radiation in thermodynamic equilibrium in a transparent medium, i.e. the energy of *black-body radiation*. This part of the formula could also have been obtained without considering fluctuations, by means of a generalization of Planck's formula for black-body radiation in a vacuum. According to this formula, the energy of black-body radiation per unit volume in the wave vector range d^3k is

$$\frac{\hbar\omega}{e^{\hbar\omega/T}-1}\cdot\frac{2d^3k}{(2\pi)^3};$$

the factor 2 takes account of the two directions of polarization. Correspondingly to obtain the spectral energy density we must replace d^3k by $4\pi k^2dk$ and substitute $k = \omega/c$. To change from a vacuum to a transparent medium, it is sufficient to write $k = n\omega/c$, i.e.

$$k^2\,dk = k^2(dk/d\omega)\,d\omega = \frac{\omega^2n^2}{c^3}\frac{d(n\omega)}{d\omega}d\omega$$

which gives the required result.

† The total energy is found by integrating with respect to ω from 0 to ∞; the factors of 2 in the square brackets arise because, by our definition of the spectral functions of the fluctuations, the mean value $\langle x^2\rangle$ is obtained by integrating $(x^2)_\omega$ with respect to $\omega/2\pi$ from $-\infty$ to ∞ (see Part 1, (122.6)).

PROBLEMS

PROBLEM 1. Find the electromagnetic field fluctuations at a large distance from a body embedded in a transparent rarefied medium with which it is thermal equilibrium; the wavelength of the radiation and the distance from the body to the point considered are large compared with the size of the body. The body has an anisotropic electric polarizability $\alpha_{ik}(\omega)$.

SOLUTION. The rarefied transparent medium is regarded as a vacuum. The required fluctuati ons are determined by the change in the vacuum Green's function due to the presence of the body; this change is small (at large distances). To calculate the change, we start from an analogy in which the vacuum function $D^R_{ik}(\omega; \mathbf{r}, \mathbf{r}')$ (for a given suffix k) may be formally regarded as the electric field $E_i(\mathbf{r}, \mathbf{r}')$ at the point $\mathbf{r}$ due to a source at the point $\mathbf{r}'$. This analogy is based on the fact that the field $E_i(\mathbf{r}, \mathbf{r}')$, like its potential $A_i(\mathbf{r}, \mathbf{r}')$, satisfies for $\mathbf{r} \neq \mathbf{r}'$ a similar equation to $D^R_{ik}(\omega; \mathbf{r}, \mathbf{r}')$, namely (75.16) with $\varepsilon = 1$. Let the body be at the point $\mathbf{r} = 0$. The field

$$E_i(0, \mathbf{r}') = D^R_{ik}(\omega; 0, \mathbf{r}') \equiv D^R_{ik}(\omega; \mathbf{r}'),$$

where $D^R_{ik}(\omega; \mathbf{r})$ is the Green's function in the vacuum in the absence of the body (given by (77.6) with $\varepsilon = 1$), polarizes the body and thus creates at the point $\mathbf{r} = 0$ a dipole moment $d_i = \alpha_{il} D^R_{lk}(\omega; 0, \mathbf{r}')$. The field created in turn by this dipole moment at the point $\mathbf{r}$ gives the required change $\delta D^R_{ik}(\omega; \mathbf{r}, \mathbf{r}')$. According to a formula in electrodynamics (see *Fields*, §72), the field created at the point $\mathbf{r}$ by a dipole moment $\mathbf{d}$ (varying with time as $e^{-i\omega t}$) is

$$E_i = d_l \left[\frac{\omega^2}{c^2} \delta_{il} + \frac{\partial^2}{\partial x_i \, \partial x_l} \right] \frac{e^{i\omega r/c}}{r},$$

where the distance r has to be large only in comparison with the size of the body, not with the wavelength. This expression may be written

$$E_i = -\frac{\omega^2}{\hbar c^2} D^R_{il}(\omega; \mathbf{r}) \, d_l$$

(the function $D^R_{il}(\omega; \mathbf{r})$ is even in $\mathbf{r}$). With the dipole moment as given above, we therefore have

$$\delta D^R_{ik}(\omega; \mathbf{r}, \mathbf{r}') = -(\omega^2/\hbar c^2) \, D^R_{il}(\omega; \mathbf{r}) \, \alpha_{lm} D^R_{mk}(\omega; \mathbf{r}').$$

The required correlation functions of the fluctuations are now found from the general formulae (76.3)–(76.6) with δD^R_{ik} instead of D^R_{ik}. The final result is

$$\delta(A^{(1)}_i A^{(2)}_k)_\omega = \frac{2\omega^2}{\hbar c^2} \left\{ \frac{1}{2} + \frac{1}{e^{\hbar\omega/T} - 1} \right\} \operatorname{im} [D^R_{il}(\omega; \mathbf{r}_1) \, \alpha_{lm} D^R_{mk}(\omega; \mathbf{r}_2)]. \tag{1}$$

The body is at $\mathbf{r} = 0$; $\mathbf{r}_1$ and $\mathbf{r}_2$ are two points remote from it. There is a contribution to the fluctuations not only from the imaginary part of the polarizability but also from its real part; the latter contribution may be regarded as the result of scattering by the body of black-body radiation occupying the transparent medium.

PROBLEM 2. The same as Problem 1, but for a body with magnetic polarizability $\alpha_{ik}(\omega)$.†

SOLUTION. In this case we regard $\operatorname{curl}_{il} D^R_{lk}(\omega; \mathbf{r}, \mathbf{r}')$ as the magnetic field $H_i(\mathbf{r}, \mathbf{r}')$ created at the point $\mathbf{r}$ by a source at the point $\mathbf{r}'$; an equation of the same form as for the function D^R_{ik} is satisfied not by the field H_i but by its potential A_1. This field magnetizes the body, creating at $\mathbf{r} = 0$ a magnetic moment

$$m_i = -\alpha_{il} \operatorname{curl}'_{lm} D^R_{mk}(\omega; 0, \mathbf{r}');$$

† The presence of a magnetic polarizability does not necessarily mean that the body consists of a magnetic material; for example, we may be considering the displacement of the magnetic field from the body because of the skin effect.

the differentiation with respect to **r** is replaced by one with respect to **r**′, using the fact that D^R_{mk} depends only on the difference **r**−**r**′. The required change in the Green's function is equal to the vector potential of the magnetic field created by this magnetic moment at the point **r**:

$$A_i = \mathrm{curl}_{il}\left[\frac{1}{r} m_l e^{i\omega r/c}\right];$$

see *Fields*, §72, Problem 1. Thus

$$\delta D^R_{ik}(\omega; \mathbf{r}, \mathbf{r}') = -\left(\mathrm{curl}_{il}\frac{e^{i\omega r/c}}{r}\right)\alpha_{lm}\,\mathrm{curl}'_{mn}\,D^R_{nk}(\omega; 0, \mathbf{r}').$$

Lastly, substituting D^R_{nk} from (77.6), we find

$$\delta D^R_{ik}(\omega; \mathbf{r}, \mathbf{r}') = \hbar\left(\mathrm{curl}_{il}\frac{e^{i\omega r/c}}{r}\right)\alpha_{lm}\,\mathrm{curl}'_{mk}\frac{e^{i\omega r'/c}}{r'}, \tag{2}$$

using the fact that $\mathrm{curl}_{mn}\nabla_n = e_{mkn}\nabla_k\nabla_n \equiv 0$.

PROBLEM 3. Determine the fluctuations of the electromagnetic field in the conditions of Problem 1, but assuming that the temperature of the medium is much less than that of the body.

SOLUTION. The field calculated in Problem 1 separates naturally, according to the two terms in the braces in (1), into zero-point fluctuations and thermal black-body radiation. The latter in turn consists of two parts; the thermal radiation of the body itself, and the field resulting from scattering of black-body radiation by the body. If the temperature of the medium is low, the second part does not appear. In solving the problem, we calculate that part separately and then subtract it from (1). Let $\mathbf{A}(\mathbf{r}) = \mathbf{A}^{(0)} + \mathbf{A}^{(s)}$, where $\mathbf{A}^{(0)}$ is the fluctuational field in the absence of the body, and $\mathbf{A}^{(s)}$ the field scattered by the body. At large distances, where $\mathbf{A}^{(s)}$ is small, we can neglect the terms quadratic in $\mathbf{A}^{(s)}$ when calculating $\delta(A_{i1}A_{k2})_\omega$. The scattering contribution is therefore

$$\delta^{(s)}(A_{i1}A_{k2})_\omega \approx (A^{(s)}_{i1}A^{(0)}_{k2})_\omega + (A^{(0)}_{i1}A^{(s)}_{k2})_\omega = (A^{(s)}_{i1}A^{(0)}_{k2})_\omega + (A^{(s)}_{k2}A^{(0)}_{i1})^*_\omega.$$

The scattered field is again given by the formula in *Fields*, §72, but the dipole moment is now to be taken simply as that induced by the black-body radiation, $d_i = \alpha_{ik}A^{(0)}_k(0)$. Again using the Green's function in the vacuum in the absence of the body, we have

$$A^{(s)}_i(\mathbf{r}_1) = -\frac{\omega^2}{\hbar c^2} D^R_{il}(\omega; \mathbf{r}_1)\,\alpha_{lm}(\omega)\,A^{(0)}_m(0),$$

so that

$$(A^{(s)}_{i1}A^{(0)}_{k2})_\omega = -\frac{\omega^2}{\hbar c^2} D^R_{il}(\omega; \mathbf{r}_1)\,\alpha_{lm}(A^{(0)}_m(0)\,A^{(0)}_k(\mathbf{r}_2))_\omega.$$

The correlation function $(A^{(0)}_{m1}A^{(0)}_{k2})_\omega$ is again taken from (76.2). Since we are interested only in the thermal radiation, the zero-point oscillations in this formula are to be omitted, with the change

$$\frac{1}{2}\coth\frac{\hbar\omega}{2T} = \frac{1}{e^{\hbar\omega/T}-1} + \frac{1}{2} \to \frac{1}{e^{\hbar\omega/T}-1}.$$

The result for the contribution of scattered black-body radiation to the correlation function is

$$\delta^{(s)}(A_{i1}A_{k2})_\omega = \frac{2\omega^2}{\hbar c^2(e^{\hbar\omega/T}-1)}[D^R_{il}(\omega; \mathbf{r}_1)\,\alpha_{lm}\,\mathrm{im}\,D^R_{mk}(\omega; \mathbf{r}_2) + D^{R*}_{kl}(\omega; \mathbf{r}_2)\,\alpha^*_{lm}\,\mathrm{im}\,D^R_{mi}(\omega; \mathbf{r}_1)]. \tag{3}$$

Lastly, in order to find the fluctuational field in a cold medium, we must subtract (3) from (1). A simple rearrangement using the symmetry of the tensors D_{ik} and α_{ik} gives

$$\delta^{(T)}(A_{i1}A_{k2})_\omega = \frac{2\omega^2}{\hbar c^2(e^{\hbar\omega/T}-1)} D^R_{il}(\omega; \mathbf{r}_1)[\mathrm{im}\,\alpha_{lm}(\omega)]\,D^{R*}_{mk}(\omega; \mathbf{r}_2), \tag{4}$$

where T is the temperature of the body. Only the thermal term has been written here; the zero-point oscillation term in (1) remains unchanged. It should be noted that the expression (4) for the thermal radiation of the body depends only on the imaginary part of the polarizability. The energy flux calculated from (4) is not zero; it gives the intensity of thermal radiation from the heated body into the surrounding cold medium.

§ 78. Current fluctuations in linear circuits

Another interesting application of the fluctuation–dissipation theorem is the problem of current fluctuations in linear circuits, first discussed by H. Nyquist (1928).

The current fluctuations are free electrical oscillations in the conductor (i.e. they occur in the absence of any externally applied e.m.f.). In a closed linear circuit the oscillations of greatest interest are, of course, those in which a non-zero total current J flows in the conductor. In what follows we shall assume that the condition for a quasi-steady state holds: the dimensions of the circuit are small compared with the wavelength $\lambda \sim c/\omega$. Then the total current J is the same at every point in the circuit, and is a function of time only.

This current J may be taken as the quantity $x(t)$ which appears in the general formulation of the fluctuation–dissipation theorem (Part 1, §124). In order to ascertain the meaning of the corresponding generalized susceptibility α, let us suppose that an external e.m.f. $\mathcal{E}$ acts in the circuit. Then the rate of dissipation of energy in the circuit is $Q = J\mathcal{E}$. A comparison with the expression $Q = -\bar{x}\dot{f}$, which serves to define the "force" f (see Part 1, (123.10)), shows that $\dot{f} = -\mathcal{E}$, or in Fourier components $\mathcal{E}_\omega = i\omega f_\omega$. On the other hand, the current and the e.m.f. in a linear circuit are related by $\mathcal{E}_\omega = Z(\omega)\, J_\omega$, where $Z(\omega)$ is the impedance of the circuit. We therefore have

$$J_\omega = \mathcal{E}_\omega/Z = i\omega f_\omega/Z,$$

and comparison with the definition of the generalized susceptibility in the relation $(\bar{x})_\omega = \alpha(\omega) f$ gives $\alpha(\omega) = i\omega/Z(\omega)$. The imaginary part of α is

$$\operatorname{im} \alpha = \operatorname{im}(i\omega/Z) = (\omega/|Z|^2)\, R(\omega),$$

where $R = \operatorname{re} Z$.

According to the fluctuation–dissipation theorem,

$$(x^2)_\omega = \hbar \coth(\hbar\omega/2T) . \operatorname{im} \alpha(\omega),$$

we now find as the spectral function of the current fluctuations

$$(J^2)_\omega = [\hbar\omega/|Z(\omega)|^2]\, R(\omega) \coth(\hbar\omega/2T). \tag{78.1}$$

This formula can be put in another form by regarding the current fluctuations as resulting from the action of the "random" e.m.f. $\mathcal{E}_\omega = ZJ_\omega$. This gives

$$(\mathcal{E}^2)_\omega = \hbar\omega R(\omega) \coth(\hbar\omega/2T). \tag{78.2}$$

In the classical case ($\hbar\omega \ll T$).

$$(\mathscr{E}^2)_\omega = 2TR(\omega). \tag{78.3}$$

We must again emphasize that these formulae are entirely independent of the nature of the phenomena responsible for the dispersion of the circuit resistance.

§ 79. Temperature Green's function of a photon in a medium

The temperature Green's function of a photon in a medium is constructed from the Matsubara operators of the electromagnetic field potentials in the same way as the time Green's function (75.2) is constructed from the Heisenberg operators:

$$\mathscr{D}_{ik} = -\langle \mathrm{T}_\tau\, \hat{A}_i^M(\tau_1, \mathbf{r}_1)\, \hat{A}_k^M(\tau_2, \mathbf{r}_2)\rangle. \tag{79.1}$$

Here we have used the fact that, since the Schrödinger operators of the field are Hermitian, the Matsubara operators $\hat{\mathbf{A}}^M$ and $\hat{\bar{\mathbf{A}}}^M$ (defined as in (37.1)) are the same. These operators, however, unlike the Heisenberg ones, are not themselves Hermitian: since the parameter τ is real, we have

$$\left[\hat{\mathbf{A}}^M(\tau, \mathbf{r})\right]^+ = \left[e^{\tau\hat{H}'/\hbar}\hat{\mathbf{A}}(\mathbf{r})\, e^{-\tau\hat{H}'/\hbar}\right]^+ = e^{-\tau\hat{H}'/\hbar}\,\hat{\mathbf{A}}(\mathbf{r})\, e^{\tau\hat{H}'/\hbar},$$

or

$$[\hat{\mathbf{A}}^M(\tau, \mathbf{r})]^+ = \hat{\mathbf{A}}^M(-\tau, \mathbf{r}).$$

Since the function (79.1) depends only on the difference $\tau = \tau_1 - \tau_2$ (cf. §37), we can write (taking, for example, $\tau > 0$)

$$\mathscr{D}_{ik}(\tau; \mathbf{r}_1, \mathbf{r}_2) = -\langle \hat{A}_i^M(\tau, \mathbf{r}_1)\, \hat{A}_k^M(0, \mathbf{r}_2)\rangle,$$
$$\mathscr{D}_{ik}(-\tau; \mathbf{r}_1, \mathbf{r}_2) = -\langle \hat{A}_k^M(\tau, \mathbf{r}_2)\, \hat{A}_i^M(0, \mathbf{r}_1)\rangle.$$

A comparison of these two expressions shows that

$$\mathscr{D}_{ik}(-\tau; \mathbf{r}_1, \mathbf{r}_2) = \mathscr{D}_{ki}(\tau; \mathbf{r}_2, \mathbf{r}_1). \tag{79.2}$$

The function $\mathscr{D}_{ik}$ can be expanded in a Fourier series in the variable τ:

$$\mathscr{D}_{ik}(\tau; \mathbf{r}_1, \mathbf{r}_2) = T \sum_{s=-\infty}^{\infty} \mathscr{D}_{ik}(\zeta_s; \mathbf{r}_1, \mathbf{r}_2)\, e^{-i\zeta_s\tau}, \tag{79.3}$$

the "frequencies" ζ_s taking values such that $\hbar\zeta_s = 2\pi sT$ (because the photons obey Bose statistics; see (37.8)). For the components of this expansion, (79.2) gives the corresponding relation

$$\mathscr{D}_{ik}(\zeta_s; \mathbf{r}_1, \mathbf{r}_2) = \mathscr{D}_{ki}(-\zeta_s; \mathbf{r}_2, \mathbf{r}_1). \tag{79.4}$$

According to the general relation (37.12), these components are related to the retarded Green's function by

$$\mathscr{D}_{ik}(\zeta_s; \mathbf{r}_1, \mathbf{r}_2) = D_{ik}^R(i\zeta_s; \mathbf{r}_2, \mathbf{r}_1)$$

for positive ζ_s. It has been shown in §75 that the functions $D^R_{ik}(\omega; \mathbf{r}_1, \mathbf{r}_2)$ may in a sense be regarded as generalized susceptibilities occurring in the general theory of the response of a macroscopic system to an external interaction. Hence there followed the symmetry property of these functions expressed (for non-magnetoactive media) by equation (75.12); because of the relation between D^R_{ik} and $\mathcal{D}_{ik}$, the latter have a similar property,

$$\mathcal{D}_{ik}(\zeta_s; \mathbf{r}_1, \mathbf{r}_2) = \mathcal{D}_{ki}(\zeta_s; \mathbf{r}_2, \mathbf{r}_1). \tag{79.5}$$

From this equation together with (79.4), it now follows that the functions $\mathcal{D}_{ik}(\zeta_s; \mathbf{r}_1, \mathbf{r}_2)$ are even functions of the discrete variable ζ_s, so that for all its values (positive and negative) we have

$$\mathcal{D}_{ik}(\zeta_s; \mathbf{r}_1, \mathbf{r}_2) = D^R_{ik}(i|\zeta_s|; \mathbf{r}_1, \mathbf{r}_2). \tag{79.6}$$

Furthermore, the function $D^R_{ik}(\omega; \mathbf{r}_1, \mathbf{r}_2)$, like any generalized susceptibility, is real on the positive imaginary ω-axis (see Part 1, §123); it therefore follows from (79.6) that the function $\mathcal{D}_{ik}(\zeta_s; \mathbf{r}_1, \mathbf{r}_2)$ is real for all values of ζ_s. Lastly, it follows from these properties that the original function $\mathcal{D}_{ik}(\tau; \mathbf{r}_1, \mathbf{r}_2)$ is real and an even function of τ:

$$\mathcal{D}_{ik}(\tau; \mathbf{r}_1, \mathbf{r}_2) = \mathcal{D}_{ik}(-\tau; \mathbf{r}_1, \mathbf{r}_2). \tag{79.7}$$

The relation (79.6) between the temperature Green's function and the retarded Green's function enables us to write down immediately the differential equation that must be satisfied by the function $\mathcal{D}_{ik}$ in an inhomogeneous medium; to do so, it is sufficient to replace ω by $i|\zeta_s|$ in equation (75.15) or (75.16). For example, with an isotropic non-magnetoactive medium having $\mu = 1$, we find the equation

$$\left[\frac{\partial^2}{\partial x_i\, \partial x_l} - \delta_{il}\triangle + \frac{\zeta_s^2}{c^2}\varepsilon(i|\zeta_s|, \mathbf{r})\,\delta_{il}\right]\mathcal{D}_{lk}(\zeta_s; \mathbf{r}, \mathbf{r}') = -4\pi\hbar\delta_{ik}\delta(\mathbf{r}-\mathbf{r}'). \tag{79.8}$$

For a homogeneous infinite medium, the function $\mathcal{D}_{ik}(\zeta_s; \mathbf{r}, \mathbf{r}')$ is expanded as a Fourier integral with respect to the difference $\mathbf{r}-\mathbf{r}'$. The components of this expansion satisfy the algebraic equations

$$\frac{1}{4\pi\hbar}\left[k_i k_l - \delta_{il}k^2 - \delta_{il}\frac{\zeta_s^2}{c^2}\varepsilon(i|\zeta_s|)\right]\mathcal{D}_{lk}(\zeta_s, \mathbf{k}) = \delta_{ik}, \tag{79.9}$$

and are given by†

$$\mathcal{D}_{ik}(\zeta, \mathbf{k}) = -\frac{4\pi\hbar}{\zeta_s^2\varepsilon(i|\zeta_s|)/c^2 + k^2}\left[\delta_{ik} + \frac{c^2 k_i k_k}{\zeta_s^2\varepsilon(i|\zeta_s|)}\right]. \tag{79.10}$$

Since the function $D_{ik}(\zeta_s, \mathbf{k})$ is expressed (in the long-wavelength range $ka \ll 1$) in terms of $\varepsilon(\omega)$, the diagram technique for calculating it becomes a

† In practical applications (cf. §80), the function $\mathcal{D}_{ik}$ always occurs as a product with ζ_s^2, and this removes the divergence at $\zeta_s = 0$.

technique for calculating the permittivity of the medium. The latter quantity also has a definite diagram significance, which will now be elucidated.

We shall represent the exact $\mathcal{D}$ function by a thick broken line, and the vacuum function $\mathcal{D}^{(0)}$ by a thin one:[†]

$$-\,-\,-\,-\,= -\mathcal{D}_{ik} \qquad -\,-\,-\,-\,= -\mathcal{D}_{ik}^{(0)} \tag{79.11}$$

The whole set of diagrams representing the $\mathcal{D}$ function can be expressed as a series exactly analogous to the series (14.3) for the function G:

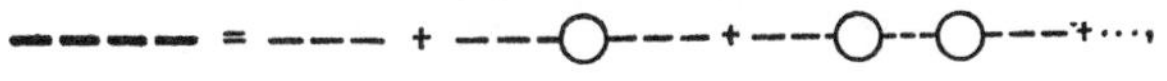

$$-\,-\,-\,-\, = \,-\,-\,-\, + \,-\,-\!\bigcirc\!-\,-\, + \,-\,-\!\bigcirc\!-\!\bigcirc\!-\,-\,+\cdots, \tag{79.12}$$

where a circle stands for the set of diagram blocks that do not fall into two parts joined only by one broken line; this set will be denoted by $-\mathcal{P}_{ik}/4\pi$. The function $\mathcal{P}_{ik}$ (analogous to the self-energy part of the Green's function of the particles) is called the *polarization operator*.

The diagram equation (79.12) is equivalent to

$$-\,-\,-\,-\, = \,-\,-\,-\,-\, + \,-\,-\!\bigcirc\!-\,-\,; \tag{79.13}$$

cf. the derivation of (14.4) from (14.3). In analytical form, this is

$$\mathcal{D}_{ik} = \mathcal{D}_{ik}^{(0)} + \mathcal{D}_{il}^{(0)}(\mathcal{P}_{lm}/4\pi)\,\mathcal{D}_{mk}, \tag{79.14}$$

where all the factors are functions of the same arguments ζ_s and $\mathbf{k}$. Multiplying this equation on the right by the inverse tensor $\mathcal{D}^{-1}$ and on the left by $[\mathcal{D}^{(0)}]^{-1}$, we can rewrite it as

$$\mathcal{D}^{-1}{}_{ik} = [\mathcal{D}^{(0)}]^{-1}{}_{ik} - \mathcal{P}_{ik}/4\pi. \tag{79.15}$$

Finally, taking $\mathcal{D}^{-1}{}_{ik}$ from the left-hand side of equation (79.9), and a similar expression with $\varepsilon = 1$ for $[\mathcal{D}^{(0)}]^{-1}{}_{ik}$, we find

$$\mathcal{P}_{ik}(\zeta_s, \mathbf{k}) = (\zeta_s^2/\hbar c^2)[\varepsilon(i\,|\,\zeta_s\,|) - 1]\,\delta_{ik}, \tag{79.16}$$

which determines the diagram significance of the function $\varepsilon(\omega) - 1$ at a discrete set of points on the positive imaginary ω-axis. The analytical continuation of $\varepsilon(i\,|\,\zeta_s\,|)$ to the whole of the upper half-plane must, in principle, take account of the facts that $\varepsilon(\omega)$ cannot have a singularity in this half-plane and that $\varepsilon(\omega) \to 1$ as $|\omega| \to \infty$.[‡]

[†] The use of broken lines to denote the $\mathcal{D}$ functions cannot cause any misunderstanding here, since this and the following section do not explicitly involve the energy of pair interaction of the particles in the medium (for which that notation was previously used).

[‡] In an anisotropic medium, we must write

$$\mathcal{P}_{ik}(\zeta_s, \mathbf{k}) = (\zeta_s^2/\hbar c^2)[\varepsilon_{ik}(i\,|\,\zeta_s\,|) - \delta_{ik}].$$

In this form, the expression remains valid when there is spatial dispersion and ε_{ik} depends on the wave vector as well as on the frequency.

In an inhomogeneous medium, the polarization operator is, like $\mathcal{D}_{ik}$, a function of the coordinates of two points. Repeating the derivation in the coordinate representation, we obtain instead of (79.14)

$$\mathcal{D}_{ik}(\mathbf{r}_1, \mathbf{r}_2) = \mathcal{D}_{ik}^{(0)}(\mathbf{r}_1, \mathbf{r}_2) + \frac{1}{4\pi}\int \mathcal{D}_{il}^{(0)}(\mathbf{r}_1, \mathbf{r}_3)\,\mathcal{P}_{lm}(\mathbf{r}_3, \mathbf{r}_4)\,\mathcal{D}_{mk}(\mathbf{r}_4, \mathbf{r}_2)\, d^3x_3\, d^3x_4;$$

the arguments ζ_s are omitted, for brevity. Applying the operator

$$\frac{\partial^2}{\partial x_{1n}\,\partial x_{1l}} - \delta_{nl}\triangle_1 + (\zeta_s^2/c^2)\,\delta_{nl}$$

to the left of this equation, and noting that $D^{(0)}$ satisfies equation (79.8) with $\varepsilon = 1$, we obtain

$$\int \mathcal{P}_{il}(\mathbf{r}_1, \mathbf{r}')\,\mathcal{D}_{lk}(\mathbf{r}', \mathbf{r}_2)\, d^3x' = [\varepsilon(\mathbf{r}_1) - 1](\zeta_s^2/\hbar c^2)\,\mathcal{D}_{ik}(\mathbf{r}_1, \mathbf{r}_2),$$

whence

$$\mathcal{P}_{ik}(\zeta_s; \mathbf{r}_1, \mathbf{r}_2) = (\zeta_s^2/\hbar c^2)\,\delta_{ik}\delta(\mathbf{r}_1 - \mathbf{r}_2)[\varepsilon(i\,|\,\zeta_s\,|, \mathbf{r}_1) - 1]. \tag{79.17}$$

The structure of a condensed medium, and hence its dielectric properties, are determined by the forces acting between its particles at distances of the order of the atomic dimensions *a*. At these distances (if the particle velocities are non-relativistic) we can neglect the retardation of the interactions, which becomes important only for the long-wave components ($ka \ll 1$) of the field. That is, in calculating the polarization operator we can neglect the long-wave part of the field. In the diagrams for the Green's function $\mathcal{D}_{ik}$ itself, however, the long-wave field occurs only through the thin broken lines on the right of (79.12).

The three-dimensional tensor $\mathcal{P}_{ik}$ considered in this section is, of course, only the spatial part of the polarization 4-tensor $\mathcal{P}_{\mu\nu}$. We emphasize, to avoid misunderstanding, that its time component $\mathcal{P}_{00}$ and mixed components $\mathcal{P}_{0i}$ are not zero. Moreover, as in quantum electrodynamics, this 4-tensor is independent of the gauge of the potentials. In non-relativistic theory the gauge invariance is obvious from the possibility just mentioned of calculating the polarization operator with only the non-retarded forces, which are independent of the gauge of the long-wave field.

The components $\mathcal{P}_{00}$ and $\mathcal{P}_{0i}$ can be found from the condition for the 4-tensor to be transverse: $\mathcal{P}_{\mu\nu}k^\mu = 0$, where $k^\mu = (i\zeta_s, \mathbf{k})$ is the wave 4-vector:

$$\left.\begin{aligned} \mathcal{P}_{00} &= -(\mathbf{k}^2/\hbar c^2)[\varepsilon(i\,|\,\zeta_s\,|) - 1], \\ \mathcal{P}_{0i} &= (i\zeta_s k_i/\hbar c^2)[\varepsilon(i\,|\,\zeta_s\,|) - 1]. \end{aligned}\right\} \tag{79.18}$$

§ 80. The van der Waals stress tensor

Although the structure of condensed bodies is essentially determined (as noted at the end of §79) by the forces acting between its particles at atomic distances, a definite contribution to the thermodynamic quantities of the body (its free energy, say) comes also from the *van der Waals forces* which act between the atoms at distances large compared with the atomic distances a. For free atoms, the energy of this interaction decreases with increasing distance as r^{-6} (see *QM*, §89), and as r^{-7} when retardation effects have become important (see *RQT*, §83). In a condensed medium, of course, the van der Waals forces do not reduce to an interaction between separate pairs of atoms. However, since their range of action is large compared with interatomic distances, we can use a macroscopic approach to the problem of their influence on the thermodynamic properties of the body.

In the macroscopic theory, the van der Waals interaction in a material medium is regarded as brought about through a long-wavelength electromagnetic field (E. M. Lifshitz 1954); this concept includes not only thermal fluctuations but also the zero-point oscillations of the field. An important property of the contribution of this interaction to the free energy is that it is not additive; it is not simply proportional to the volume of the bodies, but depends also on parameters that characterize their shape and configuration. This non-additivity, resulting from the long range of the van der Waals forces, is the property that distinguishes their contribution to the free energy from the much larger additive part. In the macroscopic picture, this property arises from the fact that any change in the electrical properties of the medium in some region causes, in accordance with Maxwell's equations, a change in the fluctuational field even outside that region. In practice, of course, the non-additivity effects are appreciable only when the characteristic dimensions are sufficiently small (though still large compared with atomic dimensions), e.g. in thin films or in bodies separated by a narrow gap.

In the calculation of the contribution of the electromagnetic fluctuations to the free energy, the important wavelengths in each case are of the order of the characteristic dimensions of the inhomogeneity of the medium (film thickness, gap width, etc.). In the macroscopic theory, this is the reason for the power-law decrease of the van der Waals forces; if fluctuations with some fixed wavelength λ_0 were important, this would give an exponential decrease, with exponent $\sim r/\lambda_0$. Since the characteristic dimensions, and therefore the characteristic wavelengths of the fluctuations, are much greater than atomic dimensions, all properties of these fluctuations and their contribution to the free energy are expressed entirely in terms of the complex permittivity of the bodies.

Our object will be to calculate the macroscopic forces acting in an inhomogeneous medium.† As a first step in the derivation, we shall determine the change

† The theory given below is due to I. E. Dzyaloshinskiĭ and L. P. Pitaevskiĭ (1959).

in the free energy of the medium due to a small change in its permittivity; the magnetic properties of the material will be neglected (the permeability $\mu = 1$). We shall suppose that the change in ε is caused by a small change $\delta\hat{H}$ in the Hamiltonian of the system. Then the change in the free energy is

$$\delta F = \langle \delta\hat{H} \rangle, \tag{80.1}$$

where the averaging is taken (with given temperature and volume of the system) over the Gibbs distribution with the unperturbed Hamiltonian $\hat{H}$. We substitute the latter in the form†

$$\hat{H} = \hat{H}_0 + \hat{V}_{lw}, \quad \hat{V}_{lw} = -\int \hat{\mathbf{j}} . \hat{\mathbf{A}}\, d^3x, \tag{80.2}$$

where $\hat{V}_{lw}$ describes the interaction of the particles with the long-wave electromagnetic field, and $\hat{H}_0$ includes all other interactions together with terms corresponding to free particles and photons. Strictly speaking, the integral in (80.2) should be regarded as cut off at a wave number $k_0 \ll 1/a$, but the cut-off parameter does not appear in the final result. The operator $\hat{\mathbf{A}}$ is the long-wave field vector potential operator; it is important that the operator $\delta\hat{H}$ responsible for the change in the permittivity does not contain $\hat{\mathbf{A}}$, since the permittivity is determined solely by the interaction of particles at atomic distances.

Let us now change in (80.1) to Matsubara operators in what may be called the long-wave representation of the interaction: in this representation, the dependence of the operators on τ is determined by all the terms in the Hamiltonian except $\hat{V}_{lw}$. By the same method as in the derivation of (38.7), we obtain

$$\left.\begin{aligned} \delta F &= \frac{1}{\langle\hat{\sigma}\rangle_0} \langle \mathrm{T}_\tau\, \delta\hat{H}^M \hat{\sigma} \rangle_0, \\ \hat{\sigma} &= \mathrm{T}_\tau \exp \int_0^{1/T} \int \hat{\mathbf{j}}^M . \hat{\mathbf{A}}^M\, d^3x\, d\tau, \end{aligned}\right\} \tag{80.3}$$

where $\langle \ldots \rangle_0$ denotes averaging over the Gibbs distribution with the Hamiltonian $\hat{H}_0$. According to the meaning of the chosen representation, the Matsubara operators are defined as

$$\hat{\mathbf{A}}^M(\tau, \mathbf{r}) = \exp(\tau\hat{H}_0)\, \hat{\mathbf{A}}(\mathbf{r}) \exp(-\tau\hat{H}_0), \tag{80.4}$$

and similarly for $\delta\hat{H}^M$ and for the ψ operators from which the particle current operator $\hat{\mathbf{j}}^M$ is constructed.‡ Since $\hat{H}_0$ does not contain the interaction of the long-wavelength photons with anything else, $\hat{\mathbf{A}}^M$ is the same as the (Matsubara) operator of the free photon field; for the ψ operators of the particles this is, of course, not so, since $\hat{H}_0$ includes the interaction between particles.

† In this section we take $\hbar = 1, c = 1$.

‡ The suffix 0 which should also be attached to the operators in this representation is omitted to simplify the notation.

Following the general construction principles of the diagram technique, we expand the exponential in (80.3) in powers of $\hat{V}_{lw}$.† In each term of the expansion, the product of the free-field operators $\hat{\mathbf{A}}^M$ is averaged in the usual way as pair contractions, using Wick's theorem. The zero-order term in the expansion, which does not contain $\hat{\mathbf{A}}^M$, gives δF_0, the change in the free energy without allowance for the long-wave fluctuations. The next term, linear in $\hat{\mathbf{A}}^M$, gives zero on averaging. In the term quadratic with respect to the field, the contraction of two operators $\langle \hat{A}_i^M \hat{A}_k^M \rangle$ gives $\mathcal{D}_{ik}^{(0)}$, the free-photon Green's function; this term may be represented by the diagram

$\delta F^{(2)} = \frac{1}{2} \cdot$  (80.5)

with the numerical factor 1/2!, which occurs in the expansion of the exponential, shown separately. The thin broken line denotes the function $\mathcal{D}^{(0)}$, and the shaded circle denotes the result of averaging all other factors. The explicit form of the latter quantity will not be given here; the only important point is that it is just $\delta \mathcal{P}_{ik}/4\pi$, where $\delta \mathcal{P}_{ik}$ is the change in the polarization operator when the Hamiltonian of the system changes by $\delta \hat{H}$. This is easily seen by considering similarly the change in the function $\mathcal{D}$. In the same representation of the operators, this function is

$$\mathcal{D}_{ik}(\tau_1, \mathbf{r}_1; \tau_2, \mathbf{r}_2) = -\frac{1}{\langle \hat{\sigma} \rangle_0} \langle \mathrm{T}_\tau \hat{A}_i^M(\tau_1, \mathbf{r}_1) \hat{A}_k^M(\tau_2, \mathbf{r}_2) \hat{\sigma} \rangle_0,$$

where now

$$\hat{\sigma} = \mathrm{T}_\tau \exp \int_0^{1/T} (-\hat{V}_{lw}^M - \delta \hat{H}^M)\, d\tau;$$

the "interaction" includes $\delta \hat{H}$ as well as $\hat{V}_{lw}$. The required change $\delta \mathcal{D}_{ik}$ is given by the linear term in the expansion of this expression in powers of $\delta \hat{H}^M$;

$$\delta \mathcal{D}_{ik} = \frac{1}{\langle \hat{\sigma} \rangle_0} \left\langle \mathrm{T}_\tau \int \delta \hat{H}\, d\tau . \hat{A}_i^M(\tau_1, \mathbf{r}_1) \hat{A}_k^M(\tau_2, \mathbf{r}_2) \exp \int \hat{\mathbf{j}}^M . \hat{\mathbf{A}}^M\, d^3x\, d\tau \right\rangle_0 . \tag{80.6}$$

In the expansion of the remaining exponential in powers of $\hat{V}_{lw}$, the zero-order term is to be omitted, as corresponding to a detached diagram (the contraction $\langle \hat{A}_i^M \hat{A}_k^M \rangle$ is separated from the other factors, which do not contain the variables $\mathbf{r}_1$ and $\mathbf{r}_2$). The first-order term contains an odd number of A operators and gives zero on averaging. Lastly, the second-order term gives in $\delta \mathcal{D}_{ik}$ an expression represented by the diagram

$\delta \mathcal{D}_{ik}^{(2)} =$  (80.7)

† It is sufficient to show the expansion of the numerator in the expression for δF. As usual, the role of the factor $\langle \hat{\sigma} \rangle_0$ in the denominator is just to exclude diagrams that separate into two or more disconnected parts.

with the same circle as in (80.5); the factor $\frac{1}{2}$ does not appear here, because there are two ways of contracting the "internal" A operators from the operators $\hat{V}_{lw}$ with the "external" $\hat{A}_i^M$ and $\hat{A}_k^M$. On the other hand, from the definition of the polarization operator, the Green's function in the approximation considered is given by the sum

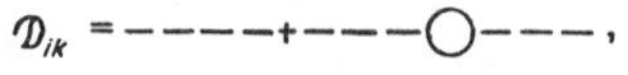

where the white circle is the polarization operator $\mathcal{P}_{ik}/4\pi$. The variation of this function therefore gives the diagram (80.7) with $\delta\mathcal{P}_{ik}/4\pi$ as the shaded circle.

All subsequent terms of the expansion in (80.3) are corrections of various orders to the broken line and the circle in the diagram (80.5). These corrections convert the broken line into the exact function $\mathcal{D}_{ik}$. The long-wave corrections to $\delta\mathcal{P}_{ik}$ are small, as already discussed, so that we can immediately take $\delta\mathcal{P}_{ik}$ to be the variation of the exact polarization operator.

In analytical form, this result is written (after changing to the Fourier expansion with respect to the variable τ)†

$$\delta F = \delta F_0 - \frac{1}{2}\sum_{s=-\infty}^{\infty} T \int \mathcal{D}_{ik}(\zeta_s; \mathbf{r}_1, \mathbf{r}_2)\frac{1}{4\pi}\,\delta\mathcal{P}_{ki}(\zeta_s; \mathbf{r}_2, \mathbf{r}_1)\, d^3x_1\, d^3x_2. \qquad (80.8)$$

According to (79.17), the change in the polarization operator is expressed (for an isotropic medium) in terms of the change in the permittivity:

$$\delta\mathcal{P}_{ki}(\zeta_s; \mathbf{r}_1, \mathbf{r}_2) = \zeta_s^2\delta_{ki}\delta(\mathbf{r}_1-\mathbf{r}_2)\delta\varepsilon(i\,|\,\zeta_s\,|, \mathbf{r}_1);$$

the delta function here eliminates one of the integrations in (80.8). Taking account also of the fact that $\mathcal{D}_{ik}$ is an even function of ζ_s, we can rewrite (80.8) as

$$\delta F = \delta F_0 - \frac{T}{4\pi}\sum_{s=0}^{\infty}{}' \int \zeta_s^2\mathcal{D}_{ll}(\zeta_s; \mathbf{r}, \mathbf{r})\, \delta\varepsilon(i\,|\,\zeta_s\,|, \mathbf{r})\, d^3x, \qquad (80.9)$$

where the summation is taken only over positive s; the prime denotes that the term with $s = 0$ has an extra factor $\frac{1}{2}$. This term is finite, since the factor ζ_s^2 cancels the divergence of $\mathcal{D}_{ll}$ at $\zeta_s = 0$.

For the writing of further formulae, it will be convenient to introduce two further functions:

$$\left.\begin{aligned} \mathcal{D}_{ik}^E(\zeta_s; \mathbf{r}, \mathbf{r}') &= -\zeta_s^2\mathcal{D}_{ik}(\zeta_s; \mathbf{r}, \mathbf{r}'), \\ \mathcal{D}_{ik}^H(\zeta_s; \mathbf{r}, \mathbf{r}') &= \operatorname{curl}_{il}\operatorname{curl}'_{km}\mathcal{D}_{lm}(\zeta_s; \mathbf{r}, \mathbf{r}'), \end{aligned}\right\} \qquad (80.10)$$

† We shall not give the general rule for determining the sign of diagrams of the type (80.5) (without free external lines). In the present case, the sign is easily found by writing explicitly the corresponding terms in the expansions in (80.3) and (80.6). It is indeed sufficient to note that this term in (80.3) contains one contraction of a pair of A operators, and in (80.6) two pairs; since the contraction of one pair gives $-\mathcal{D}_{ik}$, the diagrams (80.5) and (80.7) have opposite signs, and this leads to the minus sign in (80.8).

constructed analogously to (76.3) and (76.4). Then δF can be written

$$\delta F = \delta F_0 + \frac{T}{4\pi}\sum_{s=0}^{\infty}{}' \int \mathcal{D}_{ll}^{E}(\zeta_s; \mathbf{r}, \mathbf{r})\, \delta\varepsilon(i\,|\,\zeta_s\,|, \mathbf{r})\, d^3x. \tag{80.11}$$

We now use (80.11) to determine the forces acting in an inhomogeneous medium. The isotropy of the medium has already been assumed; we shall now suppose that it is a liquid, so that the change of state at each point (at a given temperature) can only be due to a change in the density ϱ.

Let us imagine that the medium is subjected to an isothermal small deformation with displacement $\mathbf{u}(\mathbf{r})$. The corresponding change in its free energy is

$$\delta F = -\int \mathbf{f}\cdot\mathbf{u}\, d^3x, \tag{80.12}$$

where $\mathbf{f}$ is the volume density of the forces acting on the medium. On the other hand, the same change can be determined from (80.11) by expressing the variations δF_0 and $\delta\varepsilon$ in terms of the same displacement vector. Let $P_0(\varrho, T)$ be the pressure without allowance for the van der Waals corrections for given values of ϱ and T; the corresponding density of volume forces is $\mathbf{f}_0 = -\nabla P_0$, so that

$$\delta F_0 = \int \mathbf{u}\cdot\nabla P_0\, d^3x.$$

Next, the change in the density is related to the displacement vector by the equation of continuity $\delta\varrho = -\operatorname{div}(\varrho\mathbf{u})$. The change in the permittivity is therefore

$$\delta\varepsilon = (\partial\varepsilon/\partial\varrho)\,\delta\varrho = -(\partial\varepsilon/\partial\varrho)\operatorname{div}(\varrho\mathbf{u}).$$

Substituting this in (80.11), integrating by parts over the whole volume of the body, and then comparing the resulting expression for δF with (80.12), we find

$$\mathbf{f} = -\nabla P_0 - \frac{T}{4\pi}\sum_{s=0}^{\infty}{}' \varrho\, \operatorname{grad}\left[\mathcal{D}_{ll}^{E}(\zeta_s; \mathbf{r}, \mathbf{r})\frac{\partial\varepsilon}{\partial\varrho}\right]. \tag{80.13}$$

This formula enables us, in particular, to determine at once the correction to the chemical potential of the body. To do so, we write the condition of mechanical equilibrium $\mathbf{f} = 0$, and use the fact that at constant temperature

$$dP_0(\varrho, T) = (\varrho/m)\, d\mu_0(\varrho, T),$$

where $\mu_0(\varrho, T)$ is the unperturbed chemical potential of the body (m being the mass of a particle). Then the condition becomes $\varrho\nabla\mu = 0$, where

$$\mu = \mu_0(\varrho, T) + \frac{mT}{4\pi}\sum_{s=0}^{\infty}{}' \mathcal{D}_{ll}^{E}(\zeta_s; \mathbf{r}, \mathbf{r})\frac{\partial\varepsilon}{\partial\varrho}. \tag{80.14}$$

On the other hand, the condition for the mechanical equilibrium of any inhomogeneous body is that the chemical potential should be constant throughout the body; it is therefore clear that (80.14) gives this potential.

The most complete description of the forces acting in a medium is given by the stress tensor σ_{ik}, which is related to the components of the vector **f** by

$$f_i = \partial\sigma_{ik}/\partial x_k. \tag{80.15}$$

To bring the expression (80.13) to this form, we first rewrite it as

$$f_i = -\frac{\partial P_0}{\partial x_i}+\frac{T}{4\pi}\sum{}'\frac{\partial}{\partial x_i}\left\{\left(\varepsilon(\mathbf{r})-\varrho\frac{\partial\varepsilon}{\partial\varrho}\right)\mathcal{D}^E_{ll}(\mathbf{r},\mathbf{r})\right\}$$
$$-\frac{T}{4\pi}\sum{}'\varepsilon(\mathbf{r})\frac{\partial}{\partial x_i}\mathcal{D}^E_{ll}(\mathbf{r},\mathbf{r});$$

for brevity, the arguments ζ_s will not be written in the intermediate formulae. The first two terms already have the required form. The third term may be written as

$$-\frac{T}{4\pi}\sum{}'\left\{\varepsilon(\mathbf{r}')\frac{\partial}{\partial x_i}+\varepsilon(\mathbf{r})\frac{\partial}{\partial x'_i}\right\}\mathcal{D}^E_{ll}(\mathbf{r},\mathbf{r}'),$$

separating the differentiations with respect to the two arguments of the function $\mathcal{D}_{ll}(\mathbf{r},\mathbf{r})$, and putting $\mathbf{r}=\mathbf{r}'$ at the end of the calculation. The calculation makes use of the equations (see (79.8))

$$\hat{\Lambda}_{il}\mathcal{D}_{lk}(\mathbf{r},\mathbf{r}') = -4\pi\delta_{ik}\delta(\mathbf{r}-\mathbf{r}'),$$
$$\hat{\Lambda}'_{il}\mathcal{D}_{kl}(\mathbf{r},\mathbf{r}') = -4\pi\delta_{ik}\delta(\mathbf{r}-\mathbf{r}'),$$

where

$$\hat{\Lambda}_{il} = \zeta_s^2\varepsilon(\mathbf{r})\,\delta_{il}+\mathbf{curl}_{im}\,\mathbf{curl}_{ml} = \zeta_s^2\varepsilon(\mathbf{r})\,\delta_{il}+\frac{\partial^2}{\partial x_i\,\partial x_l}-\delta_{il}\triangle$$

The resulting equation is (with $\mathbf{r}=\mathbf{r}'$)

$$\varepsilon\frac{\partial}{\partial x_i}\mathcal{D}^E_{ll} = 2\frac{\partial}{\partial x_k}[\varepsilon\mathcal{D}^E_{ik}+\mathcal{D}^H_{ik}]-\frac{\partial}{\partial x_i}\mathcal{D}^H_{ll}$$

and the final expression for the stress tensor is

$$\sigma_{ik} = -P_0\delta_{ik}-\frac{T}{2\pi}\sum_{s=0}^{\infty}{}'\left\{-\frac{1}{2}\delta_{ik}\left[\varepsilon(i\zeta_s,\mathbf{r})-\varrho\frac{\partial\varepsilon(i\zeta_s,\mathbf{r})}{\partial\varrho}\right]\times\right.$$
$$\times\mathcal{D}^E_{ll}(\zeta_s;\mathbf{r},\mathbf{r})+\varepsilon(i\zeta_s,\mathbf{r})\,\mathcal{D}^E_{ik}(\zeta_s;\mathbf{r},\mathbf{r})$$
$$\left.-\frac{1}{2}\delta_{ik}\mathcal{D}^H_{ll}(\zeta_s;\mathbf{r},\mathbf{r})+\mathcal{D}^H_{ik}(\zeta_s;\mathbf{r},\mathbf{r})\right\}. \tag{80.16}$$

The formulae obtained, however, do not yet have a direct physical significance. The reason is that the function $\mathcal{D}_{ik}(\mathbf{r},\mathbf{r}')$ tends to infinity as $1/|\mathbf{r}-\mathbf{r}'|$ when $\mathbf{r}'\to\mathbf{r}$, as is easily seen by means of equation (79.8). This divergence arises from the contribution of large wave numbers ($k\sim 1/|\mathbf{r}-\mathbf{r}'|$), and is due only to the invalidity of equation (79.8) for $k \gtrsim a$. This difficulty can be avoided by not making an explicit cut-off at large k. We note that the short-wavelength fluctuations have no connection with the effects under consideration, which are due to the inhomogeneity of the medium. Their contribution to the thermodynamic quantities at any given point in the body is the same for a homogeneous medium and for a medium that is inhomogeneous but has the same value of $\varepsilon(\mathbf{r})$ at the point. To give the formulae a definite meaning that is in fact independent of the nature of the cut-off, we must therefore subtract appropriately. The Green's function $\mathcal{D}_{ik}(\zeta_s; \mathbf{r}, \mathbf{r})$ is to be taken as the limit of the difference

$$\lim_{\mathbf{r}' \to \mathbf{r}} \{\mathcal{D}_{ik}(\zeta_s; \mathbf{r}, \mathbf{r}') - \overline{\mathcal{D}}_{ik}(\zeta_s; \mathbf{r}, \mathbf{r}')\}, \tag{80.17}$$

where $\overline{\mathcal{D}}_{ik}$ is the Green's function of an auxiliary homogeneous infinite medium whose permittivity is the same as that of the actual medium at the given point $\mathbf{r}$; this limit is not divergent. To avoid further complication of the formulae, we leave them in the previous form and treat $\mathcal{D}_{ik}$ as denoting the difference (80.17). Here $P_0(\varrho, T)$ is the pressure in an infinite homogeneous medium for given values of ϱ and T.

Both in formula (80.16) and in the equation (79.8) that determines the Green's function $\mathcal{D}_{ik}$, the properties of the medium occur only through $\varepsilon(i\zeta)$, the permittivity as a function of the imaginary frequency. In this connection it may be recalled that the function has a simple relation to the imaginary part of the permittivity for real frequencies:

$$\varepsilon(i\zeta) = 1 + \frac{2}{\pi} \int_0^\infty \frac{\omega \operatorname{im} \varepsilon(\omega)}{\omega^2 + \zeta^2} d\omega \tag{80.18}$$

(see *ECM*, §62). We may therefore say that the only macroscopic characteristic that determines the van der Waals forces in a material medium is ultimately the imaginary part of its permittivity.

Formula (80.16) has exactly the same form as the expression known in macroscopic electrodynamics for the Maxwell stress tensor in a constant electromagnetic field, the quadratic combinations of the components **E** and **H** being replaced by the corresponding functions $-\mathcal{D}^E_{ik}$ and $-\mathcal{D}^H_{ik}$. This analogy is not a very profound one, however: it does not signify that for a variable electromagnetic field as such there is a general expression for the stress tensor in an absorbing medium, containing only the permittivity as a characteristic of the medium. In the present case we have not an arbitrary electromagnetic field but the thermodynamic-equilibrium intrinsic fluctuational field in the medium.

§ 81. Forces of molecular interaction between solid bodies. The general formula

Let us apply the general formulae derived in §80 to calculate the forces of interaction between solids whose surfaces are a very short distance apart, this distance satisfying only the one condition that it is large compared with interatomic distances in the bodies. This enables us to treat the problem macroscopically, regarding the bodies as continuous media and their interaction as being brought about by the fluctuational electromagnetic field. The important fluctuations are those whose wavelengths are of the order of the characteristic dimensions of the problem, namely the width of the gap between the bodies.†

The suffixes 1 and 2 will denote quantities pertaining to the two solids, and 3 will denote those pertaining to the gap between them (Fig. 17). The gap will

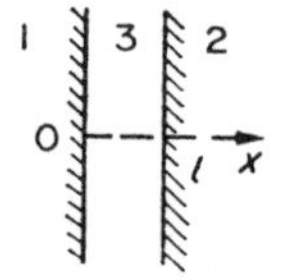

FIG. 17.

be assumed plane-parallel, with the x-axis perpendicular to its plane, so that the surfaces of bodies 1 and 2 are the planes $x = 0$ and $x = l$, where l is the gap width. The force F acting on unit area of the surface of body 2, say, is calculated as the momentum flux into the body through this surface. The flux is given by the component σ_{xx} of the electromagnetic stress tensor in the gap, taken at $x = l$. In a vacuum, $\varepsilon = 1$, and the expression for σ_{xx} from (80.16) becomes‡

$$F = \sigma_{xx}(l) = \frac{T}{4\pi} \sum_{n=0}^{\infty}{}' \{\mathcal{D}^E_{yy}(\zeta_n; l, l) + \mathcal{D}^E_{zz}(\zeta_n; l, l) - \mathcal{D}^E_{xx}(\zeta_n; l, l)$$

$$+\mathcal{D}^H_{yy}(\zeta_n; l, l) + \mathcal{D}^H_{zz}(\zeta_n; l, l) - \mathcal{D}^H_{xx}(\zeta_n; l, l)\}; \tag{81.1}$$

in this section the summation suffix will be denoted by n.

Because the problem is homogeneous in the y and z directions, the functions $\mathcal{D}_{ik}(\zeta_n; \mathbf{r}, \mathbf{r}')$ depend only on the differences $y-y'$ and $z-z'$ (the arguments $y-y'$ and $z-z'$ are not written out in (81.1)); $\mathcal{D}_{ik}(\zeta_n, \mathbf{q}; x, x')$ are the Fourier components with respect to these variables. Then

$$\mathcal{D}_{ik}(\zeta_n; \mathbf{r}, \mathbf{r}) = \int \mathcal{D}_{ik}(\zeta_n, \mathbf{q}; x, x)\, d^2q/(2\pi)^2. \tag{81.2}$$

† The results in this section and §82 are due to E. M. Lifshitz (1954).
‡ In the intermediate calculations we put $\hbar = 1, c = 1$.

For the functions $\mathcal{D}_{ik}(\zeta_n, \mathbf{q}; x, x')$, equations (79.8) become (with the y-axis parallel to the vector $\mathbf{q}$)

$$\left(w^2-\frac{d^2}{dx^2}\right)\mathcal{D}_{zz}(x, x') = -4\pi\delta(x-x'),$$

$$\left(w^2-q^2-\frac{d^2}{dx^2}\right)\mathcal{D}_{yy}(x, x')+iq\frac{d}{dx}\mathcal{D}_{xy}(x, x') = -4\pi\delta(x-x'),$$

$$w^2\mathcal{D}_{xy}(x, x')+iq\, d\mathcal{D}_{yy}(x, x')/dx = 0,$$

$$w^2\mathcal{D}_{xx}(x, x')+iq\, d\mathcal{D}_{yx}(x, x')/dx = -4\pi\delta(x-x'),$$

where $w = (\varepsilon\zeta_n^2+q^2)^{1/2}$, $\varepsilon=\varepsilon(i\zeta_n)$, and x' acts as a parameter; the components $\mathcal{D}_{xz} = \mathcal{D}_{yz} = 0$, since the equations for them prove to be homogeneous. The solution of this system reduces to that of the two equations

$$\left(w^2-\frac{d^2}{dx^2}\right)\mathcal{D}_{zz}(x, x') = -4\pi\delta(x-x'), \tag{81.3}$$

$$\left(w^2-\frac{d^2}{dx^2}\right)\mathcal{D}_{yy}(x, x') = -\frac{4\pi w^2}{\varepsilon\zeta_n^2}\delta(x-x'), \tag{81.4}$$

and $\mathcal{D}_{xy}$ and $\mathcal{D}_{xx}$ are then determined as

$$\left.\begin{aligned}\mathcal{D}_{xy}(x, x') &= -\frac{iq}{w^2}\frac{d}{dx}\mathcal{D}_{yy}(x, x'),\\ \mathcal{D}_{xx}(x, x') &= -\frac{iq}{w^2}\frac{d}{dx}\mathcal{D}_{yx}(x, x')-\frac{4\pi}{w^2}\delta(x-x').\end{aligned}\right\} \tag{81.5}$$

Here it must be taken into account that, from (79.5),

$$\mathcal{D}_{yx}(\mathbf{r}, \mathbf{r}') = \mathcal{D}_{xy}(\mathbf{r}', \mathbf{r}), \quad \text{and therefore} \quad \mathcal{D}_{yx}(\mathbf{q}; x, x') = \mathcal{D}_{xy}(-\mathbf{q}; x', x).$$

The boundary conditions, corresponding to continuity of the tangential components of the electric and magnetic fields, amount to requiring the continuity of the quantities $\mathcal{D}^E_{yk}, \mathcal{D}^E_{zk}, \mathcal{D}^H_{yk}, \mathcal{D}^H_{zk}$, or, equivalently, of the quantities

$$\mathcal{D}_{yk}, \quad \mathcal{D}_{zk}, \quad \operatorname{curl}_{yl}\mathcal{D}_{lk}, \quad \operatorname{curl}_{zl}\mathcal{D}_{lk}.$$

Using the first equation (81.5), we find that the following quantities must be continuous at the boundary:

$$\mathcal{D}_{zz}, \quad \frac{d}{dx}\mathcal{D}_{zz}, \quad \mathcal{D}_{yy}, \quad \frac{\varepsilon}{w^2}\frac{d}{dx}\mathcal{D}_{yy}. \tag{81.6}$$

Since we have in mind to calculate the stress tensor only in the region of the gap, we can immediately take $0 < x' < l$. In the range $0 < x < l$ the functions

$\mathcal{D}_{yy}$ and $\mathcal{D}_{zz}$ are determined by equations (81.3), (81.4) with $\varepsilon = 1$, $w = w_3 = (\zeta_n^2+q^2)^{1/2}$. In regions 1 $(x < 0)$ and 2 $(x > l)$, they satisfy the same equations with zero on the right (since here $x \neq x'$) and with respectively ε_1, w_1 and ε_2, w_2 as ε, w.

The subtraction needed according to (80.17) amounts to subtracting from all functions $\mathcal{D}_{ik}$ in the gap region their values for $\varepsilon_1 = \varepsilon_2 = 1$. In particular, therefore, we can immediately omit the second term on the right of the second equation (81.5), so that in the gap region

$$\mathcal{D}_{xy} = -\frac{iq}{w_3^2}\frac{d}{dx}\mathcal{D}_{yy}, \quad \mathcal{D}_{xx} = -\frac{iq}{w_3^2}\frac{d}{dx}\mathcal{D}_{yx}. \tag{81.7}$$

Before going on to solve the equations, we should make one further comment. The general solution of equations (81.3) and (81.4) is $f^-(x-x')+f^+(x+x')$. Using these equations with (81.7) and the definition of the functions $\mathcal{D}_{ik}^E$ and $\mathcal{D}_{ik}^H$, we can show that the parts of the Green's functions which depend on $x+x'$ make no contribution to the expression (81.1) for the force. We shall not discuss this point further here, as it is already obvious from physical considerations: putting $x = x'$ in a solution of the form $f^+(x+x')$, we should obtain a momentum flux in the gap that varied with the coordinate, which would contradict the law of conservation of momentum. Henceforward we shall therefore include only the expressions for the parts $\mathcal{D}_{ik}^-$ of the Green's functions that do not depend on $x+x'$.

Let us now determine the function $\mathcal{D}_{zz}$. It satisfies the equations

$$\left.\begin{aligned}
(w_1^2-d^2/dx^2)\,\mathcal{D}_{zz}(x,x') &= 0, && x<0,\\
(w_2^2-d^2/dx^2)\,\mathcal{D}_{zz}(x,x') &= 0, && x>l,\\
(w_3^2-d^2/dx^2)\,\mathcal{D}_{zz}(x,x') &= -4\pi\delta(x-x'), && 0<x<l.
\end{aligned}\right\} \tag{81.8}$$

Hence we find

$$\mathcal{D}_{zz} = Ae^{w_1x}, \quad x<0; \qquad \mathcal{D}_{zz} = Be^{-w_2x}, \quad x>l;$$
$$\mathcal{D}_{zz} = C_1e^{w_3x}+C_2e^{-w_3x}-(2\pi/w_3)\,e^{-w_3|x-x'|}, \qquad 0<x<l.$$

In the last expression we have used the fact that, according to the third equation (81.8), the derivative $d\mathcal{D}_{zz}/dx$ has a discontinuity of 4π at $x = x'$. Determining A, B, C_1 and C_2 (which are functions of x') from the boundary conditions that $\mathcal{D}_{zz}$ and $d\mathcal{D}_{zz}/dx$ are continuous, we obtain

$$\mathcal{D}_{zz}^- = \frac{4\pi}{w_3\Delta}\cosh w_3(x-x')-\frac{2\pi}{w_3}e^{-w_3|x-x'|}, \qquad 0<x<l,$$

where

$$\Delta = 1-e^{2w_3l}\frac{(w_1+w_3)(w_2+w_3)}{(w_1-w_3)(w_2-w_3)}.$$

Subtracting the value of $\mathcal{D}_{zz}^{-}$ for $w_1 = w_2 = w_3$ (and $1/\Delta = 0$), we have finally

$$\mathcal{D}_{zz}^{-} = \frac{4\pi}{w_3\Delta}\cosh w_3(x-x').$$

Similarly, solving the equation for $\mathcal{D}_{yy}$, we obtain (after the subtraction)

$$\mathcal{D}_{yy}^{-} = \frac{4\pi w_3}{\zeta_n^2\Delta_1}\cosh w_3(x-x'),$$

$$\Delta_1 = 1-e^{2w_3 l}\frac{(\varepsilon_1 w_3+w_1)(\varepsilon_2 w_3+w_2)}{(\varepsilon_1 w_3-w_1)(\varepsilon_2 w_3-w_2)}$$

and, using (81.7),

$$\mathcal{D}_{xy}^{-} = \mathcal{D}_{yx}^{-} = -(4\pi iq/\zeta_n^2\Delta_1)\sinh w_3(x-x'),$$

$$\mathcal{D}_{xx}^{-} = -(4\pi q^2/\zeta_n^2 w_3\Delta_1)\cosh w_3(x-x').$$

Now, calculating the functions $\mathcal{D}_{ik}^{E}$ and $\mathcal{D}_{ik}^{H}$, and then transforming them in accordance with (81.2), and substituting in (81.1), we obtain

$$F(l) = -\frac{T}{2\pi}\sum_{n=0}^{\infty}{}' \int_0^{\infty} w_3\left(\frac{1}{\Delta}+\frac{1}{\Delta_1}\right) q\,dq.$$

Finally, changing to a new variable of integration p with $q = \zeta_n\sqrt{(p^2-1)}$, and returning to ordinary units, we have as the expression for the force F on unit area of each of the two bodies separated by a gap of width l

$$F(l) = \frac{T}{\pi c^3}\sum_{n=0}^{\infty}{}'\zeta_n^3\int_1^{\infty} p^2\left\{\left[\frac{(s_1+p)(s_2+p)}{(s_1-p)(s_2-p)}\exp\left(\frac{2p\zeta_n l}{c}\right)-1\right]^{-1}\right.$$
$$\left.+\left[\frac{(s_1+p\varepsilon_1)(s_2+p\varepsilon_2)}{(s_1-p\varepsilon_1)(s_2-p\varepsilon_2)}\exp\left(\frac{2p\zeta_n l}{c}\right)-1\right]^{-1}\right\}dp, \qquad (81.9)$$

where $s_1 = \sqrt{(\varepsilon_1-1+p^2)}$, $s_2 = \sqrt{(\varepsilon_2-1+p^2)}$, $\zeta_n = 2\pi nT/\hbar$, ε_1 and ε_2 being functions of the imaginary frequency $\omega = i\zeta_n$; in this connection it should be remembered that $\varepsilon(i\zeta)$ is a positive real quantity that decreases monotonically from its electrostatic value ε_0 at $\zeta = 0$ to 1 at $\zeta = \infty$.† The positive values of F correspond to attraction between the bodies. The integrand in each term of the sum (81.9) is positive, and for any given p and ζ_n decreases monotonically

† Formula (81.9) has been derived on the assumption that both bodies are isotropic. Its application to crystals therefore depends on neglecting the anisotropy of their permittivity. Although this is entirely legitimate in most cases, the anisotropy of bodies causes in general a specific effect, namely a torque tending to rotate the bodies relative to one another.

as l increases.† Hence $F > 0$ and $dF/dl < 0$, i.e. the bodies (separated by a vacuum gap) attract each other with a force that decreases monotonically with increasing distance.

The general formula (81.9) is very complicated. It can be considerably simplified, however, because the influence of the temperature on the interaction force is usually quite unimportant.‡ The reason is that, because of the exponentials in the integrands in (81.9), only those terms are important in the sum which have $\zeta_n \sim c/l$ or $n \sim c\hbar/lT$. In the case $lT/c\hbar \ll 1$ the important values of n are therefore large, and in (81.9) we can change from summation to integration over $dn = \hbar d\zeta/2\pi T$. The temperature then disappears from the formula, and the result is

$$F(l) = \frac{\hbar}{2\pi^2 c^3} \int_0^\infty \int_1^\infty p^2 \zeta^3 \left\{ \left[\frac{(s_1+p)(s_2+p)}{(s_1-p)(s_2-p)} \exp\left(\frac{2p\zeta l}{c}\right) - 1 \right]^{-1} + \left[\frac{(s_1+p\varepsilon_1)(s_2+p\varepsilon_2)}{(s_1-p\varepsilon_1)(s_2-p\varepsilon_2)} \exp\left(\frac{2p\zeta l}{c}\right) - 1 \right]^{-1} \right\} dp\, d\zeta. \tag{81.10}$$

According to the above discussion, this is valid for distances $l \ll c\hbar/T$, and even at room temperatures the distances concerned are up to about 10^{-4} cm.

Formula (81.10) can be considerably further simplified in two limiting cases.

§ 82. Forces of molecular interaction between solid bodies. Limiting cases

Let us first consider the limiting case of "small" distances, by which we mean distances small in comparison with the characteristic wavelengths λ_0 of the absorption spectra of the bodies concerned. The temperatures that may be in question for condensed bodies are always small compared with the $\hbar\omega_0$ important here (for example, in the visible spectrum), and the inequality $lT/c\hbar \ll 1$ is therefore always satisfied.

Because of the exponential factor in the denominators of the integrand, the important range in the integration with respect to p is that where $p\zeta l/c \sim 1$. Here $p \gg 1$, and therefore we can put $s_1 \approx s_2 \approx p$ in determining the principal term in the integral. In this approximation, the first term in the braces in (81.10) is zero. The second term, with a new variable of integration $x = 2p\zeta l/c$, gives

$$F(l) = \frac{\hbar}{16\pi^2 l^3} \int_0^\infty \int_0^\infty x^2 \left[\frac{(\varepsilon_1+1)(\varepsilon_2+1)}{(\varepsilon_1-1)(\varepsilon_2-1)} e^x - 1 \right]^{-1} dx\, d\zeta; \tag{82.1}$$

† This is easily seen by noting that for $s = \sqrt{(\varepsilon - 1 + p^2)}$ (and $p \geqslant 1$) we have the inequalities $\varepsilon p > s > p$ when $\varepsilon > 1$.

‡ In speaking of the influence of the temperature, we are not referring to that due simply to the temperature dependence of the permittivity itself.

in this approximation, the lower limit of integration with respect to x is replaced by zero.†

In this case, the force is inversely proportional to the cube of the distance, as we should expect in accordance with the usual behaviour of the van der Waals forces between two atoms; see the next footnote. The functions $\varepsilon(i\zeta)-1$ decrease monotonically with increasing ζ and tend to zero. Hence the values of ζ beyond some ζ_0 no longer contribute significantly to the integral; the condition for l to be small means that we must have $l \ll c/\zeta_0$.

We shall show how the change can be made from the macroscopic formula (82.1) to the interaction of individual atoms in vacuum. To do so, we formally assume that both bodies are sufficiently rarefied. Macroscopically, this means that their permittivities are almost unity, i.e. that ε_1-1 and ε_2-1 are small. From (82.1) we then have with the necessary accuracy

$$F = \frac{\hbar}{64\pi^2 l^3}\int_0^\infty\int_0^\infty x^2 e^{-x}(\varepsilon_1-1)(\varepsilon_2-1)\,dx\,d\zeta$$

$$= \frac{\hbar}{32\pi^2 l^3}\int_0^\infty [\varepsilon_1(i\zeta)-1][\varepsilon_2(i\zeta)-1]\,d\zeta.$$

Expressing $\varepsilon(i\zeta)$ in terms of im $\varepsilon(\omega)$ on the real ω-axis, by (80.18), we obtain

$$F = \frac{\hbar}{8\pi^4 l^3}\int\int\int_0^\infty \frac{\omega_1\omega_2 \,\mathrm{im}\,\varepsilon_1(\omega_1)\,\mathrm{im}\,\varepsilon_2(\omega_2)}{(\omega_1^2+\zeta^2)(\omega_2^2+\zeta^2)}\,d\zeta\,d\omega_1\,d\omega_2$$

$$= \frac{\hbar}{16\pi^3 l^3}\int\int_0^\infty \frac{\mathrm{im}\,\varepsilon_1(\omega_1)\,\mathrm{im}\,\varepsilon_2(\omega_2)}{\omega_1+\omega_2}\,d\omega_1\,d\omega_2. \qquad (82.2)$$

This force corresponds to an interaction of atoms with energy

$$U(r) = -\frac{3\hbar}{8\pi^4 n_1 n_2 r^6}\int\int \frac{\mathrm{im}\,\varepsilon_1(\omega_1)\,\mathrm{im}\,\varepsilon_2(\omega_2)}{\omega_1+\omega_2}\,d\omega_1\,d\omega_2, \qquad (82.3)$$

† An integral of the form

$$\frac{1}{2}a\int_0^\infty \frac{x^2\,dx}{ae^x-1}$$

varies only slightly, from 1 to 1.2, when a varies from ∞ to 1. We can therefore write (82.1) with sufficient accuracy in practice, as

$$F = \frac{\hbar\bar{\omega}}{8\pi^2 l^3}, \quad \bar{\omega} = \int_0^\infty \frac{[\varepsilon_1(i\zeta)-1][\varepsilon_2(i\zeta)-1]}{[\varepsilon_1(i\zeta)+1][\varepsilon_2(i\zeta)+1]}\,d\zeta.$$

The quantity $\bar{\omega}$ acts as a frequency characterizing the absorption spectra of the two bodies.

where r is the distance between the atoms; n_1, n_2 are the number densities of atoms in the two bodies.† This formula agrees with London's formula in quantum mechanics, which is derived by applying ordinary perturbation theory to the dipole interaction of two atoms (see *QM*, §89, Problem). In making the comparison, it must be borne in mind that the imaginary part of $\varepsilon(\omega)$ is related to the spectral density of "oscillator strengths" $f(\omega)$ by

$$\omega \operatorname{im} \varepsilon(\omega) = (2\pi^2 e^2/m)\, n f(\omega),$$

where e and m are the electron charge and mass; see *ECM*, §62. The oscillator strengths are expressed, in the usual manner, in terms of the squared matrix elements of the dipole moments of the atoms; see *QM*, (149.10).

Let us now turn to the opposite case of "large" distances, $l \gg \lambda_0$. We shall, however, suppose that the distances are still not so large as to violate the inequality $lT/\hbar c \ll 1$.

In (81.10) we again use a new variable of integration $x = 2pl\zeta/c$, but leave p and not ζ as the second variable. Then ε_1 and ε_2 are functions of $i\zeta = ixc/2pl$. On account of the factor e^x in the denominators of the integrand, the important values of x in the integral with respect to x are ~ 1, and since $p \geq 1$ the argument of ε for large l is almost zero throughout the important range of values of the variables. Accordingly, we can replace ε_1 and ε_2 simply by their values for $\zeta = 0$, i.e. the electrostatic permittivities ε_{10} and ε_{20}. Thus we have finally

$$\left.\begin{aligned} F = \frac{\hbar c}{32\pi^2 l^4} \int_0^\infty \int_1^\infty \frac{x^3}{p^2} \Bigg\{ \left[\frac{(s_{10}+p)(s_{20}+p)}{(s_{10}-p)(s_{20}-p)} e^x - 1 \right]^{-1} \\ + \left[\frac{(s_{10}+p\varepsilon_{10})(s_{20}+p\varepsilon_{20})}{(s_{10}-p\varepsilon_{10})(s_{20}-p\varepsilon_{20})} e^x - 1 \right]^{-1} \Bigg\} dp\, dx, \\ s_{10} = \sqrt{(\varepsilon_{10}-1+p^2)}, \quad s_{20} = \sqrt{(\varepsilon_{20}-1+p^2)}. \end{aligned}\right\} \tag{82.4}$$

The law of decrease with distance as l^{-4} here corresponds to the decrease of the van der Waals forces between two atoms with allowance for retardation (see below).

Formula (82.4) reduces to a very simple expression when both bodies are metals. For metals, $\varepsilon(i\zeta) \to \infty$ as $\zeta \to 0$, and we can therefore take $\varepsilon_0 = \infty$. Putting $\varepsilon_{10} = \varepsilon_{20} = \infty$, we obtain

$$F = \frac{\hbar c}{16\pi^2 l^4} \int_0^\infty \int_1^\infty \frac{x^3\, dp\, dz}{p^2(e^x-1)} = \frac{\pi^2}{240} \frac{\hbar c}{l^4} \tag{82.5}$$

† If the potential energy of the interaction of atoms 1 and 2 is $U(r) = -ar^{-6}$, the total energy of pair interactions of all the atoms in two half-spaces separated by a gap of width l is $U_{tot} = -a\pi n_1 n_2/12l^2$. The force is $F = dU_{tot}/dl = a\pi n_1 n_2/6l^3$. This is the correspondence between (82.2) and (82.3).

(H. B. G. Casimir 1948). This force is independent of the nature of the metals, a property which does not hold at small distances, where the interaction force depends on the behaviour of the function $\varepsilon(i\zeta)$ for all values of ζ and not only at $\zeta = 0$.

Figure 18 shows a graph of the function $\phi_{ii}(\varepsilon_0)$, which gives the attractive force between two identical insulators ($\varepsilon_{10} = \varepsilon_{20} \equiv \varepsilon_0$); formula (82.4) is written as

$$F = \frac{\pi^2}{240}\frac{\hbar c}{l^4}\left(\frac{\varepsilon_0-1}{\varepsilon_0+1}\right)^2 \phi_{ii}(\varepsilon_0). \tag{82.6}$$

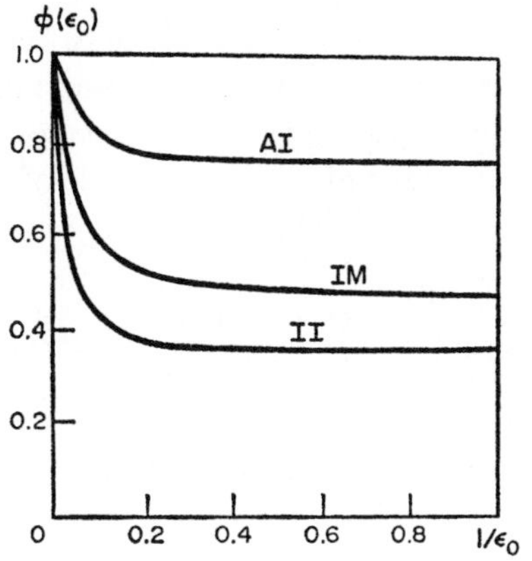

FIG. 18.

The same diagram shows the function $\phi_{im}(\varepsilon_0)$, which gives the attractive force for an insulator and a metal ($\varepsilon_{10} = \varepsilon_0$, $\varepsilon_{20} = \infty$), from the formula†

$$F = \frac{\pi^2}{240}\frac{\hbar c}{l^4}\frac{\varepsilon_0-1}{\varepsilon_0+1}\phi_{im}(\varepsilon_0). \tag{82.7}$$

In (82.4) we make the transition to the interaction of individual atoms, as was done above for formula (82.1). For small $\varepsilon_0 - 1$, we have

$$s_0 - p \approx (\varepsilon_0-1)/2p, \quad s_0 - p\varepsilon_0 \approx (\varepsilon_0-1)(-p+\tfrac{1}{2}p^{-1}),$$

and the integral (82.4) becomes

$$F = \frac{\hbar c}{32\pi^2 l^4}(\varepsilon_{10}-1)(\varepsilon_{20}-1)\int_0^\infty x^3 e^{-x}\,dx \int_1^\infty \frac{1-2p^2+2p^4}{8p^6}\,dp,$$

whence

$$F = \frac{\hbar c}{l^4}\frac{23}{640\pi^2}(\varepsilon_{10}-1)(\varepsilon_{20}-1). \tag{82.8}$$

† As $\varepsilon_0 \to 1$, the functions ϕ_{ii} and ϕ_{im} tend to 0.35 and 0.46 respectively, corresponding to the limiting forms (82.8) and (1) in the Problem. As $\varepsilon_0 \to \infty$, both functions tend to unity, corresponding to (82.5).

This force corresponds to the interaction of two atoms with energy

$$U(r) = -\frac{23\hbar c}{4\pi r^7}\alpha_1\alpha_2, \qquad (82.9)$$

where α_1 and α_2 are the static polarizabilities of the atoms ($\varepsilon_0 = 1+4\pi n\alpha$). Formula (82.9) agrees with the result of the calculation in quantum electrodynamics for the attraction of two atoms at sufficiently large distances, when retardation effects become important (see *RQT*, §85).

Lastly, let us consider distances so great that $lT/\hbar c \gg 1$, the opposite inequality to that required for the effect of the temperature to be negligible. In this case only the first term in the sum (81.9) need be retained, but we cannot immediately put $n = 0$ in this term, since there is an indeterminacy: the factor ζ_n^3 is zero, but the integral with respect to p diverges. This difficulty can be avoided by at first using instead of p a new variable of integration $x = 2p\zeta_n l/c$ (so that the factor ζ_n^3 disappears). Then putting $\zeta_n = 0$, we obtain

$$F = \frac{T}{16\pi l^3}\int_0^\infty x^2\left[\frac{(\varepsilon_{10}+1)(\varepsilon_{20}+1)}{(\varepsilon_{10}-1)(\varepsilon_{20}-1)}e^x - 1\right]^{-1} dx. \qquad (82.10)$$

Thus, at sufficiently large distances the decrease of the attractive force becomes slower, and returns to the l^{-3} law, but with a coefficient that depends on the temperature; all subsequent terms in the sum in (81.9) decrease exponentially with increasing l. The condition $lT/\hbar c \gg 1$ is essentially the classicality condition ($\hbar\omega \ll T$, where $\omega \sim c/l$). It is therefore natural that the expression (82.10) does not involve $\hbar$.†

PROBLEM

Find the law of interaction of an atom with a metal wall at "large" distances.

SOLUTION. The interaction of an individual atom with a condensed body can be found by considering only one of the bodies (labelled 2, say) as a rarefied medium. Regarding $\varepsilon_{20}-1$ as small and putting $\varepsilon_{10} = \infty$, we obtain from (82.4)

$$F = \frac{\hbar c(\varepsilon_{20}-1)}{32\pi^2 l^4}\int_0^\infty x^3 e^{-x}\,dx \int_1^\infty \frac{dp}{2p^2} = \frac{3\hbar c(\varepsilon_{20}-1)}{32\pi^2 l^4}. \qquad (1)$$

If the atom–wall interaction energy is $U = -aL^{-4}$ (where L is the distance from the atom to the wall), the energy of interaction of atoms in a half-space separated from the wall by a gap l is $U_{\text{tot}} = -an/3l^3$, and the force $F = dU_{\text{tot}}/dl = an/l^4$. Thus the value found for F corresponds to the attraction of an individual atom to the wall with the energy

$$U(L) = -3\alpha_2\hbar c/8\pi L^4 \qquad (2)$$

(H. B. G. Casimir and D. Polder 1948).

† The formulae obtained in §§81 and 82 can be generalized to include the case where the gap between the solids is filled with a liquid and the case of a thin liquid film on a solid surface; see I. E. Dzyaloshinskiĭ, E. M. Lifshitz and L. P. Pitaevskiĭ, *Soviet Physics Uspekhi* **4**, 153, 1961 (= *Advances in Physics* **10**, 165, 1961).

For the interaction of an atom with an insulating wall, the same method gives

$$U(L) = -\frac{3\hbar c\alpha_2}{8\pi L^4}\frac{\varepsilon_{10}-1}{\varepsilon_{10}+1}\phi_{ai}(\varepsilon_{10})$$

with the function ϕ_{ai} shown graphically in Fig. 18. As $\varepsilon_{10} \to 1$ this function tends to the value $23/30 = 0.77$, corresponding to (82.8).

§ 83. Asymptotic behaviour of the correlation function in a liquid

Long-wavelength electromagnetic fluctuations lead also to certain specific properties of the correlation function of the density fluctuations in a homogeneous liquid.

The correlation function $\nu(r)$ is determined (see Part 1, §116) in terms of the mean value of the product of the fluctuations of the particle number density n at two points in space by

$$\langle \delta n(\mathbf{r}_1)\, \delta n(\mathbf{r}_2)\rangle = \bar{n}\delta(\mathbf{r})+\bar{n}\nu(r), \qquad \mathbf{r} = \mathbf{r}_1-\mathbf{r}_2. \tag{83.1}$$

The correlation function is related to the interaction between particles, and its asymptotic behaviour at large distances is determined by the long-range van der Waals part of this interaction. Hence $\nu(r)$, like the van der Waals forces, decreases as an inverse power of the distance (J. E. Enderby, T. Gaskell and N. H. March 1965).

This, of course, also affects the properties of the Fourier components of the correlation function $\nu(\mathbf{k}) \equiv \nu(k)$. If the only forces acting between particles in the liquid had a range of the order of the atomic dimensions a, $\nu(r)$ would decrease exponentially with increasing distance, the exponent being $\sim r/a$.† In terms of Fourier components this means that $\nu(k)$ would be a regular function of ka and could be expanded in even powers of ka when $ka \ll 1$. The long-range forces, however, cause the occurrence in $\nu(k)$ of a term $\nu_1(k)$ that varies considerably even in the range $k \sim 1/\lambda_0$ (not $k \sim 1/a$), where λ_0 ($\gg a$) denotes characteristic wavelengths in the spectrum of the liquid. In the range $ka \ll 1$, the parameter $k\lambda_0$ may be either small or large; the function $\nu_1(k)$ has a singularity in this range.

To calculate the correlation function, we use its relation to the second variational derivative of the free energy of the body with respect to its density. By definition, this derivative is the function $\phi(r)$ that occurs in the expression

$$\delta F = \tfrac{1}{2}\int \phi(|\mathbf{r}_1-\mathbf{r}_2|)\, \delta n(\mathbf{r}_1)\, \delta n(\mathbf{r}_2)\, d^3x_1\, d^3x_2 \tag{83.2}$$

for the change in the free energy due to density fluctuations (at a given

† The liquid under consideration is at a temperature $T \sim \Theta$, where $\Theta \sim \hbar u/a$ is the "Debye temperature" of the liquid, and far from the critical point. Near the critical point, the correlation radius increases indefinitely (see Part 1, §§152, 153). It also increases at low temperatures, and for $T \ll \Theta$ it is of the order of $\hbar u/T$ (see §87 below).

temperature). The Fourier component $\phi(\mathbf{k}) \equiv \phi(k)$ of this function is related to the required function $\nu(k)$ by

$$\nu(k) = \frac{T}{\bar{n}\phi(k)} - 1; \tag{83.3}$$

see Part 1, (116.14). We must emphasize that this formula assumes the fluctuations to be classical, for which it is necessary that $\hbar\omega \ll T$, where ω is the frequency of oscillations with wave number k. With $\omega \sim ku$ (where u is the velocity of sound in the liquid) we get the condition

$$\hbar ku \ll T, \tag{83.4}$$

corresponding to distances $r \gg \hbar u/T$.

The "regular" part of the function $\phi(k)$, due to the short-range forces, can be expanded in powers of k; taking (when $ka \ll 1$) only the first term of the expansion and denoting it by b, we can write

$$\phi(k) \approx b + \phi_1(k), \tag{83.5}$$

where $\phi_1(k)$ is the "singular" part that is now of interest.† Because the van der Waals forces are relatively weak, $\phi_1(k) \ll b$, and so the result of substituting (83.5) in (83.3) can be put in the form

$$\nu(k) = \frac{T}{\bar{n}b} - 1 - \frac{T}{\bar{n}b^2}\phi_1(k). \tag{83.6}$$

Since $\nu(k)$ and $\phi_1(k)$ are linearly related, the function $\nu(r)$ at large distances is simply

$$\nu(r) = -(T/\bar{n}b^2)\,\phi_1(r). \tag{83.7}$$

The first (k-independent) term in (83.6) corresponds to a coordinate function constant$\times\delta(\mathbf{r})$ due to the short-range forces (if their range of action is regarded as negligible).

To determine $\phi_1(r)$, we start from formula (80.11) for the variation of the free energy. Putting there

$$\delta\varepsilon(i\zeta_s, \mathbf{r}) = \frac{\partial\varepsilon(i\zeta_s)}{\partial\bar{n}}\,\delta n(\mathbf{r}), \tag{83.8}$$

we see that the expression

$$-\frac{T}{4\pi\hbar c^2}\sum_{s=0}^{\infty}{}' \zeta_s^2 \mathcal{D}_{ll}(\zeta_s; \mathbf{r}, \mathbf{r})\frac{\partial\varepsilon(i\zeta_s)}{\partial\bar{n}}$$

† The constant b is expressed in terms of the thermodynamic quantities for the liquid by $b = (1/\bar{n})(\partial P/\partial\bar{n})_T$; see Part 1, §152.

is the first variational derivative of the free energy with respect to the density. For the second differentiation we must vary this expression in turn, obtaining†

$$-\frac{T}{4\pi\hbar c^2}\sum_{s=0}^{\infty}{}' \zeta_s^2 \delta \mathcal{D}_{ll}(\zeta_s; \mathbf{r}, \mathbf{r}) \frac{\partial \varepsilon(i\zeta_s)}{\partial \bar{n}}. \tag{83.9}$$

The function $\mathcal{D}$ itself satisfies equation (79.8):

$$\left[\frac{\partial^2}{\partial x_i \partial x_l} - \delta_{il}\triangle + \frac{\zeta_s^2}{c^2}\varepsilon(i\zeta_s, \mathbf{r})\,\delta_{il}\right]\mathcal{D}_{lk}(\zeta_s; \mathbf{r}, \mathbf{r}') = -4\pi\hbar\delta_{ik}\delta(\mathbf{r}-\mathbf{r}'), \tag{83.10}$$

and its variation gives the equation for the variation of $\mathcal{D}$:

$$\left[\frac{\partial^2}{\partial x_i \partial x_l} - \delta_{il}\triangle + \frac{\zeta_s^2}{c^2}\varepsilon(i\zeta_s)\,\delta_{il}\right]\delta\mathcal{D}_{ik}(\zeta_s; \mathbf{r}, \mathbf{r}') = -\frac{\zeta_s^2}{c^2}\delta\varepsilon(i\zeta_s, \mathbf{r})\,\mathcal{D}(\zeta_s; \mathbf{r}, \mathbf{r}'). \tag{83.11}$$

The solution of (83.11) may be written down at once by noticing that, from (83.10), the "unperturbed" function $\mathcal{D}_{ik}$ is the Green's function of this equation; hence

$$\delta\mathcal{D}_{ik}(\zeta_s; \mathbf{r}, \mathbf{r}') = \frac{\zeta_s^2}{4\pi\hbar c^2}\int \delta\varepsilon(i\zeta_s, \mathbf{r}'')\,\mathcal{D}_{lk}(\zeta_s; \mathbf{r}'', \mathbf{r}')\mathcal{D}_{li}(\zeta_s; \mathbf{r}'', \mathbf{r})\,d^3x'';$$

here we have also used the fact that $D_{il}(\mathbf{r}, \mathbf{r}'') = D_{li}(\mathbf{r}'', \mathbf{r})$. Finally, substituting (83.8) here and the result in (83.9), we obtain the second variational derivative

$$\phi_1(r) = -\frac{T}{(4\pi\hbar c^2)^2}\sum_{s=0}^{\infty}{}' \zeta_s^4 \left[\frac{\partial \varepsilon(i\zeta_s)}{\partial \bar{n}}\right]^2 \mathcal{D}_{lm}^2(\zeta_s; \mathbf{r}_1, \mathbf{r}_2), \tag{83.12}$$

with $r = |\mathbf{r}_1 - \mathbf{r}_2|$. This formula together with (83.7) gives the required general expression for the correlation function $\nu(r)$ when $r \gg \hbar u/T$ (M. P. Kemoklidze and L. P. Pitaevskiĭ 1970).

The condition (83.4) already assumed previously for the wave numbers is equivalent to $r \gg \hbar u/T$ for the distances. If, simultaneously with this condition, we restrict the range of values of r by an upper limit also:

$$\hbar c/T \gg r \gg \hbar u/T, \tag{83.13}$$

then large values of s are important in the sum, and the summation over discrete "frequencies" $\zeta_s = 2\pi Ts/\hbar$ can be replaced by integration over $ds = \hbar d\zeta/2\pi T$:

$$\nu(r) = \frac{T}{\bar{n}b^2\hbar c^4}\int_0^\infty \left[\frac{1}{4\pi}\frac{\partial \varepsilon(i\zeta)}{\partial \bar{n}}\right]^2 \zeta^4 \mathcal{D}_{lm}^2(\zeta; \mathbf{r}_1, \mathbf{r}_2)\frac{d\zeta}{2\pi}. \tag{83.14}$$

† Only the function $\mathcal{D}_{ll}$ is varied. Varying ε would lead to a term of the form constant $\times\delta(\mathbf{r})$ in $\phi(r)$, which does not relate to the long-range forces.

The function $\mathcal{D}_{lm}$ is obtained from (77.6) by the substitution $\omega \to i\zeta$. Carrying out the differentiation and squaring, we have

$$\left.\begin{aligned} \mathcal{D}_{lm}^2 &= \frac{2\hbar^2}{r^2}e^{-2w}\left(1+\frac{2}{w}+\frac{5}{w^2}+\frac{6}{w^3}+\frac{3}{w^4}\right), \\ w &= r\zeta\sqrt{\varepsilon(i\zeta)}/c. \end{aligned}\right\} \tag{83.15}$$

Substitution of (83.15) in (83.14) gives a fairly complicated expression, but this becomes simpler in two limiting cases.

For "small" distances ($r \ll \lambda_0$; cf. §81) the important range in the integral is $\zeta \sim c/\lambda_0$; then $r\zeta/c \ll 1$, so that we can replace the exponential factor in (83.15) by unity, and keep only the last term in the brackets. We then find

$$\nu(r) = \frac{A}{r^6}, \quad A = \frac{3\hbar T}{16\pi^3\bar{n}b^2}\int\limits_0^\infty \left[\frac{\partial\varepsilon(i\zeta)}{\partial\bar{n}}\right]^2 \frac{d\zeta}{\varepsilon^2(i\zeta)}, \qquad r \ll \lambda_0. \tag{83.16}$$

The Fourier transform of this function is†

$$\nu(k) = \pi^2 A k^3/12, \qquad k\lambda_0 \gg 1. \tag{83.17}$$

In the opposite case of "large" distances ($r \gg \lambda_0$) the important range in the integral is $\zeta \sim c/r \ll c/\lambda_0 \sim \omega_0$. We can therefore replace $\varepsilon(i\zeta)$ by its electrostatic value ε_0 and take $(\partial\varepsilon_0/\partial n)^2$ outside the integral in (83.14). The integration is then elementary (and all the terms in (83.15) make contributions of the same order of magnitude). The result is

$$\nu(r) = B/r^7, \quad B = \frac{23\hbar cT}{64\pi^3\varepsilon_0^{3/2}\bar{n}b^2}\left(\frac{\partial\varepsilon_0}{\partial\bar{n}}\right)^2, \qquad r \gg \lambda_0. \tag{83.18}$$

The Fourier transform of this function is

$$\nu(k) = -(\pi/30)\,Bk^4 \log k\lambda_0, \qquad k\lambda_0 \ll 1. \tag{83.19}$$

§ 84. Operator expression for the permittivity

In this section we shall derive a useful expression for the permittivity of a medium in terms of the commutator of the charge density operator (P. Nozières and D. Pines 1958). This formula is analogous to Kubo's formula, taking account of the specific nature of the electromagnetic field.

† By direct integration in spherical polar coordinates in **k**-space, we obtain

$$I_\nu \equiv \lim_{\lambda \to +0} \int e^{i\mathbf{k}\cdot\mathbf{r}-\lambda k}k^\nu \frac{d^3k}{(2\pi)^3} = -\frac{\Gamma(\nu+2)\sin\frac{1}{2}\pi\nu}{2\pi^2 r^{\nu+3}}.$$

The integral needed to verify (83.17) is I_3. The integral needed to verify (83.19) is $dI_\nu/d\nu$ with $\nu = 4$.

We shall consider a homogeneous medium having both time and space dispersion of the permittivity. This means that the induction $\mathbf{D}(t, \mathbf{r})$ depends on the values of the field $\mathbf{E}(t, \mathbf{r})$ not only at previous times but at other points in space. Such a dependence can be generally represented as

$$D_i(t, \mathbf{r}) = E_i(t, \mathbf{r}) + \int_0^\infty \int f_{ik}(\tau, \mathbf{r}')\, E_k(t-\tau, \mathbf{r}-\mathbf{r}')\, d^3x'\, d\tau. \tag{84.1}$$

For a monochromatic field in which $\mathbf{E}$ and $\mathbf{D} \propto \exp[i(\mathbf{k}.\mathbf{r}-\omega t)]$, this relation becomes

$$D_i = \varepsilon_{ik}(\omega, \mathbf{k})\, E_k, \tag{84.2}$$

where

$$\varepsilon_{ik}(\omega, \mathbf{k}) = \delta_{ik} + \int_0^\infty \int f_{ik}(\tau, \mathbf{r}')\, e^{i(\omega\tau - \mathbf{k}.\mathbf{r}')}\, d^3x'\, d\tau. \tag{84.3}$$

We shall take only the case where the medium is not only homogeneous but also isotropic and without natural optical activity. Then the permittivity remains a tensor, but one that contains only the vector $\mathbf{k}$. The general form of such a tensor is

$$\varepsilon_{ik} = \varepsilon_l(\omega, \mathbf{k}) \frac{k_i k_k}{k^2} + \varepsilon_t(\omega, \mathbf{k}) \left(\delta_{ik} - \frac{k_i k_k}{k^2}\right). \tag{84.4}$$

The scalar functions ε_l and ε_t are called respectively the *longitudinal permittivity* and the *transverse permittivity*. If $\mathbf{E}$ is a potential field, $\mathbf{E} = -\nabla\phi$, then for a plane wave it is parallel to the wave vector ($\mathbf{E} = -i\mathbf{k}\phi$) and then $\mathbf{D} = \varepsilon_l\mathbf{E}$. If the field is solenoidal ($\operatorname{div}\mathbf{E} = i\mathbf{k}.\mathbf{E} = 0$), $\mathbf{E}$ is perpendicular to the wave vector, and then $\mathbf{D} = \varepsilon_t\mathbf{E}$.

With this description of the properties of the medium, there is (cf. *ECM*, §83) no significance in dividing the mean microscopic current density $\overline{\varrho\mathbf{v}}$ (ϱ being the charge density) into two parts $\partial\mathbf{P}/\partial t$ and $c \operatorname{curl}\mathbf{M}$, where $\mathbf{P}$ is the electric polarization and $\mathbf{M}$ the magnetization of the medium. Thus Maxwell's equations are

$$\operatorname{curl}\mathbf{E} = -\frac{1}{c}\frac{\partial\mathbf{B}}{\partial t}, \qquad \operatorname{curl}\mathbf{B} = \frac{1}{c}\frac{\partial\mathbf{D}}{\partial t},$$

without the introduction of the vector $\mathbf{H}$ as well as the magnetic induction $\mathbf{B}$ which is the mean microscopic magnetic field. All terms resulting from the averaging of microscopic currents are assumed to be included in the definition $\mathbf{D} = \mathbf{E}+4\pi\mathbf{P}$, $\overline{\varrho\mathbf{v}} = \partial\mathbf{P}/\partial t$.

The longitudinal permittivity is important in applications, and we shall derive an operator expression for it. This is found by considering the response of the system to an external (i.e. generated by sources outside the system) potential electric field $\mathbf{E}_{\text{ex}} = -\nabla\phi_{\text{ex}}$.

The operator of the interaction of the system with this field is

$$\hat{V} = \int \hat{\varrho}(t, \mathbf{r})\, \phi_{\text{ex}}(t, \mathbf{r})\, d^3x, \tag{84.5}$$

where $\hat{\varrho}(t, \mathbf{r})$ is the charge density operator in the system. Comparing this expression with the general formula (75.8) and regarding ϕ_{ex} as a "generalized force" f, we immediately find from formulae (75.9)–(75.11) that the Fourier components with respect to time of the mean charge density are

$$\bar{\varrho}_\omega(\mathbf{r}) = -\frac{i}{\hbar}\int_0^\infty \int e^{i\omega t}\langle \hat{\varrho}(t, \mathbf{r})\, \hat{\varrho}(0, \mathbf{r}') - \hat{\varrho}(0, \mathbf{r}')\, \hat{\varrho}(t, \mathbf{r})\rangle\, \phi_\omega^{\text{ex}}(\mathbf{r}')\, d^3x'\, dt.$$

Changing also to spatial Fourier components and using the fact that, since the system is homogeneous, the mean value of the commutator depends only on $\mathbf{r}-\mathbf{r}'$, we obtain

$$\bar{\varrho}_{\omega\mathbf{k}} = \alpha(\omega, \mathbf{k})\, \phi_{\omega\mathbf{k}}^{\text{ex}}, \tag{84.6}$$

where

$$\alpha(\omega, \mathbf{k}) = -\frac{i}{\hbar}\int_0^\infty \int e^{i(\omega t - \mathbf{k}\cdot\mathbf{r})}\langle \hat{\varrho}(t, \mathbf{r})\, \hat{\varrho}(0, 0) - \hat{\varrho}(0, 0)\, \hat{\varrho}(t, \mathbf{r})\rangle\, d^3x\, dt. \tag{84.7}$$

The mean charge density is related to the polarization vector of the medium by $\bar{\varrho} = -\operatorname{div}\mathbf{P}$ (see *ECM*, §6). Hence, for the Fourier components,

$$\bar{\varrho}_{\omega\mathbf{k}} = -i\mathbf{k}\cdot\mathbf{P}_{\omega\mathbf{k}} = -i(\varepsilon_l - 1)\,\mathbf{k}\cdot\mathbf{E}_{\omega\mathbf{k}}/4\pi.$$

On the other hand, $\triangle\phi_{\text{ex}} = -4\pi\varrho_{\text{ex}}$, where ϱ_{ex} is the density of the charges that create the external field; the induction $\mathbf{D}$ is related to this charge density by $\operatorname{div}\mathbf{D} = 4\pi\varrho_{\text{ex}}$. From these two equations we find

$$\phi_{\omega\mathbf{k}}^{\text{ex}} = (4\pi/k^2)\, \varrho_{\omega\mathbf{k}}^{\text{ex}} = (i\varepsilon_l/k^2)\, \mathbf{k}\cdot\mathbf{E}_{\omega\mathbf{k}}.$$

Finally, substituting these expressions in (84.6), we obtain the required expression for the longitudinal permittivity:

$$\frac{1}{\varepsilon_l(\omega, \mathbf{k})} = 1 + \frac{4\pi}{k^2}\alpha(\omega, \mathbf{k}). \tag{84.8}$$

In (84.7) $\hat{\varrho}(t, \mathbf{r})$ should be taken, strictly speaking, as the charge density operator of all particles in the system, both electrons and nuclei. Usually, however, the electrons make the principal contribution to the permittivity throughout the important range of values of ω and $\mathbf{k}$; we can therefore take $\hat{\varrho}$ as $e(\hat{n} - \bar{n})$, where $\hat{n}$ is the electron density operator and $\bar{n}$ its mean value.

Formulae (84.7) and (84.8) can be further transformed by expressing them in terms of the matrix elements of the Fourier components of the operator $\hat{\varrho}$. To do so, we first rewrite (84.7) as

$$\alpha(\omega, \mathbf{k}) = -\frac{i}{\hbar V}\int_0^\infty e^{i\omega t}\langle \hat{\varrho}_{\mathbf{k}}(t)\,\hat{\varrho}_{-\mathbf{k}}(0) - \hat{\varrho}_{-\mathbf{k}}(0)\,\hat{\varrho}_{\mathbf{k}}(t)\rangle\,dt, \qquad (84.9)$$

where V is the volume of the system. The matrix elements of the Heisenberg operator $\hat{\varrho}_{\mathbf{k}}(t)$ are expressed in terms of those of the Schrödinger operator by

$$\left(\varrho_{\mathbf{k}}(t)\right)_{mn} = e^{i\omega_{mn}t}(\varrho_{\mathbf{k}})_{mn}.$$

Expanding the product of operators by the matrix multiplication rule and integrating according to (31.21), we have finally

$$\frac{1}{\varepsilon_l(\omega, \mathbf{k})} = 1 + \frac{4}{\hbar k^2 V}\sum_n |(\varrho_{\mathbf{k}})_{n0}|^2\left\{\frac{1}{\omega-\omega_{n0}+i0} - \frac{1}{\omega+\omega_{n0}+i0}\right\}, \qquad (84.10)$$

where the suffix 0 refers to the given state for which the permittivity is sought.

§ 85. A degenerate plasma

Let us consider a fully ionized plasma, in which the ions form a classical (Boltzmann) gas and the electron component is degenerate. For this, the temperature must satisfy the conditions $\mu_i \ll T \lesssim \mu_e$, i.e.

$$\hbar^2 n^{2/3}/m_i \ll T \lesssim \hbar^2 n^{2/3}/m_e, \qquad (85.1)$$

where μ_e and μ_i are the chemical potentials of the electrons and ions in the plasma, m_e and m_i the electron and ion masses, n the particle number density; in making estimates, we do not distinguish between n_e and n_i. We shall also suppose that the plasma is almost ideal. For this to be so, the energy of the Coulomb interaction between two particles at a distance $l \sim n^{-1/3}$ apart must be small in comparison with their mean kinetic energy ε. For ions $\varepsilon \sim T$, and for electrons $\varepsilon \sim \mu_e \sim n^{2/3}\hbar^2/m_e$. Hence we have the conditions

$$m_e e^2/\hbar^2 \ll n^{1/3} \ll T/e^2. \qquad (85.2)$$

It has been shown in Part 1, §80, that under these conditions the chief source of corrections in the thermodynamic quantities of the plasma (as compared with their values for an ideal gas) is the exchange interaction of the electrons; the energy of this interaction (per unit volume of plasma) is $\sim e^2/n^{4/3}$. The correlation correction (the main one in a classical plasma) is small in a degenerate plasma, in the ratio $\eta^{1/2}$ relative to the exchange correction, where $\eta = m_e e^2/\hbar^2 n^{1/3} \ll 1$. Nevertheless, its calculation for a degenerate plasma is of methodological interest, and affords an instructive illustration of the use of the diagram technique.

The Coulomb interaction operator of the plasma particles is

$$\hat{V} = \frac{1}{2} e^2 \sum_{a,b} \int \hat{\Psi}^+_{a\alpha} \hat{\Psi}'^+_{b\beta} \frac{z_a z_b}{|\mathbf{r}-\mathbf{r}'|} \hat{\Psi}'_{b\beta} \hat{\Psi}_{a\alpha}\, d^3x\, d^3x', \tag{85.3}$$

where the suffixes a and b label the different kinds of particle (electrons and various ions); $z_a e$ is the charge on a particle (for electrons, $z_e = -1$). Taking the ψ operators in the Matsubara representation, we obtain the interaction operator in that representation. The diagram technique for calculating the mean value $\langle \hat{V} \rangle$ (over the Gibbs distribution) is then carried through in the usual manner by changing to the interaction representation for Matsubara operators; the resulting perturbation-theory series is an expansion of $\langle \hat{V} \rangle$ in powers of e^2.

The expression (85.3) contains no "free" variables (i.e. variables over which there is no integration). In the diagram technique, this is expressed by the fact that the terms in the perturbation-theory series for $\langle \hat{V} \rangle$ are represented by diagrams having no free external lines. The broken lines in these diagrams, with 4-momenta $Q = (\zeta_s, \mathbf{q})$, will be arbitrarily associated with factors†

$$-\phi(\mathbf{q}) = -4\pi/q^2 \tag{85.4}$$

(which are independent of ζ), i.e. minus the Fourier components of the potential $\phi(\mathbf{r})$ of the field of a unit charge. The continuous lines must now be assigned, together with the 4-momentum $P = (\zeta_s, \mathbf{p})$, an additional suffix a which indicates the type of particle, and each such line is associated with a factor $-\mathcal{G}^{(0)}_{a\alpha\beta}(P)$, which is minus the Green's function of the free particles a. The continuous lines in the diagram form closed loops, each containing sections with the same a. Each vertex of the diagram (a point of intersection of a broken line with continuous lines of type a) is associated with an additional factor $z_a e$. Each fermion loop contributes an extra factor of -1. The diagrams constructed according to these rules give terms in the expansion of

$$-(2/V)\langle \hat{V} \rangle. \tag{85.5}$$

The factor V in the denominator is the volume of the system. This factor arises because the integrand in each term of the series depends only on the differences of the coordinates, and therefore one of the integrations over d^3x gives simply the volume V. The minus sign in (85.5) results from the determination of the broken lines by the rule (85.4), i.e. with the minus sign before $\phi(\mathbf{q})$. The factor 2 results from taking the factor $\frac{1}{2}$ in (85.3) to the left-hand side.

In the first order of perturbation theory, there are diagrams of two kinds:

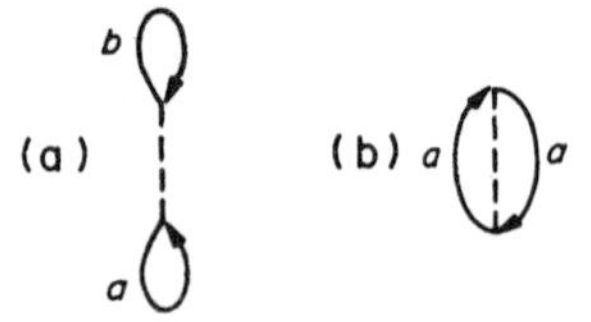

(85.6)

† In the rest of this section we put $\hbar = 1$, $c = 1$, and e (> 0) denotes the unit charge.

with all possible a and b. Those of the form (85.6a) arise from contractions of ψ operators taken at the same point in space. These diagrams correspond to the direct Coulomb interaction of particles a and b distributed uniformly in space; their contributions cancel out on summation over all pairs a, b, because the plasma is electrically neutral. Diagrams of the form (85.6b) arise from contractions of ψ operators with different arguments, and correspond to the exchange interaction of particles of one type a. The calculation of this diagram leads to the results already obtained in Part 1, §80.

In the next order, diagrams of the following kinds occur:

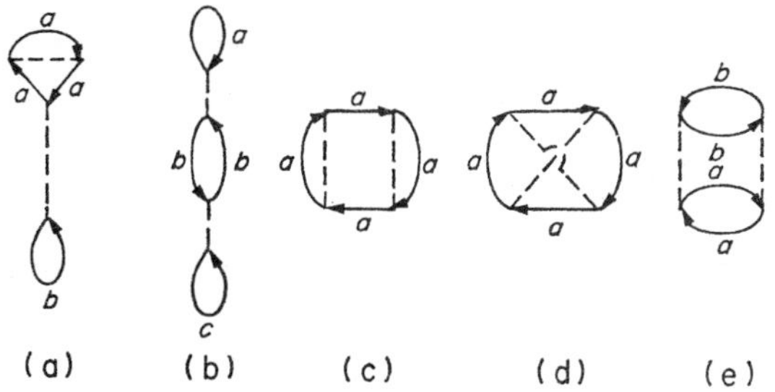

(85.7)

Diagrams (85.7a, b) are corrections to (85.6a) and for the same reason cancel on summation over all a, b, c. Diagrams (85.7c, d) are small corrections to the exchange interaction energy and are of no interest here.

Diagram (85.7e) is "anomalously large" because the corresponding integral diverges. This divergence occurs because the momenta $\mathbf{q}$ of the two broken lines in the diagram are the same (as is obvious from the conservation of momentum at the vertices). Hence the diagram contains the integral $\int d^3q/q^4$, which diverges as $1/q$ when $\mathbf{q}$ is small.

In subsequent approximations there occur (as well as correction diagrams) new "ring" diagrams with even stronger divergence. For example, the third-order diagram

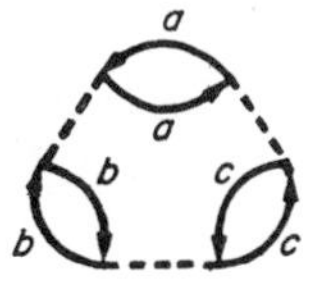

with three broken lines having the same momentum $\mathbf{q}$ contains the integral $\int q^{-6}\, d^3q$, which diverges as q^{-3}. In general, the ring diagram of order n, formed by n continuous loops joined by n broken lines, diverges as $q^{-(2n-3)}$.

The summation of the infinite sequence of ring diagrams leads, as we shall see, to an effective cut-off of the divergences at values of q of the order of smallness of e; hence all these diagrams together give a contribution to $\langle V \rangle$ that is of the order of smallness of $(e^2)^n/e^{2n-3} = e^3$. This contribution is represented graphically by the sum (over kinds of particle) of the skeleton diagrams

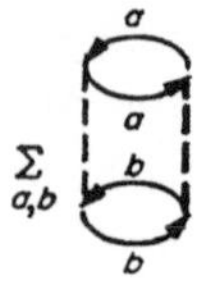

$$\sum_{a,b} \qquad\qquad \text{(85.8)}$$

where the thick broken line represents the sum of the infinite set of linear diagrams

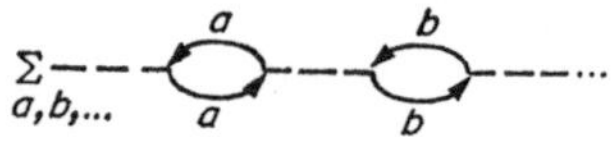

$$\sum_{a,b,\ldots} \qquad\qquad \text{(85.9)}$$

with various numbers of continuous loops.

Whereas the thin broken line represents the potential ϕ of the Coulomb field of an isolated charge, the thick one represents the potential (which we denote by Φ) of the field perturbed by the polarization of the surrounding plasma. The total contribution (85.8) therefore gives the required correlation part of the mean interaction energy in the plasma.

We use the notation $-\mathcal{P}(\zeta_s, \mathbf{q})/4\pi$ for the sum of simple continuous loops of all types of particles, and denote this quantity graphically by a white circle:

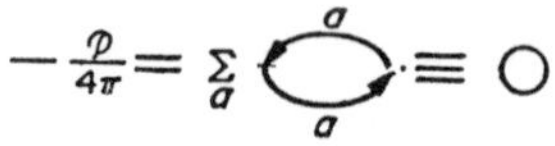

$$-\frac{\mathcal{P}}{4\pi} = \sum_a \qquad\qquad \equiv \bigcirc \qquad\qquad \text{(85.10)}$$

The argument ζ_s of this function takes "even" values $\zeta_s = 2s\pi T$, whatever the statistics obeyed by the particles a: by the law of conservation of frequencies at the vertex, this argument is equal to the difference of the frequencies of the two continuous lines, which is "even" for both "even" and "odd" terms.

With the notation (85.10), the sum (85.8) is represented by one skeleton diagram

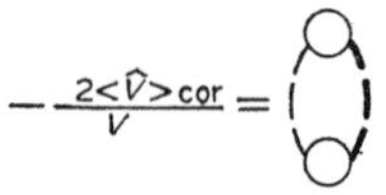

$$-\frac{2\langle \hat{V} \rangle_{\text{cor}}}{V} = \qquad\qquad \text{(85.11)}$$

The thick broken line itself satisfies the diagram equation

$$---- = ---- + ---\bigcirc--- \qquad\qquad \text{(85.12)}$$

which is exactly analogous to (14.4) and (79.13). In analytical form, this equation is

$$-\Phi(\zeta_s, \mathbf{q}) = -\phi(\mathbf{q}) - \phi(\mathbf{q}) \frac{\mathcal{P}(\zeta_s, \mathbf{q})}{4\pi} \Phi(\zeta_s, \mathbf{q}),$$

whence

$$\Phi(\zeta_s, \mathbf{q}) = \frac{4\pi}{q^2 - \mathcal{P}(\zeta_s, \mathbf{q})}. \tag{85.13}$$

It is useful to regard these formulae from a somewhat different viewpoint, in order to establish the connection with the diagrams in §79. The Coulomb interaction between charges may be treated as resulting from the exchange of virtual photons. Here, however, it is more convenient to use not the gauge (75.1) but the Coulomb gauge (see *RQT*, §77), in which $-D_{00}$ is the Fourier component of the Coulomb potential. The spatial part D_{ik} in this gauge describes the retardation and the magnetic interaction, and can be neglected in a non-relativistic plasma. We may therefore suppose that the broken lines in (85.11) correspond to the Matsubara $\mathcal{D}_{00}$, and that the function $\mathcal{P}$ is just the component $\mathcal{P}_{00}$ of the polarization operator. According to (79.18) we can therefore write

$$\mathcal{P}(\zeta_s, \mathbf{q}) = -q^2[\varepsilon_l(i\,|\zeta_s|, \mathbf{q}) - 1];$$

it is easy to see that the longitudinal permittivity ε_l occurs in (79.18) when there is spatial dispersion. Substituting this expression in (85.13), we find

$$\Phi(\zeta_s, \mathbf{q}) = \frac{4\pi}{q^2 \varepsilon_l(i\,|\zeta_s|, \mathbf{q})}, \tag{85.14}$$

i.e. the Fourier component of the potential of a unit charge in the medium, as it should be.

Expanding the diagram (85.11) by the general rules of the Matsubara technique, we find

$$\begin{aligned} \langle V \rangle_{\text{cor}} &= -\frac{1}{2} VT \sum_s \int \frac{\mathcal{P}^2(\zeta_s, \mathbf{q})}{(4\pi)^2} \phi(\mathbf{q}) \Phi(\zeta_s, \mathbf{q}) \frac{d^3q}{(2\pi)^3} \\ &= -\frac{1}{2} VT \sum_s \int \frac{\mathcal{P}^2(\zeta_s, \mathbf{q})}{q^2[q^2 - \mathcal{P}(\zeta_s, \mathbf{q})]} \frac{d^3q}{(2\pi)^3}. \end{aligned} \tag{85.15}$$

We shall see later that the term with $s = 0$ is the most important in the sum, and the corresponding integral is governed by the region of small $\mathbf{q}$. Hence, in calculating (85.15), it is in practice sufficient to know the limiting value of $\mathcal{P}(0, \mathbf{q})$ as $\mathbf{q} \to 0$. This quantity is easily determined from simple physical considerations without any direct calculation from the diagrams (85.10).

When $\zeta_s = 0$, the function $\Phi(0, \mathbf{q})$ is the Fourier transform of the potential $\Phi(r)$ of the electrostatic field of a unit charge in the plasma. The unperturbed potential $\phi(r)$ satisfies Poisson's equation with a delta function on the right-hand side: $\triangle\phi = -4\pi\delta(\mathbf{r})$. The equation for the potential Φ perturbed by the polarization of the plasma is found by adding on the right-hand side the change $\delta\varrho$ in the charge density in the plasma caused by the field itself:

$$\triangle\Phi = -4\pi[\delta(\mathbf{r}) + \delta\varrho]. \tag{85.16}$$

On the other hand, as $\mathbf{q} \to 0$ we have a field that varies only slowly through the plasma volume. In such a field, the thermodynamic equilibrium condition holds:

$$\mu_a + ez_a\Phi = \text{constant} = \mu_a^{(0)}, \tag{85.17}$$

where μ_a is the chemical potential of particles of type a, and $\mu_a^{(0)}$ its value in the absence of the field. From this condition we find for the change in the particle density n_a

$$\delta\mu_a = (\partial\mu_a/\partial n_a)_{T,V}\,\delta n_a = -ez_a\Phi$$

and hence for the change in the charge density

$$\delta\varrho = \sum_a ez_a\delta n_a = -\sum_a (ez_a)^2(\partial n_a/\partial\mu_a)_{T,V}\,\Phi.$$

Substituting this expression in (85.16), we obtain the equation

$$\triangle\Phi - \varkappa^2\Phi = -4\pi\delta(\mathbf{r}), \tag{85.18}$$

with

$$\varkappa^2 = 4\pi e^2 \sum_a z_a^2(\partial n_a/\partial\mu_a)_{T,V}. \tag{85.19}$$

It is seen from (85.18) that $1/\varkappa$ is the Debye radius of the field screening in the plasma (cf. Part 1, §78). Finally, taking the Fourier component of each side of (85.18), we find

$$\Phi(\mathbf{q}) = 4\pi/(q^2+\varkappa^2),$$

and a comparison of this with (85.13) gives

$$[\mathcal{D}(0, \mathbf{q})]_{\mathbf{q}\to 0} = -\varkappa^2. \tag{85.20}$$

Now integrating in (85.15) with this value of $\mathcal{D}$, we have

$$\langle\hat{V}\rangle_{\text{cor}} = -\frac{VT\varkappa^4}{2(2\pi)^3}\int\frac{4\pi q^2\,dq}{q^2(q^2+\varkappa^2)} = -\frac{VT\varkappa^3}{8\pi}. \tag{85.21}$$

First of all, we note that the integral converges at the lower limit and that the most important range is $q \sim \varkappa$. For the non-degenerate ion component of the plasma, $\partial n_i/\partial\mu_i = n_i/T$; for the electrons, $\partial n_e/\partial\mu_e \sim n_e/\mu_e$. It is easily seen that, from the conditions (85.2), $\varkappa \ll n^{1/3}$, and so $q \ll n^{1/3}$, i.e. $1/q$ is large compared with the distances between particles. This justifies the use of the equilibrium condition (85.17). To justify the neglect of all terms in the sum in (85.15) except that with $s = 0$, we note that by (85.14) the polarization of the plasma at non-zero frequencies is described by the permittivity $\varepsilon_l(\omega, \mathbf{q})$. According to the known asymptotic formula for high frequencies, $\varepsilon_l(\omega) \approx 1 - 4\pi n_e e^2/m_e\omega^2$, and therefore

$$\varepsilon_l(i\,|\zeta_s|) = 1 + 4\pi n_e e^2/m_e\zeta_s^2;$$

see *ECM*, §59. From the conditions (85.1), (85.2), all the non-zero frequencies $\zeta_s = 2s\pi T \gg (n_e e^2/m_e)^{1/2}$, and for them we can therefore take $\varepsilon(i|\zeta_s|) = 1$, i.e. the plasma is not polarized and $\mathcal{D}$ is small.

Formula (85.21) is expressed in terms of the thermodynamic variables T, V, μ_a. Hence the thermodynamic potential Ω of the plasma can be found by direct integration of the equation

$$(\partial\Omega/\partial e^2)_{T, V, \mu_a} = \langle \hat{V} \rangle / e^2; \tag{85.22}$$

see Part 1, (80.4). The result for the correlation part of Ω is (in ordinary units)

$$\Omega_{\mathrm{cor}} = -\frac{VT\varkappa^3}{12\pi} = -\frac{2\sqrt{\pi}VTe^3}{3}\left[\sum_a z_a^2 \left(\frac{\partial n_a}{\partial \mu_a}\right)_{V,T}\right]^{3/2} \tag{85.23}$$

(A. A. Vedenov 1959). According to the general theorem of small increments, the same formula expressed in terms of other thermodynamic variables gives the correction to other thermodynamic potentials.

For a non-degenerate plasma, all the derivatives $\partial n_a/\partial \mu_a = n_a/T$, and (85.23) then becomes

$$F_{\mathrm{cor}} = -\frac{2\sqrt{\pi}Ve^3}{3\sqrt{T}}\left(\sum_a z_a^2 n_a\right)^{3/2} \tag{85.24}$$

for the correction to the free energy, the same as in Part 1, (78.12).

For strong degeneracy of the electrons in a plasma ($T \ll \mu_e$), the derivative $\partial n_e/\partial \mu_e \sim n_e/\mu_e \ll n_e/T$. In the sum over a in (85.23) we can then neglect the electron term, and return to (85.24) with the difference that the sum is taken only over the kinds of ion in the plasma. Thus with strong degeneracy the electrons have no influence on the screening length or on the correlation part of the thermodynamic quantities in the plasma.

CHAPTER IX

HYDRODYNAMIC FLUCTUATIONS

§ 86. Dynamic form factor of a liquid

THE correlation function of density fluctuations, discussed in Part 1, §116, is a particular case of a more general function which relates the density fluctuations not only at different points in space but also at different times. In the classical theory, this function is defined as the mean value

$$\bar{n}\sigma(t; \mathbf{r}_1, \mathbf{r}_2) = \langle \delta n(t_1, \mathbf{r}_1)\, \delta n(t_2, \mathbf{r}_2) \rangle, \tag{86.1}$$

where $t = t_1 - t_2$; the factor $\bar{n} = N/V$, the mean number density of particles, is taken outside the definition of σ. For a homogeneous isotropic medium (a liquid or a gas) the function (86.1) depends on $\mathbf{r}_1$ and $\mathbf{r}_2$ only through the distance $r = |\mathbf{r}_1 - \mathbf{r}_2|$ between the two points, and this will be assumed in what follows.

In the quantum theory, the corresponding function is defined by means of the symmetrized product of time-dependent (Heisenberg) density operators as

$$\bar{n}\tilde{\sigma}(t, r) = \tfrac{1}{2}\langle \delta\hat{n}(t_1, \mathbf{r}_1)\, \delta\hat{n}(t_2, \mathbf{r}_2) + \delta\hat{n}(t_2, \mathbf{r}_2)\, \delta\hat{n}(t_1, \mathbf{r}_1) \rangle, \tag{86.2}$$

in accordance with the general definition as in Part 1, (118.4). There are certain advantages in the present case, however, in using the asymmetric definition

$$\bar{n}\sigma(t, r) = \langle \delta\hat{n}(t_1, \mathbf{r}_1)\, \delta\hat{n}(t_2, \mathbf{r}_2) \rangle \tag{86.3}$$

for which we retain the notation $\sigma(t, r)$.† Unlike $\tilde{\sigma}(t, r)$, $\sigma(t, r)$ is not an even function of t; it is evident that

$$\tilde{\sigma}(t, r) = \tfrac{1}{2}[\sigma(t, r) + \sigma(-t, r)]. \tag{86.4}$$

The Fourier transform of the function $\sigma(t, r)$ with respect to time and coordinates,

$$\sigma(\omega, \mathbf{k}) \equiv \sigma(\omega, k) = \int \int_{-\infty}^{\infty} e^{i(\omega t - \mathbf{k}.\mathbf{r})} \sigma(t, r)\, dt\, d^3x \tag{86.5}$$

† It is this function which is a directly observable quantity, for example, in inelastic scattering of neutrons in a liquid (see Problem).

is called the *dynamic form factor* of the medium. Since $\sigma(t, r)$ is isotropic, the form factor depends only on the magnitude of the wave vector. It follows from (86.4) that the Fourier transform of $\tilde{\sigma}(t, r)$ is

$$\tilde{\sigma}(\omega, k) = \tfrac{1}{2}[\sigma(\omega, k)+\sigma(-\omega, k)]. \tag{86.6}$$

The purely spatial correlation of the liquid density fluctuations is determined by (86.1) with $t = 0$: $\sigma(r) = \sigma(t = 0, r) = \tilde{\sigma}(t = 0, r)$. This function is related to $\nu(r)$, defined in Part 1 (§116) and used in §83, by $\sigma(r) = \nu(r)+\delta(\mathbf{r})$; the Fourier transforms are such that $\sigma(k) = \nu(k)+1$. The function $\sigma(k)$ or $\nu(k)$ is called the *static form factor* of the liquid. The functions $\sigma(\omega, k)$ and $\sigma(k)$ are related by the integral formula

$$\sigma(k) = \left[\int_{-\infty}^{\infty} \sigma(\omega, k)\, e^{-i\omega t}\frac{d\omega}{2\pi}\right]_{t=0} = \int_{-\infty}^{\infty} \sigma(\omega, k)\frac{d\omega}{2\pi}. \tag{86.7}$$

The Schrödinger (time-independent) density operator is given by the sum

$$\hat{n}(\mathbf{r}) = \sum_a \delta(\mathbf{r}-\mathbf{r}_a), \tag{86.8}$$

taken over all the particles in the medium; the coordinates $\mathbf{r}_a$ of the particles act as parameters; cf. (24.4). We shall need the components of the Fourier expansion of this operator with respect to the coordinates:

$$\begin{aligned} \hat{n}_{\mathbf{k}} &= \int \hat{n}(\mathbf{r})\, e^{-i\mathbf{k}.\mathbf{r}}\, d^3x \\ &= \sum_a e^{-i\mathbf{k}.\mathbf{r}_a}. \end{aligned} \tag{86.9}$$

The change to the time-dependent (Heisenberg) operator is made by the general rule

$$\hat{n}(t, \mathbf{r}) = \exp(i\hat{H}t/\hbar)\, \hat{n}(\mathbf{r}) \exp(-i\hat{H}t/\hbar), \tag{86.10}$$

where $\hat{H}$ is the Hamiltonian of the system. This operator may be represented by the expressions (86.8) and (86.9) with $\mathbf{r}_a$ replaced by $\hat{\mathbf{r}}_a(t)$, the Heisenberg operators of the particle coordinates.

According to the basic principles of statistical physics, the averaging $\langle\ldots\rangle$ can be variously interpreted, according to the thermodynamic variables in terms of which the result is to be expressed. For example, if the function σ is defined for given total energy and number of particles in the system, the averaging is taken with respect to a definite (mth) stationary state, i.e. from the appropriate diagonal matrix element. For a homogeneous system (a liquid), the dependence of the matrix elements of the operator $\delta\hat{n}(t, \mathbf{r})$ on the time and coordinates is given by

$$\langle m|\,\delta\hat{n}(t, \mathbf{r})\,|l\rangle = \langle m|\,\delta\hat{n}(0)|l\rangle \exp[i(\omega_{ml}t-\mathbf{k}_{ml}.\mathbf{r})], \tag{86.11}$$

which is exactly analogous to (8.4); the right-hand side contains the matrix element of the Schrödinger operator $\delta\hat{n}(\mathbf{r})$ taken at $\mathbf{r} = 0$. Using this formula, we write

$$\overline{n\sigma}(t, r) = \sum_l \langle m|\,\delta\hat{n}(t_1, \mathbf{r}_1)\,|l\rangle\langle l|\,\delta\hat{n}(t_2, \mathbf{r}_2)\,|m\rangle$$
$$= \sum_l |\langle m|\,\delta\hat{n}(0)\,|l\rangle|^2 \exp\,[i(\omega_{ml}t - \mathbf{k}_{ml}\cdot\mathbf{r})].$$

The Fourier transform of this function is

$$\overline{n\sigma}(\omega, k) = (2\pi)^4 \sum_l |\langle m|\,\delta\hat{n}(0)|\,l\rangle\,|^2\,\delta(\omega - \omega_{lm})\,\delta(\mathbf{k} - \mathbf{k}_{lm}). \tag{86.12}$$

The summation in these formulae is over all states of the system with a given number N_m of particles, since the operator $\delta\hat{n}$ does not affect this number.

If, however, we wish to express the form factor in terms of the temperature and chemical potential of the liquid, the expression (86.12) must also be averaged over the Gibbs distribution:

$$\overline{n\sigma}(\omega, k) = (2\pi)^4 \sum_{l,m} \exp\left(\frac{\Omega - E_m - \mu N_m}{T}\right) |\langle m|\,\delta\hat{n}(0)|\,l\rangle\,|^2 \delta(\omega - \omega_{lm})\,\delta(\mathbf{k} - \mathbf{k}_{lm}); \tag{86.13}$$

$N_l = N_m$ in each term of the sum. With a similar formula for $\sigma(-\omega, -\mathbf{k}) \equiv \sigma(-\omega, k)$, interchanging the summation indices l and m and putting in the exponential factor $E_l = E_m + \hbar\omega_{lm} = E_m + \hbar\omega$ (because of the delta function), we find

$$\sigma(-\omega, k) = \sigma(\omega, k)\, e^{-\hbar\omega/T} \tag{86.14}$$

and then, by (86.6),

$$\tilde{\sigma}(\omega, k) = \tfrac{1}{2}\,(1 + e^{-\hbar\omega/T})\,\sigma(\omega, k). \tag{86.15}$$

It follows from (86.13) or (86.12) that $\sigma(\omega, k) \geqslant 0$ for all values of the arguments. From (86.14) we have at zero temperature

$$\sigma(\omega, k) = 0 \quad \text{for} \quad \omega < 0, \quad T = 0. \tag{86.16}$$

In the macroscopic limit (N and $V \to \infty$ for a fixed ratio N/V), the "palisade" of delta functions in (86.13) is smoothed into a continuous function, but the delta-function peaks in $\sigma(\omega, k)$ remain for values $\omega = \omega(k)$ corresponding to non-decaying elementary excitations, as follows from arguments similar to those in §8. Such peaks occur, however, only for excitations without change in the number of particles.†

† For example, in a Fermi liquid $\sigma(\omega, k)$ has a delta-function singularity at $\omega = ku_0$ (u_0 being the velocity of zero sound), but does not have such singularities corresponding to the fermion branch of the spectrum; see §91.

We shall show how the form factor of a liquid may be related to quantities occurring in the general formulation of the fluctuation–dissipation theorem (D. Pines and P. Nozières 1958).

Let each particle in the liquid be subject to an external field which gives it the potential energy $U(t, \mathbf{r})$. Then the perturbation operator acting on the whole liquid is

$$\hat{V}(t) = \int \hat{n}(t, \mathbf{r})\, U(t, \mathbf{r})\, d^3x. \tag{86.17}$$

Taking a Fourier expansion of all the quantities here with respect to time, we can represent the response of the system, i.e. the mean value of the density change caused by the perturbation, as

$$\delta\bar{n}(\omega, \mathbf{r}_1) = -\int \alpha(\omega, |\mathbf{r}_1 - \mathbf{r}_2|)\, U(\omega, \mathbf{r}_2)\, d^3x_2, \tag{86.18}$$

where the function $\alpha(\omega, r)$ acts as a generalized susceptibility. The time Fourier component of the correlation function $\tilde{\sigma}(t, r)$ is, in the notation of the fluctuation–dissipation theorem,

$$\bar{n}\tilde{\sigma}(\omega, r) = (\delta n(\mathbf{r}_1)\, \delta n(\mathbf{r}_2))_\omega, \quad \mathbf{r} = \mathbf{r}_1 - \mathbf{r}_2.$$

According to that theorem, this function is expressed in terms of the generalized susceptibility by

$$\bar{n}\tilde{\sigma}(\omega, r) = \hbar \coth(\hbar\omega/2T)\, \mathrm{im}\, \alpha(\omega, r). \tag{86.19}$$

A similar formula gives the coordinate Fourier component $\tilde{\sigma}(\omega, k)$ in terms of $\alpha(\omega, k)$, and then from (86.15) we find the dynamic form factor

$$\bar{n}\sigma(\omega, k) = \frac{2\hbar}{1 - e^{-\hbar\omega/T}}\, \mathrm{im}\, \alpha(\omega, k). \tag{86.20}$$

The importance of these formulae is due mainly to the fact that they establish a relation between the dynamic form factor and a function with known general analytical properties (as regards the variable ω); for the function $\alpha(\omega, k)$ these properties are described in Part 1, §123. They also allow the use, in calculating the form factor, of the general formula (cf. (75.11)) whereby

$$\alpha(\omega, k) = \frac{i}{\hbar} \int \int_0^\infty e^{i(\omega t - \mathbf{k}.\mathbf{r})} \langle \hat{n}(t, \mathbf{r})\, \hat{n}(0, 0) - \hat{n}(0, 0)\, \hat{n}(t, \mathbf{r}) \rangle\, dt\, d^3x. \tag{86.21}$$

Expressing the density operators in terms of the ψ operators ($n = \hat{\Psi}^+\hat{\Psi}$), we can write this as a two-particle Green's function, which may be calculated by means of the diagram technique.

PROBLEM

Express in terms of the dynamic form factor the probability of inelastic scattering of slow neutrons in a liquid consisting of identical atoms (G. Placzek 1952).

SOLUTION. According to the pseudo-potential method (see *QM*, §151), the scattering of slow neutrons may be described as the result of interaction with a potential energy

$$U(\mathbf{r}) = (2\pi^2\hbar/M)\, a\hat{n}(\mathbf{r}), \tag{1}$$

where $\hat{n}(\mathbf{r})$ is the density operator (86.8), M the reduced mass of the atom and the neutron, and a the slow neutron scattering length for a single atom, i.e. minus the limit of the scattering amplitude. The transition probability from an initial state i of the liquid + neutron system to a final state f in a range $d\nu_f$ is

$$dw_{fi} = \left| \frac{1}{\hbar} \int_{-\infty}^{\infty} U_{fi}(t)\, dt \right|^2 d\nu_f; \tag{2}$$

see *QM*, (40.5). For non-diagonal matrix elements U_{fi} in (1), $\delta\hat{n}$ may be written in place of $\hat{n}$. The wave function of the initial state of the neutron (with momentum $\mathbf{p}$ and energy ε) is normalized to one particle in the volume V, and that of the final state (momentum $\mathbf{p}'$ and energy ε') by the delta function of $\mathbf{p}/2\pi$. Then $d\nu_f = d^3p'/(2\pi\hbar)^3$, and the perturbation matrix element is

$$U_{fi}(t) = \frac{2\pi\hbar^2 a}{M\sqrt{V}} \int \delta n_{fi}(t, \mathbf{r})\, e^{i(\mathbf{k}\cdot\mathbf{r} - \omega t)}\, d^3x,$$

where $\hbar\mathbf{k} = \mathbf{p} - \mathbf{p}'$, $\hbar\omega = \varepsilon - \varepsilon'$, and $\delta n_{fi}(t, \mathbf{r})$ is the matrix element with respect to the wave functions of the liquid. We substitute this expression in dw_{fi} and sum the transition probability over all possible final states of the liquid. The squared modulus of the integral is written as a double integral (over $dt\, dt'\, d^3x\, d^3x'$), and we use the fact that

$$\begin{aligned} \sum_f \delta n_{fi}(t, \mathbf{r})\, \delta n_{fi}(t', \mathbf{r}')^* &= \sum_f \delta n_{if}(t', \mathbf{r}')\, \delta n_{fi}(t, \mathbf{r}) \\ &= \langle i |\, \delta\hat{n}(t', \mathbf{r}')\, \delta\hat{n}(t, \mathbf{r})\, | i \rangle \\ &= \bar{n}\sigma(t' - t, \mathbf{r}' - \mathbf{r}); \end{aligned}$$

σ is expressed as a function of the total energy of the liquid in the state i. The integration over $d(t' - t)d^3(x' - x)$ gives $\sigma(\omega, k)$, and a further integration (say, over $dt\, d^3x$) gives just the volume V and the total time interval t. Omitting the factor t, we obtain as the scattering probability per unit time

$$w = \frac{4\pi^2\hbar^2}{M^2} \bar{n} a^2 \sigma'(\omega, k) \frac{d^3p'}{(2\pi\hbar)^3}. \tag{3}$$

This expression remains valid, of course, after being averaged over the Gibbs distribution, i.e. when the form factor is expressed in terms of the temperature.

The property (86.16) of the form factor, as applied to the scattering of neutrons, expresses the fact that at $T = 0$ a liquid can only gain energy, not lose it. The relation (86.14) expresses the principle of detailed balancing, since scattering processes with energy and momentum transfer $(\omega, \mathbf{k})$ and $(-\omega, -\mathbf{k})$ are inverse to each other.

§ 87. Summation rules for the form factor

The dynamic form factor satisfies certain integral (with respect to the frequencies ω) relations called *summation rules*.

The derivation of one of these is based on the commutation rule between the operators $\hat{n}_{\mathbf{k}}(t)$ and $\dot{\hat{n}}_{\mathbf{k}}(t)$. The commutator of Heisenberg operators taken at the same instant is the same as that of the Schrödinger operators $\hat{n}_{\mathbf{k}}$ and $\dot{\hat{n}}_{\mathbf{k}}$. The operator $\dot{\hat{n}}_{\mathbf{k}}$ is determined by (86.9), and the required commutator is

$$\dot{\hat{n}}_{\mathbf{k}}\hat{n}_{\mathbf{k}}^{+}-\hat{n}_{\mathbf{k}}^{+}\dot{\hat{n}}_{\mathbf{k}} = -(i\hbar/m)\,k^2N, \tag{87.1}$$

where m is the mass of a liquid particle.†

We start from the expression for the component of the Fourier expansion of the function $\sigma(t, r)$ with respect to the coordinates only:

$$\bar{n}\sigma(t, k) = \int e^{-i\mathbf{k}.(\mathbf{r}_1-\mathbf{r}_2)}\langle\delta\hat{n}(t_1, \mathbf{r}_1)\,\delta\hat{n}(t_2, \mathbf{r}_2)\rangle\,d^3(x_1-x_2).$$

Since the integrand depends only on $\mathbf{r}_1-\mathbf{r}_2$, we replace the integration over $d^3(x_1-x_2)$ by one over $d^3x_1\,d^3x_2/V$; carrying out this integration within the averaging, we obtain

$$\sigma(t, k) = (1/N)\,\langle\delta\hat{n}_{\mathbf{k}}(t_1)\,\delta\hat{n}_{-\mathbf{k}}(t_2)\rangle. \tag{87.2}$$

We calculate the derivative $\partial\sigma(t, k)/\partial t$ at $t = 0$. Since $\sigma(t, k)$ depends only on the difference $t = t_1-t_2$,

$$\frac{\partial\sigma(t, k)}{\partial t} = \frac{1}{2}\left(\frac{\partial\sigma}{\partial t_1}-\frac{\partial\sigma}{\partial t_2}\right)$$

and, after substitution of (87.2),

$$\partial\sigma(t, k)/\partial t = (1/2N)\,\langle\delta\dot{\hat{n}}_{\mathbf{k}}(t_1)\,\delta\hat{n}_{-\mathbf{k}}(t_2)-\delta\hat{n}_{\mathbf{k}}(t_1)\,\delta\dot{\hat{n}}_{-\mathbf{k}}(t_2)\rangle.$$

Each of the two terms here depends only on the absolute value of the vector $\mathbf{k}$, and we therefore replace $\mathbf{k}$ by $-\mathbf{k}$ in the second term. Then, putting $t_1=t_2$, and noting that $\hat{n}_{-\mathbf{k}} = \hat{n}_{\mathbf{k}}^{+}$, we find that the difference in the angle brackets is equal to the commutator (87.1). Hence

$$[\partial\sigma(t, k)/\partial t]_{t=0} = -(i\hbar/2m)k^2.$$

On the other hand, expressing $\sigma(t, k)$ as a Fourier integral with respect to frequencies, we have

$$[\partial\sigma(t, k)/\partial t]_{t=0} = \left[\frac{\partial}{\partial t}\int_{-\infty}^{\infty} e^{-i\omega t}\sigma(\omega, k)\,\frac{d\omega}{2\pi}\right]_{t=0} = -i\int_{-\infty}^{\infty}\omega\sigma(\omega, k)\,\frac{d\omega}{2\pi}.$$

Comparison of the two expressions for the derivative gives the required relation

$$\int_{-\infty}^{\infty}\omega\sigma(\omega, k)\,\frac{d\omega}{2\pi} = \frac{\hbar k^2}{2m} \tag{87.3}$$

† The calculation of this commutator is the same as the calculation used in *QM* §149 in deriving the summation rule (149.5); the number of electrons Z is here replaced by the total number of liquid particles N.

(G. Placzek 1952). It must be emphasized that this is valid for all k. When the classical limit ($\hbar \to 0$) of this expression is taken, the integral on the left must be written as

$$\int_0^\infty \omega[\sigma(\omega, k)-\sigma(-\omega, k)]\frac{d\omega}{2\pi}$$

and we must put, in accordance with (86.14),

$$\sigma(\omega, k)-\sigma(-\omega, k) \approx (\hbar\omega/T)\,\sigma(\omega, k).$$

The factor $\hbar$ then cancels on each side, leaving

$$\int_0^\infty \omega^2\sigma(\omega, k)\frac{d\omega}{2\pi} = Tk^2/2m.$$

We can apply formula (87.3) to a Bose liquid at $T = 0$, and consider the range of small k. As $k \to 0$, the principal contribution to the integral comes from the delta-function peak in the form factor $\sigma(\omega, k)$, which arises in (86.13) from transitions creating one phonon; since there are no phonons in the ground state of the liquid, there are no transitions annihilating a phonon at $T = 0$. This term has the form $A\delta(\omega-uk)$, where $\hbar uk$ is the phonon energy (u is the velocity of sound). Substituting it as $\sigma(\omega, k)$ in (87.3), we find the coefficient A, and the result is

$$\sigma(\omega, k) = (\pi\hbar k/mu)\,\delta(\omega-uk). \tag{87.4}$$

Integration of this according to formula (86.7) gives the static form factor

$$\sigma(k) = \hbar k/2mu \tag{87.5}$$

(R. P. Feynman 1954).† Since this formula relates to the range of small k, its Fourier transformation gives the asymptotic expression for the correlation function at large r:

$$\nu(r) = -\hbar/2\pi^2 mur^4; \tag{87.6}$$

this may be verified by means of the integral given in the last footnote to §83. At $T = 0$, formula (87.6) is valid at arbitrarily large distances. At low but finite temperatures, it is valid up to distances $r \sim \hbar u/T$, where the fluctuations cease to be purely quantum ones. At still greater distances, formula (87.6) is replaced

† Formula (87.5), written in the form $\sigma(k) = \hbar^2k^2/2m\varepsilon(k)$, where $\varepsilon(k)$ is the quasi-particle energy, is strictly valid only as $k \to 0$. As k increases, there are increasingly important contributions to $\sigma(k)$ from transitions creating several quasi-particles. If this contribution is nevertheless neglected, we may suppose that the formula gives the relation between the form factor and the energy of the quasi-particles in a Bose liquid. The maximum of σ at $k \sim 1/a$ (where a is the interatomic distance in the liquid) corresponds to the "roton" minimum on the curve of $\varepsilon(k)$.

by an exponential decrease (if the contribution of the van der Waals forces is neglected; see §83).†

Another summation rule can be obtained from the relation established in §86 between the form factor and a generalized susceptibility $\alpha(\omega, k)$. This relation is given by (86.20), which for $T = 0$ (and $\omega > 0$) becomes

$$\bar{n}\sigma(\omega, k) = 2\hbar \,\mathrm{im}\,\alpha(\omega, k). \tag{87.7}$$

According to the Kramers–Kronig formulae (see Part 1, (123.15)),

$$\mathrm{re}\,\alpha(\omega, k) = \frac{1}{\pi}\mathrm{P}\int_{-\infty}^{\infty} \frac{\mathrm{im}\,\alpha(\omega', k)}{\omega' - \omega}\, d\omega'.$$

Putting here $\omega = 0$, and noting that $\alpha(0, k)$ is real,‡ we get

$$\alpha(0, k) = \frac{2}{\pi}\int_0^{\infty} \frac{1}{\omega}\,\mathrm{im}\,\alpha(\omega, k)\frac{d\omega}{2\pi}. \tag{87.8}$$

In the limit $k \to 0$,

$$\begin{aligned}\alpha(0, k \to 0) &= (\partial \bar{n}/\partial \mu)_{T=0} \\ &= \bar{n}(\partial \bar{n}/\partial P)_{T=0}.\end{aligned} \tag{87.9}$$

This follows because, in a static weak field U varying slowly in space, the equilibrium condition is $\mu + U =$ constant, so that the switching-on of the external field is equivalent to changing the chemical potential by $-U$. In the limit $k \to 0$, we therefore have from (86.18)

$$\begin{aligned}\delta\bar{n} &= -(\partial\bar{n}/\partial\mu)\, U \\ &\approx -U\int \alpha(0, \mathbf{r}_1 - \mathbf{r}_2)\, d^3(x_1 - x_2) \\ &= -U\alpha(0, k = 0),\end{aligned}$$

whence (87.9) follows.

Thus, combining the formulae (87.7)–(87.9), we find the following summation rule for the form factor of a liquid at $T = 0$:

$$\frac{1}{\pi\hbar}\int_0^{\infty} \sigma(\omega, k \to 0)\frac{d\omega}{\omega} = \frac{\partial\bar{n}}{\partial P} \tag{87.10}$$

(D. Pines and P. Nozières 1958).

† The correlation function (87.6) is negative (corresponding to repulsion between particles), unlike that of an ideal Bose gas, which is positive (see Part 1, §117). In this connection it may be recalled (§25) that in a slightly non-ideal Bose gas the energy spectrum has the phonon form only at $k \ll mu/\hbar$ (with $\hbar/mu \gg a$). The corresponding distances $r \sim 1/k \gg \hbar/mu$, so that in the transition to an ideal gas ($u \to 0$) the range of applicability of (87.6) moves away to infinity.

‡ The quantity $\alpha(\omega = 0, \mathbf{r})$ is real because of the general properties of the generalized susceptibility. Then the Fourier component $\alpha(\omega = 0, \mathbf{k})$ is real because $\alpha(\omega, \mathbf{r})$ is an even function of $\mathbf{r}$.

PROBLEMS

PROBLEM 1. Find the correlation function $\nu(r)$ in a Bose liquid at distances $r \gtrsim \hbar u/T$ and temperatures $T \ll T_\lambda$.

SOLUTION. The required correlation function is determined by the form factor for $k \sim 1/r \lesssim T/\hbar u \ll 1/a$, for which the liquid has a phonon energy spectrum. When $T \neq 0$, $\sigma(\omega, k)$ contains a term in $\delta(\omega + ku)$ corresponding to phonon absorption, as well as one in $\delta(\omega - ku)$ corresponding to phonon emission. The coefficients in these terms can be found from (86.14) and (87.3):

$$\sigma(\omega, k) = \frac{\pi\hbar k}{mu}[1 - e^{-\hbar ku/T}]^{-1}\{\delta(\omega - ku) + e^{-\hbar ku/T}\,\delta(\omega + ku)\}. \tag{1}$$

Integration of this expression gives

$$\sigma(k) = \frac{\hbar k}{2mu}\coth\frac{\hbar ku}{2T} \tag{2}$$

and hence

$$\nu(r) = \int e^{i\mathbf{k}\cdot\mathbf{r}}\sigma(k)\frac{d^3k}{(2\pi)^3}$$

$$= \frac{\hbar}{8\pi^2 imur}\int_{-\infty}^{\infty} e^{ikr}k^2\coth\frac{\hbar ku}{2T}\,dk.$$

By completing the contour of integration with respect to k by an infinite semicircle in the upper half-plane of complex k, we reduce the integral to a sum of residues at the poles (which lie on the imaginary axis). For $r \gg \hbar u/T$, the main contribution to the integral comes from the residue at the pole $\hbar ku/2T = i\pi$:

$$\nu(r) = -\frac{2\pi T^3}{mu^4\hbar^2 r}\exp\left(-\frac{2\pi Tr}{\hbar u}\right). \tag{3}$$

With the condition $aT/\hbar u \ll 1$, the characteristic decay length for the function $\nu(r)$ is much greater than the interatomic distances over which effects decay that are due to the direct interaction between atoms. In formula (3), $\hbar$ is essentially involved, and the correlation described by this formula is therefore a quantum effect. In the derivation, the contribution from the van der Waals forces has been neglected. It follows from the results of §83 that this contribution has a power-law form and predominates at sufficiently large distances. The distances at which (3) is replaced by (83.16) depend on the specific relation between the coefficients, but a range of applicability of formula (3) always exists at sufficiently low temperatures, since at the limit of this range with $r \sim \hbar u/T$, according to (3) $\nu \propto T^4$, and according to (83.16) $\nu \propto T^7$.

PROBLEM 2. Find the limiting form (at large distances) of the correlation function for the fluctuations of the condensate wave function in a Bose superfluid (P. C. Hohenberg and P. C. Martin 1965).

SOLUTION. In the long-wave limit, the strongest fluctuations are those of the condensate wave function Φ, since they involve only the relatively small energy of the macroscopic superfluid motion. The corresponding contribution to the total thermodynamic potential Ω of the liquid (in a volume V with given T and μ) is

$$\delta\Omega = \int \frac{1}{2}\varrho_s \mathbf{v}_s^2\, dV = \frac{\hbar^2\varrho_s}{2m^2}\int (\nabla\Phi)^2\, dV.$$

Expressing $\delta\Phi$ as a Fourier series,

$$\delta\Phi = \sum_{\mathbf{k}} \delta\Phi_{\mathbf{k}}\, e^{i\mathbf{k}\cdot\mathbf{r}}, \qquad \delta\Phi_{-\mathbf{k}} = \delta\Phi_{\mathbf{k}}^*,$$

we obtain

$$\delta\Omega = \frac{\hbar^2}{2m^2}\varrho_s V \sum_{\mathbf{k}} k^2 |\delta\Phi_{\mathbf{k}}|^2$$

and hence the mean square fluctuations

$$\langle |\delta\Phi_{\mathbf{k}}|^2\rangle = Tm^2/V\hbar^2\varrho_s k^2; \tag{4}$$

the calculations are exactly similar to those in Part 1, § 146. The contribution of these fluctuations to the one-time correlation function

$$G(r) = \langle \delta\Xi(0)\,\delta\Xi(\mathbf{r})\rangle$$

is

$$G(r) \approx n_0\langle\delta\Phi(0)\,\delta\Phi(\mathbf{r})\rangle = Tn_0m^2/4\pi\hbar^2\varrho_s r.$$

Thus $G(r)$ decreases by a power law at large distances. The contribution from fluctuations of the condensate density n_0 decreases exponentially there. The two contributions are comparable at $r \sim r_c$; at distances $r \ll r_c$, they jointly follow (at temperatures near the λ-point) the law

$$G(r) \propto r^{-(1+\zeta)}, \tag{6}$$

where ζ is the appropriate critical index. The correlation radius r_c may be defined as the distance at which the asymptotic form (5) is replaced by (6):

$$r_c^{\zeta} \propto \varrho_s/n_0.$$

With the critical indices β and ν, which describe the temperature dependence of n_0 and r_c according to (28.1) and (28.3), we find that

$$\varrho_s \propto (T_\lambda - T)^{2\beta-\nu\zeta}.$$

With the known relations between the critical indices α, β, ν, ζ (see Part 1, §§ 148, 149), we can easily see that this result agrees with (28.4).

§ 88. Hydrodynamic fluctuations

In the preceding sections, we have considered density fluctuations in a liquid for any frequencies ω and wave vectors $\mathbf{k}$. Here, of course, the actual form of the correlation function could not be found in the general case. However, this can be done in the hydrodynamic limit, where the wavelength of the fluctuations is large compared with the characteristic microscopic dimensions (interatomic distances in a liquid, mean free path in a gas).

The calculation of the one-time correlation functions of fluctuations of density, temperature, velocity etc. in a liquid at rest calls for no special study: these fluctuations (in the classical or non-quantum limit) are described by the usual thermodynamic formulae, which are valid for any medium in thermal equilibrium. The correlations between fluctuations at the same time at different points in space are propagated to distances of the order of interatomic distances (here we neglect the weak long-range van der Waals forces). But, in hydrodynamics, such distances are regarded as infinitesimal. In the hydrodynamic limit, therefore, the fluctuations at the same time at different points are uncorrelated. This statement follows formally from the additivity of a thermodynamic quantity, the minimum work $R_{\min}$ needed to cause the fluctuation. Since the probability of the fluctuation is proportional to $\exp(-R_{\min}/T)$, by representing $R_{\min}$ as a sum of terms pertaining to various physically infinitesimal volumes we find that the probabilities of fluctuations in these volumes are independent of one another.

Using this independence, we can immediately rewrite the known formulae (see Part 1, §112) for the mean square fluctuations of the thermodynamic quantities at a given point in space as formulae for the correlation functions. For example, from the formula $\langle(\delta T)^2\rangle = T^2/\varrho c_v V$ for the temperature fluctuations in the volume V (where ϱ is the density, and c_v the specific heat per unit mass of the medium), we first write

$$\langle \delta T(\mathbf{r}_a)\delta T(\mathbf{r}_b)\rangle = (T^2/\varrho c_v V_a)\,\delta_{ab},$$

where the fluctuations relate to two small volumes V_a and V_b. Then, as the volumes tend to zero, we obtain†

$$\langle \delta T(\mathbf{r}_1)\,\delta T(\mathbf{r}_2)\rangle = (T^2/\varrho c_v)\,\delta(\mathbf{r}_1-\mathbf{r}_2). \tag{88.1}$$

Similarly, we have the following formulae for the fluctuations of other thermodynamic quantities:

$$\langle \delta\varrho(\mathbf{r}_1)\,\delta\varrho(\mathbf{r}_2)\rangle = \varrho T(\partial\varrho/\partial P)_T\,\delta(\mathbf{r}_1-\mathbf{r}_2), \tag{88.2}$$

$$\langle \delta P(\mathbf{r}_1)\,\delta P(\mathbf{r}_2)\rangle = \varrho T(\partial P/\partial\varrho)_S\,\delta(\mathbf{r}_1-\mathbf{r}_2) = \varrho T u^2\delta(\mathbf{r}_1-\mathbf{r}_2), \tag{88.3}$$

$$\langle \delta s(\mathbf{r}_1)\,\delta s(\mathbf{r}_2)\rangle = (c_p/\varrho)\,\delta(\mathbf{r}_1-\mathbf{r}_2), \tag{88.4}$$

where P is the pressure and s the entropy per unit mass of the medium; the fluctuations of the pairs ϱ, T and P, s are independent. We can also write a formula for the fluctuations of the macroscopic velocity $\mathbf{v}$ of the liquid (which is zero in equilibrium):

$$\langle \delta v_i(\mathbf{r}_1)\,\delta v_k(\mathbf{r}_2)\rangle = (T/\varrho)\,\delta_{ik}\,\delta(\mathbf{r}_1-\mathbf{r}_2). \tag{88.5}$$

A problem specific to hydrodynamics is that of the time correlations of the fluctuations, as is that of fluctuations in a moving liquid. The solution of these

† This and subsequent formulae for one-time correlations in gases are valid for fluctuations with wavelengths large only in comparison with intermolecular distances but not necessarily with the mean free path. The latter condition is, however, required for different-time correlation functions in the hydrodynamic approximation (since the microscopic mechanism of propagation of perturbations in gases is determined by the mean free path of the particles).

problems calls for the consideration of dissipative processes (viscosity and thermal conduction) in the liquid.

The construction of a general theory of fluctuation phenomena in hydrodynamics amounts to setting up the "equations of motion" for the fluctuating quantities. This can be done by adding the appropriate terms to the hydrodynamic equations (L. D. Landau and E. M. Lifshitz 1957).

The equations of hydrodynamics, written in the form

$$\partial\varrho/\partial t+\mathrm{div}\,(\varrho\mathbf{v}) = 0, \tag{88.6}$$

$$\varrho\frac{\partial v_i}{\partial t}+\varrho v_k\frac{\partial v_i}{\partial x_k} = -\frac{\partial P}{\partial x_i}+\frac{\partial\sigma'_{ik}}{\partial x_k}, \tag{88.7}$$

$$\varrho T\left(\frac{\partial s}{\partial t}+\mathbf{v}.\nabla s\right) = \frac{1}{2}\sigma'_{ik}\left(\frac{\partial v_i}{\partial x_k}+\frac{\partial v_k}{\partial x_i}\right)-\mathrm{div}\,\mathbf{q}, \tag{88.8}$$

with no specific form of the stress tensor σ'_{ik} and the heat flux vector $\mathbf{q}$, simply express the conservation of mass, momentum and energy. In this form they are therefore valid for any motion, including fluctuational changes in the state of a liquid. Then ϱ, P, $\mathbf{v}$, ... are to be interpreted as the sum of the values of ϱ_0, P_0, $\mathbf{v}_0$, ... in the basic motion and their fluctuations $\delta\varrho$, δP, $\delta\mathbf{v}$, ... (the equations can, of course, always be linearized with respect to the latter).

The usual expressions for the stress tensor and the heat flux relate them respectively to the velocity gradients and the temperature gradient. When there are fluctuations in a liquid, there are also spontaneous local stresses and heat fluxes unconnected with these gradients; we denote these "random" quantities by s_{ik} and $\mathbf{g}$. Then

$$\sigma'_{ik} = \eta\left(\frac{\partial v_i}{\partial x_k}+\frac{\partial v_k}{\partial x_i}-\frac{2}{3}\delta_{ik}\,\mathrm{div}\,\mathbf{v}\right)+\zeta\delta_{ik}\,\mathrm{div}\,\mathbf{v}+s_{ik}, \tag{88.9}$$

$$\mathbf{q} = -\varkappa\nabla T+\mathbf{g}, \tag{88.10}$$

where η and ζ are the viscosity coefficients, and $\varkappa$ the thermal conductivity.

The problem now is to establish the properties of s_{ik} and $\mathbf{g}$ as regards their correlation functions. For simplicity, the arguments will be given for the normal case in hydrodynamics, that of non-quantum fluctuations; this means that the frequencies of the fluctuations are assumed to satisfy the condition $\hbar\omega \ll T$. The viscosities and the thermal conductivity are assumed non-dispersive, i.e. independent of the fluctuation frequency.

In the general theory of fluctuations given in Part 1, §§119–122, a discrete sequence of fluctuating quantities x_1, x_2, ..., is considered, whereas here we have a continuous sequence, the values of ϱ, P, ... at each point in the liquid. This unimportant difficulty is avoided by dividing the volume of the body into small but finite portions ΔV and considering certain mean values of the quantities in each portion; the change to infinitesimal volume elements is made in the final formulae.

We shall take formulae (88.9) and (88.10) as the equations

$$\dot{x}_a = -\sum_b \gamma_{ab} X_b + y_b \qquad (88.11)$$

of the general theory of quasi-stationary fluctuations; see Part 1, (122.20). The quantities $\dot{x}_a$ are taken to be the components of the tensor σ'_{ik} and the vector $\mathbf{q}$ in each portion ΔV. Then the quantities y_a are s_{ik} and $\mathbf{g}$:

$$\left.\begin{aligned} \dot{x}_a &\to \sigma'_{ik}, q_i, \\ y_a &\to s_{ik}, g_i. \end{aligned}\right\} \qquad (88.12)$$

The significance of the thermodynamically conjugate quantities X_a is determined by means of the formula for the rate of change of the total entropy S of the liquid. We find in the usual way (cf. *FM*, §49) from (88.8)–(88.10)

$$\dot{S} = \int \left\{ \frac{\sigma'_{ik}}{2T} \left(\frac{\partial v_i}{\partial x_k} + \frac{\partial v_k}{\partial x_i} \right) - \frac{\mathbf{q}\cdot\nabla T}{T^2} \right\} dV. \qquad (88.13)$$

Replacing this integral by a sum over the portions ΔV and then comparing with the expression

$$\dot{S} = -\sum_a \dot{x}_a X_a,$$

we find that

$$X_a \to -\frac{1}{2T} \left(\frac{\partial v_i}{\partial x_k} + \frac{\partial v_k}{\partial x_i} \right) \Delta V, \quad \frac{1}{T^2} \frac{\partial T}{\partial x_i} \Delta V. \qquad (88.14)$$

It is now easy to find the coefficients γ_{ab}, which immediately give the required correlations by

$$\langle y_a(t_1) y_b(t_2) \rangle = (\gamma_{ab} + \gamma_{ba})\, \delta(t_1 - t_2); \qquad (88.15)$$

see Part 1, (122.21a).

First of all, we note that in formulae (88.9) and (88.10) there are no terms which would relate σ'_{ik} to the temperature gradient, or $\mathbf{q}$ to the velocity gradients. This means that the corresponding coefficients $\gamma_{ab} = 0$, and from (88.15)

$$\langle s_{ik}(t_1, \mathbf{r}_1)\, g_l(t_2, \mathbf{r}_2) \rangle = 0, \qquad (88.16)$$

i.e. the values of s_{ik} and $\mathbf{g}$ are uncorrelated.

Next, the coefficients relating the values of q_i to those of $(\Delta V/T^2)\, \partial T/\partial x_i$ are zero, if these quantities are taken in different portions ΔV, and are $\gamma_{ik} = \varkappa T^2 \delta_{ik}/\Delta V$ if taken in the same ΔV. With these values of γ_{ab}, we obtain from (88.15), after taking the limit $\Delta V \to 0$,

$$\langle g_i(t_1, \mathbf{r}_1)\, g_k(t_2, \mathbf{r}_2) \rangle = 2\varkappa T^2 \delta_{ik} \delta(\mathbf{r}_1 - \mathbf{r}_2)\, \delta(t_1 - t_2). \qquad (88.17)$$

Similarly we obtain formulae for the correlation functions of the random stress tensor:

$$\langle s_{ik}(t_1,\mathbf{r}_1)\, s_{lm}(t_2,\mathbf{r}_2)\rangle = 2T[\eta(\delta_{il}\delta_{km}+\delta_{im}\delta_{kl})+(\zeta-\tfrac{2}{3}\eta)\,\delta_{ik}\delta_{lm}]\,\delta(\mathbf{r}_1-\mathbf{r}_2)\,\delta(t_1-t_2). \tag{88.18}$$

Formulae (88.16)–(88.18) in principle solve this problem of calculating the hydrodynamic fluctuations in any particular case. The solution procedure is as follows. Regarding s_{ik} and $\mathbf{g}$ as given functions of coordinates and time, we formally solve the linearized equations (88.6)–(88.8) for $\delta\varrho$, $\delta\mathbf{v}$, ..., with the necessary hydrodynamic boundary conditions. This gives the quantities expressed as linear functionals of s_{ik} and $\mathbf{g}$. Correspondingly, any quantity quadratic in $\delta\varrho$, $\delta\mathbf{v}$, ... is expressed in terms of quadratic functionals of s_{ik} and $\mathbf{g}$, and the mean value is calculated from (88.16)–(88.18); the auxiliary quantities s_{ik} and $\mathbf{g}$ do not appear in the result.

We can also write out formulae (88.16)–(88.18) in Fourier components with respect to frequency, and we shall do this in a form which generalizes them to the case of quantum fluctuations. According to the general rules of the fluctuation–dissipation theorem, such a generalization is obtained by including an extra factor $(\hbar\omega/2T)\coth(\hbar\omega/2T)$ (which is unity in the classical limit $\hbar\omega \ll T$). In the presence of dispersion of the viscosity and thermal conductivity, the quantities η, ζ and $\varkappa$ are complex functions of the frequency; in the formulae for the fluctuations, they are replaced by the real parts of those functions:

$$(s_{ik}^{(1)}g_l^{(2)})_\omega = 0, \tag{88.19}$$

$$(g_i^{(1)}g_k^{(2)})_\omega = \delta_{ik}\delta(\mathbf{r}_1-\mathbf{r}_2)\,\hbar\omega T\coth(\hbar\omega/2T)\,\mathrm{re}\,\varkappa(\omega), \tag{88.20}$$

$$(s_{ik}^{(1)}s_{lm}^{(2)})_\omega = \hbar\omega\delta(\mathbf{r}_1-\mathbf{r}_2)\coth(\hbar\omega/2T)\times$$
$$\times[(\delta_{il}\delta_{km}+\delta_{im}\delta_{kl}-\tfrac{2}{3}\,\delta_{ik}\delta_{lm})\,\mathrm{re}\,\eta(\omega)+\delta_{ik}\delta_{lm}\,\mathrm{re}\,\zeta(\omega)]. \tag{88.21}$$

§ 89. Hydrodynamic fluctuations in an infinite medium

In this section we shall consider hydrodynamic fluctuations in an infinite liquid at rest. This problem can, of course, be solved by the method given in §88. Here, however, we shall use a different method, and thus exemplify an alternative way of solving problems of hydrodynamic fluctuations.

This other way employs the general theory of quasi-stationary fluctuations in its earlier stage, before the introduction of the random forces. The relevant general formulae are as follows (see Part 1, §122).

Let

$$\dot{x}_a = -\sum_q \lambda_{ab}x_b \tag{89.1}$$

be the macroscopic "equations of motion" for the set of quantities $x_a(t)$ which describe the non-equilibrium state of the system (in equilibrium, all the $x_a = 0$). These equations are valid if the x_a are large compared with their mean fluctuations (but also so small that the equations of motion may be linearized). We can then say that similar equations are satisfied (when $t > 0$) by the correlation functions of the fluctuations:

$$\frac{d}{dt}\langle x_a(t)\, x_c(0)\rangle = -\sum_b \lambda_{ab}\langle x_b(t)\, x_c(0)\rangle, \quad t > 0. \tag{89.2}$$

The initial condition for these is formed by the equations

$$[\langle x_a(t)\, x_c(0)\rangle]_{t=+0} = \langle x_a x_c\rangle, \tag{89.3}$$

where $\langle x_a x_c\rangle$ is the one-time correlation function, assumed known. In the range $t < 0$, the correlation functions are continued by the rule

$$\langle x_a(t)\, x_c(0)\rangle = \pm\langle x_a(-t)\, x_c(0)\rangle, \tag{89.4}$$

the upper sign relating to the case where x_a and x_c are both even or both odd under time reversal, and the lower sign to the case where one is even and the other odd. The solution of equation (89.2) with the condition (89.3) is obtained by means of a one-sided Fourier transformation: multiplying the equation by $e^{i\omega t}$ and integrating with respect to t from 0 to ∞ (with integration by parts on the left-hand side), we obtain the equations

$$-i\omega(x_a x_c)^{(+)}_\omega = -\sum_b \lambda_{ab}(x_b x_c)^{(+)}_\omega + \langle x_a x_c\rangle \tag{89.5}$$

for the quantities (functions of frequency)

$$(x_a x_b)^{(+)}_\omega = \int_0^\infty e^{i\omega t}\langle x_a(t)\, x_b(0)\rangle\, dt. \tag{89.6}$$

The ordinary Fourier components of the correlation function are expressed in terms of the quantities (89.6) by

$$\begin{aligned}(x_a x_b)_\omega &= \int_{-\infty}^{\infty} e^{i\omega t}\langle x_a(t)\, x_b(0)\rangle\, dt \\ &= (x_a x_b)^{(+)}_\omega \pm [(x_a x_b)^{(+)}_\omega]^* \\ &= (x_a x_b)^{(+)}_\omega + (x_b x_a)^{(+)}_{-\omega},\end{aligned} \tag{89.7}$$

where the signs $\pm$ correspond to those in (89.4).

Proceeding to the stated problem of fluctuations in a liquid at rest, we first linearize the hydrodynamic equations (88.6)–(88.8) with σ'_{ik} and $\mathbf{q}$ from (88.9) and (88.10) (without the final terms). Putting $\varrho = \varrho_0 + \delta\varrho$, $\mathbf{v} = \delta\mathbf{v}, \ldots$, and omitting the non-linear terms we find

$$\frac{\partial\delta\varrho}{\partial t} + \varrho \operatorname{div}\mathbf{v} = 0, \tag{89.8}$$

$$\varrho\, \partial\mathbf{v}/\partial t = -\nabla\delta P + \eta\triangle\mathbf{v} + (\zeta + \tfrac{1}{3}\eta)\nabla \operatorname{div}\mathbf{v}, \tag{89.9}$$

$$\frac{\partial\delta s}{\partial t} = \frac{\varkappa}{\varrho T}\triangle\delta T; \tag{89.10}$$

the suffix 0 to the constant quantities ϱ_0 etc. is omitted after the linearization. In equations (89.8)–(89.10) it will be convenient to divide the velocity at once into the potential ("longitudinal") and rotational ("transverse") parts $\mathbf{v}^{(l)}$ and $\mathbf{v}^{(t)}$ defined by

$$\mathbf{v} = \mathbf{v}^{(l)} + \mathbf{v}^{(t)}, \tag{89.11}$$

$$\operatorname{div} \mathbf{v}^{(t)} = 0, \quad \operatorname{curl} \mathbf{v}^{(l)} = 0.$$

In (89.8) only the longitudinal velocity occurs:

$$\frac{\partial \delta\varrho}{\partial t} + \varrho \operatorname{div} \mathbf{v}^{(l)} = 0, \tag{89.12}$$

and (89.9) separates into the two equations

$$\frac{\partial \mathbf{v}^{(t)}}{\partial t} = \frac{\eta}{\varrho} \triangle \mathbf{v}^{(t)}, \tag{89.13}$$

$$\varrho \frac{\partial \mathbf{v}^{(l)}}{\partial t} = -\nabla \delta P + \left(\zeta + \frac{4}{3}\eta\right) \nabla \operatorname{div} \mathbf{v}^{(l)}. \tag{89.14}$$

The equation for the transverse velocities is independent of the other equations. Accordingly, we also have one equation for the correlation function of its fluctuations:

$$\frac{\partial}{\partial t}\langle v_i^{(t)}(t, \mathbf{r})\, v_k^{(t)}(0, 0)\rangle - \nu \triangle \langle v_i^{(t)}(t, \mathbf{r})\, v_k^{(t)}(0, 0)\rangle = 0, \tag{89.15}$$

where $\nu = \eta/\varrho$ is the kinematic viscosity. Taking the one-sided Fourier transform, we obtain

$$-i\omega(v_i^{(t)}(\mathbf{r})\, v_k^{(t)}(0))_\omega^{(+)} - \nu \triangle (v_i^{(t)}(\mathbf{r})\, v_k^{(t)}(0))_\omega^{(+)} = \langle v_i^{(t)}(\mathbf{r})\, v_k^{(t)}(0)\rangle,$$

where the right-hand side is the one-time correlation function; or, with the Fourier components with respect to the coordinates,

$$(v_i^{(t)}\, v_k^{(t)})_{\omega\mathbf{k}} = (v_i^{(t)}\, v_k^{(t)})_{\mathbf{k}}/(\nu k^2 - i\omega).$$

The one-time correlation function of the velocity fluctuations is given by (88.5); changing to Fourier components and separating the transverse part, we have

$$(v_i^{(t)}\, v_k^{(t)})_{\mathbf{k}} = (T/\varrho)(\delta_{ik} - k_i k_k/k^2). \tag{89.16}$$

Substituting in the preceding formula, we have finally†

$$(v_i^{(t)}\, v_k^{(t)})_{\omega\mathbf{k}} = 2 \operatorname{re} (v_i^{(t)}\, v_k^{(t)})_{\omega\mathbf{k}}^{(+)} = \frac{2T}{\varrho}\left(\delta_{ik} - \frac{k_i k_k}{k^2}\right) \frac{\nu k^2}{\omega^2 + \nu^2 k^4}\,. \tag{89.17}$$

For the other variables we have a system of coupled equations (89.10), (89.12), (89.14). This becomes simpler, however, in the limiting cases of high

† It is easy to see that, on integrating (89.17) with respect to $\omega/2\pi$, we return to the one-time correlation function, as is to be expected.

or low frequencies. The reason is that the perturbations of longitudinal velocity and pressure are propagated in the liquid with the velocity of sound u, and those of entropy according to the equation of thermal conduction. The latter mechanism requires a time $\sim 1/\chi k^2$ for propagation of the perturbation to a distance $\sim 1/k$ (where $\chi = \varkappa/\varrho c_p$ is the thermometric conductivity of the medium). Hence, for frequencies that satisfy (with a given value of the wave number) the condition

$$\chi k^2 \ll \omega \sim ku, \tag{89.18}$$

we may suppose that only $\mathbf{v}^{(l)}$ and P fluctuate at constant entropy. On the other hand, if

$$\chi k^2 \sim \omega \ll ku, \tag{89.19}$$

there are isobaric fluctuations of entropy.†

Let us first consider the high-frequency range (89.18) and determine the fluctuations of pressure, for example.

Equation (89.14), written for the correlation functions, has the form

$$\frac{\partial}{\partial t}\langle \mathbf{v}^{(l)}(t, \mathbf{r})\, \delta P(0, 0)\rangle = -\operatorname{grad}\langle \delta P(t, \mathbf{r})\, \delta P(0, 0)\rangle +$$

$$+\left(\zeta+\frac{4}{3}\eta\right)\operatorname{grad}\operatorname{div}\langle \mathbf{v}^{(l)}(t, \mathbf{r})\, \delta P(0, 0)\rangle,$$

and the initial condition for it is the vanishing of the one-time correlation of $\mathbf{v}^{(l)}$ and δP. With a one-sided Fourier transformation with respect to time and a complete transformation with respect to coordinates, we hence have

$$-i\omega\varrho(\mathbf{v}^{(l)}\,\delta P)^{(+)}_{\omega\mathbf{k}} = -i\mathbf{k}(\delta P^2)^{(+)}_{\omega\mathbf{k}} - (\zeta+\tfrac{4}{3}\eta)\,\mathbf{k}(\mathbf{k}\,.\,\mathbf{v}^{(l)}\delta P)^{(+)}_{\omega\mathbf{k}}. \tag{89.20}$$

Next, in equation (89.12) we write

$$\delta\varrho = \left(\frac{\partial\varrho}{\partial P}\right)_s \delta P + \left(\frac{\partial\varrho}{\partial s}\right)_P \delta s = \frac{1}{u^2}\,\delta P - \varrho^2\left(\frac{\partial T}{\partial P}\right)_s \delta s,$$

and $\partial\delta s/\partial t$ is expressed by means of (89.10) in the form

$$\frac{\partial \delta s}{\partial t} = \frac{\varkappa}{\varrho T}\left(\frac{\partial T}{\partial P}\right)_s \triangle\delta P;$$

the term in $\triangle\delta s$ on the right is neglected in comparison with $\partial\delta s/\partial t$, since $\chi k^2 \ll \omega$. This leads to the equation

$$\frac{1}{u^2}\frac{\partial\delta P}{\partial t} - \frac{\varkappa\varrho}{T}\left(\frac{\partial T}{\partial P}\right)_s^2 \triangle\delta P + \varrho\operatorname{div}\mathbf{v}^{(l)} = 0.$$

† The inequality $\chi k^2 \ll ku$ is always satisfied in the hydrodynamic region. For example, in gases $u \sim v_T$ and $\chi \sim v_T l$, where v_T is the mean thermal velocity of the particles and l their mean free path. Hence the inequality $\chi k \ll u$ is equivalent to the necessary condition $kl \ll 1$.

The corresponding equation for the correlation functions is again obtained by replacing δP and $\mathbf{v}^{(l)}$ by $\langle \delta P(t, \mathbf{r})\, \delta P(0, 0)\rangle$ and $\langle \mathbf{v}^{(l)}(t, \mathbf{r})\, \delta P(0, 0)\rangle$ respectively; the initial condition is (88.3). After the Fourier transformations, this equation becomes

$$\left[-\frac{i\omega}{u^2}+\frac{k^2\varkappa\varrho}{T}\left(\frac{\partial T}{\partial P}\right)_s^2\right](\delta P^2)^{(+)}_{\omega\mathbf{k}}+i\varrho(\mathbf{k}\cdot\mathbf{v}^{(l)}\delta P)^{(+)}_{\omega\mathbf{k}} = \varrho T. \qquad (89.21)$$

From the two equations (89.20) and (89.21) we can find

$$(\delta P^2)_{\omega\mathbf{k}} = 2\,\mathrm{re}\,(\delta P^2)^{(+)}_{\omega\mathbf{k}} = 2\,\mathrm{re}\,\frac{k^2\varrho T u^4(i+2\gamma_T\omega/uk^2)}{\omega(\omega^2-k^2u^2+2i\omega u\gamma)}, \qquad (89.22)$$

where

$$\gamma = \frac{k^2}{2\varrho u}\left[\zeta+\frac{4}{3}\eta+\frac{\varkappa u^2\varrho^2}{T}\left(\frac{\partial T}{\partial P}\right)_s^2\right] \qquad (89.23)$$

is the coefficient of absorption of sound in the medium (see *FM*, §77), and γ_T is the part of this due to thermal conduction. The final result for frequencies near the values $\omega = \pm ku$, where the fluctuations are especially great, is

$$(\delta P^2)_{\omega\mathbf{k}} = \varrho T u^3\gamma/[(\omega\mp ku)^2+u^2\gamma^2]. \qquad (89.24)$$

This formula is valid when $|\omega\mp ku| < u\gamma$.†

In the low-frequency range (89.19) it is sufficient, as already mentioned, to consider the fluctuations of entropy, neglecting those of pressure. This means that in (89.10) we can put

$$\delta T \approx (\partial T/\partial s)_P\, \delta s = (T/c_p)\, \delta s,$$

the specific heat c_p being taken per unit mass. The required correlation function therefore satisfies an equation of the same type as (89.15), and the initial condition is (88.4). The result is

$$(\delta s^2)_{\omega\mathbf{k}} = \frac{2c_p}{\varrho}\,\frac{\chi k^2}{\omega^2+\chi^2k^4}. \qquad (89.25)$$

PROBLEMS

PROBLEM 1. Find the correlation function of the fluctuations of the number of solute particles in a weak solution.

SOLUTION. The number density n of solute particles satisfies the diffusion equation

$$\partial n/\partial t = D\,\triangle n,$$

† It may be recalled (see *FM*, §77) that the hydrodynamic sound-absorption coefficient is always small in gases (the inequality $\gamma \ll k$ necessarily follows from the condition $kl \ll 1$), and is small in liquids where there is no significant dispersion of sound.

where D is the diffusion coefficient. In a weak solution, the simultaneous values of the density at different points are uncorrelated (just as there is no one-time correlation for the density of an ideal gas); the one-time correlation function is therefore

$$\langle \delta n(\mathbf{r}_1)\, \delta n(\mathbf{r}_2)\rangle = \bar{n}\delta(\mathbf{r}_1-\mathbf{r}_2).$$

Similarly to (89.25), we find

$$(\delta n^2)_{\omega\mathbf{k}} = 2\bar{n}k^2D/(\omega^2+k^4D^2).$$

Here, thermal diffusion is neglected, and the fluctuations of n may therefore be treated independently of those of temperature.

PROBLEM 2. Find the correlation function of pressure fluctuations in a liquid having a large and dispersive second viscosity $\zeta(\omega)$ (due to slow relaxation of some parameter).

SOLUTION. The presence of slow relaxation processes causes a second viscosity of the form

$$\zeta(\omega) = \frac{\tau\varrho}{1-i\omega\tau}(u_\infty^2-u_0^2),$$

where τ is the relaxation time, u_0 the equilibrium velocity of sound, and u_∞ the velocity of sound for a constant value of the relaxation parameter; see *FM*, §78. Equations (89.20) and (89.21), and therefore (89.22), are valid also in the presence of dispersion. Putting $\zeta = \zeta(\omega)$ and neglecting the terms arising from η and $\varkappa$, we obtain after some calculation

$$(\delta P^2)_{\omega\mathbf{k}} = \frac{2T\tau\varrho u_0^4(u_\infty^2-u_0^2)}{(u_0^2-\omega^2/k^2)^2+\omega^2\tau^2(u_\infty^2-\omega^2/k^2)^2}.$$

§ 90. Operator expressions for the transport coefficients

The formulae (88.20) and (88.21) may be viewed differently by reading them "from right to left", i.e. by regarding them as expressions for the thermal conductivity and the viscosity. The correlation functions on the left may then be expressed, in accordance with their definition, in terms of the operators of certain quantities having microscopic significance; we thereby obtain the transport coefficients of the liquid, expressed in terms of these operators.

First of all, we must note that the absence of any correlation between the fluctuations of the "random" energy and momentum fluxes at different points in space (the delta function $\delta(\mathbf{r}_1-\mathbf{r}_2)$ in (88.20) and (88.21)) is a consequence of the hydrodynamic approximation, which is valid only for small values of the wave vector. In order to express this condition explicitly, we write the formulae in Fourier components with respect to the spatial coordinates (which amounts to substituting unity for the factors $\delta(\mathbf{r}_1-\mathbf{r}_2)$) and take the limit $\mathbf{k}\to 0$. For example, formula (88.20) contracted with respect to the suffixes i and k,

$$(\mathbf{g}^{(1)}\cdot\mathbf{g}^{(2)})_\omega = 3\delta(\mathbf{r}_1-\mathbf{r}_2)\,\hbar\omega T\coth(\hbar\omega/2T)\,.\,\mathrm{re}\,\varkappa(\omega),$$

is written as

$$\mathrm{re}\,\varkappa(\omega) = \frac{1}{3\hbar\omega T}\tanh\frac{\hbar\omega}{2T}\lim_{\mathbf{k}\to 0}(g^2)_{\omega\mathbf{k}}. \tag{90.1}$$

It is easy to see that in this formula we can replace the "random" heat flux $\mathbf{g}$ by the total energy flux, which we shall denote by $\mathbf{Q}$. The latter, as we know from hydrodynamics, consists of the convective energy flux and the heat flux $\mathbf{q}$:

$$\mathbf{Q} = (\tfrac{1}{2}v^2+w)\,\varrho\mathbf{v}+\mathbf{q} \approx \varrho w\mathbf{v}-\varkappa\,\nabla T+\mathbf{g}, \tag{90.2}$$

where w is the heat function per unit mass of the liquid; in the last expression, we omit the term containing a higher power of the fluctuation velocity $\mathbf{v}$. For small $\mathbf{k}$, however, the fluctuations of the actual physical quantities $\mathbf{v}$, T, ϱ, etc., contain an extra power of $\mathbf{k}$ in comparison with those of the random fluxes, and therefore the fluctuations of $\mathbf{g}$ are the same as those of $\mathbf{Q}$ in the limit $\mathbf{k} \to 0$. This is immediately evident from the fact that the fluxes $\mathbf{g}$ and s_{ik} appear in the equation of motion (88.6)–(88.8) of the hydrodynamic fluctuations only as spatial derivatives, but the physical quantities mentioned occur also as time derivatives; when the Fourier components are taken, the latter quantities are therefore of order k/ω relative to the former.

Unlike $\mathbf{g}$, the total energy flux $\mathbf{Q}$ has a direct mechanical significance and corresponds to a definite quantum-mechanical operator $\hat{\mathbf{Q}}(t, \mathbf{r})$ expressible in terms of the operators of dynamic variables of particles in the medium. With the definition of the correlation functions in terms of the (Heisenberg) operators of the corresponding quantities, we thus arrive at the formula

$$\operatorname{re}\varkappa(\omega) = \frac{1}{6\hbar\omega T}\tanh\frac{\hbar\omega}{2T}\times$$

$$\times \lim_{\mathbf{k}\to 0}\int\int_{-\infty}^{\infty} e^{i(\omega t-\mathbf{k}\cdot\mathbf{r})}\langle\hat{\mathbf{Q}}(t,\mathbf{r})\,\hat{\mathbf{Q}}(0,0)+\hat{\mathbf{Q}}(0,0)\,\hat{\mathbf{Q}}(t,\mathbf{r})\rangle\,dt\,d^3x \tag{90.3}$$

(M. S. Green 1954).

A more useful representation of the function $\varkappa(\omega)$ is, however, obtained by means of a formula expressing the correlation function in terms of the commutator of the corresponding operators.

If $x_a(\mathbf{r})$ and $x_b(\mathbf{r})$ are two fluctuating quantities (equal to zero in equilibrium and behaving in the same way under time reversal), their correlation function may be written, according to (76.1) and (75.11), as

$$(x_a^{(1)}\,x_b^{(2)})_\omega = \coth\frac{\hbar\omega}{2T}\cdot\operatorname{re}\int_0^{\infty} e^{i\omega t}\langle[\hat{x}_a(t,\mathbf{r}_1),\ \hat{x}_b(0,\mathbf{r}_2)]\rangle\,dt,$$

where $[\ldots]$ denotes the commutator. Changing to a Fourier expansion with respect to the coordinates $\mathbf{r} = \mathbf{r}_1-\mathbf{r}_2$, we obtain

$$(x_a x_b)_{\omega\mathbf{k}} = \coth\frac{\hbar\omega}{2T}\cdot\operatorname{re}\int\int_0^{\infty} e^{i(\omega t-\mathbf{k}\cdot\mathbf{r})}\langle[\hat{x}_a(t,\mathbf{r}),\ \hat{x}_b(0,0)]\rangle\,dt\,d^3x. \tag{90.4}$$

Applying this formula to the correlation function $(\mathbf{Q}^2)_{\omega\mathbf{k}}$ and substituting in (90.1), we have

$$\operatorname{re} \varkappa(\omega) = \frac{1}{3\omega T} \lim_{\mathbf{k}\to 0} \operatorname{re} \int \int_0^\infty e^{i(\omega t - \mathbf{k}\cdot\mathbf{r})} \langle [\hat{\mathbf{Q}}(t, \mathbf{r}), \hat{\mathbf{Q}}(0, 0)] \rangle \, dt \, d^3x.$$

The two sides of this equation contain the real parts of functions of ω that tend to zero as $\omega \to \infty$ and have no singularity in the upper half-plane of the complex variable ω. If the real parts of such functions are equal on the real axis of ω, it follows that the functions themselves are equal, and we arrive at the final formula

$$\varkappa(\omega) = \frac{1}{3\omega T} \lim_{\mathbf{k}\to 0} \int \int_0^\infty e^{i(\omega t - \mathbf{k}\cdot\mathbf{r})} \langle [\hat{\mathbf{Q}}(t, \mathbf{r}), \hat{\mathbf{Q}}(0, 0)] \rangle \, dt \, d^3x. \qquad (90.5)$$

In order to derive the static value of the thermal conductivity, we must then also take the limit $\omega \to 0$.

Similarly, formula (88.21) may be transformed into an operator expression for the viscosity coefficients.

If we use the total momentum flux $\sigma_{ik} = -P\delta_{ik} + \sigma'_{ik}$ (with σ'_{ik} from (88.9)) then in the limit $\mathbf{k} \to 0$ the fluctuations of all terms except s_{ik} become zero, and so in this limit we can replace the correlation function $(s_{ik}s_{lm})_{\omega\mathbf{k}}$ by $(\sigma_{ik}\sigma_{lm})_{\omega\mathbf{k}}$. The result is

$$\eta(\omega)\left(\delta_{il}\delta_{km} + \delta_{im}\delta_{kl} - \frac{2}{3}\delta_{ik}\delta_{lm}\right) + \zeta(\omega)\,\delta_{ik}\delta_{lm}$$

$$= \frac{1}{\omega} \lim_{\mathbf{k}\to 0} \int \int_0^\infty e^{i(\omega t - \mathbf{k}\cdot\mathbf{r})} \langle [\hat{\sigma}_{ik}(t, \mathbf{r}), \hat{\sigma}_{lm}(0, 0)] \rangle \, dt \, d^3x, \qquad (90.6)$$

where $\hat{\sigma}_{ik}(t, \mathbf{r})$ is the momentum flux density operator (H. Mori 1958). Contracting this equation with respect to the pairs of suffixes i, k and l, m or i, l and k, m, we obtain separate expressions for 9ζ or for $10\eta + 3\zeta$ respectively.

§ 91. Dynamic form factor of a Fermi liquid

Formulae (87.4)–(87.6) for the form factor at $T = 0$ are not applicable to a Fermi liquid, since their derivation assumes the existence (for small ω and $\mathbf{k}$) of only the phonon branch of the spectrum of elementary excitations. The hydrodynamic theory of fluctuations developed in §§88 and 89 is also inapplicable to a Fermi liquid. It requires the fulfilment of the condition $kl \ll 1$ (where l is the quasi-particle mean free path), which is certainly not satisfied in a Fermi liquid, since $l \propto T^{-2}$ and tends to infinity as $T \to 0$. Hence the transport equation must be used to calculate the form factor of a Fermi liquid.

Here it is convenient to start from equations (86.17)–(86.20), which give the relation between the form factor and the generalized susceptibility with respect to the action of a field $U(t, \mathbf{r})$ on the liquid. In Fourier components also with respect to the coordinates, the definition (86.18) becomes

$$\delta\bar{n}(\omega, \mathbf{k}) = -\alpha(\omega, \mathbf{k})\, U_{\omega\mathbf{k}}. \tag{91.1}$$

We shall consider only the case $T = 0$. Then the dynamic form factor is expressed in terms of $\alpha(\omega, \mathbf{k})$ by

$$\bar{n}\sigma(\omega, \mathbf{k}) = \left\{ \begin{array}{ll} 2\hbar \operatorname{im} \alpha(\omega, \mathbf{k}), & \omega > 0, \\ 0, & \omega < 0. \end{array} \right\} \tag{91.2}$$

The density perturbation $\delta\bar{n}(\omega, \mathbf{k})$ is calculated by means of the transport equation, in which the collision integral may be neglected (as $T \to 0$). These calculations only differ from those for zero sound in §4 by the addition of the term

$$U(t, \mathbf{r}) = U_{\omega\mathbf{k}}\, e^{i(\mathbf{k}\cdot\mathbf{r}-\omega t)}$$

in the quasi-particle energy. Correspondingly, the derivative $\partial\varepsilon/\partial\mathbf{r}$ (4.3) contains an added term $\partial U/\partial\mathbf{r} = i\mathbf{k}U$, and on the left of the transport equation (4.8) there is a term

$$-i\mathbf{k}U\,.\,\partial n_0/\partial\mathbf{p} = i\mathbf{k}\,.\,\mathbf{v}U\delta(\varepsilon-\varepsilon_F).$$

The solution of the transport equation is sought in the form

$$\left. \begin{array}{l} \delta n(\mathbf{p}) = \delta n_{\omega\mathbf{k}}(\mathbf{p})\, e^{i(\mathbf{k}\cdot\mathbf{r}-\omega t)}, \\ \delta n_{\omega\mathbf{k}}(\mathbf{p}) = -\delta(\varepsilon-\varepsilon_F)\dfrac{\pi^2\hbar^2}{2m^* p_F}\chi(\mathbf{n}), \qquad \mathbf{n} = \mathbf{p}/p. \end{array} \right\} \tag{91.3}$$

This is the Fourier component of the perturbation of the quasi-particle momentum distribution. The required change in the density of the total number of quasi-particles (the number density of actual particles) is given by the integral

$$\delta\bar{n}(\omega, \mathbf{k}) = \int \delta n_{\omega\mathbf{k}}(\mathbf{p})\,.\,2\, d^3p/(2\pi\hbar)^3 = -\frac{1}{2\hbar}\int \chi(\mathbf{n})\frac{do}{4\pi}\,.\,U_{\omega\mathbf{k}}.$$

The definition of the function $\chi(\mathbf{n})$ in (91.3) differs from that of $\nu(\mathbf{n})$ in (4.9) as regards normalization: here, it is chosen so that formula (91.2) becomes

$$\bar{n}\sigma(\omega, \mathbf{k}) = \operatorname{im} \int \chi(\mathbf{n})\, do/4\pi, \qquad \omega > 0. \tag{91.4}$$

For $\chi(\mathbf{n})$ itself, we get the equation

$$(\omega - v_F\mathbf{k}\,.\,\mathbf{n})\,\chi(\mathbf{n}) - v_F\mathbf{k}\,.\,\mathbf{n}\int F(\vartheta)\,\chi(\mathbf{n}')\, do'/4\pi = -\mathbf{k}\,.\,\mathbf{n}\,.\,2p_F^2/\pi^2\hbar^2, \tag{91.5}$$

which differs on the right-hand side from (4.11).

Equation (91.5) does not explicitly contain imaginary quantities. The presence of an imaginary part of its solution $\chi(\mathbf{n})$ is therefore due only to the passages round poles in the integrals that arise during the process of solution. The rule for these passages depends on the requirement that the field $U \propto e^{-i\omega t}$ applied to the system is applied adiabatically from $t = -\infty$ onwards; for this purpose, its frequency ω must be replaced by $\omega + i0$.

The specific form of the solution depends on the form of the quasi-particle interaction function $F(\vartheta)$. We shall illustrate the process of solution and the resulting properties for the simplest example, in which $F \equiv F_0$, a constant.

In this case, the solution of equation (91.5) has the form

$$\chi(\mathbf{n}) = Cv_F\mathbf{k}.\mathbf{n}/(v_F\mathbf{k}.\mathbf{n}-\omega-i0), \tag{91.6}$$

where C is a constant. This constant is determined by substituting (91.6) back into (91.5), which gives

$$C(1+I) = 2m^*p_F/\pi^2\hbar^2, \tag{91.7}$$

where

$$I = \int \frac{\mathbf{k}.\mathbf{n}'v_F}{\mathbf{k}.\mathbf{n}'v_F-\omega-i0}\cdot\frac{do'}{4\pi}.$$

The integrand depends only on the angle between $\mathbf{n}'$ and $\mathbf{k}$, and an obvious substitution gives

$$I(s) = \frac{1}{2}\int_{-1}^{1}\frac{x\,dx}{x-s-i0} = 1-\frac{1}{2}s\log\left|\frac{s+1}{s-1}\right| + \begin{cases}\frac{1}{2}is\pi, & s<1,\\ 0, & s>1,\end{cases} \tag{91.8}$$

where $s = \omega/kv_F$; the imaginary part of the integral is determined by the rule (8.11).

Substituting the function $\chi(\mathbf{n})$ from (91.6)–(91.8) in (91.4), we get the dynamic form factor

$$\bar{n}\sigma(\omega, k) = \frac{2m^*p_F}{\pi^2\hbar^2}\,\mathrm{im}\,\frac{I(s)}{1+F_0I(s)} \tag{91.9}$$

(A. A. Abrikosov and I. M. Khalatnikov 1958). According to (91.8), this is non-zero for $s < 1$, i.e. for all $\omega < kv_F$.

If $F_0 > 0$, zero sound can be propagated in a Fermi liquid with a velocity u_0 determined by (4.15):

$$1+F_0I(s_0) = 0, \quad s_0 = u_0/v_F.$$

For values of s close to s_0, the expression (91.9) becomes

$$\text{constant} \times \mathrm{im}\, 1/(s-s_0);$$

according to the comment made above, $s = \omega/kv_F$ is to be taken as $s+i0$. This means that $\sigma(\omega, k)$ contains also a delta-function term having the form constant$\times\delta(s-s_0)$, or

$$\sigma(\omega, k) = \text{constant} \times k\delta(\omega-ku_0). \tag{91.10}$$

This term is the contribution to the form factor from the zero-sound branch of the energy spectrum of the Fermi liquid; it is exactly analogous to the phonon contribution (87.4) to the form factor of a Bose liquid.

The existence of such a term does not, of course, depend on the assumption that F is constant; it is a general property of a Fermi liquid in which zero sound can be propagated. Only the value of the constant coefficient in (91.10) depends on the law of interaction of quasi-particles. Equation (91.5), with zero on the right-hand side, is the equation of zero sound; the solution of the inhomogeneous equation therefore has a pole at $\omega/k = u_0$.

It is clear from the form of (91.5) that its solution depends on the parameters ω and k only through the ratio ω/k. The dynamic form factor will therefore also be a function of this ratio. The static form factor

$$\sigma(k) = \int_{-\infty}^{\infty} \sigma(\omega, k)\frac{d\omega}{2\pi}$$

will consequently have the form

$$\sigma(k) = \text{constant} \times k. \tag{91.11}$$

This means that the one-time spatial correlation function of the density fluctuations at $T = 0$ in a Fermi liquid obeys the law $\nu(r) \propto r^{-4}$, as in a Bose liquid.

Lastly, we may note that the dynamic form factor of an ideal Fermi gas is obtained from (91.9) by taking the limit $F_0 \to 0$:

$$\sigma(\omega, k) = m^2\omega/\pi\hbar^2\bar{n}k, \qquad 0 < \omega < kv_F.$$

The static form factor is then

$$\sigma(k) = \int_0^{kv_F} \sigma(\omega, k)\frac{d\omega}{2\pi} = \frac{p_F^2 k}{(2\pi\hbar)^2\bar{n}},$$

in agreement with the result in Part 1, §117, Problem 1.

INDEX